# Zinc Signaling in Physiology and Pathogenesis

Special Issue Editors

**Toshiyuki Fukada**
**Taiho Kambe**

MDPI • Basel • Beijing • Wuhan • Barcelona • Belgrade

**MDPI**

*Special Issue Editors*
Toshiyuki Fukada
Tokushima Bunri University
Japan

Taiho Kambe
Kyoto University
Japan

*Editorial Office*
MDPI
St. Alban-Anlage 66
Basel, Switzerland

This edition is a reprint of the Special Issue published online in the open access journal *IJMS* (ISSN 1422-0067) from 2017–2018 (available at: http://www.mdpi.com/journal/ijms/special_issues/zinc_signaling).

For citation purposes, cite each article independently as indicated on the article page online and as indicated below:

Lastname, F.M.; Lastname, F.M. Article title. *Journal Name* **Year**, *Article number*, page range.

**First Edition 2018**

**ISBN 978-3-03842-821-3 (Pbk)**
**ISBN 978-3-03842-822-0 (PDF)**

Cover photo courtesy of Koh-ei Toyoshima and Takashi Tsuji. It depicts mouse epidermal hair follicles which express zinc transporter ZIP10 (Bum-Ho Bin et al. PNAS 114, 12243-12248, 2017)

# Table of Contents

# About the Special Issue Editors

**Toshiyuki Fukada**, Professor. I received my BS and MS from Hiroshima University in 1990 and 1992, respectively. After working for Mochida Pharmaceutical Co., Ltd. for two years, I entered Osaka University Graduate School of Medicine and received my Ph.D. in 1998. After postdoctoral training at the Cold Spring Harbor Laboratory, New York, from 1999 to 2003, I joined the Research Center for Allergy and Immunology at Riken Yokohama Institute as a senior investigator, until 2014. Then, I moved to Showa University School of Dentistry as an Assistant Professor, and joined Tokushima Bunri University as a full Professor from April 2015. My research goal is to understand the physiological and pathophysiological role of zinc signaling in vivo. Zinc signaling is a novel platform in the life sciences, and I would thus like to explore zinc signaling at the molecular level. In particular, I am interested in the mechanisms regarding how each transporter-mediated zinc signals and controls its downstream target molecules specifically, which may eventually affect the health and disease status in mammals. To address these questions, I mainly employ molecular and physiological methodologies using genetically modified mice to study zinc transporters and human genetics.

**Taiho Kambe**, Associate Professor. I received my BS in Agriculture from Kyoto University in 1995, and my Ph.D. in Agriculture from Kyoto University in 2001. I served as an Assistant Professor at Kyoto University from 1998. After postdoctoral training at the University of Missouri, Columbia, from 2006 to 2007 and at the University of Kansas Medical Center from 2007 to 2008, I rejoined the faculty at Kyoto University. My research interests focus on elucidating the cellular and physiological functions of ZIP and ZNT zinc transporters. In particular, I am very interested in when, where, and how ZIPs and ZNTs monitor zinc concentrations and regulate the influx and efflux of zinc across biological membranes. Moreover, I am investigating when, where, and how zinc-containing proteins capture and dissociate zinc mediated by ZIPs and ZNTs. I am also trying to establish a new strategy to improve zinc nutrition in humans.

# Preface to "Zinc Signaling in Physiology and Pathogenesis"

Zinc, an essential trace element, plays indispensable roles in multiple cellular processes. It regulates a great number of protein functions, including transcription factors, enzymes, adapters, and growth factors, as a structural and/or catalytic factor. Recent studies have highlighted another function of zinc as an intra- and intercellular signaling mediator, which is now recognized as the zinc signal. Indeed, zinc regulates cellular signaling pathways, which enables the conversion of extracellular stimuli into intracellular signals, and controls various intracellular and extracellular events. Thus, zinc mediates communication between cells. The zinc signal is essential for physiology, and its dysregulation causes a variety of diseases such as diabetes, cancer, osteoarthritis, dermatitis, and dementia. This indicates that the zinc signal is an emerging topic that will assist in our understanding of the nature of physiology and pathophysiology.

This special issue, "Zinc Signaling in Physiology and Pathogenesis" has two main goals. The first is to update the current information available about the crucial role of zinc signaling in biological processes on both a molecular and a physiological level. This will assist in addressing the questions underlying this unique phenomenon and discerning its future direction through the publishing of review articles by experts, as well as original papers. The second aim is to feature the 5th Meeting of the International Society for Zinc Biology 2017 (ISZB-2017) in collaboration with Zinc-Net (COST Action TD1304), held in Cyprus. As Lowe and Moran reported, ISZB-2017 was held in conjunction with the final dissemination meeting of the Network for the Biology of Zinc (Zinc-Net) at the University of Central Lancashires Cyprus campus in June 2017, with over 160 participants, 17 scientific symposia, four plenary speakers, and two poster discussion sessions [1]. Much of the research presented at this meeting had never been presented or published before, and the most of the authors featured in this special issue presented their research at this meeting. This means that this issue contains the most up-to-date information on zinc signaling and related biology. Twelve review articles by such invited authors, which have been included in this issue, are mentioned below.

From a molecular and biochemical point of view, the first article by Maret provides an overview of the regulatory functions of zinc signaling through its interaction with Ca2+, redox, and phosphorylation signaling, thus enabling the transmission of information within cells and communication between cells [2]. The article by Kambe et al. summarizes the various zinc transporters, i.e., the family of zinc transporters (ZNTs) and Zrt- and Irt-like proteins (ZIPs), and discusses the roles of these transporters in the early secretory pathway [3]. Further, Takagishi et al. review an update of zinc transporters and zinc signaling. They focus on the recent progress in determining the roles of SLC39A/ZIP family members in vivo [4]. Sunuwar et al. focuses on ZnR/GPR39, a G-protein coupled receptor that senses changes in the concentration of extracellular zinc, reviewing its physiological role in skin and the colon, as well as its implication in cancer [5]. In addition, Subramanian Vignesh and Deepe provide an overview of the current understanding of a family of metal-binding proteins, metallothioneins (MTs), especially focusing on their role in immunity [6].

From a viewpoint of physiology and medicine, Takeda and Tamano describe the impact of synaptic zinc signaling on cognition and its decline [7]. Portbury and Adlard highlight the role of zinc signaling in the central nervous system, and its potential implications in brain diseases such as cognitive decline, depression, and Alzheimers disease [8]. Fukunaka and Fujitani emphasize the contribution of zinc homeostasis on the pathophysiology of metabolic diseases, by focusing on the zinc transporters ZnT8 and ZIP13 [9]. Maywald et al. review the critical role of zinc homeostasis in the immune system. In addition, they describe the molecular mechanisms and targets that are affected by altered zinc homeostasis and illustrate several types of zinc signaling that are involved in the immune system [10]. Pyle et al. describe the possible molecular relationship between tuberculosis and zinc homeostasis. They also review the protective role of the zinc transporter ZIP8 in macrophages in Mycobacterium tuberculosis infection [11]. Turan and Tuncay review the current understanding of the physiological role of zinc signaling on heart functions and related diseases [12]. Cherasse and Urade suggest the potent connection between zinc status and sleep, and investigate its molecular mechanisms [13].

In addition to the reviews mentioned above, five research articles are included in this special issue. Hence, we must emphasize that this special issue will provide new insights into the role of zinc signaling, mediated by zinc transporters and zinc-binding proteins, in health and disease from a molecular to a physiological level. We hope that this will present our readers with novel opportunities to raise new ideas and connections to resolve persisting questions in the future. Finally, we would like to express our heartfelt gratitude to all of the authors and referees for their tremendous efforts in supporting this special issue. Without their valuable assistance, we would not have had even a glance of this timely and successfully publication with its useful updates on zinc signaling biology.

**Toshiyuki Fukada and Taiho Kambe**

*Special Issue Editors*

### References

1. Lowe, N.M.; Moran, V.H. Report of the International Society for Zinc Biology 5th Meeting, in Collaboration with Zinc-Net (COST Action TD1304)-UCLan Campus, Pyla, Cyprus. Int. J. Mol. Sci. 2017, 18, 2518.
2. Maret, W. Zinc in Cellular Regulation: The Nature and Significance of Zinc Signals. Int. J. Mol. Sci. 2017, 18, 2285.
3. Kambe, T.; Matsunaga, M.; Takeda, T.A. Understanding the Contribution of Zinc Transporters in the Function of the Early Secretory Pathway. Int. J. Mol. Sci. 2017, 18, 2179.
4. Takagishi, T.; Hara, T.; Fukada, T. Recent Advances in the Role of SLC39A/ZIP Zinc Transporters In Vivo. Int. J. Mol. Sci. 2017, 18, 2708.
5. Sunuwar, L.; Gilad, D.; Hershfinkel, M. The zinc sensing receptor, ZnR/GPR39, in health and disease. Int. J. Mol. Sci. 2018, 19, 439.
6. Subramanian Vignesh, K.; Deepe, G.S., Jr. Metallothioneins: Emerging Modulators in Immunity and Infection. Int. J. Mol. Sci. 2017, 18, 2197.
7. Takeda, A.; Tamano, H. The Impact of Synaptic Zn2+ Dynamics on Cognition and Its Decline. Int. J. Mol. Sci. 2017, 18, 2411.

8.    Portbury, S.D.; Adlard, P.A. Zinc Signal in Brain Diseases. Int. J. Mol. Sci. 2017, 18, 2506.

9.    Fukunaka, A.; Fujitani, Y. Role of Zinc Homeostasis in the Pathogenesis of Diabetes and Obesity. Int. J. Mol. Sci. 2018, 19, 476.

10.    Maywald, M.; Wessels, I.; Rink, L. Zinc Signals and Immunity. Int. J. Mol. Sci. 2017, 18, 2222.

11.    Pyle, C.J.; Azad, A.K.; Papp, A.C.; Sadee, W.; Knoell, D.L.; Schlesinger, L.S. Elemental Ingredients in the Macrophage Cocktail: Role of ZIP8 in Host Response to Mycobacterium tuberculosis. Int. J. Mol. Sci. 2017, 18, 2375.

12.    Turan, B.; Tuncay, E. Impact of Labile Zinc on Heart Function: From Physiology to Pathophysiology. Int. J. Mol. Sci. 2017, 18, 2395.

13.    Cherasse, Y.; Urade, Y. Dietary Zinc Acts as a Sleep Modulator. Int. J. Mol. Sci. 2017, 18, 2334.

International Journal of
*Molecular Sciences*

MDPI

*Conference Report*

# Report of the International Society for Zinc Biology 5th Meeting, in Collaboration with Zinc-Net (COST Action TD1304)—UCLan Campus, Pyla, Cyprus

Nicola M. Lowe [1,*,†] and Victoria Hall Moran [2,†]

1   International Institute of Nutritional Sciences, and Applied Food Safety Studies,
    Faculty of Health and Wellbeing, University of Central Lancashire, Preston PR1 2HE, UK
2   School of Community Health & Midwifery, Faculty of Health and Wellbeing,
    University of Central Lancashire, Preston PR1 2HE, UK; vlmoran@uclan.ac.uk
*   Correspondence: nmlowe@uclan.ac.uk; Tel.: +44-(0)1772-893-599
†   These authors contributed equally to this work.

Received: 30 October 2017; Accepted: 22 November 2017; Published: 24 November 2017

**Abstract:** From 18 to 22 June 2017, the fifth biennial meeting of the International Society for Zinc Biology was held in conjunction with the final dissemination meeting of the Network for the Biology of Zinc (Zinc-Net) at the University of Central Lancashire, Cyprus campus. The meeting attracted over 160 participants, had 17 scientific symposia, 4 plenary speakers and 2 poster discussion sessions. In this report, we give an overview of the key themes of the meeting and some of the highlights from the scientific programme.

**Keywords:** zinc; conference report; Zinc-Net; ISZB; COST Action

## 1. Introduction

The meeting was jointly hosted as a partnership between the International Society for Zinc Biology (ISZB) and the Network for the Biology of Zinc (Zinc-Net), a European Union Framework Programme, Horizon 2020-funded Collaboration in Science and Technology (COST) Action. It was the fifth biennial meeting of the ISZB, which is now celebrating a decade of achievements. The society was established to allow unique interactions between scientists interested in zinc biology and has been successful in fostering links between the chemical, biological and clinical fields of zinc biology [1]. Zinc-Net was launched in 2013 and has been funded for 4 years. It is comprised of over 200 scientists from 26 different European countries, as well as Australia. The overall mission of the network was to create a multi-disciplinary research platform that brings together expertise from research groups throughout the COST countries and beyond, as well as to stimulate and accelerate new, innovative and high-impact scientific research [2]. The clear synergy between these two organisations created a stimulating forum for lively scientific debate over the 5 days of the conference.

## 2. Meeting Overview

### 2.1. Zinc-Net Celebration Symposium

The meeting was the final major event of the 4 year COST Action, Zinc-Net. Therefore, it began with a symposium that celebrated the achievements of Zinc-Net, with presentations from each of the theme leaders in Chemical Biology, Biomarker Discovery and Clinical Coordination. A key note lecture by Professor Janet King (Chair of the International Zinc Nutrition Consultative Group, Children's Hospital Oakland Research Institute, Oakland, CA, USA), and Professor Nicola Lowe

(Chair of the Zinc-Net Management Committee, University of Central Lancashire, Preston, UK), with comments by Professor Mukhtiar Zaman (Lady Reading Hospital, Khyber Pakhtunkhwa, Pakistan), placed zinc research into a global context, highlighting the extent of zinc deficiency worldwide, with the highest prevalence of over 40% of the population in low- and middle-income countries. A display of posters showcasing some of the short-term scientific missions (laboratory exchange visits) that had been undertaken by early career researchers in the network were available to view [3].

## 2.2. Keynote Lectures

The first keynote lecture of the main conference was given by the Founding President of the ISZB, Professor Glen K. Andrews (University of Kansas School of Medicine, Kansas City, MO, USA). He was introduced by the current president, Dr. Kathryn Taylor (Cardiff University, Wales, UK). Professor Andrews's lecture recognised the contributions of key scientists whose research was of fundamental importance to the advancement of the field of zinc. These included the discovery of the metallothionein (MT) and the recognition that these proteins have a wide species distribution and are inducible by metals, the elucidation of a zinc-sensing mechanism of transcriptional regulation of MT genes, and the discovery of the Slc30a (ZnT) and Slc39a (ZIP) families of zinc transporters. These findings informed further studies of the mechanisms of zinc-dependent regulation of the ZIP proteins and their important physiological role. Each of these discoveries led to thousands of subsequent research publications and provided the foundation for much of the current research into the biology of zinc [4].

The second keynote lecture, by Professor Hidenori Ichijo (Graduate School of Pharmaceutical Sciences, U-Tokyo, Japan) discussed the physiological and pathophysiological roles of copper/zinc superoxide dismutase (SOD1) under conditions of zinc deficiency. The importance of understanding the molecular mechanism of zinc homeostasis in living organisms is highlighted by zinc's essentiality in a wide variety of biological processes and the consequences of zinc deficiency in human health and disease. Ichijo's group recently reported that SOD1 is one of the key factors to regulate cellular zinc homeostasis under zinc-deficient conditions. In zinc deficiency, SOD1 adopts an abnormal conformation and evokes endoplasmic reticulum (ER) stress through specific interaction with Derlin-1, a component of ER-associated degradation (ERAD) machinery, leading to the restoration of cellular homeostasis. Intriguingly, they found that wild-type SOD1 under zinc-deficient conditions and over 100 types of amyotrophic lateral sclerosis (ALS)-linked SOD1 mutants share the common aberrant conformation, suggesting that wild-type SOD1 has a potential to exert neuronal toxicity under stress conditions. Professor Ichijo's group have performed genome-wide siRNA screening to identify mediators of the conformational alteration in wild-type SOD1 under conditions of zinc deficiency, and he discussed the physiological and pathophysiological implications [5].

Professor Stephen J. Lippard's (Massachusetts Institute of Technology, Cambridge, MA, USA) keynote discussed the role of zinc probes as tools in the study of mobile $Zn^{2+}$; these serve as signaling agents in a number of biological processes, including neurotransmission. A new class of hybrid fluorescent sensors that facilitate the tunability of small molecule probes and the targetability of protein-based sensors was described, which can detect exogenous $Zn^{2+}$ and endogenous mobile $Zn^{2+}$ in response to reactive nitrogen species in live cells. The functional role of $Zn^{2+}$ in the olfactory system was discussed, including results obtained from fluorescence imaging and electrophysiology recordings of live animals exposed to a variety of odours. Synaptically released mobile $Zn^{2+}$ attenuates excitatory postsynaptic currents carried by *N*-methyl-D-aspartate (NMDA) receptors in the olfactory bulb, thus attenuating sensory input gain [6].

Despite increasing access to sufficient food for all and significant achievements in reducing global hunger, micronutrient deficiencies ("hidden hunger"), especially zinc deficiency, still remain a major public health problem in the world. An inadequate daily intake of zinc is the major reason for the problem, particularly in the developing world, where extensive amounts of cereals are consumed with very low concentrations of bioavailable zinc. In the final keynote lecture, Professor Ismail Cakmak (Sabanci University, Istanbul, Turkey) described several agricultural strategies that are known to improve grain-zinc concentration, including conventional plant breeding, genetic engineering, and plant nutrition-based agronomy. In recent years, there has been an increase in the number of published reports showing that the maintenance of the high pool of zinc in the leaf tissue during the reproductive growth stage is required to achieve desirable concentrations of zinc in grains for human nutrition. Field experiments conducted in different countries under the HarvestZinc project [4] on maize, wheat and rice demonstrated that a foliar spray of zinc and other micronutrients such as iodine and selenium results in substantial increases in concentrations of those micronutrients, in both the whole grain and the endosperm. In addition, foods made from cereal grains that have been biofortified agronomically with micronutrients, such as bread and cookies, also had elevated micronutrient concentrations, evidencing the stability of the micronutrients in products. Consuming agronomically biofortified foods is expected to result in a significant contribution to human nutrition with the potential to impact on micronutrient deficiencies worldwide [7].

## 3. Research Themes of the Meeting

The 17 scientific symposia that comprised the meeting can be grouped into three main themes: zinc in health and disease, zinc signalling, and zinc proteins and transporters. A summary of each is reported below and further up-to-date information on key topics in zinc research has been published by Rink [8] and zinc signalling by Fukada and Kambe [9].

### 3.1. Zinc in Health and Disease

The role of zinc in immunity and infectious disease was explored and debated. Zinc is recognized as an important metal ion in relation to nutritional immunity, a process by which the immune system withholds micronutrients from potential invaders. An understanding of the underlying mechanisms by which host immune defences manipulate metal levels to attack invading microbes by metal-restriction and/or exposure to metal-excess may have considerable clinical significance. Giving *Campylobacter jejuni* (Cj), a common cause of acute human gastroenteritis, as an example of an important foodborne pathogen that targets different host niches with different metal challenges, how Cj adapts to different metal stresses within its different hosts was described. Exploring the functions and mechanisms of the zinc handling systems in *Campylobacter* will expand our understanding of how they contribute to infections. Zinc deficiency is linked to an increased susceptibility to bacterial infection, such as *Streptococcus pyogenes* (Group A *Streptococcus*—GAS), a Gram-positive human pathogen responsible for a wide spectrum of diseases ranging from pharyngitis and impetigo, to severe invasive diseases including necrotizing fasciitis and streptococcal toxic shock-like syndrome. It was demonstrated that zinc homeostasis is an important contributor to GAS pathogenesis and innate immune defence against infection. Strategies to manipulate zinc homeostasis in order to reduce GAS infection were discussed.

Zinc-based therapeutics were explored in relation to cognitive disorders, cancer and cardiovascular disease. One of the critical cell processes that becomes dysregulated with age and also in disease, and which participates both directly and indirectly in cognitive function, is metal homeostasis and the neurochemistry of metalloproteins. This is particularly true for zinc, for which 10–15% of brain zinc exists in a chelatable form, primarily within synaptic vesicles at glutamatergic synapses, highlighting its potential importance in synaptic plasticity/cognition. Zinc dyshomeostasis has been implicated in dementia and autism spectrum disorders. Taken together with other supporting data in the literature, this demonstrates a critical role for zinc in cognitive function, and that it may be a therapeutic target for improving functional outcomes in health and disease.

A significant body of evidence has shown that zinc plays important roles in metabolism and the development of metabolic disease. Zinc has a role in insulin secretion, insulin signalling and subsequent glucose metabolism. Low zinc status also promotes inflammatory stress, and using mouse models of atherosclerosis, vascular inflammation and plaque formation have been shown to be enhanced by marginal zinc deficiency. Increased intestinal permeability plays an important role in the onset of a variety of chronic inflammatory conditions and metabolic diseases, and zinc has been found to improve gut barrier integrity in vitro. In humans, a low-zinc diet is associated with a decrease in fatty acid desaturase enzyme 1 (FADS1) activity, lowered arachidonic acid incorporation into lipid subclasses, and an increase in DNA strand breaks, suggesting that FADS1 activity and DNA strand breaks respond to small changes in dietary zinc that may be provided by food fortification programmes.

### 3.2. Zinc Signalling

The roles of zinc in modulating cellular function in various disease states emerged as a key theme of this meeting. These included new advances in understanding how disrupted $Zn^{2+}$ homeostasis in chronic heart failure is linked to dysregulated intracellular $Ca^{2+}$ responses, resulting in leakage of $Ca^{2+}$ from the sarcoplasmic reticulum in cardiac tissue. Research describing the pathophysiological role of zinc in neurological disorders provided insights into possible novel therapeutic approaches. Examples included the role of zinc in triggering neuronal apoptosis and blocking optic nerve regeneration after injury, as well as the role of extracellular zinc in the modulation of the cytokine-induced pro-inflammatory response following brain ischaemia, which may contribute to impaired memory function. New research examining the pathophysiological role of extracellular $Zn^{2+}$ in cognitive decline with aging was received with interest. Highlights within this theme also included new understandings of the molecular mechanisms for maintaining cellular zinc homeostasis and the use of novel zinc sensors, which can be activated by UV light or enzymes for the study of the dynamics of cellular "free" zinc.

### 3.3. Zinc Proteins and Transporters

One of the most exciting areas of research over the last decade has been the discovery of the zinc transporter families ZIP and ZnT and the elucidation of their role in the control of cellular zinc homeostasis. Within this theme, the relationship between the loss of ZnT2 function within paneth cells and intestinal dysbiosis was discussed. This research revealed that genetic polymorphisms that influence the ZnT2 transporter function might lead to clinically relevant shifts in the intestinal microbiome of preterm infants, which is a fascinating new area of research relating to infant nutrition. Similarly, ZIP7 plays a critical role in ER function within connective tissue cells, such that a loss of the function of this transporter results in inhibited cell proliferation, preventing proper dermis formation. Highlights within this theme included potential new therapies for the treatment of cancer, linked to the inhibition of mitosis through the selective blocking of ZIP transporters. Additionally, the unexpected association between genetic mutations in ZIP13, zinc homeostasis and beige adipocyte biogenesis may contribute to new therapies for obesity and metabolic syndrome.

Other zinc binding proteins also shared the limelight in this theme. Notably, the influence of zinc binding on protein folding and aggregation may have deleterious consequences if intracellular zinc homeostasis becomes imbalanced. Also discussed was the mechanism of the activation of the zinc-requiring ectoenzymes, defined as secretory, membrane bound, and organelle-resident enzymes, which play pivotal roles in numerous biological responses. One such example is tissue non-specific alkaline phosphatase, which is activated in a two-step process involving ZnT transporters.

## 4. Focus on Early Career Researchers

The conference was attended by over 160 scientists, including early career researchers (ECRs), many of whom presented posters in one of the two evening poster sessions. Both Zinc-Net and ISZB place a strong emphasis on providing training opportunities, capacity building and support for the next generation of zinc biologists. In a competitive process, ECRs were invited to present their research in two special symposia showcasing the work of these up and coming young scientists in this exciting field.

## 5. Final Remarks

Much of the research presented at this meeting had never been presented or published before. The meeting had a strict embargo on the photographing of slides without the presenters' permission, and the abstracts, although made available to all participants, were not to be published in proceedings of the meeting. However, it was gratifying to observe that the atmosphere within the meeting was extremely open and collegiate, with new collaborations initiated and many animated discussions during the social, networking and poster events. The zinc research community is clearly thriving, and it is exciting to be a part of it.

**Acknowledgments:** We gratefully acknowledge the networking support from the COST Action TD1304 and our commercial sponsors, STREM Chemicals Inc., The Japanese Society for Zinc Nutritional Therapy, and the Royal Society of Chemistry, Metallomics Journal.

**Author Contributions:** Nicola M. Lowe and Victoria Hall Moran contributed equally to this work.

**Conflicts of Interest:** The authors declare no conflict of interest.

## Abbreviations

| | |
|---|---|
| FADS1 | Fatty acid desaturase enzyme 1 |
| NMDA | N-Methyl-D-aspartate |
| ERAD | ER-associated degradation |
| ISZB | International Society of Zinc Biology |
| COST | Collaboration in Science and Technology |
| SOD | Superoxide dismutase |
| ECR | Early career researcher |
| MT | Metallothionein |
| ER | Endoplasmic reticulum |
| Cj | *Campylobacter jejuni* |

## References

1. International Society for Zinc Biology. Available online: https://iszb.org/ (accessed on 10 November 2017).
2. The Network for the Biology of Zinc. Available online: http://www.cost.eu/COST_Actions/fa/TD1304 (accessed on 10 November 2017).
3. Zinc-Net. Available online: http://zinc-net.com/?page_id=1879 (accessed on 10 November 2017).
4. HarvestZinc. Available online: www.harvestzinc.org (accessed on 10 November 2017).
5. Andrews, G.K. Cellular Zinc Sensors: MTF-1 regulation of gene expression. In *Zinc Biochemistry, Physiology, and Homeostasis: Recent Insights and Current Trends*; Maret, W., Ed.; Springer Science: Berlin, Germany, 2013; pp. 37–51, ISBN 9401737282.
6. Zastrow, M.L.; Radford, R.J.; Chyan, W.; Anderson, C.T.; Zhang, D.Y.; Loas, A.; Tzounopoulos, T.; Lippard, S.J. Reaction-Based Probes for Imaging Mobile Zinc in Live Cells and Tissues. *ACS Sens.* **2016**, *1*, 32–39. [CrossRef] [PubMed]
7. Cakmak, I. Enrichment of cereal grains with zinc: Agronomic or genetic biofortification? *Plant Soil* **2008**, *302*, 1–17. [CrossRef]
8. Zinc in Human Health. *Biomedical and Health Research*; Rink, L., Ed.; IOS Press: Amsterdam, The Netherlands, 2011; Volume 76, ISBN1 978-1-60750-815-1 (print), ISBN2 978-1-60750-816-8 (online).
9. Fukada, T.; Kambe, T. *Zinc Signals in Cellular Functions and Disorders*, 1st ed.; Springer: Tokyo, Japan, 2014; ISBN 978-4-431-55113-3.

International Journal of
*Molecular Sciences*

MDPI

*Review*

# Zinc in Cellular Regulation: The Nature and Significance of "Zinc Signals"

Wolfgang Maret

Metal Metabolism Group, Departments of Biochemistry and Nutritional Sciences, School of Life Course Sciences, Faculty of Life Sciences and Medicine, King's College London, Franklin-Wilkins Bldg, 150 Stamford St., London SE1 9NH, UK; wolfgang.maret@kcl.ac.uk; Tel.: +44-(0)-20-7848-4264; Fax: +44-(0)-20-7848-4195

Received: 27 September 2017; Accepted: 26 October 2017; Published: 31 October 2017

**Abstract:** In the last decade, we witnessed discoveries that established $Zn^{2+}$ as a second major signalling metal ion in the transmission of information within cells and in communication between cells. Together with $Ca^{2+}$ and $Mg^{2+}$, $Zn^{2+}$ covers biological regulation with redox-inert metal ions over many orders of magnitude in concentrations. The regulatory functions of zinc ions, together with their functions as a cofactor in about three thousand zinc metalloproteins, impact virtually all aspects of cell biology. This article attempts to define the regulatory functions of zinc ions, and focuses on the nature of zinc signals and zinc signalling in pathways where zinc ions are either extracellular stimuli or intracellular messengers. These pathways interact with $Ca^{2+}$, redox, and phosphorylation signalling. The regulatory functions of zinc require a complex system of precise homeostatic control for transients, subcellular distribution and traffic, organellar homeostasis, and vesicular storage and exocytosis of zinc ions.

**Keywords:** zinc; homeostasis; signalling; regulation

## 1. Zinc in Enzymatic Catalysis, Protein Structure, and Regulation of Proteins

It is ingrained in our understanding of the scientific literature that zinc has catalytic, structural, and regulatory functions in proteins. However, while the first two functions are well-established, validated examples of regulatory molecular functions are much more difficult to pinpoint. Catalytic and structural functions occur in an estimated 3000 human zinc metalloproteins, a number that translates into about every tenth protein being a zinc protein [1]. This myriad of functions demonstrates the major role of zinc in cell biology [2], which is now reinforced by the emerging roles of zinc ions in cellular regulation [3]. Many earlier postulates about regulatory functions of zinc in proteins are based on outdated premises, were not linked to specific molecular actions in physiological events, and involved zinc/protein interactions that occur with micromolar affinities. Compelling arguments put such interactions outside the physiological range of cellular zinc ion concentrations. They are based on experimental results and the requirement to control zinc in a range of concentrations that avoids interference with the biochemistry of the other essential metal ions. Measurements of zinc binding constants of zinc proteins, and total and available ("free") zinc concentrations now provide a different view of the physiological significance of zinc/protein interactions. Affinities of structural and catalytic zinc in cellular proteins are in the picomolar to femtomolar range, and at present, evidence is lacking that these sites are regulated via zinc binding and release [4,5]. As a consequence of this high affinity, the "free" zinc ion concentration is in the picomolar range, as shown experimentally, despite the total cellular zinc concentration being in the range of hundreds of micromolar [6]. This large difference between total and "free" zinc is a distinctive chemical property of zinc in biology [7]. It rules out low affinity zinc binding sites as being physiologically significant for cellular regulation, and suggests that regulatory sites of zinc in cellular proteins must have binding constants commensurate with these chemical and biological constraints. In fact, a couple of examples of zinc inhibiting enzymes with picomolar affinity are now known [8–10]. Also, it was reported about 50 years ago that picomolar concentrations of zinc (II) ions inhibit phosphoglucomutase by replacing the magnesium ion required for activity [11]. The authors discussed the physiological significance of the finding, but their reports received very little, if any recognition in the field of metallobiochemistry [12]. In terms of its rather high affinity for binding sites, $Zn^{2+}$ is different from $Mg^{2+}$ and $Ca^{2+}$, and makes an additional range available for biological regulation using all three redox-inert metal ions (Figure 1). What is needed for regulation is a change in the concentration of "free" zinc ions (zinc transients). This article discusses how such changes can be effected, and controlled, in a zinc-buffered biological environment. In the cell, zinc transients occur from a basal level of tens to hundreds of pM free zinc (about 0.2 mM total cellular zinc). They leave a gap of three orders of magnitude before calcium transients can set in, which are above a basal level of 100 nM free calcium (about 2 mM total cellular calcium). Like calcium, zinc needs a system to control these transients, and is cytotoxic if not properly controlled. Regarding magnesium, recently described circadian rhythms of total magnesium concentrations in cells provide strong evidence for regulatory functions [13].

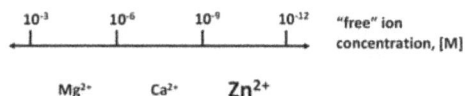

**Figure 1.** Biological regulation with the three redox-inert metals ions: $Mg^{2+}$, $Ca^{2+}$, and $Zn^{2+}$. Zinc extends the range of regulation with metal ions. The regulatory function of each metal ion is in a specific range of concentrations, thus avoiding overlap with the signalling functions of the other metal ions. However, there are interactions among the metabolism of the metal ions.

## 2. Control of Cellular Zinc Homeostasis

Knowledge about the proteins that control cellular zinc now provides a basis for understanding how zinc ions can regulate cellular processes. In humans, at least twenty-four membrane transporters (14 Zrt, Irt-like proteins (ZIP) zinc importers and 10 zinc transporters (ZnT) zinc exporters), about a dozen metallothioneins (MTs), and a zinc-sensing transcription factor, metal-response element (MRE)-binding transcription factor-1 (MTF-1), are involved in controlling cellular zinc [14]. A few of these transporters have a role—or an additional role—in $Mn^{2+}$ and $Fe^{2+}$ transport, while MT also has a function in copper metabolism [15]. The number of membrane transporters is much higher than what one would expect for simple control of homeostasis of a metabolite. Cellular iron, for example, is essentially controlled by one importer (DMT1) and one exporter protein (ferroportin). It turned out that many additional transporter proteins are necessary for the subcellular distribution of zinc, the control of organellar zinc homeostasis, and the generation and control of zinc transients. Most of the ZIP transporters have been found on the plasma membrane, but some have roles intracellularly. The large number of ZIP transporters on the plasma membrane is likely a reflection of the importance of securing the correct supply of zinc for the cell under various conditions. Remarkably, zinc ions accumulate in cellular vesicles for storage and/or release, and in secretory vesicles for exocytosis. The occurrence of zinc ions in cellular vesicles is a characteristic feature of cellular zinc biology, and resolves the long-standing issue why a storage protein akin to ferritin, which stores several thousand iron ions in its core, was never found for zinc. There is some ambiguity in the nature of some of these vesicles. Vesicular stores that accumulate zinc added to cultured cells have been called zincosomes [16]. Others are secretory vesicles from which zinc ions are exocytosed for various purposes. These vesicles differ from intracellular vesicular/organellar stores of zinc, as ZIP transporters do not seem to counteract the action of the ZnT transporters (ZnT2, 3, and 8) loading these vesicles with zinc ions. The "lethal milk" mouse has a truncated form of ZnT4, and presents with low zinc concentration in the milk and developmental defects of the mammary gland [17]. Zinc transporters

ZnT4–7 are involved in loading zinc-requiring ectoenzymes with zinc at the *trans* Golgi network (TGN)/early secretory pathway [18]. Zinc is also translocated to lysosomes, mitochondria, and nuclei, but very little is known about the transporters involved. Zincosomes and related vesicles require additional components for control of zinc homeostasis in order to link cellular uptake and store loading. Thus, there is an at least three-tiered system for the homeostatic control of cellular zinc: import and export through the plasma membrane, including cytosolic binding proteins such as metallothioneins and sensors such as MTF-1 (tier 1); intracellular storage and release of zinc generating zinc transients (tier 2); and allocation of zinc for exocytosis in some cells, somehow linked to re-uptake (tier 3) (Figure 2). It appears that a hierarchy is associated with these three tiers, namely, that the capacity of loading vesicles is compromised, first, under zinc deficiency, as suggested by investigations with cultured cells [19].

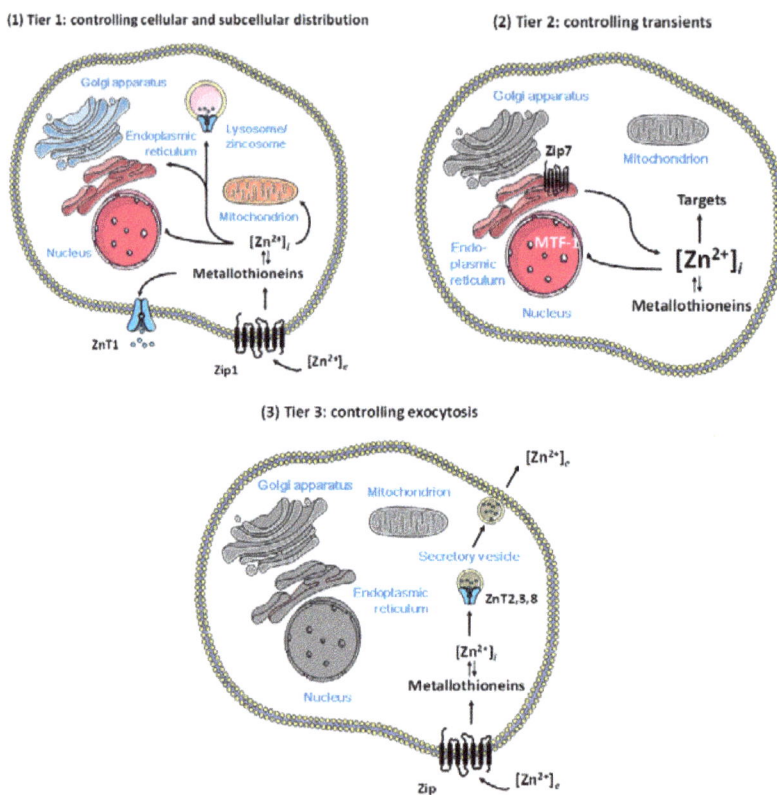

**Figure 2.** Control of cellular zinc homeostasis including regulation of zinc and functions of zinc in regulation. (**1**) Tier 1: cellular homeostasis through import and export through the plasma membrane and subcellular distribution; (**2**) Tier 2: vesicular storage and/or release of zinc associated with zinc importers and exporters on the vesicles; (**3**) Tier 3: loading of secretory vesicles with zinc associated only with zinc exporters on the vesicles. Regulatory functions of zinc require additional complexity in homeostatic control, namely, molecules that control the spatiotemporal fluctuations of zinc ions in a metal-buffered environment, above the fluxes in re-distributing zinc for supply as a cofactor in catalytic and structural sites of proteins. With this distinction, we assume that the functions of bona fide zinc proteins are not regulated by zinc coming on and off from their catalytic and structural sites. (Figures were composed from Servier Medical Art templates (http://smart.servier.com/)).

There are at least three pathways for transient release of zinc ions within and from cells. The proteins participating in the control of homeostasis (ZIPs, ZnTs, metallothioneins, and MTF-1) restore the steady-state after the transients have occurred (Figure 3). A breakthrough that triggered the discovery of these pathways, and the functions associated with them, came with the synthesis and availability of fluorescent chelating agents for measuring zinc ions, similar to the ones employed in the field of calcium biology to measure calcium fluxes [20].

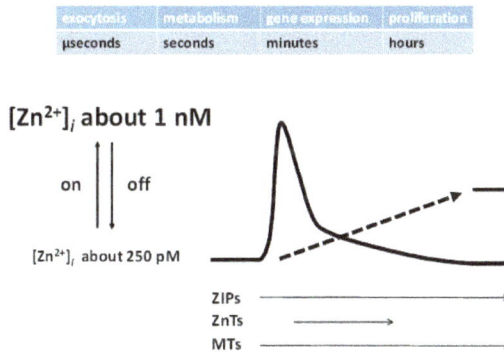

**Figure 3.** Cellular zinc transients and their control. Several pathways generate rapid zinc transients (zinc signals) with biological responses on various time scales. The amplitude of these zinc signals is about a few nanomolar globally, from picomolar basal "free" zinc ion concentrations, but may be higher in microdomains. Proteins involved in the control of zinc homeostasis serve as mufflers (ZIPs, ZnTs) and buffers (MTs) to restore the steady-state. A much slower process (arrow) leads to an overall change of the zinc-buffering characteristics of the cell, establishing different basal levels of "free" zinc ion concentrations.

## 3. The Paradigms for Using Zinc Ions in Information Transfer/Regulation/Communication

A long train of discoveries followed the original observation of a histochemically stainable pool of zinc ions in the brain over half a century ago [21]. It culminated in the demonstrations that (i) the zinc ions localize to presynaptic vesicles (boutons) of specialized neurons; (ii) the zinc transporter ZnT3 loads the vesicles with zinc; and (iii) the zinc ions are exocytosed upon electrical stimulation of such neurons. Subsequently, three types of "zinc signals" in the brain have been defined: the synaptic signal, described as a type of chemical transmitter/neuromodulator; the transmembrane signal, where the zinc ions are translocated through the plasma membrane; and the intracellular zinc signal, with release of zinc ions from a source within cells [22]. All of these signals have now been shown to have physiological roles, not only in the brain, but importantly, in many other tissues, thus establishing at least three general pathways for the generation of zinc ion transients.

### 3.1. Release of Zinc Ions from Cells by Vesicular Exocytosis

Every cell has the capacity to export an excess of zinc. It appears that zinc is sequestered first in cellular vesicular stores, and that export from the cell occurs only after the capacity of the stores is exhausted. However, a number of cells, including neurons with glutamatergic/zincergic synaptic vesicles, have the additional capacity to secrete zinc ions by calcium-dependent exocytosis for a specific purpose [23,24]. In neurons, in addition to postsynaptic effects, zinc has a presynaptic effect, re-entering the neuron and affecting phosphorylation signalling via inhibition of protein tyrosine phosphatases (PTPs) [25,26]. In cells secreting zinc via exocytosis, there is an additional demand for zinc, as zinc brought into the secretory vesicles can reach relatively high concentrations (about mM) and needs to be replenished after secretion. High zinc concentrations have a function within the vesicles, e.g., a structural role together with calcium in the insulin hexamer in the dense granules of β-cells in the endocrine pancreas or inhibiting proteases in zymogen granules of acinar cells in the exocrine pancreas. Zinc ions are co-secreted from these vesicles, and paracrine, autocrine, and even endocrine roles for the zinc ions $[Zn^{2+}]_e$ have been suggested, e.g., in β-cells, in addition to their function in preventing amyloidosis of the secreted amylin and insulin [27]. For an endocrine effect, proof is lacking that the secreted cellular zinc increases the rather high total zinc concentration of blood sufficiently, in order for the liver to detect a zinc signal emanating from the pancreas [28]. It would seem that a specific complex of zinc needs to carry the information of the endocrine signal. Zinc-secreting cells also include prostate epithelial cells (zinc in prostate fluid), mammary gland epithelial cells (zinc in milk), intestinal Paneth cells, cells of the immune system (mast cells, granulocytes, neutrophils), and platelets. It is not clear whether zinc is the only cargo in any of these exocytotic vesicles. In fact, it is highly unlikely because secretory vesicles are biochemically quite complex, as shown for the dense granules of β-cells, which contain more than 300 different proteins [29]. It is also not clear how zinc is made available in the cytosol, where the concentration of free zinc is only picomolar, for loading the vesicles, and what the chemical forms of zinc are, in addition to protein-bound zinc in the vesicles.

Zinc in secretory vesicles → $Ca^{2+}$-dependent exocytosis → increase of $[Zn^{2+}]_e$ → multiple targets

Another important biological event is the release of zinc ions from fertilized mouse oocytes [30]. These "zinc sparks" are preceded by calcium oscillations after fertilization/chemical-induced activation of the oocyte. The secreted zinc ions have a role in hardening the zona pellucida (glycoprotein matrix) to avoid polyspermy—yet another function of the secreted zinc on proteins.

### 3.2. Intracellular Release of Zinc Ions through Reactive Species Modifying Proteins

This pathway is based on the recognition that zinc coordination environments with thiolate (cysteine) sulphur, such as in MTs, are redox-active, despite zinc ions being redox-inert in biology. Oxidation of the sulphur ligand donor controls the dissociation of zinc ions [31]. It was the first pathway to show a way of releasing zinc intracellularly and increasing the free zinc ion concentrations, $[Zn^{2+}]_i$, overcoming the paradox of how zinc can be made available when it is bound with such high affinity to proteins. The list of oxidants is extensive, and includes reducible sulphur and selenium compounds [32,33]. In addition, reactive species such as nitric oxide, and electrophiles such as carbonyls react with the sulphur ligand donor [34]. The release of zinc has been shown in many cellular systems, e.g., nitric oxide signals target MTs and release zinc [35,36], and cell-permeable disulphides increase the available cellular zinc ion concentration [37]. The majority of investigations has addressed pathophysiological (oxidative stress releasing zinc), pharmacological (drugs releasing zinc), and toxicological (toxins releasing zinc) events. The released zinc ions first have a cytoprotective, and then a cytotoxic effect [38]. Many of the agents investigated react with MTs with concomitant release of zinc, which then activates MTF-1 to induce gene expression. In this pathway, MT is a signal transducer, generating zinc signals in response to redox signals. The physiological significance seems to be that signalling with many growth factors generates reactive species that can then generate cellular zinc transients $[Zn^{2+}]_i$ for the purpose of modulating signal transduction.

Reactive species/redox signalling → MTs → increase of $[Zn^{2+}]_i$ → multiple targets and MTF-1-dependent gene expression

### 3.3. Intracellular Release of Zinc Ions through Channels of Vesicular Stores

It was summarized, at the turn of the millennium, that zinc has functions at all levels of cellular signalling, and that growth factor signalling affects intracellular zinc re-distribution and functions [39]. The underlying idea, though, was that the extracellular zinc taken up by the cell is responsible for the observed effects.

A key discovery that expanded the possibilities of zinc as a signalling ion was that cross-linking of the high affinity immunoglobulin E receptor (Fcε receptor I) in mast cells causes release of zinc ions intracellularly from the perinuclear area, including the endoplasmic reticulum (ER). The response was dependent on calcium influx, caused mitogen-activated protein kinase (MAPK)/extracellular signal-regulated kinase (ERK) activation, and was called a "zinc wave". While the targets of zinc were not identified, it demonstrated that zinc serves the role of a second messenger in modulating the signalling of plasma membrane receptors [40].

Another key observation was that zinc ions are released from an ER store through a specific pathway that includes the casein kinase 2-mediated phosphorylation and opening of the ZIP7 channel [41]. The released zinc affects phosphorylation signalling, and it appears to do so primarily by inhibiting PTPs [42].

Growth factor signals → casein kinase 2 (CK2) → ZIP7 phosphorylation → increase of $[Zn^{2+}]_i$ → inhibition of protein tyrosine phosphatases (PTPs) and enhanced phosphorylation signalling

## 4. Targets of Zinc Signals (Sensors, Effectors, and Stimuli)

Zinc binds to and stimulates a zinc-sensing receptor (ZnR) that triggers intracellular calcium release [43]. This receptor turned out to be the G-protein coupled receptor GPR39 [44–46], which activates inositol triphosphate $(IP_3)$ signalling and calcium release. What is not resolved, as yet, is where the zinc signal originates from, i.e., how a zinc signal can occur in a zinc-buffered extracellular milieu, whether it is part of the exocytotic pathway of zinc ions, and whether it originates from the same cell or another cell. GPR39 is an orphan receptor, and it is not clear whether or not zinc is the primary agonist or merely a modulator.

$[Zn^{2+}]_e$ → GPR39 → $IP_3$ → increase of $[Ca^{2+}]_i$ → phosphorylation signalling
(ERK/protein kinase B (AKT))

With the exception of ZnR/GPR39 and activation of MTF-1, the investigated effects downstream of the signalling zinc ions are, in most cases, not the zinc-dependent proteins themselves. Therefore, the direct molecular targets of the signalling zinc ions remain poorly defined. This is an important missing piece of information that is affecting the wider acceptance of $Zn^{2+}$ as a second messenger, and so is the issue in which chemical form zinc ions transmit the signal, in particular, since a zinc-binding messenger protein, such as the calcium signal transducer calmodulin, has not been identified. In none of the cases of zinc signalling is the chemical identity of the zinc signal known. It is an important issue as the bioinorganic chemistry of zinc is quite different from that of calcium, and the term "free" does not apply in the same way. Any ligand of zinc could impact the specificity of the zinc signal, and thus, its information content [7]. Links between signals and effectors (targets) have been established, though. In neuronal zinc signalling, the NMDA (*N*-methyl-D-aspartate) receptor, a calcium channel, is a well characterized target. Zinc inhibits this receptor with nanomolar affinities at a structurally characterized site [47]. In the case of GPR39, calcium signalling is a downstream event of the zinc signal, too. Calcium signals can also be upstream of the zinc signal, e.g., when $Ca^{2+}$/calmodulin activates nitric oxide synthase, and nitric oxide releases zinc from proteins. The intracellularly released "zinc wave" modulates phosphorylation signalling. Here, it seems that protein tyrosine phosphatases (PTPs) are targeted. Their inhibition enhances the phosphorylation signalling, i.e., the effect of kinases. PTPs are not described as metalloproteins, but zinc inhibits them with picomolar to nanomolar affinities with a specific mechanism that is independent of their redox modulation [48]. In the case of ZIP7 channel-released zinc, the phosphorylation signal is upstream and downstream of the zinc signal. There are other targets of zinc transients, and we are far from having a complete list of them [10]. The zinc signals downstream of redox signals seem to be a response to redox or carbonyl stress, and are linked to repair and defence pathways, and to initiating a response that restores the redox balance. PTPs are targets, but there is the potential for a much greater number of targets, depending on the intensity of the stress signal, and hence, the amplitudes of the zinc signal. Here, too, like in calcium and phosphorylation signalling, zinc transients/signals are upstream and downstream of redox effects, as zinc ions have pro-antioxidant and pro-oxidant effects, depending on their intracellular concentrations [49]. The observation of both upstream and downstream effects of zinc ion signals in signal transduction by calcium, redox, and phosphorylation signalling, suggests that the regulatory function of zinc signals is one of feedforward or feedback on these pathways and other pathways [50,51]. Zinc inhibition of phosphodiesterases controlling cAMP and cGMP signalling have been described, expanding this network of zinc regulation even further [39]. It appears that zinc, as the redox-inert metal ion with the highest affinity, is well suited for a regulatory position higher up in the hierarchy.

It remains crucial to quantitate the amplitudes and duration of zinc signals, and match them to the affinities of targets of zinc, in order to tease apart the biological significance of the signalling effects. Proteins targeted by transient zinc signals need to be differentiated from proteins affected by long-lasting changes in cellular zinc concentrations. In one case, the effect of the signal is readily reversible, while it will be long-lasting when the cell is brought into a different physiological state. Importantly, such a transition does not necessarily require a change in total zinc concentrations, but a change in the zinc buffering capacity, such as an increase or decrease in the concentrations of MTs. The very high affinity of some PTPs for zinc suggests that they are inhibited tonically, and then activated [9]. It is not known whether this occurs via a general mechanism, with generation of higher chelating capacity of the zinc-buffering species, or by specific molecules that bind to PTPs and remove the inhibitory zinc ion.

## 5. Genomic Effects and the Roles of Metallothioneins (MTs) and MTF-1 in Buffering and Muffling Zinc

Zinc signalling and zinc regulation are two different aspects. The discussed pathways of non-genomic zinc signalling eventually lead to genomic (transcriptional) responses via their effects on classical signal transduction pathways. These responses are part of zinc regulation. MTF-1 can sense zinc signals directly and induce gene transcription. More than 1000 genes have cis-acting metal response elements (MREs), but only 43 were identified as putative MTF-1 targets [52,53]. Among these targets, about 50% are transcription factors, and 19 are genes involved in development, demonstrating the role of zinc in developmental programmes as genetic ablation of MTF-1 is embryonically lethal in the mouse. Other target genes are γ-glutamate-cysteine ligase heavy chain needed for glutathione biosynthesis, and the zinc exporter ZnT1, and most, but not all MTs. In addition to the constitutively expressed ZnT1 and MT proteins buffering zinc signals/transients, the induced ZnT1 and MT proteins can adjust the zinc buffering capacity, and hence, the free zinc ion concentrations (Figure 4). Two effects are responsible for such adjustments. One is zinc buffering by MTs, and the other is a process called muffling, which refers to transporting metals into and out of the cytosol, and is a typical component of biological buffering mechanisms [54]. Knockdown of MTF-1 results in more genes becoming zinc-responsive and in unmasking repression of transcriptional responses of genes that have a zinc transcriptional regulatory element (ZTRE). It suggests that MTF-1 is high up in a hierarchy of sensors, and that its control of the expression of ZnT1 and MTs buffers the transcriptomic response to zinc [55]. A postulated mechanism of how MTF-1 functions is that a 10–50-fold difference in affinity of its zinc fingers is used for zinc sensing [56]. This direct effect of zinc on gene expression, via the zinc sensor MTF-1, is in addition to its indirect effects as a structural cofactor in hundreds of transcription factors with zinc fingers and related motifs.

The biochemistry of both MTF-1 and MTs shows that the control of zinc homeostasis interacts with numerous other cellular systems. A host of factors induce MT expression, not only with the purpose of sequestering any surplus of zinc, but also for providing enough metabolically available zinc for cellular processes. The rapid zinc binding, the mechanisms of releasing zinc, and the affinities of MTs for zinc match the requirements for controlling cellular zinc. Only now, 60 years after its discovery, with an understanding of both the quantitative aspects of zinc metabolism and the complexity of cellular homeostatic control of zinc, the role of the particular zinc/thiolate cluster chemistry of MTs in zinc metabolism, believed to be elusive, is becoming evident [15].

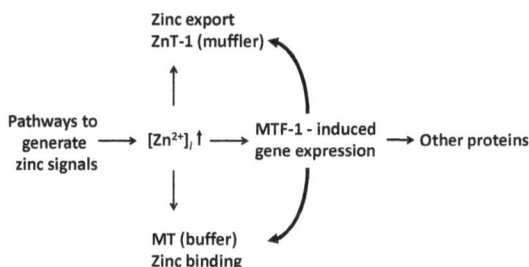

**Figure 4.** Genomic effects as a consequence of sensing zinc signals. MTF-1 induces the expression of the zinc exporter ZnT1 (muffler) and MT (buffer), which adjust the cellular zinc buffering capacity, and other proteins. In addition, constitutively expressed zinc transporters and MTs control the zinc signals.

## 6. Definition of Signalling/Regulation with Regard to Zinc

It was the intent of this short account to introduce the paradigms of zinc signals and zinc signalling for some of the regulatory roles of zinc. The terms "signals" and "signalling" have been applied to describe different scenarios where zinc (II) ion transients are detectable. With some ambiguity in the use of the terms, the credibility of the field will suffer without a clear definition of the role of zinc in cellular signalling.

First, the terms zinc signals and zinc signalling should be used only where zinc ions participate as messengers in cellular signalling. The criteria for a second messenger, namely that changes of a short-lived metabolite—and a mechanism to terminate the response—lead to a rapid alteration of the activity of (an) enzyme(s), are fulfilled in the above pathways. Of course, with metal ions, the metabolite itself is not short-lived—the availability of the metal ion controls the signal. As pointed out in calcium biology, where $IP_3$ is the second messenger that releases calcium from the ER, calcium is actually the third messenger. This distinction is rarely made, and calcium is generally referred to as a classical second messenger. In the same line of thought: zinc will be the fourth messenger in the signalling cascade when the calcium signal is upstream of the zinc signal. The other signalling functions discussed refer to extracellular zinc ions. Here, $Zn^{2+}$ is a first messenger, a stimulus like a hormone binding to a receptor, when it binds to GPR39. For such an action, a zinc signal must be generated in the extracellular environment which has a buffered steady-state concentration of zinc.

Second, the terms should be reserved for transient physiological signals and not be used for non-physiological concentrations of zinc that are the result of an imbalance or a breakdown of homeostatic control. The two situations are part of a spectrum of actions, and can be distinguished only when it is known that the homeostatic system is unable to cope with the zinc concentrations.

Third, like calcium signals, zinc signals have spatiotemporal characteristics with relatively short transients. I suggest that the terms "late" zinc signals and signalling should not be used because they are a consequence of the short-lived zinc transients, and refer to slow transcriptional, or other responses to zinc. For example, cells adjust to different physiological states requiring a re-programming that may involve gene expression. Resting, proliferating, differentiating, and apoptotic cells all have a different basal "free" zinc ion concentration [7].

## 7. Conclusions

The most important message is that zinc is a major cellular regulatory ion in the series of biologically redox-inert metal ions $Na^+$, $K^+$, $Mg^{2+}$, $Ca^{2+}$, $Zn^{2+}$, making zinc regulation not a specific topic with isolated modes of actions, but a general one in cellular biochemistry/biology. The regulatory functions are in addition to its roles as a catalytic and structural cofactor in the already impressive number of zinc metalloproteins. Cellular zinc is controlled by a large network of specific proteins that interact with virtually all pathways controlling metabolism and cell fate. A most recent example is the identification of ZIP9 as an androgen receptor that is coupled to G-proteins, and mediates non-classical responses to androgens [57].

Recent developments show the field of zinc biology to be as important as the fields of iron or calcium biology. Zinc shares relatively high abundance and cellular concentrations with both. Zinc is not a trace element, but rather a mineral and major constituent of the cell. While iron biochemistry has been a forerunner in metallobiochemistry, zinc biochemistry turns out to be an even more pervasive topic. Zinc has been classified as a type 2 nutrient, such as magnesium, with a general function in metabolism, as opposed to type 1 nutrients, such as iron with more specific functions [58]. With iron, it shares the many functions in metalloproteins, and with calcium, it shares the signalling capacity. Like calcium, genetic and environmental factors that affect homeostatic control can cause zinc disorders and dysregulation, which are the cause of many diseases [59].

The physicochemical properties that make zinc ideally suited for the wide range of biological functions have been pointed out repeatedly, and the potential of the field was predicted in a now classic review published 25 years ago [60]. Likewise, the anticipation that zinc will be the calcium of the 21st century, in terms of its signalling capacities, seems to be gaining traction [61]. The time has come for the implications of zinc serving major regulatory functions in the cell to be recognized outside the relatively small community of zinc biologists.

**Conflicts of Interest:** The author declares no conflict of interest.

## Abbreviations

| | |
|---|---|
| PTP | protein tyrosine phosphatase |
| ZnT | zinc transporters |
| ZIP | Zrt, Irt-like proteins (zinc transporters) |
| MT | metallothionein |
| MTF-1 | (MRE)-binding transcription factor 1 |
| ER | endoplasmic reticulum |
| $IP_3$ | inositol triphosphate |
| ZnR | Zinc-sensing receptor |
| GPR | G-protein-coupled receptor |

## References

1. Andreini, C.; Banci, L.; Bertini, I.; Rosato, I. Counting the zinc-proteins encoded in the human genome. *Proteome Res.* **2006**, *5*, 196–201. [CrossRef] [PubMed]
2. Maret, W. Zinc biochemistry: From a single zinc enzyme to a key element of life. *Adv. Nutr.* **2013**, *4*, 82–91. [CrossRef] [PubMed]
3. Fukada, T.; Kambe, T. (Eds.) *Zinc Signals in Cellular Functions and Disorders*; Springer: Tokyo, Japan, 2014.
4. Maret, W. Zinc and sulfur: A critical biological partnership. *Biochemistry* **2004**, *43*, 3301–3309. [CrossRef] [PubMed]
5. Kochańczyk, T.; Drozd, A.; Krężel, A. Relationship between the architecture of zinc coordination and zinc binding affinity in proteins—Insights into zinc regulation. *Metallomics* **2015**, *7*, 244–257. [CrossRef] [PubMed]
6. Krężel, A.; Maret, W. Zinc buffering capacity of a eukaryotic cell at physiological pZn. *J. Biol. Inorg. Chem.* **2006**, *11*, 1049–1062. [CrossRef] [PubMed]
7. Krężel, A.; Maret, W. The biological inorganic chemistry of zinc ions. *Arch. Biochem. Biophys.* **2016**, *611*, 3–19. [CrossRef] [PubMed]
8. Hogstrand, C.; Verbost, P.M.; Wendelaar Bonga, S.E. Inhibition of human $Ca^{2+}$-ATPase by $Zn^{2+}$. *Toxicology* **1999**, *133*, 139–145. [CrossRef]

9.   Wilson, M.; Hogstrand, C.; Maret, W. Picomolar concentrations of free zinc (II) ions regulate receptor protein tyrosine phosphatase beta activity. *J. Biol. Chem.* **2012**, *287*, 9322–9326. [CrossRef] [PubMed]

10.  Maret, W. Inhibitory zinc sites in enzymes. *Biometals* **2013**, *26*, 197–204. [CrossRef] [PubMed]

11.  Peck, E.J.; Ray, W.J. Metal complexes of phosphoglucomutase in vivo. *J. Biol. Chem.* **1971**, *246*, 1160–1167. [PubMed]

12.  Ray, W.J. Role of bivalent cations in the phosphoglucomutase system. *J. Biol. Chem.* **1969**, *244*, 3740–3747. [PubMed]

13.  Feeney, K.A.; Hansen, L.L.; Putker, M.; Olivares-Yanez, C.; Day, J.; Eades, L.J.; Larrondo, L.F.; Hoyle, N.P.; O'Neill, J.S.; van Ooijen, G. Daily magnesium fluxes regulate cellular timekeeping and energy balance. *Nature* **2016**, *532*, 375–379. [CrossRef] [PubMed]

14.  Kambe, T.; Tsuji, T.; Hashimoto, A.; Itsumura, N. The physiological, biochemical, and molecular roles of zinc transporters in zinc homeostasis and metabolism. *Physiol. Rev.* **2015**, *95*, 749–784. [CrossRef] [PubMed]

15.  Krężel, A.; Maret, W. The functions of metamorphic metallothioneins in zinc and copper metabolism. *Int. J. Mol. Sci.* **2017**, *18*, 1237. [CrossRef] [PubMed]

16.  Haase, H.; Beyersmann, D. Uptake and distribution of labile and total Zn (II) in C6 rat glioma cells investigated with fluorescent probes and atomic absorption. *Biometals* **1999**, *12*, 247–254. [CrossRef] [PubMed]

17.  McCormick, N.H.; Lee, S.; Hennigar, S.R.; Kelleher, S.L. ZnT4 (SLC30A4)-null ("lethal milk") mice have defects in mammary gland secretion and hallmarks of precocious involution during lactation. *Am. J. Physiol. Regul. Integr. Comp. Physiol.* **2016**, *310*, R33–R40. [CrossRef] [PubMed]

18.  Tsuji, T.; Kurokawa, Y.; Chiche, J.; Pouysségur, J.; Sato, H.; Fukuzawa, H.; Nagao, M.; Kambe, T. Dissecting the process of activation of cancer-promoting zinc-requiring ectoenzymes by zinc metalation mediated by ZNT transporters. *J. Biol. Chem.* **2017**, *292*, 2159–2173. [CrossRef] [PubMed]

19.  Li, Y.; Maret, W. Transient fluctuations of intracellular zinc ions in cell proliferation. *Exp. Cell. Res.* **2009**, *315*, 2463–2470. [CrossRef] [PubMed]

20.  Maret, W. Analyzing free zinc (II) ion concentrations in cell biology with fluorescent chelating molecules. *Metallomics* **2015**, *7*, 202–211. [CrossRef] [PubMed]

21.  Maske, H. Über den topochemischen Nachweis von Zink im Ammonshorn verschiedener Säugetiere. *Naturwissenschaften* **1955**, *42*, 424. [CrossRef]

22.  Frederickson, C.J.; Bush, A.I. Synaptically released zinc: Physiological functions and pathological effects. *Biometals* **2001**, *14*, 353–366. [CrossRef] [PubMed]

23.  Danscher, G.; Stoltenberg, M. Zinc-specific autometallographic in vivo selenium methods: Tracing of zinc-enriched (ZEN) pathways, and pools of zinc ions in a multitude of other ZEN cells. *J. Histochem. Cytochem.* **2005**, *53*, 141–153. [CrossRef] [PubMed]

24.  Maret, W. Molecular aspects of zinc signals. In *Zinc Signals in Cellular Functions and Disorders*; Fukuda, T., Kambe, T., Eds.; Springer: Tokyo, Japan, 2014; pp. 7–26.

25.  Pan, E.; Zhang, X.A.; Huang, Z.; Krezel, A.; Tinberg, C.E.; Lippard, S.J.; McNamara, J.O. Vesicular zinc promotes presynaptic and inhibits postsynaptic long-term potentiation of mossy fiber-CA3 synapse. *Neuron* **2011**, *71*, 1116–1126. [CrossRef] [PubMed]

26.  Sindreu, C.; Palmiter, R.D.; Storm, D.R. Zinc transporter ZnT-3 regulates presynaptic Erk1/2 signaling and hippocampus-dependent memory. *Proc. Natl. Acad. Sci. USA* **2011**, *108*, 3366–3370. [CrossRef] [PubMed]

27.  Maret, W. Zinc in pancreatic islet biology, insulin sensitivity, and diabetes. *Prev. Nutr. Food Sci.* **2017**, *22*, 1–8. [CrossRef] [PubMed]

28.  Tamaki, M.; Fujitani, Y.; Hara, A.; Uchida, T.; Tamura, Y.; Takeno, K.; Kawaguchi, M.; Watanabe, T.; Ogihara, T.; Fukunaka, A.; et al. The diabetes-susceptible gene SLC30A8/ZnT8 regulates hepatic insulin clearance. *J. Clin. Investig.* **2013**, *123*, 4513–4524. [CrossRef] [PubMed]

29.  Schvartz, D.; Brunner, Y.; Couté, Y.; Foti, M.; Wollheim, C.B.; Sanchez, J.-C. Improved characterization of the insulin secretory granule proteomes. *J. Proteom.* **2012**, *75*, 4620–4631. [CrossRef] [PubMed]

30.  Que, E.L.; Duncan, F.E.; Bayer, A.R.; Philips, S.J.; Roth, E.W.; Bleher, R.; Gleber, S.C.; Vogt, S.; Woodruff, T.K.; O'Halloran, T.V. Zinc sparks induce physiochemical changes in the egg zona pellucida that prevent polyspermy. *Integr. Biol. (Camb.)* **2017**, *9*, 135–144. [CrossRef] [PubMed]

31.  Maret, W.; Vallee, B.L. Thiolate ligands in metallothionein confer redox activity on zinc clusters. *Proc. Natl. Acad. Sci. USA* **1998**, *95*, 3478–3482. [CrossRef] [PubMed]

32. Maret, W. Zinc Coordination environments in proteins as redox sensors and signal transducers. *Antioxid. Redox Signal.* **2006**, *8*, 1419–1441. [CrossRef] [PubMed]

33. Jacob, C.; Maret, W.; Vallee, B.L. Selenium redox biochemistry of zinc/sulfur coordination sites in proteins and enzymes. *Proc. Natl. Acad. Sci. USA* **1999**, *96*, 1910–1914. [CrossRef] [PubMed]

34. Hao, Q.; Maret, W. Aldehydes release zinc from proteins. A pathway from oxidative stress/lipid peroxidation to cellular functions of zinc. *FEBS J.* **2006**, *273*, 4300–4310. [CrossRef] [PubMed]

35. Pearce, L.L.; Gandley, R.E.; Han, W.; Wasserloos, K.; Stitt, M.; Kannai, A.J.; McLaughlin, M.K.; Pitt, B.R.; Levitan, E.S. Role of metallothionein in nitric oxide signaling as revealed by a green fluorescent fusion protein. *Proc. Natl. Acad. Sci. USA* **2000**, *97*, 477–482. [CrossRef] [PubMed]

36. Spahl, D.U.; Berendji-Grün, D.; Suschek, C.V.; Kolb-Bachofen, V.; Kröncke, K.D. Regulation of zinc homeostasis by inducible NO synthase-derived NO: Nuclear metallothionein translocation and intranuclear $Zn^{2+}$ release. *Proc. Natl. Acad. Sci. USA* **2003**, *100*, 13952–13957. [CrossRef] [PubMed]

37. Aizenman, E.; Stout, A.K.; Hartnett, K.A.; Dineley, K.E.; McLaughlin, B.; Reynolds, I.J. Induction of neuronal apoptosis by thiol oxidation: Putative role of intracellular zinc release. *J. Neurochem.* **2000**, *75*, 1878–1888. [CrossRef] [PubMed]

38. Maret, W. Metallothionein redox biology in the cytoprotective and cytotoxic functions of zinc. *Exp. Gerontol.* **2008**, *43*, 363–369. [CrossRef] [PubMed]

39. Beyersmann, D.; Haase, H. Functions of zinc in signaling, proliferation and differentiation of mammalian cells. *Biometals* **2001**, *14*, 331–341. [CrossRef] [PubMed]

40. Yamasaki, S.; Sakata-Sogawa, K.; Hasegawa, A.; Suzuki, T.; Kabu, K.; Sato, E.; Kurosaki, T.; Yamashita, S.; Tokunaga, M.; Nishida, K.; et al. Zinc is a novel intracellular second messenger. *J. Cell. Biol.* **2007**, *177*, 637–645. [CrossRef] [PubMed]

41. Taylor, K.M.; Hiscox, S.; Nicholson, R.I.; Hogstrand, C.; Kille, P. Protein kinase CK2 triggers cytosolic zinc signaling pathways by phosphorylation of zinc channel ZIP7. *Sci. Signal.* **2012**, *5*, ra11. [CrossRef] [PubMed]

42. Haase, H.; Maret, W. Intracellular zinc fluctuations modulate protein tyrosine phosphatase activity in insulin/insulin-like growth factor-1 signaling. *Exp. Cell. Res.* **2003**, *291*, 289–298. [CrossRef]

43. Hershfinkel, M.; Moran, A.; Grossman, N.; Sekler, I. A zinc-sensing receptor triggers the release of intracellular Ca2+ and regulates ion transport. *Proc. Natl. Acad. Sci. USA* **2001**, *98*, 11749–11754. [CrossRef] [PubMed]

44. Holst, B.; Egerod, K.L.; Schild, E.; Vickers, S.P.; Cheetham, S.; Gerlach, L.-O.; Storjohann, L.; Stidsen, C.E.; Jones, R.; Beck-Sickinger, A.G.; et al. GPR39 signaling is stimulated by zinc ions but not by obstetatin. *Endocrinology* **2007**, *148*, 13–20. [CrossRef] [PubMed]

45. Yasuda, S.-I.; Miyazaki, T.; Munechika, K.; Yamashita, M.; Ikeda, Y.; Kamizono, A. Isolation of $Zn^{2+}$ as an endogenous agonist of GPR39 from fetal bovine serum. *J. Recept. Signal Transduct. Res.* **2007**, *27*, 235–246. [CrossRef] [PubMed]

46. Popovics, P.; Steward, A.J. GPR39: A $Zn^{2+}$-activated G protein-coupled receptor that regulates pancreatic, gastrointestinal and neuronal functions. *Cell. Mol. Life Sci.* **2011**, *68*, 85–95. [CrossRef] [PubMed]

47. Romero-Hernandez, A.; Simorowski, N.; Karakas, E.; Furukawa, H. Molecular basis for subtype specificity and high-affinity zinc inhibition in the GluN1-GluN2A NMDA receptor amino-terminal domain. *Neuron* **2016**, *92*, 1324–1336. [CrossRef] [PubMed]

48. Bellomo, E.; Singh, B.K.; Massarotti, A.; Hogstrand, C.; Maret, W. The metal face of PTP-1B. *Coord. Chem. Rev.* **2016**, *327–328*, 70–83. [CrossRef] [PubMed]

49. Hao, Q.; Maret, W. Imbalance between pro-oxidant and pro-antioxidant functions of zinc in disease. *J. Alzheimer's Dis.* **2005**, *8*, 161–170. [CrossRef]

50. Maret, W. The Function of zinc metallothionein: A link between cellular zinc and redox state. *J. Nutr.* **2000**, *130*, 1455S–1458S. [PubMed]

51. Maret, W. Crosstalk of the group IIa and IIb metals calcium and zinc in cellular signaling. *Proc. Natl. Acad. Sci. USA* **2001**, *98*, 12325–12327. [CrossRef] [PubMed]

52. Günther, V.; Lindert, U.; Schaffner, W. The taste of heavy metals: Gene regulation by MTF-1. *Biochim. Biophys. Acta* **2012**, *1823*, 1416–1425. [CrossRef] [PubMed]

53. Hogstrand, C.; Fu, D. Zinc. In *Binding, Transport and Storage of Metal Ions in Biological Cells*; Maret, W., Wedd, A., Eds.; The Royal Society of Chemistry: Cambridge, UK, 2014; pp. 666–694.

54. Colvin, R.A.; Holmes, W.R.; Fontaine, C.P.; Maret, W. Cytosolic zinc buffering and muffling: Their role in intracellular zinc homeostasis. *Metallomics* **2010**, *2*, 306–317. [CrossRef] [PubMed]

55. Hardyman, J.E.; Tyson, J.; Jackson, K.A.; Aldridge, C.; Cockell, S.J.; Wakeling, L.A.; Valentine, R.A.; Ford, D. Zinc sensing by metal-responsive transcription factor 1 (MTF1) controls metallothionein and ZnT1 expression to buffer the sensitivity of the transcriptome response to zinc. *Metallomics* **2016**, *8*, 337–343. [CrossRef] [PubMed]

56. Laity, J.H.; Andrews, G.K. Understanding the mechanisms of zinc-sensing by metal-responsive element binding transcription factor-1 (MTF-1). *Arch. Biochem. Biophys.* **2007**, *463*, 201–210. [CrossRef] [PubMed]

57. Thomas, P.; Converse, A.; Berg, H.A. ZIP9, a novel membrane androgen receptor and zinc transporter protein. *Gen. Comp. Endocrinol.* **2017**. [CrossRef] [PubMed]

58. King, J.C. Zinc: An essential but elusive nutrient. *Am. J. Clin. Nutr.* **2011**, *94*, 679S–684S. [CrossRef] [PubMed]

59. Hogstrand, C.; Maret, W. Genetics of human zinc deficiencies. In *Encyclopedia of Life Sciences (ELS)*; John Wiley & Sons, Ltd.: Chichester, UK, 2016.

60. Vallee, B.L.; Falchuk, K.H. The biochemical basis of zinc physiology. *Physiol. Rev.* **1993**, *73*, 79–118. [PubMed]

61. Frederickson, C.J.; Koh, J.-Y.; Bush, A.I. The neurobiology of zinc in health and disease. *Nat. Rev. Neurosci.* **2005**, *6*, 449–462. [CrossRef] [PubMed]

International Journal of
*Molecular Sciences*

MDPI

*Review*

# Understanding the Contribution of Zinc Transporters in the Function of the Early Secretory Pathway

Taiho Kambe * , Mayu Matsunaga and Taka-aki Takeda

Division of Integrated Life Science, Graduate School of Biostudies, Kyoto University, Kyoto 606-8502, Japan; matsunaga.mayu.76x@st.kyoto-u.ac.jp (M.M.); takeda.takaaki.77s@st.kyoto-u.ac.jp (T.T.)
* Correspondence: kambe1@kais.kyoto-u.ac.jp; Tel.: +81-75-753-6273

Received: 8 September 2017; Accepted: 15 October 2017; Published: 19 October 2017

**Abstract:** More than one-third of newly synthesized proteins are targeted to the early secretory pathway, which is comprised of the endoplasmic reticulum (ER), Golgi apparatus, and other intermediate compartments. The early secretory pathway plays a key role in controlling the folding, assembly, maturation, modification, trafficking, and degradation of such proteins. A considerable proportion of the secretome requires zinc as an essential factor for its structural and catalytic functions, and recent findings reveal that zinc plays a pivotal role in the function of the early secretory pathway. Hence, a disruption of zinc homeostasis and metabolism involving the early secretory pathway will lead to pathway dysregulation, resulting in various defects, including an exacerbation of homeostatic ER stress. The accumulated evidence indicates that specific members of the family of Zn transporters (ZNTs) and Zrt- and Irt-like proteins (ZIPs), which operate in the early secretory pathway, play indispensable roles in maintaining zinc homeostasis by regulating the influx and efflux of zinc. In this review, the biological functions of these transporters are discussed, focusing on recent aspects of their roles. In particular, we discuss in depth how specific ZNT transporters are employed in the activation of zinc-requiring ectoenzymes. The means by which early secretory pathway functions are controlled by zinc, mediated by specific ZNT and ZIP transporters, are also subjects of this review.

**Keywords:** ZNT/Solute carrier family 30 member (SLC30A); ZIP/SLC39A; early secretory pathway; ER stress; unfolded protein response (UPR); zinc-requiring ectoenzymes; tissue non-specific alkaline phosphatase (TNAP); metallation

## 1. Introduction

Zinc is an essential trace element that is required for a large variety of cellular processes [1,2]. Approximately 10% of the eukaryotic proteome requires zinc for cellular activity [3,4], and thus any disturbance in zinc homeostasis can result in disease, including cancer, neuronal degeneration, chronic inflammation, hypertension, osteoarthritis, and age-related macular degeneration. A diverse range of symptoms is also found in cases of zinc deficiency [1,2,5–8]. The biological functions of zinc can be grouped into three major categories, structural, catalytic, and regulatory. However, the molecular basis of how zinc engages in such diverse functions is still far from being completely understood [1,2].

At the cellular level, zinc plays a pivotal role in the function of a variety of subcellular compartments, one of which is the early secretory pathway constituted by the endoplasmic reticulum (ER), the Golgi apparatus, and other intermediate organelles, such as the ER-Golgi intermediate compartment. Zinc homeostasis in the lumen of these compartments requires a transport system to translocate zinc across biological membranes. In vertebrates, Zn transporters (ZNTs)/Solute carrier family 30 member (SLC30A) and Zrt- and Irt-like proteins (ZIPs)/SLC39A are widely recognized as being critical transporters in zinc metabolism under physiological conditions [1,2,9]. Both of these proteins are clearly important for zinc metabolism involved in early secretory pathways. This review outlines the functions of ZNT and ZIP transporters in the regulation and function of secretory pathways, in particular, the early secretory pathway, focusing on several recent aspects of the molecular processes underlying the ER stress response, as well as the activation of zinc-requiring ectoenzymes. Zinc transporters also play important roles in secretory granules/vesicles that contain high amounts of zinc, such as insulin granules, synaptic vesicles, and secretory vesicles involved in milk secretion; these are discussed in this review for comparison. Further details of these transporters can be found in other comprehensive reviews of zinc transporters [1,10–12].

## 2. Brief Overview of the Properties of ZNT and ZIP Transporters

In mammals, there are nine ZNT and 14 ZIP transporters that play distinct roles in the maintenance of systemic, cellular, and subcellular zinc homeostasis. These transporters act in a cell or tissue-specific manner, and are developmentally regulated [1,2,9] (Figure 1). ZNTs transport zinc from the cytosol into either the lumen of intracellular compartments or the extracellular milieu, whereas ZIPs transport zinc in the opposite direction. Zinc transport by ZNT and ZIP transporters is coordinately controlled through precisely timed increases or decreases in their expression, and by their precise subcellular localization [2,13]. A growing body of evidence has shown that cooperative zinc transport across biological membranes mediated by both transporters contributes to the control of expression, localization, and functional activity of target proteins [1]. The molecular features of ZNT and ZIP transporters have been extensively summarized in other review papers [1,2], and thus only their main features are outlined briefly here.

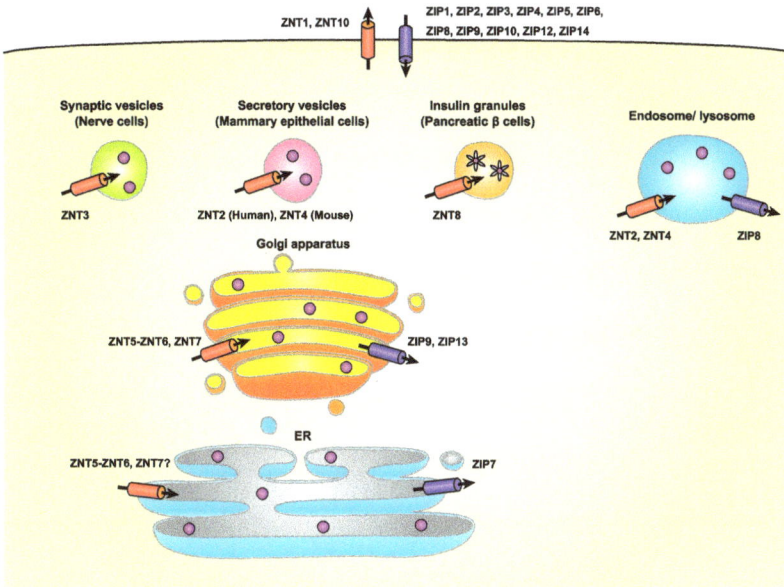

**Figure 1.** Subcellular localization of ZNT and ZIP transporters. ZNT transporters move cytosolic zinc into the lumen of vesicles involved in the early secretory pathway, including the endoplasmic reticulum (ER), Golgi apparatus, as well as into cytoplasmic vesicles/granules such as synaptic and secretory vesicles and insulin granules, in which specific ZNT proteins are localized. ZNT5 and ZNT6 form heterodimers to transport zinc. ZIP transporters move zinc in the opposite direction. In contrast to the specific localization of ZIP9, ZIP13, and ZIP7 in the Golgi apparatus and the ER, the subcellular location of ZNT5-ZNT6 heterodimers has not been definitively determined.

Based on the three-dimensional structure of the *Escherichia coli* homolog YiiP, ZNT transporters are predicted to have six transmembrane (TM) helices (TM helices I-VI) [14–18]. ZNT transporters function as zinc/proton exchangers [19,20], and can form homodimers or heterodimers [21–24]. With respect to the zinc transport mechanism used by YiiP, two models have been proposed, the alternative access mechanism model, in which the TM helices form inward- and outward-facing conformations [17,18], and the allosteric mechanism model, in which cytosolic zinc binding induces a scissor-like movement of the homodimers and interlocks the TM helices at the dimer interface [15,16]. ZNT transporters likely transport zinc using either of the two proposed mechanisms. As has been observed for YiiP, which has an intramembranous zinc-binding site formed by TM helices II and V, ZNT transporters are also thought to have a conserved intramembranous zinc-binding site, which is indispensable for zinc transport activity [19,20,25,26]. The intramembranous zinc-binding site in most ZNT transporters consists of two His and two Asp residues in TM helices II and V [19,27]. Interestingly, ZNT10 has an Asn residue in TM helix II instead of His, which confers the ability to transport manganese [28], as has also been seen for the homologous bacterial protein [29]. The nine ZNT transporters belong to the cation diffusion facilitator (CDF) family of transporters, which are classified into three subgroups, namely Zn-CDF, Zn/Fe-CDF, and Mn-CDF [14,30]. All of the ZNT transporters are classified as being Zn-CDF members (although ZNT10 is a manganese transporter), and, based on their sequence similarities, can be further subdivided into four groups: (i) ZNT1 and ZNT10, (ii) ZNT2, ZNT3, ZNT4, and ZNT8, (iii) ZNT5 and ZNT7, and (iv) ZNT6, [14,31,32] (Figure 2). Of interest to this review, some characteristics of the transporters, such as subcellular localization, are conserved in the members of the same group [32] (Figure 1).

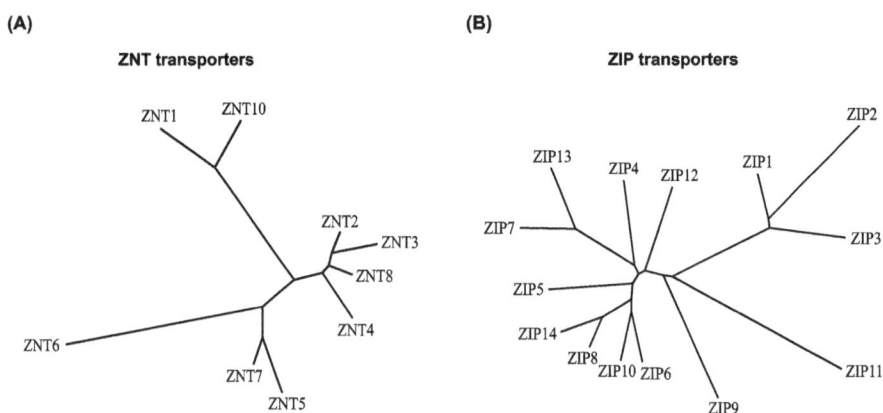

**Figure 2.** Phylogeny of ZNT and ZIP transporters. The neighbor-joining phylogenetic tree was constructed using ClustalW (http://clustalw.ddbj.nig.ac.jp/index.php?lang=en) protein alignment. (**A**) ZNT and (**B**) ZIP transporters. Subfamilies and subgroups are designated according to the text.

Computational analysis suggested that ZIP transporters have eight TM helices [33], and this was confirmed by the first three-dimensional structure reported for a ZIP transporter homolog in bacteria (*Bordetella bronchiseptica*) [34]. The structure shows that the ZIP transporter has a novel 3 + 2 + 3 TM architecture with a binuclear metal center, in which two His residues, one each in TM helices IV and V, form two intramembranous zinc binding sites [34]. As in ZNT transporters, ZIP transporters form functional homo- or heterodimeric complexes, which are essential for their zinc transport ability, although no dimer formation was seen in the crystal structure [34–36]. ZIP transporters may function as selective electrodiffusion channels [37], or as zinc/bicarbonate symport transporters [38–40]. However, their definitive mode of transport has not yet been completely elucidated. Phylogenetic analysis classifies the fourteen ZIP transporter members into four subfamilies, namely ZIP I (ZIP9), ZIP II

(ZIP1-ZIP3), LIV-1 (ZIP4-ZIP8, ZIP10, ZIP12-ZIP14), and gufA (ZIP11) [1,41]. In the LIV-1 subfamily, features of their extracellular domains further classify the proteins into four subgroups as follows: (i) ZIP4 and ZIP12, (ii) ZIP8 and ZIP14, (iii) ZIP5, ZIP6, and ZIP10, and (iv) ZIP7 and ZIP13 [42] (Figure 2). Members of the LIV-1 subfamily have an extended extracellular N-terminus, whose structure has been solved only in the case of ZIP4 [42]. The extracellular portion of ZIP4 can form homodimers without the need for TM helices [42], which may facilitate dimerization. Interestingly, in ZIP transporters belonging to subgroup (iii), a prion-like domain is present in the extracellular N-terminal portion proximal to the first TM helix; hence, there is an evolutionary link between these transporters and the prion protein [43]. ZIP8 and ZIP14 in subgroup (ii) have the ability to transport manganese [39,40], because they have a Glu residue in TM helix V rather than a His residue [1,44]. The His residue is therefore involved in metal substrate specificity, because of its contribution to forming intramembranous zinc binding sites [34].

Over the last two decades, the physiological roles of ZNT and ZIP transporters and their involvement in disease pathology have been clarified at the molecular level, as has been described elsewhere [1,2,8,9], and a deeper understanding will likely come in the future.

### 3. Regulation of Zinc Homeostasis by Zinc Transporters in the Early Secretory Pathway

Approximately one-third of all the cellular proteins in eukaryotes are targeted to the ER, and thus the early secretory pathway [45], in which nascent proteins are folded, assembled, and modified during their trafficking to final destinations. Importantly, a considerable proportion of the secretome requires zinc as a structural and catalytic cofactor. Moreover, resident chaperones require zinc for modulation and potentiation of their functions [46–48]. Hence, any disruption of zinc homeostasis in the early secretory pathway can cause and exacerbate ER stress [49,50], and trigger the unfolded protein response (UPR) in cells. Therefore, elaborate regulatory mechanisms are used to control zinc homeostasis in the early secretory pathway. Accumulating evidence clearly shows that both ZNT and ZIP transporters play crucial roles in this process [49–53], and this is summarized in this section.

With the exception of ZNT1 and ZNT10 (which are members of ZNT subgroup (i) described above [28,54]), ZNT transporters are mainly localized to intracellular compartments. Of these, ZNT5, ZNT6, and ZNT7 (members of ZNT subgroups (iii), and (iv) described above) are involved in the early secretory pathway [12,55]. ZNT5 has been shown to be mainly localized to coat protein complex II (COPII) vesicles and the Golgi apparatus [56], whereas ZNT6 is localized to the Golgi apparatus [57], although these ZNTs can also form heterodimers as functional complexes [21,22] (Figure 1). The actual subcellular localization of these heterodimers has however been poorly investigated. ZNT7 is also located in the Golgi apparatus [58], and a recent study indicates that it is also localized to the sarco(endo)plasmic reticulum (S(E)R) [52] (Figure 1). These ZNT transporters are employed as zinc entry routes in the early secretory pathway, suggesting that a lack of them would be potentially to elicit an ER stress response. In fact, it has been clearly shown that cells lacking these ZNT transporters do exhibit an exacerbated ER stress responses [50,59] (Figure 3). A similar exacerbation of ER stress is found in yeast lacking ZNT orthologs [49,60], which highlights the fact that the important role these ZNT transporters play in maintaining zinc homeostasis in the secretory pathway is well-conserved among subgroups. However, there remains the interesting questions of how and where these ZNT transporters transport zinc, and their association with ER stress, because ZNT5, ZNT6, and ZNT7 all appear to be principally localized to the Golgi apparatus [57,58,61]. Recent studies have shown that another ZNT transporter, either ZNT3 or ZNT10, may play a protective role in ER stress-induced toxicities [62,63], although their contributions to the early secretory pathway have not yet been clarified.

In contrast to ZNT transporters, most of which are located in intracellular compartments, most, but not all, ZIP transporters are found on the plasma membrane. Of the 14 ZIP transporters, ZIP7, ZIP9, and ZIP13 (the ZIPI subfamily and the LIV-1 subgroup (iii) described above) are involved in the early secretory pathway [64–67] (Figure 1), although recent reports also indicate that ZIP9 is also found localized on the plasma membrane where it serves as a membrane androgen receptor [68] and that ZIP13 can also be found in intracellular vesicles [69]. It has also been suggested that ZIP11 can localize to the Golgi apparatus [70], but this has not yet been thoroughly established. Accordingly, this protein is not further discussed here. ZIP7, ZIP9, and ZIP13 are known to release zinc, which is stored in the early secretory pathway, into the cytosol in response to various stimuli, thus contributing to the signaling function of zinc [66,71–73]. Importantly, ZIP7, ZIP9, and ZIP13 are also thought to contribute to homeostatic maintenance of the secretory pathway. In this regard, ZIP7, which is located in the ER, plays an indispensable role in the proper regulation of ER function, through the fine-tuning of zinc homeostasis [51,53]. The loss of ZIP7 probably increases zinc levels in the ER, which triggers zinc-dependent aggregation of protein disulfide isomerase, leading to ER stress [53] (Figure 3). This critical functional role of ZIP7 in ER homeostasis contributes to self-renewal processes of intestinal epithelium [51] and appropriate epidermal development [53]. ZIP7 is also involved in the induction of ER stress by mediating the redistribution of zinc into the cytosol from the S(E)R in cardiomyocytes under hyperglycemic conditions [52]. Consistent with the involvement of ZIP7 in ER stress, the yeast ZIP7 homolog, yKE4, has been shown to be involved in ER stress responses [74]. Similarly, Catsup, a Drosophila ZIP7 ortholog, which plays a crucial role in catecholamine synthesis, is also involved in the ER stress response [75,76]. Along with ZIP7, Golgi-localized ZIP9 is also thought to contribute to secretory homeostasis [67], and likewise, ZIP13 contributes directly to zinc homeostasis in the early secretory pathway by mobilizing zinc from the Golgi apparatus, or indirectly by releasing zinc from intracellular vesicles [69]. Recently, although it is localized to the plasma membrane [77,78], ZIP14 has also been shown to play a significant role in the adaptation to ER stress [79,80], suggesting that zinc homeostasis in the early secretory pathway might be indirectly controlled by ZIP transporters that are located in other subcellular regions.

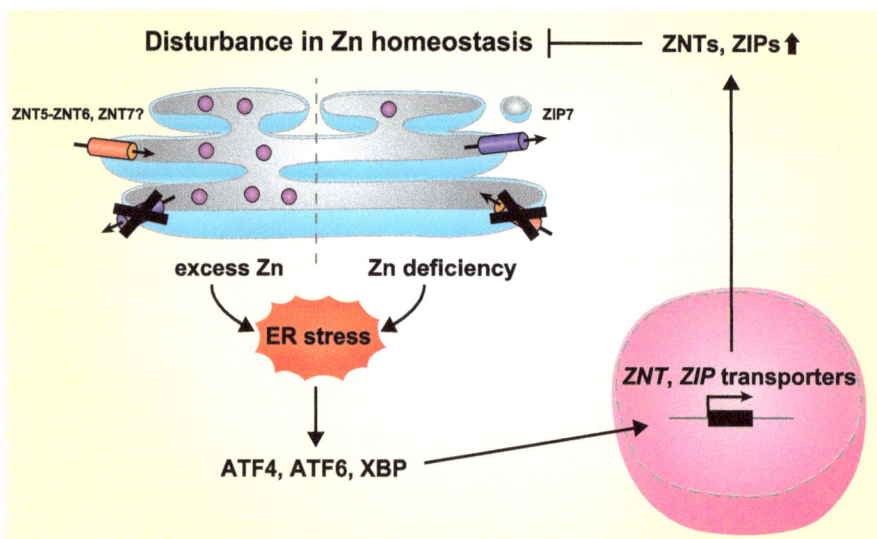

**Figure 3.** Model of feedback regulation for the maintenance of zinc homeostasis in the ER (in the early secretory pathway). A disturbance in zinc homeostasis, such as zinc deficiency or zinc overload, in the ER (and perhaps in the early secretory pathway) induces homeostatic ER stress. The unfolded protein response (UPR) leads to the activation of transcription factors such as ATF4, ATF6, and XBP1, and increases the transcription of several ZNT and ZIP transporter genes. These activities of ZNT and ZIP transporters then contribute to the maintenance of zinc homeostasis in the ER (and in the early secretory pathway), and thus attenuate homeostatic ER stress. Zn: zinc.

With respect to the direction of zinc transport mediated by ZNT and ZIP transporters, both decreases and increases in zinc levels in the early secretory pathway exacerbate its proper functioning and thus either increases or decreases in zinc levels will result in the homeostatic ER stress response. The molecular basis underlying this phenomenon may be explained by changes in the activity of chaperone proteins that are either positively or negatively regulated by zinc [46–48], although this has not yet been completely elucidated.

## 4. Regulation of Expression of ZNT and ZIP Transporters by ER Stress

Based on the crucial functions of zinc mobilized by ZNT and ZIP transporters in the early secretory pathway, it is easy to imagine that the transcription of ZNT and ZIP genes would be regulated by homeostatic ER stress. In fact, *ZNT5* transcription increases in response to inducers of ER stress, and its promoter harbors a UPR element, which serves as the binding site for the transcription factor XBP-1 [50] (Figure 3). *ZIP14* transcription is also induced by inducers of ER stress [79,80], and its promoter also has several ER stress response elements, to which the transcription factors ATF6 and ATF4 bind [79,80]. Moreover, the treatment with inducers of ER stress or *N,N,N′,N′*-tetrakis(2-pyridylmethyl)ethylenediamine, a zinc chelator which also causes ER stress, has been shown to induce the expression of ZIP3, ZIP7, ZIP9, ZIP13, and ZIP14, as well as ZNT3, ZNT6, ZNT7, and ZNT10 [51,62,63,79,80], although the elements responsible for the induction have not yet been identified in the promoter regions of these genes. Several homologues of both types of transporter have been shown to be increased by inducers of ER stress [81]. Thus, the fine-tuning of zinc homeostasis by these zinc transporters in the early secretory pathway, as well as the regulation of their expression triggered by homeostatic ER stress, are important control mechanisms in maintaining homeostasis [12].

## 5. Importance of ZNT Transporters in the Activation of Ectoenzymes in the Early Secretory Pathway

Zinc-requiring ectoenzymes, which are defined here as secretory, membrane-bound, and organelle-resident enzymes, have attracted considerable attention because they play crucial roles in various physiological functions, and in a number of pathological processes, such as cancer progression, and metastasis [1,55]. Thus, they are regarded as potential therapeutic targets in the treatment of diseases [82–86]. Moreover, the activities of some zinc-requiring ectoenzymes, e.g., alkaline phosphatases (ALP), may be used as clinical markers to reflect systemic zinc status [87,88]. These enzymes are synthesized in the early secretory pathway, at which point they acquire zinc for their activity, before being trafficked to the plasma membrane via the constitutive secretory pathway [12,55]. How zinc is made available to zinc-requiring ectoenzymes is largely unknown, but the importance of ZNT transporters has been partially clarified in the activation of specific enzymes. This section addresses these specific ectoenzyme activation processes in detail.

### 5.1. ZNT Transporters Involved in Zinc-Requiring Ectoenzyme Activation

Zinc-requiring ectoenzymes likely become active by coordinating with zinc at their active site (i.e., they become metallated) during the secretory process. Zinc coordination is generally achieved by the interaction of zinc with three or four amino acids, including His, Asp, and Glu residues [4,89,90], which must undergo precise regulation for the conversion from the apo- to holo-forms. When compared with cytosolic zinc-requiring enzymes, zinc-requiring ectoenzymes require more complicated and elaborate regulatory processes involving zinc mobilization, because their activation process requires at least two types of zinc transporters that involve two biological membranes. In the first process, zinc transport from the extracellular milieu to the cytosol (i.e., ZIP transporters) occurs, and in the second process zinc transport from the cytosol to the lumen (i.e., ZNT transporters) occurs. Information regarding the identity of the ZIP transporters involved is lacking, while information relating to the ZNT transporters is accumulating. Three ZNT transporters, constituting two independent complexes, have been shown to be indispensable in ectoenzyme metallation; one complex is formed by ZNT5 and ZNT6 as a heterodimer, in which ZNT6 operates as an auxiliary subunit, and the other is formed by ZNT7 homodimers [21,91,92]. Both of these ZNT complexes can specifically activate several zinc-requiring enzymes, such as tissue-nonspecific ALP (TNAP) and placental ALP, as well as autotaxin (ATX) [24,56,93]. These three enzymes have similar active site geometry with a bimetallic core, which consists of two zinc ions, one of which is coordinated by one Asp and two His residues, and the other coordinated by one His and two Asp residues (Table 1), although ALPs and ATX catalyze different enzymatic reactions and have different biological roles. Based on these enzymes, it could be hypothesized that specific regulation mechanisms are operative in this bimetallic core enzyme family during conversion from the apo- to holo-enzymes. Both ZNT complexes, however, can activate other zinc-requiring ectoenzymes, such as matrix metalloproteinase (MMP)-9 and probably MMP-2 [24], and thus, they likely play a critical role in the activation of many zinc-requiring ectoenzymes in the early secretory pathway.

**Table 1.** Relationship between properties of zinc-requiring enzymes and ZNT transporters *.

| Enzyme | Active Site | Zn Coordination Residues | ZNTs Involved in Activation | Defects Caused by Loss of ZNTs |
|---|---|---|---|---|
| ALP | Bimetallic center | Asp, His, His for $Zn_1$ His, Asp, Asp for $Zn_2$ | ZNT5-ZNT6, ZNT7 | Loss of enzyme activity Protein destabilization |
| ATX | Bimetallic center | Asp, His, His for $Zn_1$, His, Asp, Asp for $Zn_2$ | ZNT5-ZNT6, ZNT7 ZNT4, | Loss of enzyme activity |
| CAIX | Mononuclear Zn | His, His, His | ZNT5ZNT6,ZNT7 | Decreases in enzyme activity |
| MMP-2 MMP-9 | Mononuclear Zn ** | His, His, His | ZNT5-ZNT6, ZNT7 | Loss of enzyme activity Protein destabilization *** |

* The structures of the zinc-requiring ectoenzymes are predicted based on homologous proteins; ** MMP-2 and MMP-9 have an additional zinc for stability of protein structure; *** Zinc supplementation partially restores both enzyme activity and protein stability. Zn: zinc.

However, some enzymes can be metallated by zinc and become activated through different pathways involving these zinc transporters. For example, carbonic anhydrase IX (CAIX) can acquire zinc via ZNT4 homodimers, in addition to ZNT5-ZNT6 heterodimers and ZNT7 homodimers [24]. ZNT4 homodimers may also be involved in carbonic anhydrase VI maturation [94]. These findings are interesting for two reasons. The first is that ZNT4 homodimers have multifunctional roles depending on their subcellular localization. For example, ZNT4 was originally reported to be localized to late endosomes [95], where it has a role in reducing cytosolic zinc toxicity [25,96], but it has also been shown to be involved in secretory pathways involving the *trans*-Golgi network, cytosolic vesicles, and probably other secretory vesicles. In addition, it has been shown to be involved in zinc secretion into breast milk in mice [97,98]. The second is that ZNT4 homodimers can become functionally equivalent to ZNT5-ZNT6 heterodimers or ZNT7 homodimers in the activation of specific ectoenzymes, including CAIX, in the early secretory pathway. The involvement of ZNT4 in CAIX activation is specific because ZNT2 expression failed to result in CAIX activation [24].

*5.2. Insight into the Activation of TNAP and Other Ectoenzymes by ZNT5-ZNT6 Heterodimers and ZNT7 Homodimers*

What affects the metallation of TNAP via ZNT5-ZNT6 heterodimers or ZNT7 homodimers? This question remains to be fully resolved but some important insights have been made to date. First, the number of zinc ions and their coordination manner at the active site does not seem to affect the activation process mediated via ZNT5-ZNT6 heterodimers or ZNT7 homodimers. The ALP protein possesses two zinc ions at the active site (zinc bimetallic core), which are coordinated by His, His, and Asp residues or Asp, Asp, and His residues, as described above [55], whereas MMP-9 has a single zinc ion at the active site, which is coordinated by three His residues (MMP-9 has another zinc ion that acts as a structural component) [55] (Table 1). CAIX, which can be activated by ZNT4 homodimers,

ZNT5-ZNT6 heterodimers and ZNT7 homodimers [24], as described above, possesses a single zinc ion coordinated by three His residues at the active site, supporting this notion.

Second, a specific motif may be significantly involved in TNAP activation via ZNT5-ZNT6 heterodimers and ZNT7 homodimers [55,99]. In cells lacking both ZNT complexes, the TNAP protein is destabilized, although it is not destabilized by zinc deficiency [93]. These data show that both ZNT complexes can stabilize the TNAP protein, in addition to supplying it with zinc. In other words, the TNAP activation process can be separated into two steps: the TNAP protein is first stabilized by ZNT5-ZNT6 heterodimers or ZNT7 homodimers in the early secretory pathway and then is metallated by zinc supplied by both ZNT complexes [93] (Figure 4). In this two-step mechanism, the Pro-Pro (PP)-motif in luminal loop 2 of ZNT5 (which corresponds to luminal loop 7, because of the fact that ZNT5 has extra N-terminal TM helices [61]) and ZNT7, which is highly conserved in ZNT5 and ZNT7 homologs across multiple species, is thought to be important [100]. In model structures of ZNT5 and ZNT7, the PP-motif is located just above the intramembranous zinc-binding site in the TM helices [100], suggesting that a unique cooperative mechanism might operate between the intramembranous zinc-binding site and the PP-motif in the activation of TNAP. In contrast, a similar two-step activation mechanism does not seem to operate in the activation of ATX, because the ATX protein is not destabilized in cells lacking both ZNT5-ZNT6 heterodimers or ZNT7 homodimers, and so the PP-motif plays a somewhat minor role [24]. This discrepancy between TNAP and ATX may be explained by differences in the degree of their glycosylation, although this needs to be clarified in future studies.

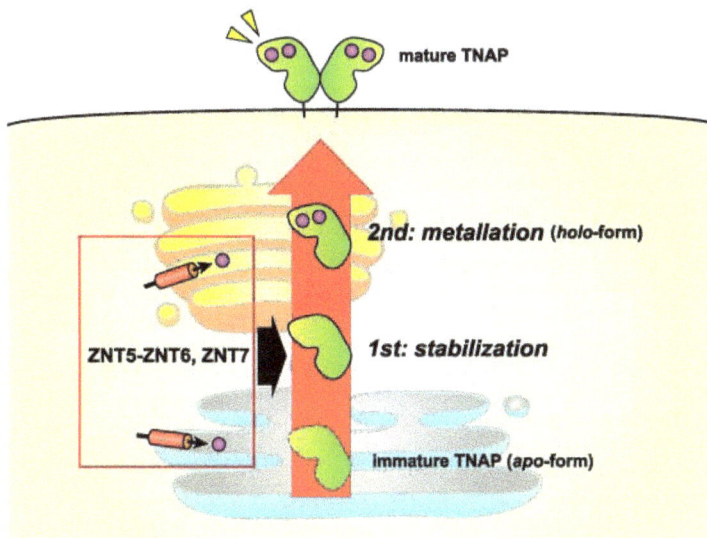

**Figure 4.** ZNT5-ZNT6 heterodimers and ZNT7 homodimers function to activate tissue-nonspecific ALP (TNAP) in a two-step mechanism. TNAP is specifically activated in a two-step mechanism involving ZNT5-ZNT6 heterodimers and ZNT7 homodimers as follows: first, the apo-form of TNAP is stabilized by either ZNT5-ZNT6 heterodimers or ZNT7 homodimers; second, the apo-form of TNAP is converted to the *holo*-form by zinc metallation. The PP-motifs in ZNT5 and ZNT7 likely play important roles in this process (see text). TNAP possesses a bimetallic core, is dimeric, and is localized to the plasma membrane via a glycophosphatidylinositol anchor. The subcellular localizations of ZNT5-ZNT6 heterodimers or ZNT7 homodimers have not been well defined. Zn: zinc.

*Znt5* or *Znt7* knockout (KO) mice show various phenotypes [101–105], but the association of those phenotypes with the early secretory pathway functions is unclear to date. One exception is the phenotype of osteopenia [101], which may be associated with reduced TNAP activity caused by a lack of ZNT5-ZNT6 heterodimers.

## 6. Importance of ZNT Transporters in Zinc-Related Regulated Secretory Pathway: After the Early Secretory Pathway

There are a number of cells that accumulate large amounts of zinc in cytoplasmic vesicles/granules (Figure 1). One can think that zinc, which is transported to the early secretory pathway, traffics those vesicles/granules through the secretory pathway and gets accumulated there. However, this is not the case. In fact, specific ZNT transporters are localized to the specific vesicles/granules and perform specific functions. In this section, representative vesicles/granules and the ZNT transporters involved in their function are briefly summarized to emphasize these points. Other aspects of this have been extensively reviewed elsewhere [1,10–12].

Insulin granules in pancreatic islet β-cells require high amounts of zinc in order to form insulin-zinc crystals, a process in which ZNT8 plays an indispensable role [106–110]. Nevertheless, a clear and crucial role for ZNT8 in regulating glucose homeostasis is yet to be established (*Znt8* KO mice are largely glucose tolerant), and thus the physiological relevance of zinc accumulation in insulin secretory granules remains unclear. Although, an interesting hypothesis is that zinc, which is secreted in concert with insulin, suppresses the insulin clearance in the liver by inhibiting clathrin-dependent insulin endocytosis [110]. Alterations in ZNT8 function are thought to lead to an increase in the risk of type 2 diabetes [111,112], because the R325W polymorphism in ZNT8 is associated with an increased risk of type 2 diabetes [113], and the R-form (i.e., the increased-risk form of ZNT8) likely alters its zinc transport activity [111,112]. In addition, another study has shown that haploinsufficiency of ZNT8 is protective against type 2 diabetes [114]. The relationship between ZNT8 and type 2 diabetes therefore requires further investigation [115]. The transporters ZNT5 and ZNT7 are also relatively highly expressed in pancreatic β-cells, [61,116,117], and both may be associated with β-cell function: loss-of-function of *Znt5* is associated with attenuation of the incidence of diabetes and mortality [103], whereas loss-of-function of *Znt7* impairs glucose tolerance and reduces glucose-stimulated increases in plasma insulin levels, hepatic glycogen levels, and pancreatic insulin content [104,105,118]. Moreover, loss-of-function of *Znt7* results in a markedly-reduced zinc content in β-cells, which is made more profound by the combined loss of function of *Znt8* [118]. ZNT5 and ZNT7 may therefore contribute to β-cell function in the early secretory pathway, but not in the insulin granules themselves.

Synaptic vesicles present in a subset of glutamatergic neurons in the hippocampus and neocortex also accumulate high amounts of zinc, which is mediated by ZNT3: *Znt3* KO mice lack synaptic zinc [119]. Zinc secreted from the synaptic vesicles into the extracellular space, as a result of ZNT3 activity, acts as a signaling molecule by modulating neuronal transmission and plasticity through binding to multiple ion channels, transporters, and receptors on postsynaptic neurons involved in neurotransmission [120–122]. The importance of synaptic zinc has been confirmed in knock-in mice studies, using glycine and *N*-methyl-D-aspartate receptors, in which the zinc-binding sites in each receptor were mutated [123,124]. A disturbance of synaptic zinc homeostasis or a dysfunction in ZNT3 has been suggested to result in neurodegenerative diseases [125–127].

Because zinc is essential for the growth and health of neonates, breast milk contains high amounts of zinc, considerably higher than the levels found in serum. The transporter ZNT2 is responsible for supplying zinc to the breast milk produced by the mammary epithelial cells in humans. Mothers with missense or nonsense mutations in the *ZNT2* gene secrete zinc-deficient milk (75–95% reduction), and thus infants exclusively breast-fed by mothers carrying the mutation experience transient neonatal zinc deficiency (TNZD; OMIM 608118) [87,128–131]. Zinc-deficient milk is produced by mothers with a heterozygous mutation in the *ZNT2* gene, and their infants also suffer from TNZD, suggesting that having one active copy of the *ZNT2* gene is not sufficient to provide zinc levels in breast milk adequate to support normal infant growth. One report has also suggested the involvement of ZNT5-ZNT6 heterodimers in the pathogenesis of TNZD [132], but there are no reports indicating that low-zinc breast milk can be attributed to mutations in the *ZNT4* gene in humans, although Znt4 is involved in low-zinc breast milk in mice [97,98].

Zinc that accumulated in granules/vesicles can be released in response to various stimuli and thereby regulate a number of diverse processes [110,133,134]. This phenomenon can be divided into two classes, the first being zinc secretion into the extracellular environment (e.g., zinc "sparks") [133,134], as described for synaptic zinc, and the second being zinc release from intracellular stores into the cytosol (e.g., zinc "wave") [135]. The latter phenomenon is strongly associated with the signaling functions of zinc and thus contributes to driving major signaling pathways [71,72,136]. The function of zinc in signaling has been extensively reviewed [2,41,137].

## 7. Perspectives

This review focuses on crucial functions of specific ZNT and ZIP zinc transporters in the early secretory pathway. As we have shown, these two classes of zinc transporter are doubtless key molecules required for the proper function of the early secretory pathway. However, there are many unsolved and fundamental questions that remain to be addressed. Specifically, how do both zinc deficiency and elevation in the early secretory pathway cause and exacerbate ER stress? How do those ZNT and ZIP transporters properly regulate zinc metabolism in a spatiotemporal manner in the early secretory pathway? Moreover, how is zinc coordinated in zinc-requiring ectoenzymes in the early secretory pathway? Are zinc chaperones required to facilitate zinc metallation of the large number of nascent proteins found in the early secretory pathway? Even the most fundamental question as to what the actual zinc concentration is in the early secretory pathway has not yet been definitively addressed, because the proposed zinc concentrations in the ER and the Golgi are controversial [138,139]. Moreover, clarification of a functional relationship(s) between the early secretory pathway and constitutive secretory or the regulated secretory pathway is required from the perspective of zinc metabolism. The answers to these questions can help our understanding of zinc in the early secretory pathway, and provide information that should be useful for the treatment of numerous diseases.

**Acknowledgments:** This work was supported by Grants-in-Aid for Scientific Research (B) from the Japan Society for the Promotion of Science (KAKENHI, Grant No. 15H04501 to Taiho Kambe). Taka-aki Takeda is a Research Fellow (DC1) of the Japan Society for the Promotion of Science.

**Author Contributions:** Taiho Kambe wrote the paper. Mayu Matsunaga and Taka-aki Takeda edited the manuscript and described the figures.

**Conflicts of Interest:** The authors declare no conflict of interest.

## Abbreviations

| | |
|---|---|
| ALP | Alkaline phosphatase |
| ATX | Autotaxin |
| CAIX | Carbonic anhydrase IX |
| CDF | Cation diffusion facilitator |
| ER | Endoplasmic reticulum |
| KO | Knockout |
| MMP | Matrix metalloproteinase |
| PP | Pro-Pro |
| S(E)R | Sarco(endo)plasmic reticulum |
| SLC | Solute carrier |
| TM | Transmembrane |
| TNAP | Tissue non-specific alkaline phosphatase |
| TNZD | Transient neonatal zinc deficiency |
| UPR | Unfolded protein response |
| ZIP | Zrt- and Irt-like protein |
| ZNT | Zn transporter |

## References

1. Kambe, T.; Tsuji, T.; Hashimoto, A.; Itsumura, N. The Physiological, Biochemical, and Molecular Roles of Zinc Transporters in Zinc Homeostasis and Metabolism. *Physiol. Rev.* **2015**, *95*, 749–784. [CrossRef] [PubMed]
2. Hara, T.; Takeda, T.A.; Takagishi, T.; Fukue, K.; Kambe, T.; Fukada, T. Physiological roles of zinc transporters: Molecular and genetic importance in zinc homeostasis. *J. Physiol. Sci.* **2017**, *67*, 283–301. [CrossRef] [PubMed]
3. Andreini, C.; Banci, L.; Bertini, I.; Rosato, A. Zinc through the three domains of life. *J. Proteome Res.* **2006**, *5*, 3173–3178. [CrossRef] [PubMed]
4. Maret, W.; Li, Y. Coordination dynamics of zinc in proteins. *Chem. Rev.* **2009**, *109*, 4682–4707. [CrossRef] [PubMed]
5. Maret, W.; Sandstead, H.H. Zinc requirements and the risks and benefits of zinc supplementation. *J. Trace Elem. Med. Biol.* **2006**, *20*, 3–18. [CrossRef] [PubMed]
6. Devirgiliis, C.; Zalewski, P.D.; Perozzi, G.; Murgia, C. Zinc fluxes and zinc transporter genes in chronic diseases. *Mutat. Res.* **2007**, *622*, 84–93. [CrossRef] [PubMed]
7. Age-Related Eye Disease Study Research Group. A randomized, placebo-controlled, clinical trial of high-dose supplementation with vitamins C and E, beta carotene, and zinc for age-related macular degeneration and vision loss: AREDS report No. 8. *Arch. Ophthalmol.* **2001**, *119*, 1417–1436.
8. Kambe, T.; Hashimoto, A.; Fujimoto, S. Current understanding of ZIP and ZnT zinc transporters in human health and diseases. *Cell. Mol. Life Sci.* **2014**, *71*, 3281–3295. [CrossRef] [PubMed]
9. Lichten, L.A.; Cousins, R.J. Mammalian zinc transporters: Nutritional and physiologic regulation. *Annu. Rev. Nutr.* **2009**, *29*, 153–176. [CrossRef] [PubMed]
10. Kelleher, S.L.; McCormick, N.H.; Velasquez, V.; Lopez, V. Zinc in specialized secretory tissues: Roles in the pancreas, prostate, and mammary gland. *Adv. Nutr.* **2011**, *2*, 101–111. [CrossRef] [PubMed]
11. Hennigar, S.R.; Kelleher, S.L. Zinc networks: The cell-specific compartmentalization of zinc for specialized functions. *Biol. Chem.* **2012**, *393*, 565–578. [CrossRef] [PubMed]
12. Kambe, T. An overview of a wide range of functions of ZnT and Zip zinc transporters in the secretory pathway. *Biosci. Biotechnol. Biochem.* **2011**, *75*, 1036–1043. [CrossRef] [PubMed]
13. Kambe, T. Regulation of zinc transport. In *Encyclopedia of Inorganic and Bioinorganic Chemistry*; Culotta, V., Scott, R.A., Eds.; John Wiley & Sons, Ltd.: Hoboken, NJ, USA, 2013; pp. 301–309. [CrossRef]
14. Kambe, T. Molecular Architecture and Function of ZnT Transporters. *Curr. Top. Membr.* **2012**, *69*, 199–220. [PubMed]
15. Lu, M.; Fu, D. Structure of the zinc transporter YiiP. *Science* **2007**, *317*, 1746–1748. [CrossRef] [PubMed]
16. Lu, M.; Chai, J.; Fu, D. Structural basis for autoregulation of the zinc transporter YiiP. *Nat. Struct. Mol. Biol.* **2009**, *16*, 1063–1067. [CrossRef] [PubMed]
17. Coudray, N.; Valvo, S.; Hu, M.; Lasala, R.; Kim, C.; Vink, M.; Zhou, M.; Provasi, D.; Filizola, M.; Tao, J.; et al. Inward-facing conformation of the zinc transporter YiiP revealed by cryoelectron microscopy. *Proc. Natl. Acad. Sci. USA* **2013**, *110*, 2140–2145. [CrossRef] [PubMed]
18. Gupta, S.; Chai, J.; Cheng, J.; D'Mello, R.; Chance, M.R.; Fu, D. Visualizing the kinetic power stroke that drives proton-coupled zinc(II) transport. *Nature* **2014**, *512*, 101–104. [CrossRef] [PubMed]
19. Ohana, E.; Hoch, E.; Keasar, C.; Kambe, T.; Yifrach, O.; Hershfinkel, M.; Sekler, I. Identification of the $Zn^{2+}$ binding site and mode of operation of a mammalian $Zn^{2+}$ transporter. *J. Biol. Chem.* **2009**, *284*, 17677–17686. [CrossRef] [PubMed]
20. Shusterman, E.; Beharier, O.; Shiri, L.; Zarivach, R.; Etzion, Y.; Campbell, C.R.; Lee, I.H.; Okabayashi, K.; Dinudom, A.; Cook, D.I.; et al. ZnT-1 extrudes zinc from mammalian cells functioning as a $Zn^{(2+)}/H^{(+)}$ exchanger. *Metallomics* **2014**, *6*, 1656–1663. [CrossRef] [PubMed]
21. Fukunaka, A.; Suzuki, T.; Kurokawa, Y.; Yamazaki, T.; Fujiwara, N.; Ishihara, K.; Migaki, H.; Okumura, K.; Masuda, S.; Yamaguchi-Iwai, Y.; et al. Demonstration and characterization of the heterodimerization of ZnT5 and ZnT6 in the early secretory pathway. *J. Biol. Chem.* **2009**, *284*, 30798–30806. [CrossRef] [PubMed]
22. Golan, Y.; Berman, B.; Assaraf, Y.G. Heterodimerization, altered subcellular localization, and function of multiple zinc transporters in viable cells using bimolecular fluorescence complementation. *J. Biol. Chem.* **2015**, *290*, 9050–9063. [CrossRef] [PubMed]

23. Zhao, Y.; Feresin, R.G.; Falcon-Perez, J.M.; Salazar, G. Differential Targeting of SLC30A10/ZnT10 Heterodimers to Endolysosomal Compartments Modulates EGF-Induced MEK/ERK1/2 Activity. *Traffic* **2016**, *17*, 267–288. [CrossRef] [PubMed]

24. Tsuji, T.; Kurokawa, Y.; Chiche, J.; Pouyssegur, J.; Sato, H.; Fukuzawa, H.; Nagao, M.; Kambe, T. Dissecting the Process of Activation of Cancer-promoting Zinc-requiring Ectoenzymes by Zinc Metalation Mediated by ZNT Transporters. *J. Biol. Chem.* **2017**, *292*, 2159–2173. [CrossRef] [PubMed]

25. Fujimoto, S.; Itsumura, N.; Tsuji, T.; Anan, Y.; Tsuji, N.; Ogra, Y.; Kimura, T.; Miyamae, Y.; Masuda, S.; Nagao, M.; et al. Cooperative functions of ZnT1, metallothionein and ZnT4 in the cytoplasm are required for full activation of tnap in the early secretory pathway. *PLoS ONE* **2013**, *8*, e77445. [CrossRef] [PubMed]

26. Golan, Y.; Itsumura, N.; Glaser, F.; Berman, B.; Kambe, T.; Assaraf, Y.G. Molecular Basis of Transient Neonatal Zinc Deficiency: Novel ZNT2 Mutations Disrupting ZINC Binding and Permeation. *J. Biol. Chem.* **2016**, *291*, 13546–13559. [CrossRef] [PubMed]

27. Hoch, E.; Lin, W.; Chai, J.; Hershfinkel, M.; Fu, D.; Sekler, I. Histidine pairing at the metal transport site of mammalian ZnT transporters controls $Zn^{2+}$ over $Cd^{2+}$ selectivity. *Proc. Natl. Acad. Sci. USA* **2012**, *109*, 7202–7207. [CrossRef] [PubMed]

28. Nishito, Y.; Tsuji, N.; Fujishiro, H.; Takeda, T.; Yamazaki, T.; Teranishi, F.; Okazaki, F.; Matsunaga, A.; Tuschl, K.; Rao, R.; et al. Direct comparison of manganese detoxification/efflux proteins and molecular characterization of ZnT10 as a manganese transporter. *J. Biol. Chem.* **2016**, *291*, 14773–14787. [CrossRef] [PubMed]

29. Martin, J.E.; Giedroc, D.P. Functional Determinants of Metal Ion Transport and Selectivity in Paralogous Cation Diffusion Facilitator Transporters CzcD and MntE in Streptococcus pneumoniae. *J. Bacteriol.* **2016**, *198*, 1066–1076. [CrossRef] [PubMed]

30. Montanini, B.; Blaudez, D.; Jeandroz, S.; Sanders, D.; Chalot, M. Phylogenetic and functional analysis of the Cation Diffusion Facilitator (CDF) family: Improved signature and prediction of substrate specificity. *BMC Genom.* **2007**, *8*, 107. [CrossRef] [PubMed]

31. Gustin, J.L.; Zanis, M.J.; Salt, D.E. Structure and evolution of the plant cation diffusion facilitator family of ion transporters. *BMC Evol. Biol.* **2011**, *11*, 76. [CrossRef] [PubMed]

32. Kambe, T.; Suzuki, T.; Nagao, M.; Yamaguchi-Iwai, Y. Sequence similarity and functional relationship among eukaryotic ZIP and CDF transporters. *Genom. Proteom. Bioinform.* **2006**, *4*, 1–9. [CrossRef]

33. Antala, S.; Ovchinnikov, S.; Kamisetty, H.; Baker, D.; Dempski, R.E. Computation and Functional Studies Provide a Model for the Structure of the Zinc Transporter hZIP4. *J. Biol. Chem.* **2015**, *290*, 17796–17805. [CrossRef] [PubMed]

34. Zhang, T.; Liu, J.; Fellner, M.; Zhang, C.; Sui, D.; Hu, J. Crystal structures of a ZIP zinc transporter reveal a binuclear metal center in the transport pathway. *Sci. Adv.* **2017**, *3*, e1700344. [CrossRef] [PubMed]

35. Bin, B.H.; Fukada, T.; Hosaka, T.; Yamasaki, S.; Ohashi, W.; Hojyo, S.; Miyai, T.; Nishida, K.; Yokoyama, S.; Hirano, T. Biochemical characterization of human ZIP13 protein: A homo-dimerized zinc transporter involved in the spondylocheiro dysplastic Ehlers-Danlos syndrome. *J. Biol. Chem.* **2011**, *286*, 40255–40265. [CrossRef] [PubMed]

36. Taylor, K.M.; Muraina, I.A.; Brethour, D.; Schmitt-Ulms, G.; Nimmanon, T.; Ziliotto, S.; Kille, P.; Hogstrand, C. Zinc transporter ZIP10 forms a heteromer with ZIP6 which regulates embryonic development and cell migration. *Biochem. J.* **2016**, *473*, 2531–2544. [CrossRef] [PubMed]

37. Lin, W.; Chai, J.; Love, J.; Fu, D. Selective electrodiffusion of zinc ions in a Zrt-, Irt-like protein, ZIPB. *J. Biol. Chem.* **2010**, *285*, 39013–39020. [CrossRef] [PubMed]

38. Gaither, L.A.; Eide, D.J. Functional expression of the human hZIP2 zinc transporter. *J. Biol. Chem.* **2000**, *275*, 5560–5564. [CrossRef] [PubMed]

39. He, L.; Girijashanker, K.; Dalton, T.P.; Reed, J.; Li, H.; Soleimani, M.; Nebert, D.W. ZIP8, member of the solute-carrier-39 (SLC39) metal-transporter family: Characterization of transporter properties. *Mol. Pharmacol.* **2006**, *70*, 171–180. [CrossRef] [PubMed]

40. Girijashanker, K.; He, L.; Soleimani, M.; Reed, J.M.; Li, H.; Liu, Z.; Wang, B.; Dalton, T.P.; Nebert, D.W. Slc39a14 gene encodes ZIP14, a metal/bicarbonate symporter: Similarities to the ZIP8 transporter. *Mol. Pharmacol.* **2008**, *73*, 1413–1423. [CrossRef] [PubMed]

41. Fukada, T.; Kambe, T. Molecular and genetic features of zinc transporters in physiology and pathogenesis. *Metallomics* **2011**, *3*, 662–674. [CrossRef] [PubMed]

42. Zhang, T.; Sui, D.; Hu, J. Structural insights of ZIP4 extracellular domain critical for optimal zinc transport. *Nat. Commun.* **2016**, *7*, 11979. [CrossRef] [PubMed]

43. Ehsani, S.; Huo, H.; Salehzadeh, A.; Pocanschi, C.L.; Watts, J.C.; Wille, H.; Westaway, D.; Rogaeva, E.; St George-Hyslop, P.H.; Schmitt-Ulms, G. Family reunion—The ZIP/prion gene family. *Prog. Neurobiol.* **2011**, *93*, 405–420. [CrossRef] [PubMed]

44. Jenkitkasemwong, S.; Wang, C.Y.; Mackenzie, B.; Knutson, M.D. Physiologic implications of metal-ion transport by ZIP14 and ZIP8. *Biometals* **2012**, *25*, 643–655. [CrossRef] [PubMed]

45. Vembar, S.S.; Brodsky, J.L. One step at a time: Endoplasmic reticulum-associated degradation. *Nat. Rev. Mol. Cell Biol.* **2008**, *9*, 944–957. [CrossRef] [PubMed]

46. Leach, M.R.; Cohen-Doyle, M.F.; Thomas, D.Y.; Williams, D.B. Localization of the lectin, ERp57 binding, and polypeptide binding sites of calnexin and calreticulin. *J. Biol. Chem.* **2002**, *277*, 29686–29697. [CrossRef] [PubMed]

47. Saito, Y.; Ihara, Y.; Leach, M.R.; Cohen-Doyle, M.F.; Williams, D.B. Calreticulin functions in vitro as a molecular chaperone for both glycosylated and non-glycosylated proteins. *EMBO J.* **1999**, *18*, 6718–6729. [CrossRef] [PubMed]

48. Solovyov, A.; Gilbert, H.F. Zinc-dependent dimerization of the folding catalyst, protein disulfide isomerase. *Protein Sci.* **2004**, *13*, 1902–1907. [CrossRef] [PubMed]

49. Ellis, C.D.; Wang, F.; MacDiarmid, C.W.; Clark, S.; Lyons, T.; Eide, D.J. Zinc and the Msc2 zinc transporter protein are required for endoplasmic reticulum function. *J. Cell Biol.* **2004**, *166*, 325–335. [CrossRef] [PubMed]

50. Ishihara, K.; Yamazaki, T.; Ishida, Y.; Suzuki, T.; Oda, K.; Nagao, M.; Yamaguchi-Iwai, Y.; Kambe, T. Zinc transport complexes contribute to the homeostatic maintenance of secretory pathway function in vertebrate cells. *J. Biol. Chem.* **2006**, *281*, 17743–17750. [CrossRef] [PubMed]

51. Ohashi, W.; Kimura, S.; Iwanaga, T.; Furusawa, Y.; Irie, T.; Izumi, H.; Watanabe, T.; Hijikata, A.; Hara, T.; Ohara, O.; et al. Zinc Transporter SLC39A7/ZIP7 Promotes Intestinal Epithelial Self-Renewal by Resolving ER Stress. *PLoS Genet.* **2016**, *12*, e1006349. [CrossRef] [PubMed]

52. Tuncay, E.; Bitirim, V.C.; Durak, A.; Carrat, G.R.J.; Taylor, K.M.; Rutter, G.A.; Turan, B. Hyperglycemia-Induced Changes in ZIP7 and ZnT7 Expression Cause $Zn^{2+}$ Release from the Sarco(endo)plasmic Reticulum and Mediate ER Stress in the Heart. *Diabetes* **2017**, *66*, 1346–1358. [CrossRef] [PubMed]

53. Bin, B.H.; Bhin, J.; Seo, J.; Kim, S.Y.; Lee, E.; Park, K.; Choi, D.H.; Takagishi, T.; Hara, T.; Hwang, D.; et al. Requirement of Zinc Transporter SLC39A7/ZIP7 for Dermal Development to Fine-Tune Endoplasmic Reticulum Function by Regulating Protein Disulfide Isomerase. *J. Investig. Dermatol.* **2017**, *137*, 1682–1691. [CrossRef] [PubMed]

54. Leyva-Illades, D.; Chen, P.; Zogzas, C.E.; Hutchens, S.; Mercado, J.M.; Swaim, C.D.; Morrisett, R.A.; Bowman, A.B.; Aschner, M.; Mukhopadhyay, S. SLC30A10 is a cell surface-localized manganese efflux transporter, and parkinsonism-causing mutations block its intracellular trafficking and efflux activity. *J. Neurosci.* **2014**, *34*, 14079–14095. [CrossRef] [PubMed]

55. Kambe, T.; Takeda, T.A.; Nishito, Y. Activation of zinc-requiring ectoenzymes by ZnT transporters during the secretory process: Biochemical and molecular aspects. *Arch. Biochem. Biophys.* **2016**, *611*, 37–42. [CrossRef] [PubMed]

56. Suzuki, T.; Ishihara, K.; Migaki, H.; Matsuura, W.; Kohda, A.; Okumura, K.; Nagao, M.; Yamaguchi-Iwai, Y.; Kambe, T. Zinc transporters, ZnT5 and ZnT7, are required for the activation of alkaline phosphatases, zinc-requiring enzymes that are glycosylphosphatidylinositol-anchored to the cytoplasmic membrane. *J. Biol. Chem.* **2005**, *280*, 637–643. [CrossRef] [PubMed]

57. Huang, L.; Kirschke, C.P.; Gitschier, J. Functional characterization of a novel mammalian zinc transporter, ZnT6. *J. Biol. Chem.* **2002**, *277*, 26389–26395. [CrossRef] [PubMed]

58. Kirschke, C.P.; Huang, L. ZnT7, a Novel Mammalian Zinc Transporter, Accumulates Zinc in the Golgi Apparatus. *J. Biol. Chem.* **2003**, *278*, 4096–4102. [CrossRef] [PubMed]

59. Kambe, T. Methods to evaluate zinc transport into and out of the secretory and endosomal-lysosomal compartments in DT40 cells. *Methods Enzymol.* **2014**, *534*, 77–92. [CrossRef] [PubMed]

60. Ellis, C.D.; Macdiarmid, C.W.; Eide, D.J. Heteromeric protein complexes mediate zinc transport into the secretory pathway of eukaryotic cells. *J. Biol. Chem.* **2005**, *280*, 28811–28818. [CrossRef] [PubMed]

61. Kambe, T.; Narita, H.; Yamaguchi-Iwai, Y.; Hirose, J.; Amano, T.; Sugiura, N.; Sasaki, R.; Mori, K.; Iwanaga, T.; Nagao, M. Cloning and characterization of a novel mammalian zinc transporter, zinc transporter 5, abundantly expressed in pancreatic beta cells. *J. Biol. Chem.* **2002**, *277*, 19049–19055. [CrossRef] [PubMed]

62. Kurita, H.; Okuda, R.; Yokoo, K.; Inden, M.; Hozumi, I. Protective roles of SLC30A3 against endoplasmic reticulum stress via ERK1/2 activation. *Biochem. Biophys. Res. Commun.* **2016**, *479*, 853–859. [CrossRef] [PubMed]

63. Go, S.; Kurita, H.; Yokoo, K.; Inden, M.; Kambe, T.; Hozumi, I. Protective function of SLC30A10 induced via PERK-ATF4 pathway against 1-methyl-4-phenylpyridinium. *Biochem. Biophys. Res. Commun.* **2017**. [CrossRef] [PubMed]

64. Taylor, K.M.; Morgan, H.E.; Johnson, A.; Nicholson, R.I. Structure-function analysis of HKE4, a member of the new LIV-1 subfamily of zinc transporters. *Biochem. J.* **2004**, *377*, 131–139. [CrossRef] [PubMed]

65. Huang, L.; Kirschke, C.P.; Zhang, Y.; Yu, Y.Y. The *ZIP7* gene (*Slc39a7*) encodes a zinc transporter involved in zinc homeostasis of the Golgi apparatus. *J. Biol. Chem.* **2005**, *280*, 15456–15463. [CrossRef] [PubMed]

66. Fukada, T.; Civic, N.; Furuichi, T.; Shimoda, S.; Mishima, K.; Higashiyama, H.; Idaira, Y.; Asada, Y.; Kitamura, H.; Yamasaki, S.; et al. The zinc transporter SLC39A13/ZIP13 is required for connective tissue development; its involvement in BMP/TGF-beta signaling pathways. *PLoS ONE* **2008**, *3*, e3642. [CrossRef]

67. Matsuura, W.; Yamazaki, T.; Yamaguchi-Iwai, Y.; Masuda, S.; Nagao, M.; Andrews, G.K.; Kambe, T. SLC39A9 (ZIP9) regulates zinc homeostasis in the secretory pathway: Characterization of the ZIP subfamily I protein in vertebrate cells. *Biosci. Biotechnol. Biochem.* **2009**, *73*, 1142–1148. [CrossRef] [PubMed]

68. Thomas, P.; Pang, Y.; Dong, J.; Berg, A.H. Identification and characterization of membrane androgen receptors in the ZIP9 zinc transporter subfamily: II. Role of human ZIP9 in testosterone-induced prostate and breast cancer cell apoptosis. *Endocrinology* **2014**, *155*, 4250–4265. [CrossRef] [PubMed]

69. Jeong, J.; Walker, J.M.; Wang, F.; Park, J.G.; Palmer, A.E.; Giunta, C.; Rohrbach, M.; Steinmann, B.; Eide, D.J. Promotion of vesicular zinc efflux by ZIP13 and its implications for spondylocheiro dysplastic Ehlers-Danlos syndrome. *Proc. Natl. Acad. Sci. USA* **2012**, *109*, E3530–E3538. [CrossRef] [PubMed]

70. Kelleher, S.L.; Velasquez, V.; Croxford, T.P.; McCormick, N.H.; Lopez, V.; Macdavid, J. Mapping the zinc-transporting system in mammary cells: Molecular analysis reveals a phenotype-dependent zinc-transporting network during lactation. *J. Cell. Physiol.* **2012**, *227*, 1761–1770. [CrossRef] [PubMed]

71. Hogstrand, C.; Kille, P.; Nicholson, R.I.; Taylor, K.M. Zinc transporters and cancer: A potential role for ZIP7 as a hub for tyrosine kinase activation. *Trends Mol. Med.* **2009**, *15*, 101–111. [CrossRef] [PubMed]

72. Taylor, K.M.; Hiscox, S.; Nicholson, R.I.; Hogstrand, C.; Kille, P. Protein Kinase CK2 Triggers Cytosolic Zinc Signaling Pathways by Phosphorylation of Zinc Channel ZIP7. *Sci. Signal.* **2012**, *5*, ra11. [CrossRef] [PubMed]

73. Taniguchi, M.; Fukunaka, A.; Hagihara, M.; Watanabe, K.; Kamino, S.; Kambe, T.; Enomoto, S.; Hiromura, M. Essential Role of the Zinc Transporter ZIP9/SLC39A9 in Regulating the Activations of Akt and Erk in B-Cell Receptor Signaling Pathway in DT40 Cells. *PLoS ONE* **2013**, *8*, e58022. [CrossRef] [PubMed]

74. Kumanovics, A.; Poruk, K.E.; Osborn, K.A.; Ward, D.M.; Kaplan, J. *YKE4* (YIL023C) encodes a bidirectional zinc transporter in the endoplasmic reticulum of *Saccharomyces cerevisiae*. *J. Biol. Chem.* **2006**, *281*, 22566–22574. [CrossRef] [PubMed]

75. Stathakis, D.G.; Burton, D.Y.; McIvor, W.E.; Krishnakumar, S.; Wright, T.R.; O'Donnell, J.M. The catecholamines up (Catsup) protein of *Drosophila melanogaster* functions as a negative regulator of tyrosine hydroxylase activity. *Genetics* **1999**, *153*, 361–382. [PubMed]

76. Groth, C.; Sasamura, T.; Khanna, M.R.; Whitley, M.; Fortini, M.E. Protein trafficking abnormalities in *Drosophila* tissues with impaired activity of the ZIP7 zinc transporter Catsup. *Development* **2013**, *140*, 3018–3027. [CrossRef] [PubMed]

77. Taylor, K.M.; Morgan, H.E.; Johnson, A.; Nicholson, R.I. Structure-function analysis of a novel member of the LIV-1 subfamily of zinc transporters, ZIP14. *FEBS Lett.* **2005**, *579*, 427–432. [CrossRef] [PubMed]

78. Liuzzi, J.P.; Lichten, L.A.; Rivera, S.; Blanchard, R.K.; Aydemir, T.B.; Knutson, M.D.; Ganz, T.; Cousins, R.J. Interleukin-6 regulates the zinc transporter Zip14 in liver and contributes to the hypozincemia of the acute-phase response. *Proc. Natl. Acad. Sci. USA* **2005**, *102*, 6843–6848. [CrossRef] [PubMed]

79. Homma, K.; Fujisawa, T.; Tsuburaya, N.; Yamaguchi, N.; Kadowaki, H.; Takeda, K.; Nishitoh, H.; Matsuzawa, A.; Naguro, I.; Ichijo, H. SOD1 as a Molecular Switch for Initiating the Homeostatic ER Stress Response under Zinc Deficiency. *Mol. Cell* **2013**, *52*, 75–86. [CrossRef] [PubMed]

80. Kim, M.H.; Aydemir, T.B.; Kim, J.; Cousins, R.J. Hepatic ZIP14-mediated zinc transport is required for adaptation to endoplasmic reticulum stress. *Proc. Natl. Acad. Sci. USA* **2017**, *114*, E5805–E5814. [CrossRef] [PubMed]

81. Shen, X.; Ellis, R.E.; Sakaki, K.; Kaufman, R.J. Genetic Interactions Due to Constitutive and Inducible Gene Regulation Mediated by the Unfolded Protein Response in *C. elegans. PLoS Genet.* **2005**, *1*, e37. [CrossRef] [PubMed]

82. Overall, C.M.; Lopez-Otin, C. Strategies for MMP inhibition in cancer: Innovations for the post-trial era. *Nat. Rev. Cancer* **2002**, *2*, 657–672. [CrossRef] [PubMed]

83. Supuran, C.T. Carbonic anhydrases: Novel therapeutic applications for inhibitors and activators. *Nat. Rev. Drug Discov.* **2008**, *7*, 168–181. [CrossRef] [PubMed]

84. Reich, R.; Hoffman, A.; Veerendhar, A.; Maresca, A.; Innocenti, A.; Supuran, C.T.; Breuer, E. Carbamoylphosphonates control tumor cell proliferation and dissemination by simultaneously inhibiting carbonic anhydrase IX and matrix metalloproteinase-2. Toward nontoxic chemotherapy targeting tumor microenvironment. *J. Med. Chem.* **2012**, *55*, 7875–7882. [CrossRef] [PubMed]

85. Reich, R.; Hoffman, A.; Suresh, R.R.; Shai, O.; Frant, J.; Maresca, A.; Supuran, C.T.; Breuer, E. Carbamoylphosphonates inhibit autotaxin and metastasis formation in vivo. *J. Enzym. Inhib. Med. Chem.* **2015**, *30*, 767–772. [CrossRef] [PubMed]

86. Neri, D.; Supuran, C.T. Interfering with pH regulation in tumours as a therapeutic strategy. *Nat. Rev. Drug Discov.* **2011**, *10*, 767–777. [CrossRef] [PubMed]

87. Itsumura, N.; Kibihara, Y.; Fukue, K.; Miyata, A.; Fukushima, K.; Tamagawa-Mineoka, R.; Katoh, N.; Nishito, Y.; Ishida, R.; Narita, H.; et al. Novel mutations in *SLC30A2* involved in the pathogenesis of transient neonatal zinc deficiency. *Pediatr. Res.* **2016**, *80*, 586–594. [CrossRef] [PubMed]

88. Hashimoto, A.; Nakagawa, M.; Tsujimura, N.; Miyazaki, S.; Kizu, K.; Goto, T.; Komatsu, Y.; Matsunaga, A.; Shirakawa, H.; Narita, H.; et al. Properties of Zip4 accumulation during zinc deficiency and its usefulness to evaluate zinc status: A study of the effects of zinc deficiency during lactation. *Am. J. Physiol. Regul. Integr. Comp. Physiol.* **2016**, *310*, R459–R468. [CrossRef] [PubMed]

89. Vallee, B.L.; Falchuk, K.H. The biochemical basis of zinc physiology. *Physiol. Rev.* **1993**, *73*, 79–118. [PubMed]

90. Kochanczyk, T.; Drozd, A.; Krezel, A. Relationship between the architecture of zinc coordination and zinc binding affinity in proteins–insights into zinc regulation. *Metallomics* **2015**, *7*, 244–257. [CrossRef] [PubMed]

91. Suzuki, T.; Ishihara, K.; Migaki, H.; Nagao, M.; Yamaguchi-Iwai, Y.; Kambe, T. Two different zinc transport complexes of cation diffusion facilitator proteins localized in the secretory pathway operate to activate alkaline phosphatases in vertebrate cells. *J. Biol. Chem.* **2005**, *280*, 30956–30962. [CrossRef] [PubMed]

92. Lasry, I.; Golan, Y.; Berman, B.; Amram, N.; Glaser, F.; Assaraf, Y.G. In Situ dimerization of multiple wild type and mutant zinc transporters in live cells using bimolecular fluorescence complementation. *J. Biol. Chem.* **2014**, *289*, 7275–7292. [CrossRef] [PubMed]

93. Fukunaka, A.; Kurokawa, Y.; Teranishi, F.; Sekler, I.; Oda, K.; Ackland, M.L.; Faundez, V.; Hiromura, M.; Masuda, S.; Nagao, M.; et al. Tissue Nonspecific Alkaline Phosphatase Is Activated via a Two-step Mechanism by Zinc Transport Complexes in the Early Secretory Pathway. *J. Biol. Chem.* **2011**, *286*, 16363–16373. [CrossRef] [PubMed]

94. McCormick, N.H.; Kelleher, S.L. ZnT4 provides zinc to zinc-dependent proteins in the trans-Golgi network critical for cell function and Zn export in mammary epithelial cells. *Am. J. Physiol. Cell Physiol.* **2012**, *303*, C291–C297. [CrossRef] [PubMed]

95. Murgia, C.; Vespignani, I.; Cerase, J.; Nobili, F.; Perozzi, G. Cloning, expression, and vesicular localization of zinc transporter Dri 27/ZnT4 in intestinal tissue and cells. *Am. J. Physiol.* **1999**, *277*, G1231–G1239. [PubMed]

96. Kukic, I.; Lee, J.K.; Coblentz, J.; Kelleher, S.L.; Kiselyov, K. Zinc-dependent lysosomal enlargement in TRPML1-deficient cells involves MTF-1 transcription factor and ZnT4 (Slc30a4) transporter. *Biochem. J.* **2013**, *451*, 155–163. [CrossRef] [PubMed]

97. Huang, L.; Gitschier, J. A novel gene involved in zinc transport is deficient in the lethal milk mouse. *Nat. Genet.* **1997**, *17*, 292–297. [CrossRef] [PubMed]

98. McCormick, N.H.; Lee, S.; Hennigar, S.R.; Kelleher, S.L. ZnT4 (*SLC30A4*)-null ("lethal milk") mice have defects in mammary gland secretion and hallmarks of precocious involution during lactation. *Am. J. Physiol. Regul. Integr. Comp. Physiol.* **2016**, *310*, R33–R40. [CrossRef] [PubMed]

99. Kimura, T.; Kambe, T. The Functions of Metallothionein and ZIP and ZnT Transporters: An Overview and Perspective. *Int. J. Mol. Sci.* **2016**, *17*, 336. [CrossRef] [PubMed]

100. Fujimoto, S.; Tsuji, T.; Fujiwara, T.; Takeda, T.A.; Merriman, C.; Fukunaka, A.; Nishito, Y.; Fu, D.; Hoch, E.; Sekler, I.; et al. The PP-motif in luminal loop 2 of ZnT transporters plays a pivotal role in TNAP activation. *Biochem. J.* **2016**, *473*, 2611–2621. [CrossRef] [PubMed]

101. Inoue, K.; Matsuda, K.; Itoh, M.; Kawaguchi, H.; Tomoike, H.; Aoyagi, T.; Nagai, R.; Hori, M.; Nakamura, Y.; Tanaka, T. Osteopenia and male-specific sudden cardiac death in mice lacking a zinc transporter gene, *Znt5*. *Hum. Mol. Genet.* **2002**, *11*, 1775–1784. [CrossRef] [PubMed]

102. Nishida, K.; Hasegawa, A.; Nakae, S.; Oboki, K.; Saito, H.; Yamasaki, S.; Hirano, T. Zinc transporter Znt5/Slc30a5 is required for the mast cell-mediated delayed-type allergic reaction but not the immediate-type reaction. *J. Exp. Med.* **2009**, *206*, 1351–1364. [CrossRef] [PubMed]

103. Sheline, C.T.; Shi, C.; Takata, T.; Zhu, J.; Zhang, W.; Sheline, P.J.; Cai, A.L.; Li, L. Dietary zinc reduction, pyruvate supplementation, or zinc transporter 5 knockout attenuates beta-cell death in nonobese diabetic mice, islets, and insulinoma cells. *J. Nutr.* **2012**, *142*, 2119–2127. [CrossRef] [PubMed]

104. Huang, L.; Yu, Y.Y.; Kirschke, C.P.; Gertz, E.R.; Lloyd, K.K. *Znt7 (Slc30a7)*-deficient mice display reduced body zinc status and body fat accumulation. *J. Biol. Chem.* **2007**, *282*, 37053–37063. [CrossRef] [PubMed]

105. Huang, L.; Kirschke, C.P.; Lay, Y.A.; Levy, L.B.; Lamirande, D.E.; Zhang, P.H. *Znt7*-null mice are more susceptible to diet-induced glucose intolerance and insulin resistance. *J. Biol. Chem.* **2012**, *287*, 33883–33896. [CrossRef] [PubMed]

106. Pound, L.D.; Sarkar, S.A.; Benninger, R.K.; Wang, Y.; Suwanichkul, A.; Shadoan, M.K.; Printz, R.L.; Oeser, J.K.; Lee, C.E.; Piston, D.W.; et al. Deletion of the mouse *Slc30a8* gene encoding zinc transporter-8 results in impaired insulin secretion. *Biochem. J.* **2009**, *421*, 371–376. [CrossRef] [PubMed]

107. Nicolson, T.J.; Bellomo, E.A.; Wijesekara, N.; Loder, M.K.; Baldwin, J.M.; Gyulkhandanyan, A.V.; Koshkin, V.; Tarasov, A.I.; Carzaniga, R.; Kronenberger, K.; et al. Insulin storage and glucose homeostasis in mice null for the granule zinc transporter ZnT8 and studies of the type 2 diabetes-associated variants. *Diabetes* **2009**, *58*, 2070–2083. [CrossRef] [PubMed]

108. Lemaire, K.; Ravier, M.A.; Schraenen, A.; Creemers, J.W.; Van de Plas, R.; Granvik, M.; Van Lommel, L.; Waelkens, E.; Chimienti, F.; Rutter, G.A.; et al. Insulin crystallization depends on zinc transporter ZnT8 expression, but is not required for normal glucose homeostasis in mice. *Proc. Natl. Acad. Sci. USA* **2009**, *106*, 14872–14877. [CrossRef] [PubMed]

109. Wijesekara, N.; Dai, F.F.; Hardy, A.B.; Giglou, P.R.; Bhattacharjee, A.; Koshkin, V.; Chimienti, F.; Gaisano, H.Y.; Rutter, G.A.; Wheeler, M.B. Beta cell-specific *Znt8* deletion in mice causes marked defects in insulin processing, crystallisation and secretion. *Diabetologia* **2010**, *53*, 1656–1668. [CrossRef] [PubMed]

110. Tamaki, M.; Fujitani, Y.; Hara, A.; Uchida, T.; Tamura, Y.; Takeno, K.; Kawaguchi, M.; Watanabe, T.; Ogihara, T.; Fukunaka, A.; et al. The diabetes-susceptible gene *SLC30A8*/ZnT8 regulates hepatic insulin clearance. *J. Clin. Investig.* **2013**, *123*, 4513–4524. [CrossRef] [PubMed]

111. Merriman, C.; Huang, Q.; Rutter, G.A.; Fu, D. Lipid-tuned Zinc Transport Activity of Human ZnT8 Protein Correlates with Risk for Type-2 Diabetes. *J. Biol. Chem.* **2016**, *291*, 26950–26957. [CrossRef] [PubMed]

112. Li, L.; Bai, S.; Sheline, C.T. *hZnT8* (Slc30a8) Transgenic Mice That Overexpress the R325W Polymorph Have Reduced Islet $Zn^{2+}$ and Proinsulin Levels, Increased Glucose Tolerance After a High-Fat Diet, and Altered Levels of Pancreatic Zinc Binding Proteins. *Diabetes* **2017**, *66*, 551–559. [CrossRef] [PubMed]

113. Sladek, R.; Rocheleau, G.; Rung, J.; Dina, C.; Shen, L.; Serre, D.; Boutin, P.; Vincent, D.; Belisle, A.; Hadjadj, S.; et al. A genome-wide association study identifies novel risk loci for type 2 diabetes. *Nature* **2007**, *445*, 881–885. [CrossRef] [PubMed]

114. Flannick, J.; Thorleifsson, G.; Beer, N.L.; Jacobs, S.B.; Grarup, N.; Burtt, N.P.; Mahajan, A.; Fuchsberger, C.; Atzmon, G.; Benediktsson, R.; et al. Loss-of-function mutations in *SLC30A8* protect against type 2 diabetes. *Nat. Genet.* **2014**, *46*, 357–363. [CrossRef] [PubMed]

115. Chabosseau, P.; Rutter, G.A. Zinc and diabetes. Arch Biochem Biophys. *Arch. Biochem. Biophys.* **2016**, *611*, 79–85. [CrossRef] [PubMed]

116. Huang, L.; Yan, M.; Kirschke, C.P. Over-expression of ZnT7 increases insulin synthesis and secretion in pancreatic beta-cells by promoting insulin gene transcription. *Exp. Cell Res.* **2010**, *316*, 2630–2643. [CrossRef] [PubMed]

117. Bellomo, E.A.; Meur, G.; Rutter, G.A. Glucose regulates free cytosolic $Zn^{2+}$ concentration, Slc39 (ZiP), and metallothionein gene expression in primary pancreatic islet beta-cells. *J. Biol. Chem.* **2011**, *286*, 25778–25789. [CrossRef] [PubMed]

118. Syring, K.E.; Boortz, K.A.; Oeser, J.K.; Ustione, A.; Platt, K.A.; Shadoan, M.K.; McGuinness, O.P.; Piston, D.W.; Powell, D.R.; O'Brien, R.M. Combined Deletion of *Slc30a7* and *Slc30a8* Unmasks a Critical Role for ZnT8 in Glucose-Stimulated Insulin Secretion. *Endocrinology* **2016**, *157*, 4534–4541. [CrossRef] [PubMed]

119. Cole, T.B.; Wenzel, H.J.; Kafer, K.E.; Schwartzkroin, P.A.; Palmiter, R.D. Elimination of zinc from synaptic vesicles in the intact mouse brain by disruption of the ZnT3 gene. *Proc. Natl. Acad. Sci. USA* **1999**, *96*, 1716–1721. [CrossRef] [PubMed]

120. Sensi, S.L.; Paoletti, P.; Koh, J.Y.; Aizenman, E.; Bush, A.I.; Hershfinkel, M. The neurophysiology and pathology of brain zinc. *J. Neurosci.* **2011**, *31*, 16076–16085. [CrossRef] [PubMed]

121. Takeda, A.; Nakamura, M.; Fujii, H.; Tamano, H. Synaptic $Zn^{2+}$ homeostasis and its significance. *Metallomics* **2013**, *5*, 417–423. [CrossRef] [PubMed]

122. Vergnano, A.M.; Rebola, N.; Savtchenko, L.P.; Pinheiro, P.S.; Casado, M.; Kieffer, B.L.; Rusakov, D.A.; Mulle, C.; Paoletti, P. Zinc dynamics and action at excitatory synapses. *Neuron* **2014**, *82*, 1101–1114. [CrossRef] [PubMed]

123. Hirzel, K.; Muller, U.; Latal, A.T.; Hulsmann, S.; Grudzinska, J.; Seeliger, M.W.; Betz, H.; Laube, B. Hyperekplexia phenotype of glycine receptor alpha1 subunit mutant mice identifies $Zn^{2+}$ as an essential endogenous modulator of glycinergic neurotransmission. *Neuron* **2006**, *52*, 679–690. [CrossRef] [PubMed]

124. Nozaki, C.; Vergnano, A.M.; Filliol, D.; Ouagazzal, A.M.; Le Goff, A.; Carvalho, S.; Reiss, D.; Gaveriaux-Ruff, C.; Neyton, J.; Paoletti, P.; et al. Zinc alleviates pain through high-affinity binding to the NMDA receptor NR2A subunit. *Nat. Neurosci.* **2011**, *14*, 1017–1022. [CrossRef] [PubMed]

125. Adlard, P.A.; Parncutt, J.M.; Finkelstein, D.I.; Bush, A.I. Cognitive loss in zinc transporter-3 knock-out mice: A phenocopy for the synaptic and memory deficits of Alzheimer's disease? *J. Neurosci.* **2010**, *30*, 1631–1636. [CrossRef] [PubMed]

126. Barnham, K.J.; Bush, A.I. Biological metals and metal-targeting compounds in major neurodegenerative diseases. *Chem. Soc. Rev.* **2014**, *43*, 6727–6749. [CrossRef] [PubMed]

127. Kaneko, M.; Noguchi, T.; Ikegami, S.; Sakurai, T.; Kakita, A.; Toyoshima, Y.; Kambe, T.; Yamada, M.; Inden, M.; Hara, H.; et al. Zinc transporters ZnT3 and ZnT6 are downregulated in the spinal cords of patients with sporadic amyotrophic lateral sclerosis. *J. Neurosci. Res.* **2015**, *93*, 370–379. [CrossRef] [PubMed]

128. Chowanadisai, W.; Lonnerdal, B.; Kelleher, S.L. Identification of a mutation in SLC30A2 (ZnT-2) in women with low milk zinc concentration that results in transient neonatal zinc deficiency. *J. Biol. Chem.* **2006**, *281*, 39699–39707. [CrossRef] [PubMed]

129. Lasry, I.; Seo, Y.A.; Ityel, H.; Shalva, N.; Pode-Shakked, B.; Glaser, F.; Berman, B.; Berezovsky, I.; Goncearenco, A.; Klar, A.; et al. A Dominant Negative Heterozygous G87R Mutation in the Zinc Transporter, ZnT-2 (SLC30A2), Results in Transient Neonatal Zinc Deficiency. *J. Biol. Chem.* **2012**, *287*, 29348–29361. [CrossRef] [PubMed]

130. Itsumura, N.; Inamo, Y.; Okazaki, F.; Teranishi, F.; Narita, H.; Kambe, T.; Kodama, H. Compound Heterozygous Mutations in *SLC30A2/ZnT2* Results in Low Milk Zinc Concentrations: A Novel Mechanism for Zinc Deficiency in a Breast-Fed Infant. *PLoS ONE* **2013**, *8*, e64045. [CrossRef] [PubMed]

131. Miletta, M.C.; Bieri, A.; Kernland, K.; Schoni, M.H.; Petkovic, V.; Fluck, C.E.; Eble, A.; Mullis, P.E. Transient Neonatal Zinc Deficiency Caused by a Heterozygous G87R Mutation in the Zinc Transporter ZnT-2 (SLC30A2) Gene in the Mother Highlighting the Importance of $Zn^{2+}$ for Normal Growth and Development. *Int. J. Endocrinol.* **2013**, *2013*, 259189. [CrossRef] [PubMed]

132. Kumar, L.; Michalczyk, A.; McKay, J.; Ford, D.; Kambe, T.; Hudek, L.; Varigios, G.; Taylor, P.E.; Ackland, M.L. Altered expression of two zinc transporters, SLC30A5 and SLC30A6, underlies a mammary gland disorder of reduced zinc secretion into milk. *Genes Nutr.* **2015**, *10*, 487. [CrossRef] [PubMed]

133. Kim, A.M.; Bernhardt, M.L.; Kong, B.Y.; Ahn, R.W.; Vogt, S.; Woodruff, T.K.; O'Halloran, T.V. Zinc sparks are triggered by fertilization and facilitate cell cycle resumption in mammalian eggs. *ACS Chem. Biol.* **2011**, *6*, 716–723. [CrossRef] [PubMed]

134. Que, E.L.; Bleher, R.; Duncan, F.E.; Kong, B.Y.; Gleber, S.C.; Vogt, S.; Chen, S.; Garwin, S.A.; Bayer, A.R.; Dravid, V.P.; et al. Quantitative mapping of zinc fluxes in the mammalian egg reveals the origin of fertilization-induced zinc sparks. *Nat. Chem.* **2015**, *7*, 130–139. [CrossRef] [PubMed]

135. Yamasaki, S.; Hasegawa, A.; Hojyo, S.; Ohashi, W.; Fukada, T.; Nishida, K.; Hirano, T. A Novel Role of the L-Type Calcium Channel alpha(1D) Subunit as a Gatekeeper for Intracellular Zinc Signaling: Zinc Wave. *PLoS ONE* **2012**, *7*, e39654. [CrossRef] [PubMed]

136. Nimmanon, T.; Ziliotto, S.; Morris, S.; Flanagan, L.; Taylor, K.M. Phosphorylation of zinc channel ZIP7 drives MAPK, PI3K and mTOR growth and proliferation signalling. *Metallomics* **2017**, *9*, 471–481. [CrossRef] [PubMed]

137. Fukada, T.; Yamasaki, S.; Nishida, K.; Murakami, M.; Hirano, T. Zinc homeostasis and signaling in health and diseases: Zinc signaling. *J. Biol. Inorg. Chem.* **2011**, *16*, 1123–1134. [CrossRef] [PubMed]

138. Chabosseau, P.; Tuncay, E.; Meur, G.; Bellomo, E.A.; Hessels, A.; Hughes, S.; Johnson, P.R.; Bugliani, M.; Marchetti, P.; Turan, B.; et al. Mitochondrial and ER-targeted eCALWY probes reveal high levels of free $Zn^{2+}$. *ACS Chem. Biol.* **2014**, *9*, 2111–2120. [CrossRef] [PubMed]

139. Qin, Y.; Dittmer, P.J.; Park, J.G.; Jansen, K.B.; Palmer, A.E. Measuring steady-state and dynamic endoplasmic reticulum and Golgi $Zn^{2+}$ with genetically encoded sensors. *Proc. Natl. Acad. Sci. USA* **2011**, *108*, 7351–7356. [CrossRef] [PubMed]

International Journal of
*Molecular Sciences*

MDPI

*Review*

# Recent Advances in the Role of SLC39A/ZIP Zinc Transporters In Vivo

Teruhisa Takagishi [1], Takafumi Hara [1] and Toshiyuki Fukada [1,2,3,*]

1   Faculty of Pharmaceutical Sciences, Tokushima Bunri University, Tokushima 770-8514, Japan;
    t.takagishi@ph.bunri-u.ac.jp (T.T.); t-hara@ph.bunri-u.ac.jp (T.H.)
2   Division of Pathology, Department of Oral Diagnostic Sciences, School of Dentistry, Showa University,
    Tokyo 142-8555, Japan
3   RIKEN Center for Integrative Medical Sciences, Yokohama, Kanagawa 230-0042, Japan
*   Correspondence: fukada@ph.bunri-u.ac.jp; Tel.: +81-88-602-8593

Received: 8 November 2017; Accepted: 8 December 2017; Published: 13 December 2017

**Abstract:** Zinc (Zn), which is an essential trace element, is involved in numerous mammalian physiological events; therefore, either a deficiency or excess of Zn impairs cellular machineries and influences physiological events, such as systemic growth, bone homeostasis, skin formation, immune responses, endocrine function, and neuronal function. Zn transporters are thought to mainly contribute to Zn homeostasis within cells and in the whole body. Recent genetic, cellular, and molecular studies of Zn transporters highlight the dynamic role of Zn as a signaling mediator linking several cellular events and signaling pathways. Dysfunction in Zn transporters causes various diseases. This review aims to provide an update of Zn transporters and Zn signaling studies and discusses the remaining questions and future directions by focusing on recent progress in determining the roles of SLC39A/ZIP family members in vivo.

**Keywords:** zinc transporter; SLC39A/ZIP; zinc signaling; physiology; diseases

## 1. Introduction

Zn is an essential micronutrient required for growth, development, immunity, and many other physiological processes. The total amount of Zn in the human body is 2–3 g, with ~60% in the skeletal muscle, ~30% in bone, and ~5% in both, the liver and skin, while the remaining 5% is in other tissues (Figure 1) [1]. Approximately 10% of human proteins may bind to Zn [2], reflecting the indispensability of Zn in numerous physiological processes. Therefore, either a deficiency or excess of Zn is detrimental [1].

Maintenance of intracellular Zn homeostasis mainly depends on two families of Zn transporters: Zrt- and Irt-like proteins (ZIPs), also known as solute carrier family 39A (SLC39A), and Zinc transporters (ZnTs), also known as SLC30A proteins, and metallothioneins (MTs) [1]. ZIPs are known to function in the uptake of Zn across the cytoplasm from the extracellular environment or regulate the release of Zn into the cytosol from intracellular organelles, including the endoplasmic reticulum (ER), mitochondria, and Golgi apparatus; ZnTs acts in the efflux of Zn from the cytoplasm to the extracellular environment or the uptake of Zn into intracellular compartments from the cytosol [1]. Vignesh and Deep describe MTs in detail in this IJMS special issue [3].

ZIPs and ZnTs are involved in many cellular responses, including cytokine- and growth factor-meditated signaling, and the regulation of enzymes, receptors, and transcription factors belonging to cellular signaling pathways [4]. Numerous Zn transporters regulate Zn homeostasis and have crucial functions in physiology; dysfunctions that are caused by mutations result in inherited diseases [1]. Moreover, single-nucleotide polymorphisms, which are related to disease pathology, in each transporter gene have been identified [5–8]. Thus, impaired Zn transporter function is strongly linked to clinical human diseases, and numerous studies have examined these membrane transporters for their great potential as drug targets.

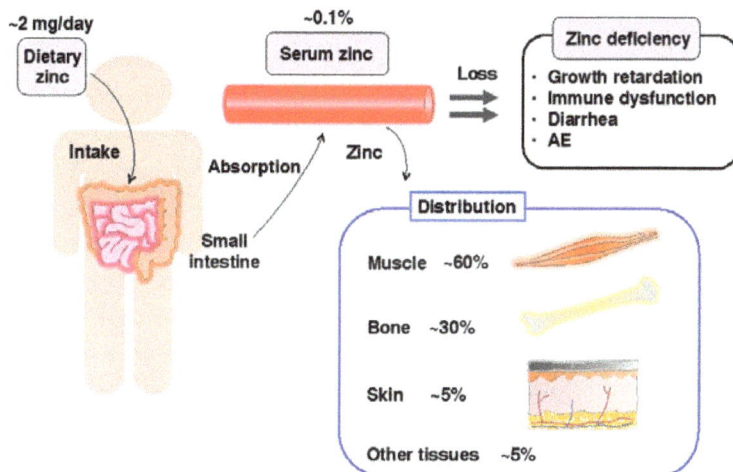

**Figure 1.** Scheme for Zinc (Zn) storage and distribution in the body. Dietary Zn is absorbed in the small intestine and distributed to the peripheral tissues, including skeletal muscle (60%), bone (30%), skin (5%), and other tissues (5%). Zn deficiency causes various abnormalities in humans and animal models, such as growth retardation, immune dysfunctions, diarrhea, and skin diseases, including acrodermatitis enteropathica (AE).

In this review, we provide the updated information related to Zn transporters, focusing on ZIP family members and their roles in Zn homeostasis, cellular functions, signal transduction, development, and human diseases. We also discuss the remaining questions by reviewing recent progress in studies of Zn transporters and Zn signaling.

## 2. Overview of Mammalian Zrt- and Irt-like Protein (ZIP) Transporters

Zn regulates a broad range of cellular functions; therefore, the dysregulation of Zn homeostasis causes various abnormalities in mammalian models [1,9,10]. Under physiological conditions, ZnTs reduce the intracellular availability of Zn by accelerating Zn efflux from the cell or into intracellular vesicles, while ZIP transporters import Zn into the cytosol from the extracellular space or intracellular compartments (Figure 2). Some ZIPs and ZnTs have been shown to be involved in the development of human diseases. Moreover, gene deficient (knockout, KO) mouse studies of ZIP and ZnT family members have revealed many unique phenotypes (Table 1), indicating that each Zn transporter-mediated Zn signaling exerts profound effects on non-overlapping molecular events to coordinate physiological conditions. Thus, Zn homeostasis is tightly regulated by the coordination of both transporters. We first provide an overview of all the ZIP transporters, followed by updates of selected ZIP transporters.

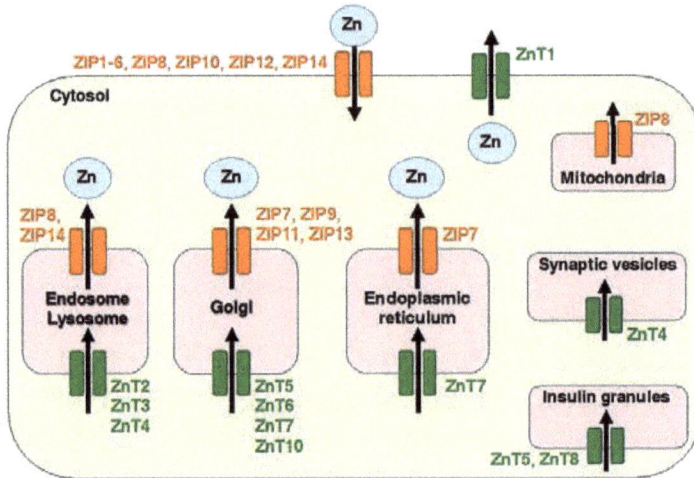

**Figure 2.** Cellular localization of Zinc transporters (ZnTs) and Zrt- and Irt-like proteins (ZIPs). The diagram shows the localization of ZIPs (orange) and ZnTs (green). The black arrow shows the direction of Zn transport in the plasma membrane and each organelle. ZIPs and ZnTs regulates the flux of Zn ion in the extra- or intra-cellular environment and tightly controls cellular Zn homeostasis in numerous cell types.

## 2.1. ZIP1

ZIP1 is a prototypic ZIP transporter that transports Zn into the cytosol and is ubiquitously expressed in human tissues [11]. *Zip1*-KO mice are sensitive to dietary Zn deficiency during pregnancy [12]. Previous studies showed that the downregulation of ZIP1 in malignant cells is accompanied by a decrease in Zn [13]. Recently, Furuta et al. observed increased ZIP1 expression in mouse astrocytes under oxidative stress conditions [14]. However, the role of ZIP1-mediated Zn signaling and the relationship between ZIP1 abnormalities and human disease remain unclear.

**Table 1.** Physiological properties of SLC39A/ Zrt- and Irt-like protein (ZIP) transporters.

| Genes/Proteins | Expression | Subcellular Location | Physiological Functions | Genetic Mutation Study in Mice | References |
|---|---|---|---|---|---|
| *Slc39a1* / ZIP1 | Ubiquitous | Plasma membrane | Abnormal embryonic development | Knockout (KO) | [12] |
| *Slc39a2* / ZIP2 | Liver, ovary, skin, dendritic cell | Plasma membrane | Abnormal embryonic development | KO | [15] |
| *Slc39a3* / ZIP3 | Widely distributed | Plasma membrane | Abnormal embryonic and T-cell development | KO | [12] |
| *Slc39a4* / ZIP4 | Small intestine, epidermis | Plasma membrane | Embryonic lethality | KO | [16,17] |
| *Slc39a5* / ZIP5 | Small intestine, kidney, pancreas | Plasma membrane | Intestinal Zn excretion; pancreatic Zn accumulation | KO | [18] |
| *Slc39a6* / ZIP6 | Widely distributed | Plasma membrane | Abnormal gonad formation and E-cadherin expression Glial cell migration in *Drosophila* | - | [19,20] |
| *Slc39a7* / ZIP7 | Widely distributed, colon, skin | Endoplasmic reticulum (ER) and Golgi apparatus | Impaired melanin synthesis, fibroblast growth factor receptor (FGFR) and Notch signaling in *Drosophila* Colon epithelial cell differentiation and proliferation in mouse Skin dermis development | KO | [21–23] |
| *Slc39a8* / ZIP8 | Widely distributed | Plasma membrane, lysosome | Cdm mouse: Resistance to cadmium-induced testicular damage, embryonic lethality | KO | [24,25] |
| *Slc39a9* / ZIP9 | Widely distributed | Golgi apparatus | Expressed in breast and prostate cancer cell lines Apoptosis regulation | - | [26,27] |
| *Slc39a10* / ZIP10 | Widely distributed, renal cell, carcinoma B cell | Plasma membrane | B cell development and function. Epidermal development Breast cancer progression | KO | [28–30] |
| *Slc39a12* / ZIP12 | Brain, pulmonary vascular smooth muscle | Plasma membrane | Neuronal differentiation Attenuation of pulmonary hypertension in a hypoxic atmosphere | KO (Rat) | [31] |
| *Slc39a13* / ZIP13 | Hard and connective tissues | Golgi apparatus, vesicles | Growth retardation, abnormal hard and connective tissue development, and adipocyte browning Growth retardation and impaired G protein-coupled receptor (GPCR) signaling | KO | [32,33] |
| *Slc39a14* / ZIP14 | Widely distributed, liver, bone, and cartilage | Plasma membrane, endosome | Growth retardation, abnormal chondrocyte differentiation Adipokineuction Impaired the phosphodiesterase (PDE) activity through GPCR-mediated cAMP-CREB signaling Hypertrophic adiposity Endotoxemia Glucose metabolism Impaired ER stress | KO | [34–38] |

## 2.2. ZIP2

ZIP2 is known to exist at the plasma membrane in human leukemia cells and functions as an importer of Zn, which increases Zn cellular levels [39]. *Zip2*-KO mice are sensitive to dietary Zn deficiency during pregnancy [15], as are *Zip1*-KO mice [12]. Gene expression analysis revealed high levels of ZIP2 expression in the epidermis, and RNAi knockdown of *ZIP2* gene expression inhibited the differentiation of keratinocytes [40]. Moreover, *Zip2*-KO mice exhibited skin blistering during early embryogenesis [15]. These results indicate that, in the skin, ZIP2 is involved in the differentiation of keratinocytes [40]; thus, ZIP2 is a potential therapeutic target for skin epidermis diseases.

## 2.3. ZIP3

ZIP3 is localized at the plasma membrane in mammary epithelial cells [41], and it functions as an importer of Zn [42]. *Zip3*-KO mice are more likely to show abnormal development during Zn-deficient pregnancy [12,42]. The absence of ZIP3 is evident in early and progressive malignancy; previous studies showed that ZIP3 expression is regulated by Ras-responsive-element-binding-protein (RREB1) in the normal ductal/acinar epithelium [43], indicating that the RREB1/ZIP3 pathway is involved in regulating oncogenesis.

## 2.4. ZIP4

ZIP4 plays an indispensable role in Zn absorption in the small intestine, and it is expressed at the apical membrane of enterocytes [44,45]. Homozygous *Zip4*-KO mice are embryonic lethal during early development, and heterozygous offspring are hypersensitive to Zn deficiency, displaying developmental defects, such as exencephalia, anophthalmia, and growth retardation [16]. Loss-of function mutations in ZIP4 cause acrodermatitis enteropathica (AE), a congenital disease that is characterized by extreme Zn deficiency if it is left untreated without supplemental Zn (OMIM 201100) (Table 2) [46,47]. It has been suggested that dietary Zn is mostly absorbed in the duodenum, ileum, and jejunum by active transport through ZIP4 [48]. However, the molecular mechanisms of dermatitis that is caused by ZIP4 mutation remain unclear. A more recent study investigated whether ZIP4 is cell-autonomously essential for maintaining human epidermal homeostasis [17]. In normal skin, Zn in the basal layer is transported to cells via ZIP4 and sufficiently supplied to Zn-binding proteins, including ΔNp63, which are essential for epidermal differentiation; thus, epidermis-localized ZIP4 has cell-autonomous functions to develop the epidermis. Taken together, ZIP4 has dual roles: ZIP4 increases Zn mass in the body via intestinal ZIP4, and is involved in the development of epidermal tissues by epidermis-localized ZIP4 [17].

**Table 2.** Hereditary human diseases of SLC39A/ZIP transporters.

| Genes/Proteins | Mutation Type | OMIM Gene Locus/Phenotype | Chromosomal Location | Disease | References |
|---|---|---|---|---|---|
| *Slc39a4*/ZIP4 | Mutation | 607059/201100 | 8q24.3 | Acrodermatitis enteropathica (AE) | [17] |
| *Slc39a5*/ZIP5 | Mutation | 608730/615946 | 12q13.3 | Nonsymptomatic high myopia | [49] |
| *Slc39a8*/ZIP8 | Mutation, Single nucleotide polymorphism (SNP) | 608732/616721 | 4q24 | Cerebellar Atrophy Syndrome, Congenital disorder of glycosylation type II | [50,51] |
| *Slc39a13*/ZIP13 | Mutation | 608735/612350 | 11p11.2 | Spondylocheiro dysplastic Ehlers-Danlos syndrome (SCD-EDS) | [32,33] |
| *Slc39a14*/ZIP14 | Mutation | 608736/617013 | 8q21.3 | Childhood-onset parkinsonism-dystonia, Hypermanganesemia with dystonia 2 | [52] |

## 2.5. ZIP5

ZIP5 is homeostatically expressed in acinar cells and enterocytes, localized to the basolateral surface, and functions as a specific transporter of Zn [53]. A lack of ZIP5 results in Zn accumulation in the liver and failure to accumulate excess Zn in the pancreas [18]. A study of pancreas-specific *Zip5*-KO mice revealed that ZIP5 in pancreatic acinar cells plays a key role in Zn accumulation/retention and protects cells from Zn-induced acute pancreatitis [18]. Although ZIP5 function is required for the survival of mammary gland epithelial cells in culture, homozygous KO mice did not show visible phenotypes. More recently, Feng et al. detected mutations in ZIP5 in patients with high myopia (Table 2) [49]; however, the relationship between ZIP5 and its pathophysiology are not understood. Therefore, ZIP5 may play a unique role in polarized cells by sensing Zn status via serosal-to-mucosal transport of Zn.

### 2.6. ZIP6

ZIP6 localizes to the plasma membrane and functions to import Zn across the cell membrane into cells [54]. The expression of ZIP6 was shown to be associated with estrogen receptor-positive breast cancer, metastatic ability, and cancer progression [20,55,56]. ZIP6 is known to be involved in the epithelial-mesenchymal transition (EMT) and cell migration. During gastrulation in zebrafish, STAT3 transactivates the expression of ZIP6, which promotes nuclear translocation of the transcriptional factor Snail and represses E-cadherin expression [19]. Several studies have revealed similarities and a functional relationship between ZIP6 and ZIP10, suggesting that these Zn transporters interact to conduct biological activities [54,57,58]. It was also shown that ZIP10 is transcriptionally regulated by signal transducer and STAT3 and STAT5, and suppresses apoptosis during the early development of B lymphocytes, and ZIP10 is also overexpressed in human lymphoma [28], as described below. Thus, both, ZIP6 and ZIP10 may be associated with the aggressive behavior of malignant cells, which is regulated by STAT3/5 signaling.

### 2.7. ZIP7

ZIP7 is localized to the Golgi apparatus [59] and ER [60], and plays a critical role in maintaining the intracellular balance of Zn and regulates both cell growth and differentiation pathways involving HER2, EGFR, Src, and IGF1R signaling [1,21,61]. ZIP7 has been shown to be consistently overexpressed in numerous breast cancers with poor prognosis and contributes to the tamoxifen resistance of breast cancer cells [21,62,63].

It has been reported that ZIP7 is involved in growth factor signaling-dependent and/or phosphorylation-mediated signaling pathways [63]. Taylor and colleagues reported that a Zn gate in the ER releases Zn from intracellular stores in response to phosphorylation by casein kinase 2 (CK2), which promotes the activation of tyrosine kinases AKT and ERKs, followed by the regulation of cell migration and proliferation [63]. These findings suggest that ZIP7 acts as a multifunctional protein in regulating a wide range of cellular processes, including ER stress during development and adult tissue homeostasis. In fact, Zn is required for normal ER function, which is supported by the observation that Zn deficiency in the ER lumen causes ER stress [64,65]. In *Drosophila*, Catsup, a member of the ZIP7 protein family, mediates Zn release from the ER and Golgi [66], indicating the possible involvement of ZIP7 in ER functions in vivo.

Recent investigations demonstrated that mice with an intestinal epithelium-specific *Zip7* deletion exhibited ER stress in proliferative progenitor cells, leading to disrupted epithelial proliferation and intestinal stemness (Figure 3A) [23]. Moreover, connective tissue-specific *Zip7*-KO mice exhibited an inhibition of protein disulfide isomerase (PDI), leading to ER dysfunction, which revealed dysgenesis of the dermis and hard connective tissue, including the bone and teeth [22]. Thus, ZIP7 plays an important role in maintaining intestinal epithelial homeostasis and skin dermis development by regulating ER function(s) (Figure 3B) [22,23]. These findings are discussed in detail in subsequent sections.

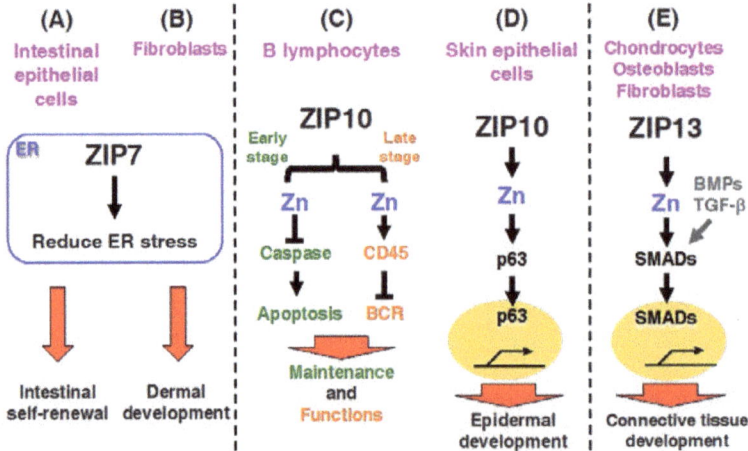

**Figure 3.** Biological relevance of ZIP7, ZIP10, and ZIP13. (**A,B**) ZIP7 is expressed in the endoplasmic reticulum (ER) membrane of various cells including dermal fibroblasts and intestinal epithelial cells, maintains Zn levels in the ER, and contributes to reducing ER stress. (**A**) In intestinal epithelial cells, ZIP7 promotes intestine epithelial self-renewal by resolving the upregulation of ER stress. Therefore, ZIP7 is a new regulator of intestinal epithelium homeostasis by regulating ER function; (**B**) In the dermal fibroblast ER, ZIP7 contributes dermal development. ZIP7 dysfunction induces ER stress caused by Zn-dependent protein disulfide isomerase (PDI) aggregation. PDI aggregation in dermal fibroblast disturbs adequate protein folding, which impairs dermal development; (**C,D**) ZIP10 contributes to the development and functions of B cells and skin epidermis; (**C**) ZIP10 inhibits caspase activity in progenitor B cells and promotes B cell development in the early stage (green color). ZIP10 also modulates B cell receptor (BCR) signaling in the late stage (orange color). Thus, ZIP10 is crucially involved in B cell-mediated immunity; (**D**) In skin epithelial cells, ZIP10 up-regulates p63 transactivation, which promotes epidermal and hair follicle development (yellow circle: nucleus). Therefore, the ZIP10-Zn-p63 signaling axis plays an important role in maintaining the skin epidermis; (**E**) ZIP13 is expressed in chondrocytes, osteoblasts, and fibroblasts and contributes to connective tissue development. ZIP13-mediated Zn signaling is required for Smad proteins activation in bone morphogenetic protein (BMP)/transforming growth factor beta (TGF-β) signaling, which regulates connective tissue development.

## 2.8. ZIP8

ZIP8 is localized to the plasma membrane and apical surface of polarized cells, mitochondria, and lysosomes [67,68]. *Zip8* mRNA expression is a transcriptional target of nuclear factor (NF)-κB, and ZIP8 negatively regulates proinflammatory responses through Zn-mediated downregulation of IκB kinase (IKK) activity, thereby inhibiting NF-κB activity (Figure 4B) [69]. Clinical studies revealed highly elevated serum Zn levels in osteoarthritis (OA) [70]. Kim et al. found that ZIP8 expression is specifically upregulated in OA cartilage of humans and mice, resulting in increased levels of intracellular Zn and the activation of a catabolic cascade by upregulating matrix-degrading enzymes, whereas upregulation of MT1 and MT2 proteins by metal responsive transcription factor (MTF1) forms a negative feedback loop and causes destruction during OA pathogenesis (Figure 4A) [24]. Thus, ZIP8 may be a potent therapeutic target for treating OA.

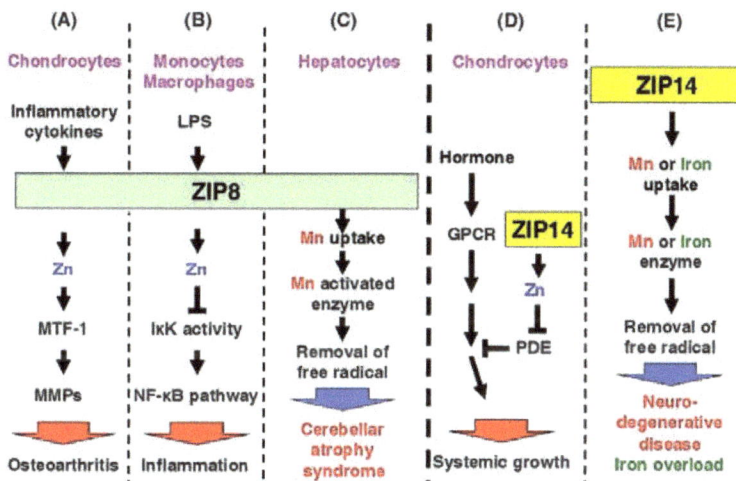

**Figure 4.** Physiological control by multiple metal transport through ZIP8 and ZIP14. (**A–C**) ZIP8 is involved in inflammatory responses and pathophysiology. ZIP8 expression is induced by (**A**) inflammatory cytokines and endotoxin in chondrocyte and (**B**) monocytes and macrophages, respectively. (**A**) In chondrocytes, ZIP8-mediated Zn activates MTF-1 and increases MMP expression, followed by cartilage degeneration of osteoarthritis; (**B**) In monocytes and macrophages, ZIP8-mediated Zn decreases IKKβ activity and NF-κB signaling and promotes inflammatory responses; (**C**) Mn (red) is transported by ZIP8. Loss of function of mutated ZIP8 reduces Mn uptake followed by a decrease in Mn-activated enzymes, resulting in cerebellar atrophy syndrome; (**D**) ZIP14 is required for systemic growth and modulates G protein-coupled receptor signaling by inhibiting hormone-stimulated phosphodiesterase (PDE) in chondrocytes; (**E**) Mn (red) and iron (green) are transported by ZIP14. Loss of function of mutated ZIP14 decreases Mn and iron uptake followed by a decrease in either Mn- or iron -activated enzymes, which results in neurodegenerative disease or iron overload disorders.

Interestingly, the mutation of *ZIP8* causes human pathogenesis, including Crohn's disease and cerebellar atrophy syndrome (Table 2) [50,51]. In fact, ZIP8 possesses higher affinity for Mn than for Zn in cells [67]; moreover, mice with liver-specific *Zip8*-KO mice showed decreased activity of arginase and β-1,4-galactosyltransferase, which are Mn-dependent enzymes (Figure 4C) [71]. Therefore, ZIP8 can regulate both Mn and Zn homeostasis. A loss-of-function mutation in ZIP8 induces the dysfunction of Mn and Zn homeostasis, resulting in human diseases, such as cerebellar atrophy syndrome (Figure 4C).

### 2.9. ZIP9

Previous studies showed that ZIP9 regulates cytosolic Zn levels, resulting in the activation of B cell receptor (BCR) signaling by enhancing Akt and Erk phosphorylation [26]. Notably, ZIP9 is expressed in breast cancer and prostate cancer cell lines [72], and ZIP9 acts as a membrane androgen receptor (mAR) that is independent of nuclear androgen receptors [27]. Testosterone treatment increases intracellular Zn concentrations, thereby upregulating a gene related to apoptosis [72]. These findings suggest that ZIP9 is important for various cellular functions, particularly in some types of cancer cells, where it regulates Zn homeostasis and/or hormone functions.

### 2.10. ZIP10

ZIP10 is mainly localized to the plasma membrane, and it functions as a cell surface Zn importer [28,73]. As described above, ZIP10 forms a functional heteromeric complex with ZIP6 [58]. Recently, ZIP6 and ZIP10 were found to control EMT by inactivating GSK-3 and downregulating

E-cadherin in breast cancer cells and renal carcinoma cells [20,74]. ZIP10 is transcriptionally regulated by STAT proteins in early B cells, and is overexpressed in lymphoma, indicating that ZIP10 is involved in the initiation or development of cancers [28]. Interestingly, Bin et al. showed that ZIP10 is required for skin epithelium development, such as the epidermis and hair follicles (Figure 3D) [30]. Together with the requirement of ZIP10 in B cell functions and skin developments, updates on the roles of ZIP10 are described in the next section.

### 2.11. ZIP11

ZIP11 is localized to the nucleus and Golgi apparatus [75,76]. Recently, Martin et al. suggested that ZIP11 plays an important role in the Zn homeostasis required to maintain mucosal integrity, function, and pH within the mouse stomach and colon [75]. Another study suggested that ZIP11 modulates the risk of bladder cancer and renal cell carcinoma [77]. However, the physiological and cellular functions of ZIP11 are not well-defined.

### 2.12. ZIP12

ZIP12 is highly expressed in human, mouse, and *Xenopus tropicalis* brain tissue [78]. Inactivation of ZIP12 caused developmental arrest and lethality during neurulation in *Xenopus tropicalis* [78]. ZIP12 was shown to play an important role in neuronal differentiation involving the activation of cAMP response element binding protein (CREB) signaling, neurite outgrowth, and tubulin polymerization [78]. A recent study detected ZIP12 expression in pulmonary vascular smooth muscle cells under hypoxic conditions. The inhibition of ZIP12 suppressed cell proliferation and increased intracellular labile Zn in hypoxic-cells [31]. Genetic disruption of ZIP12 in rat attenuates hypoxia-associated pulmonary hypertension in hypoxic environments [31]. Thus, inhibition of ZIP12 may be useful for treating pulmonary hypertension.

Interestingly, a recent study showed that increased ZIP12 expression in the dorsolateral prefrontal cortex causes schizophrenia [79], as described below.

### 2.13. ZIP13

ZIP13 is expressed in hard and connective tissues and, it is mainly localized to the Golgi apparatus [32,33]. Interestingly, a recent study showed that *Drosophila* ZIP13 (dZIP13) transports not only Zn, but also Fe [80]. The amino acid sequence of mammalian ZIP family members determined by the Protein Basic Local Alignment Search Tool (BLASTP) search, revealed that dZIP13 is highly homologous to human ZIP13 [80]. Both the gut and rest of the body exhibited Fe reduction after dZIP13 knockdown, and Fe increase when dZIP13 was overexpressed. dZIP13 affects Fe absorption, as described above, and it is known that dietary Fe absorption is mediated by ferritin in *Drosophila* [81]. Thus, these results suggest that knockdown of dZIP13 inhibits Fe transport into the secretory pathway to be available to ferritin, reducing Fe export from the gut for systemic use, while the overexpression of dZIP13 increases Fe concentrations in the body by facilitating Fe transport into the secretion pathway, making less Fe available in the cytosol of the gut cells. Taken together, dZIP13 potently mediates Fe export to the secretory pathway [80].

Previous studies reported that ZIP13 is involved in bone morphogenetic protein (BMP)/transforming growth factor β (TGF-β)-mediated Smad localization to the nucleus (Figure 3E) [32]. It was demonstrated that bone, tooth, and connective tissues development and systemic growth are impaired in *Zip13*-KO mice, as well as in patients with the loss of functions of ZIP13 proteins [32,82,83]. These patients exhibited significantly decreased white fat mass. Recently, Fukunaka et al. demonstrated that ZIP13-mediated Zn transport plays a critical role in suppressing adipocyte browning by reducing C/EBP-β proteins [84], which are discussed in this issue by Fukunaka and Fujitani. The molecular mechanism of the pathogenesis induced by the mutations is described below.

*2.14. ZIP14*

ZIP14 is localized to the plasma membrane and endosome, and expressed in the small intestine, liver, pancreas, and heart [85,86]. Recent studies have shown that ZIP14 is highly expressed in the various cancers in human including the colorectal cancer, hepatocellular cancer, and prostate cancer [87–90]. In *Zip14*-KO mice with dwarf body sizes, osteopenia, and impaired skeletal growth, cellular and molecular investigations revealed that ZIP14 modulates G protein-coupled receptor-mediated cAMP-CREB signaling by suppressing basal phosphodiesterase (PDE) activity (Figure 4D) [38]. Moreover, studies with *Zip14*-KO mice have indicated that ZIP14-mediated Zn transports involved in the metabolic endotoxemia, acute and chronic inflammation, intestinal barrier function, hypertrophic adiposity, and impaired glucose metabolism and ER stress [34–37,91,92].

In addition to Zn transportation, ZIP14 has been reported to transport metals such as Fe and Mn in vivo [93,94]. Fe is required for vital metabolic processes in cells; however, excess Fe has toxic effect in cells and can initiate Fe-overload disorders, such as hereditary hemochromatosis, resulting in liver cirrhosis, diabetes, and heart failure [95]. Fe uptake is also known to be regulated by two principle pathways, transferrin (Tf)-Fe via the Tf-receptor (TfR) pathway and nontransferrin-bound Fe (NTBI) through divalent metal transporters, such as DMT1, which is required for intestinal Fe uptake. A previous study using a cell culture system showed that SLC39A14/ZIP14 transport is involved not only in the uptake of Zn, but also in that of Fe in hepatocytes (Figure 4E) [96]. Moreover, a tissue expression array showed that *Zip14* mRNA is ubiquitously expressed at high levels in the liver, pancreas, and heart [86]. Therefore, Jenkitkasemwong et al. evaluated the role of ZIP14 in NTBI uptake in vivo [94]. *Zip14*-KO mice showed decreased $^{59}$Fe-NTBI uptake in hepatocytes. The authors crossed *Zip14*-KO mice with *Hfe*-KO and *Hfe2*-KO mice to develop an animal model of hemochromatosis in order to determine if ZIP14 is required for tissue Fe accumulation in Fe overload. Analysis of single- or double-KO mice revealed that ZIP14 deficiency in hemochromatotic mice greatly diminished Fe overloading in the liver and prevented Fe deposition in hepatocytes. These findings suggest that ZIP14 is required for NTBI uptake into hepatocytes. Thus, ZIP14 is essential for the developments of hepatic Fe overload in hemochromatosis and for Fe loading of hepatocytes (Figure 4E) [94].

In addition to Zn transportation, ZIP14 has also been reported to transport Fe and manganese (Mn) [93,96]. A recent study demonstrated that ZIP14 is a potent candidate molecule for inducing hemochromatosis [94], and, more recently, ZIP14 was reported to transport Mn in humans [52]. Its loss of function causes similar symptoms as parkinsonism-dystonia with neurodegeneration and hypermanganesemia in childhood (Table 2) [52], indicating that although ZIP14 may be a therapeutic target, further investigations to clarify the molecular basis of ZIP14 are needed, as described below.

## 3. Updates on the Role of ZIP Transporters in Pathophysiology and Human Diseases

In this section, we describe updated information on the role of ZIP transporters, mainly focusing on the pathophysiology. Recently, many studies of ZIP transporters have been conducted, which have improved the understanding of their crucial involvement in physiological events and showing that the loss of their functions causes diseases. Among the ZIP members, we selected ZIP7, ZIP10, ZIP12, ZIP13, and ZIP14 as ZIP transporters, of which investigations in vivo have been remarkably and rapidly progressed, so that advanced information of these molecules are reviewed below.

*3.1. ZIP7*

3.1.1. ZIP7 Contributes to Intestinal Epithelial Homeostasis

Although ZIP7 has attracted much interest in numerous research fields and many studies have been performed in primary cells and cell lines, as described above [21,59,61], the in vivo functions of ZIP7 remained unclear because of the lack of a *Zip7*-KO animal model. A recent investigation demonstrated that ZIP7 is highly expressed in transit-amplifying (TA) cells and Paneth cells at the intestinal crypt [23]. Ohashi et al. generated *Zip7*-conditional KO (*Zip7*-cKO) mice lacking the *Zip7* gene specifically in intestinal epithelium cells [23]. They demonstrated that *Zip7*-cKO mice, which died within a week with the loss of intestinal stem cells and epithelial integrity, showed a loss of the proliferating compartment under increased ER stress.

ER stress triggers a signaling reaction known as the unfolded protein response (UPR), which plays a crucial role in regulating the proliferation of the intestinal epithelium [97,98]. However, excessive UPR induces ER stress, leading to the activation of apoptosis signaling in the *Zip7*-KO TA cell population. Collectively, these findings suggest that TA cells enhanced UPR signaling and maintained cell proliferation in the lower region of the intestinal crypt. ZIP7 upregulated by UPR signaling maintained Zn homeostasis under ER stress, which promoted epithelial proliferation. This mechanism plays an important role in maintaining intestinal stemness, and it is highly sensitive to the death of neighboring cells induced by ER-stress. Thus, ZIP7 may be a novel regulator of intestinal epithelium homeostasis by maintaining ER function [23].

### 3.1.2. ZIP7 Is Required for Dermal Development in Skin

Skin is the first area that manifests Zn deficiency [99]. However, the molecular mechanisms by which Zn homeostasis affects skin development remain largely unknown. A recent study by Bin et al. further confirmed that ZIP7 is a critical molecule for regulating ER functions in the dermis, and thus it is necessary for proper skin formation [22]. Connective tissue-specific *Zip7*-cKO mice exhibited growth retardation, decreased hair follicles, abnormal incisor teeth, and sunken and down-slanting eyes. Microarray experiments analyzing *ZIP7* expression profiles in human mesenchymal stem cells indicated that the upregulated genes were mainly involved in the response to ER stress, while downregulated genes were mainly involved in cell cycle-related processes and differentiation processes. Moreover, deletion of ZIP7 downregulated PDI activity by increasing ER Zn levels, and induced overexpression of UPR genes [22]. These findings indicate that ZIP7 is a key regulator for resolving ER stress to normalize ER functions; therefore, control of ZIP7 may unlock therapeutic opportunities for overcoming human diseases that are arising from ER dysfunction.

### 3.2. ZIP10

### 3.2.1. ZIP10 Is Necessary for the Development and Functioning of B Lymphocytes

Zn deficiency leads to lymphopenia and the attenuation of both cellular and humoral immunity, resulting in an increased susceptibility to various pathogens [100,101]; however, little is known about how Zn regulates immune function. Miyai and Hojyo et al. investigated the expression profile of ZIP transporters and found that ZIP10 was highly expressed in B lymphocytes, particularly in cells in the early stages of B cell development, such as in pro-B cells [28,102]. They evaluated the physiological role of ZIP10 in early B cells by generating B cell-specific KO mice by using *Mb1-cre* mice, which exhibited fewer peripheral B cells and decreased pro-B cell survival. *Zip10* ablation in pro-B cells in vitro enhanced the activities of caspase-3, -8, -9, and -12, resulting in increased apoptotic cell death, which was mimicked by chemically chelating intracellular Zn; these negative effects were reversed by Zn supplementation [28] (Figure 3C left). Moreover, they demonstrated that activated STAT3 and STAT5 regulate the ZIP10 expression upon cytokine stimulation [28]. Because it is well-known that JAK-STAT signaling induced by cytokine stimulation controls pro-B cell survival and development [103,104], these findings clearly demonstrate that "the JAK/STAT-ZIP10-Zn signaling axis" is crucial for the survival of pro-B cells during their development [28,102].

Additionally, Hojyo et al. found another role for ZIP10 in late stage B cells by using mice, in which *Zip10* was deficient in antigen-presenting cells; these mice exhibited severely decreases germinal center (GC) formation, which was similar to the abnormalities observed in Zn-deficient mice [29,102]. *Zip10*-cKO late stage B cells showed dramatically decreased proliferation after BCR cross-linking in vitro. BCR signaling is initiated by Lyn, a Src-family protein tyrosine kinase, and Lyn activates Syk, which is involved in the activation of cell proliferation and survival signaling pathways [105,106].

*Zip10*-cKO B cells showed hyperactivated BCR signaling, which reduced cell proliferation (Figure 3C right). Furthermore, CD45R protein tyrosine phosphatase activity was downregulated in *Zip10*-cKO B cells. Thus, the deletion of ZIP10 in late stage B cells led to dysregulated BCR signaling due to reduced CD45R protein tyrosine phosphatase activity and impaired proliferation, as well as decreased GC formation, indicating that the ZIP10-mediated Zn stream is required for proper B cell signaling.

Together, these findings indicate that ZIP10 is indispensable for both, the homeostasis and functioning of B cells; therefore, ZIP10-mediated Zn homeostasis is relevant to B cell-immunity, explaining the importance of Zn in acquired immunity [10].

### 3.2.2. ZIP10 Is Necessary for Epidermal Homeostasis

The most recent study by Bin et al. revealed that ZIP10 is essential for epidermal formation [30]. As described above, Zn appears to be primarily essential for the differentiation, proliferation, and survival of epidermal keratinocytes in the skin [107]. However, the molecular relationship between Zn homeostasis and cells forming the skin epidermis is not well-understood. ZIP10 was found to be highly expressed in the outer root sheath of hair follicles [30]. Epithelium tissue-specific *Zip10*-cKO mice that were generated by using *Keratine14-cre* mice exhibited severe hypoplasia in the stratified epithelia, decreased hair follicles, and thymus atrophy, as the thymus medullae predominantly expresses ZIP10, which plays a role in maintaining this structure [30]. Moreover, the loss of ZIP10 interferes with the functions of p63, a master epidermal regulator containing a DNA-binding domain and Zn binding site [108], indicating the relevance of ZIP10-mediated Zn signaling in p63 function. Thus, ZIP10 plays important roles in epithelial tissue development via, at least in part, the ZIP10-Zn-p63 signaling axis, highlighting the physiological significance of Zn regulation in maintaining the skin epidermis (Figure 3D). These results provide insight for generating new therapeutic approaches by targeting hair follicle-localizing ZIP10. Furthermore, ZIP10-specific agonist may be useful as a trichogenous agent.

### 3.3. ZIP12

ZIP12 Contributes to Cortical Functions

It has been reported that ZIP12 is related to the pathogenesis of pulmonary hypertension [31]. However, the relationship between ZIP12 and human disease remains unclear. Scarr et al. demonstrated that in schizophrenia patients, ZIP12 expression is increased in the cortex according to gene expression profiling [79]. Moreover, Zn uptake analysis showed that two variants of ZIP12 have Zn transport functions in cells. However, total cortical Zn levels were not altered in brain tissues from schizophrenia patients, which may be because of changes in Zn homeostasis that are controlled by ZnTs. These results suggest that the increased expression of ZIP12 in brain tissues induces an imbalance in Zn homeostasis, causing the onset of schizophrenia. If ZIP12-specific antagonists are identified, this may be alternative approach for developing more effective drugs for schizophrenia when compared to existing drugs.

### 3.4. ZIP13

Molecular Mechanisms of ZIP13 Pathogenic Mutated Proteins

The first identified genetic disease that was associated with a ZIP family member was AE associated with a mutation in ZIP4 [46,47]; however, the pathophysiological role of ZIP family members except for ZIP4 in human diseases were unknown until 2008. Fukada et al. found that *Zip13*-KO mice show delayed growth and abnormalities in hard and connective tissue development [32], and a loss of function mutations were found in a novel variant of human Ehlers–Danlos syndrome (EDS): Spondylocheirodysplastic Ehlers–Danlos syndrome (SCD-EDS; OMIM 612350), demonstrating the importance of ZIP13 in the development and homeostasis of hard and connective tissues (Table 2) [32,83]. Moreover, they also found that SCD-EDS is attributed to a homozygous loss-of-function mutation in *ZIP13* gene, and genetic analysis showed that the pathogenic mutation was a glycine to aspartic acid substitution at position 64 (G64D) in the ZIP13 protein, which is encoded by *SLC39A13* [32]. Another mutant ZIP13 protein contains a deletion of amino acid residues 162–164 (phenylalanine–eucine–alanine) in *ZIP13*, which was also reported in SCD-EDS patients [83]. Bin et al. revealed that human ZIP13 protein forms a dimer [33,109]. Both mutant ZIP13 proteins are readily degraded by the valosin-containing protein-linked ubiquitin (Ub)-proteasome pathway, presumably by their misfolding during the protein maturation process, resulting in ZIP13 proteins with reduced functions [82]. Thus, ZIP13 mutants are susceptible to Ub-proteasome pathways, and the maintenance of Zn homeostasis via ZIP13 is impaired in cells expressing mutant ZIP13, leading to severe SCD-EDS pathogenesis in *Zip13*-KO mice [33,109] (Figure 3E). However, whether this also occurs in mammals and whether this conclusion explains the onset of SCD-EDS remain unclear. Further studies are required to resolve the complexity of ZIP13-mediated mammalian in vivo physiology and its molecular mechanisms.

### 3.5. ZIP14

ZIP14 Mediates Manganese Homeostasis

Mn is also an essential element for humans and is normally present in various tissues, including the brain, liver, and kidney. Mn imbalance impairs brain functions and causes disorders such as parkinsonism dystonia [110]. Previous studies showed that Mn transport is regulated by several transporter proteins, including DMT1 [111], ferroportin [112], SLC39A8/ZIP8 [71], and SLC30A10/ZnT10 [113], in addition to SLC39A14/ZIP14 [93]. Among these, the molecular details of ZIP14 in Mn transport are currently being examined. The expression and Mn transport function of ZIP14 is regulated by interleukin-6 in human neuroblastoma cells [114]. Recently, Tuschl and colleagues found that loss-of-function mutations in ZIP14 cause hypermanganesemia and progressive parkinsonism [52]. Mutation in *Zip14* by CRISPR/Cas9 genome editing resulted in reduced Mn disturbance in *Zip14*-mutated zebrafish [52], clearly demonstrating the relevant role of ZIP14-mediated Mn homeostasis in maintaining health.

The most recent reports described the physiological role of ZIP14 in Mn uptake in vivo [93,115]. Interestingly, *Zip14*-KO mice began to show signs of dystonia with a progressive inability to coordinate their motor activities [93], similar to PD-like motor disability in patients with a mutation in ZIP14 [52]. ZIP14 has been shown to be highly expressed in the liver [94]. Therefore, hepatocyte-specific *Zip14*-cKO mice that were generated using *Albumin-cre* mice showed significantly reduced hepatic Mn levels, but not normal levels of Mn in other tissues such as the brain, kidney, and pancreas [93]. In contrast, upon consuming a high-Mn diet, hepatocyte-specific *Zip14*-cKO mice showed increased Mn levels in the serum and brain, but not in the liver. Based on these findings, hepatic ZIP14 regulates the uptake, transport, and storage of Mn. Our results provide insight into Mn homeostasis in humans. Thus, ZIP14 is a potent and major Mn importer in the liver, and recent studies have revealed the clinical effects of disrupting Mn homeostasis (Figure 4E) [52].

### 4. Conclusions and Perspectives

As evident from previous and current studies on the role of Zn in the physiological and pathophysiological events described above, information that is related to the biological relevance of Zn as an essential trace element is accumulating rapidly. Recent studies involving mice and humans suggested that Zn transporters have various physiological functions in tissue development and homeostasis. As described in the sections above, some studies have revealed the relation between the functions of Zn transporters and specific diseases. Despite the current progress in our understanding of the physiological functions of Zn transporters, many questions remain regarding the role of Zn transporters in health and disease.

Zn transporters are expressed in various tissues and cell types, and they are localized in distinct subcellular compartments; these proteins transport not only Zn, but also Fe, Mn, cadmium, and other trace elements into subcellular compartments, indicating that the functions of Zn transporters are complex and diverse. Thus, studies clarifying the transportation mechanisms, not only of Zn but also of other metal ions, are required to understand the relation between the homeostatic mechanisms of metal ions and various diseases. This information will improve our understanding of the wide-range of functions of Zn ion and its transporters in diverse organisms. These studies may reveal how Zn ion regulates biological functions, and provide information related to its homeostasis (Figure 5) [116].

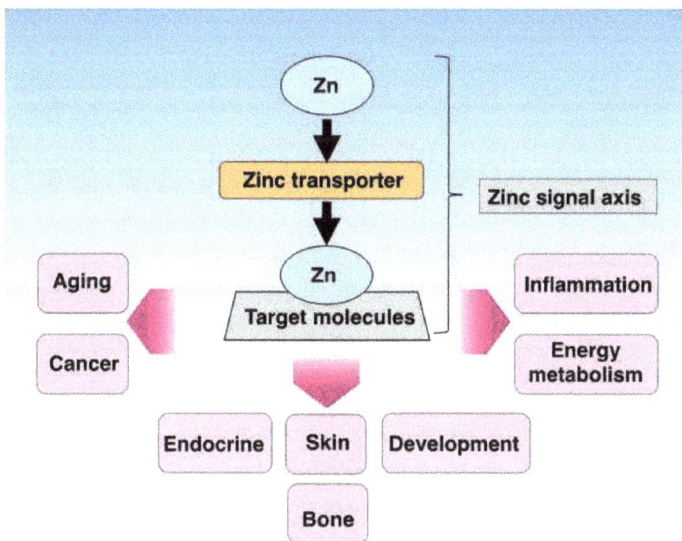

**Figure 5.** Summary of Zn signal axis in physiology and pathogenesis. Each Zn transporter regulates specific target molecules and cellular responses known as the "Zn-signal axis" [117] which transduces their signals to various physiological processes such as the bone and skin development, endocrine system, and aging. Dysfunction of the Zn signal results in impaired numerous cellular events, leading to various pathophysiological conditions such as inflammation, energy metabolism disorders, and cancer.

According to reports on the involvement of Zn transporters in tissue development and homeostasis, the dysfunction of Zn transporters is crucial not only to disease progression, but also to disease onset. Thus, investigating the functions of Zn transporters in tissue development and homeostasis using pluripotent stem cell lines from patients with abnormal Zn-homeostasis is important for addressing these fundamental questions regarding useful disease models. If Zn transporter dysfunctions determine the onset of diseases, modulating their function by specific compounds will provide crucial clues for the development of therapeutic strategies. In addition, no studies have examined the specific compounds that are regulating Zn transporter functions, but such compounds may have great potential for the development of novel therapeutic strategies.

Finally, further intensive research effort to examine Zn transporters would reveal critical molecular mechanisms that are related to mammalian health and disease.

**Acknowledgments:** This work was supported by KAKENHI (17H04011) of the Ministry of Education, Culture, Sports, Science, and Technology (MEXT), Sumitomo Foundation, Naito Foundation, Nestlé Nutrition Council Japan, Mitsubishi Foundation, Vehicle Racing Commemorative Foundation, and Takeda Science Foundation to Toshiyuki Fukada.

**Author Contributions:** Teruhisa Takagishi and Takafumi Hara contributed to the design, drafting, and revising of the manuscript. Toshiyuki Fukada helped design, draft, and revise the manuscript and granted final approval of the version to be published.

**Conflicts of Interest:** The authors declare no conflict of interest.

## Abbreviations

| | |
|---|---|
| AE | Acrodermatitis enteropathica |
| EMT | Epithelial mesenchymal transition |
| ER | Endoplasmic reticulum |
| KO | Knockout |
| NTBI | Nontransferrin-bound iron |
| PDI | Protein disulfide isomerase |
| SCD-EDS | Spondylocheirodysplastic Ehlers-Danlos syndrome |
| SLC | Solute carrier |
| TGF-$\beta$ | Transforming growth factor beta |
| UPR | Unfolded protein response |
| ZIP | Zrt- and Irt-like protein |
| ZnT | Zn transporter |

## References

1. Hara, T.; Takeda, T.; Takagishi, T.; Fukue, K.; Kambe, T.; Fukada, T. Physiological roles of zinc transporters: Molecular and genetic importance in zinc homeostasis. *J. Physiol. Sci.* **2017**, *67*, 283–301. [CrossRef] [PubMed]
2. Andreini, C.; Banci, L.; Bertini, I.; Rosato, A. Counting the zinc-proteins encoded in the human genome. *J. Proteome Res.* **2006**, *5*, 196–201. [CrossRef] [PubMed]
3. Subramanian Vignesh, K.; Deepe, G.S. Metallothioneins: Emerging modulators in immunity and infection. *Int. J. Mol. Sci.* **2017**, *18*, E2197. [CrossRef] [PubMed]
4. Fukada, T.; Yamasaki, S.; Nishida, K.; Murakami, M.; Hirano, T. Zinc homeostasis and signaling in health and diseases. *J. Biol. Inorg. Chem.* **2011**, *16*, 1123–1134. [CrossRef] [PubMed]
5. Howson, J.M.M.; Krause, S.; Stevens, H.; Smyth, D.J.; Wenzlau, J.M.; Bonifacio, E.; Hutton, J.; Ziegler, A.G.; Todd, J.A.; Achenbach, P. Genetic association of zinc transporter 8 (ZnT8) autoantibodies in type 1 diabetes cases. *Diabetologia* **2012**, *55*, 1978–1984. [CrossRef] [PubMed]
6. Xu, J.; Wang, J.; Chen, B. SLC30A8 (ZnT8) variations and type 2 diabetes in the Chinese Han population. *Genet. Mol. Res.* **2012**, *11*, 1592–1598. [CrossRef] [PubMed]
7. Sørgjerd, E.P.; Skorpen, F.; Kvaløy, K.; Midthjell, K.; Grill, V. Prevalence of ZnT8 antibody in relation to phenotype and SLC30A8 polymorphism in adult autoimmune diabetes. Results from the HUNT study, Norway. *Autoimmunity* **2012**, *46*, 74–79. [CrossRef] [PubMed]
8. Li, M.; Wu, D.D.; Yao, Y.G.; Huo, Y.X.; Liu, J.W.; Su, B.; Chasman, D.I.; Chu, A.Y.; Huang, T.; Qi, L.; et al. Recent positive selection drives the expansion of a schizophrenia risk nonsynonymous variant at SLC39A8 in Europeans. *Schizophr. Bull.* **2016**, *42*, 178–190. [PubMed]
9. Hojyo, S.; Fukada, T. Zinc transporters and signaling in physiology and pathogenesis. *Arch. Biochem. Biophys.* **2016**, *611*, 43–50. [CrossRef] [PubMed]
10. Hojyo, S.; Fukada, T. Roles of zinc signaling in the immune system. *J. Immunol. Res.* **2016**, *2016*. [CrossRef] [PubMed]
11. Gaither, L.A.; Eide, D.J. The human ZIP1 transporter mediates zinc uptake in human K562 erythroleukemia cells. *J. Biol. Chem.* **2001**, *276*, 22258–22264. [CrossRef] [PubMed]
12. Dufner-Beattie, J.; Huang, Z.L.; Geiser, J.; Xu, W.; Andrews, G.K. Mouse ZIP1 and ZIP3 genes together are essential for adaptation to dietary zinc deficiency during pregnancy. *Genesis* **2006**, *44*, 239–251. [CrossRef] [PubMed]
13. Franklin, R.B.; Feng, P.; Milon, B.; Desouki, M.M.; Singh, K.K.; Kajdacsy-Balla, A.; Bagasra, O.; Costello, L.C. hZIP1 zinc uptake transporter down regulation and zinc depletion in prostate cancer. *Mol. Cancer* **2005**, *4*, 32–44. [CrossRef] [PubMed]

14. Furuta, T.; Ohshima, C.; Matsumura, M.; Takebayashi, N.; Hirota, E.; Mawaribuchi, T.; Nishida, K.; Nagasawa, K. Oxidative stress upregulates zinc uptake activity via Zrt/Irt-like protein 1 (ZIP1) in cultured mouse astrocytes. *Life Sci.* **2016**, *151*, 305–312. [CrossRef] [PubMed]

15. Peters, J.L.; Dufner-Beattie, J.; Xu, W.; Geiser, J.; Lahner, B.; Salt, D.E.; Andrews, G.K. Targeting of the mouse Slc39a2 (Zip2) gene reveals highly cell-specific patterns of expression, and unique functions in zinc, iron, and calcium homeostasis. *Genesis* **2007**, *45*, 339–352. [CrossRef] [PubMed]

16. Dufner-Beattie, J.; Weaver, B.P.; Geiser, J.; Bilgen, M.; Larson, M.; Xu, W.; Andrews, G.K. The mouse acrodermatitis enteropathica gene Slc39a4 (Zip4) is essential for early development and heterozygosity causes hypersensitivity to zinc deficiency. *Hum. Mol. Genet.* **2007**, *16*, 1391–1399. [CrossRef] [PubMed]

17. Bin, B.H.; Bhin, J.; Kim, N.H.; Lee, S.H.; Jung, H.S.; Seo, J.; Kim, D.K.; Hwang, D.; Fukada, T.; Lee, A.Y.; et al. An acrodermatitis enteropathica-associated Zn transporter, ZIP4, regulates human epidermal homeostasis. *J. Investig. Dermatol.* **2017**, *137*, 874–883. [CrossRef] [PubMed]

18. Geiser, J.; De Lisle, R.C.; Andrews, G.K. The zinc transporter Zip5 (Slc39a5) regulates intestinal zinc excretion and protects the pancreas against zinc toxicity. *PLoS ONE* **2013**, *8*, e82149. [CrossRef] [PubMed]

19. Yamashita, S.; Miyagi, C.; Fukada, T.; Kagara, N.; Che, Y.-S.; Hirano, T. Zinc transporter LIVI controls epithelial-mesenchymal transition in zebrafish gastrula organizer. *Nature* **2004**, *429*, 298–302. [CrossRef] [PubMed]

20. Hogstrand, C.; Kille, P.; Ackland, M.L.; Hiscox, S.; Taylor, K.M. A mechanism for epithelial-mesenchymal transition and anoikis resistance in breast cancer triggered by zinc channel ZIP6 and STAT3 (signal transducer and activator of transcription 3). *Biochem. J.* **2013**, *455*, 229–237. [CrossRef] [PubMed]

21. Taylor, K.M.; Vichova, P.; Jordan, N.; Hiscox, S.; Hendley, R.; Nicholson, R.I. ZIP7-mediated intracellular zinc transport contributes to aberrant growth factor signaling in antihormone-resistant breast cancer cells. *Endocrinology* **2008**, *149*, 4912–4920. [CrossRef] [PubMed]

22. Bin, B.-H.; Bhin, J.; Seo, J.; Kim, S.-Y.; Lee, E.; Park, K.; Choi, D.-H.; Takagishi, T.; Hara, T.; Hwang, D.; et al. Requirement of zinc transporter SLC39A7/ZIP7 for dermal development to fine-tune endoplasmic reticulum function by regulating protein disulfide isomerase. *J. Investig. Dermatol.* **2017**, *137*, 1682–1691. [CrossRef] [PubMed]

23. Ohashi, W.; Kimura, S.; Iwanaga, T.; Furusawa, Y.; Irié, T.; Izumi, H.; Watanabe, T.; Hijikata, A.; Hara, T.; Ohara, O.; et al. Zinc transporter SLC39A7/ZIP7 promotes intestinal epithelial self-renewal by resolving ER stress. *PLoS Genet.* **2016**, *12*, 1–22. [CrossRef] [PubMed]

24. Kim, J.H.; Jeon, J.; Shin, M.; Won, Y.; Lee, M.; Kwak, J.S.; Lee, G.; Rhee, J.; Ryu, J.H.; Chun, C.H.; et al. Regulation of the catabolic cascade in osteoarthritis by the zinc-ZIP8-MTF1 axis. *Cell* **2014**, *156*, 730–743. [CrossRef] [PubMed]

25. Gálvez-Peralta, M.; He, L.; Jorge-Nebert, L.F.; Wang, B.; Miller, M.L.; Eppert, B.L.; Afton, S.; Nebert, D.W. ZIP8 zinc transporter: Indispensable role for both multiple-organ organogenesis and hematopoiesis in utero. *PLoS ONE* **2012**, *7*, e36055. [CrossRef] [PubMed]

26. Taniguchi, M.; Fukunaka, A.; Hagihara, M.; Watanabe, K.; Kamino, S.; Kambe, T.; Enomoto, S.; Hiromura, M. Essential role of the zinc transporter ZIP9/SLC39A9 in regulating the activations of Akt and Erk in B-cell receptor signaling pathway in DT40 cells. *PLoS ONE* **2013**, *8*, 1–10. [CrossRef] [PubMed]

27. Thomas, P.; Converse, A.; Berg, H.A. ZIP9, a novel membrane androgen receptor and zinc transporter protein. *Gen. Comp. Endocrinol.* **2017**, 30341–30346. [CrossRef] [PubMed]

28. Miyai, T.; Hojyo, S.; Ikawa, T.; Kawamura, M.; Irie, T.; Ogura, H.; Hijikata, A.; Bin, B.-H.; Yasuda, T.; Kitamura, H.; et al. Zinc transporter SLC39A10/ZIP10 facilitates antiapoptotic signaling during early B-cell development. *Proc. Natl. Acad. Sci. USA* **2014**, *111*, 11780–11785. [CrossRef] [PubMed]

29. Hojyo, S.; Miyai, T.; Fujishiro, H.; Kawamura, M.; Yasuda, T.; Hijikata, A.; Bin, B.-H.; Irie, T.; Tanaka, J.; Atsumi, T.; et al. Zinc transporter SLC39A10/ZIP10 controls humoral immunity by modulating B-cell receptor signal strength. *Proc. Natl. Acad. Sci. USA* **2014**, *111*, 11786–11791. [CrossRef] [PubMed]

30. Bin, B.-H.; Bhin, J.; Takaishi, M.; Toyoshima, K.; Kawamata, S.; Ito, K.; Hara, T.; Watanabe, T.; Irié, T.; Takagishi, T.; et al. Requirement of zinc transporter ZIP10 for epidermal development: Implication of the ZIP10-p63 axis in epithelial homeostasis. *Proc. Natl. Acad. Sci. USA* **2017**, *114*, 12243–12248. [CrossRef] [PubMed]

31. Zhao, L.; Oliver, E.; Maratou, K.; Atanur, S.S.; Dubois, O.D.; Cotroneo, E.; Chen, C.-N.; Wang, L.; Arce, C.; Chabosseau, P.L.; et al. The zinc transporter ZIP12 regulates the pulmonary vascular response to chronic hypoxia. *Nature* **2015**, *524*, 356–360. [CrossRef] [PubMed]

32. Fukada, T.; Civic, N.; Furuichi, T.; Shimoda, S.; Mishima, K.; Higashiyama, H.; Idaira, Y.; Asada, Y.; Kitamura, H.; Yamasaki, S.; et al. The zinc transporter SLC39A13/ZIP13 is required for connective tissue development; its involvement in BMP/TGF-beta signaling pathways. *PLoS ONE* **2008**, *3*, e3642. [CrossRef]

33. Bin, B.H.; Fukada, T.; Hosaka, T.; Yamasaki, S.; Ohashi, W.; Hojyo, S.; Miyai, T.; Nishida, K.; Yokoyama, S.; Hirano, T. Biochemical characterization of human ZIP13 protein: A homo-dimerized zinc transporter involved in the spondylocheiro dysplastic Ehlers-Danlos syndrome. *J. Biol. Chem.* **2011**, *286*, 40255–40265. [CrossRef] [PubMed]

34. Aydemir, T.B.; Chang, S.M.; Guthrie, G.J.; Maki, A.B.; Ryu, M.S.; Karabiyik, A.; Cousins, R.J. Zinc transporter ZIP14 functions in hepatic zinc, iron and glucose homeostasis during the innate immune response (endotoxemia). *PLoS ONE* **2012**, *7*. [CrossRef]

35. Aydemir, T.B.; Troche, C.; Kim, M.H.; Cousins, R.J. Hepatic ZIP14-mediated zinc transport contributes to endosomal insulin receptor trafficking and glucose metabolism. *J. Biol. Chem.* **2016**, *291*, 23939–23951. [CrossRef] [PubMed]

36. Troche, C.; Beker Aydemir, T.; Cousins, R.J. Zinc transporter Slc39a14 regulates inflammatory signaling associated with hypertrophic adiposity. *Am. J. Physiol. Endocrinol. Metab.* **2016**, *310*, E258–E268. [CrossRef] [PubMed]

37. Kim, M.-H.; Aydemir, T.B.; Kim, J.; Cousins, R.J. Hepatic ZIP14-mediated zinc transport is required for adaptation to endoplasmic reticulum stress. *Proc. Natl. Acad. Sci. USA* **2017**, *114*, E5805–E5814. [CrossRef] [PubMed]

38. Hojyo, S.; Fukada, T.; Shimoda, S.; Ohashi, W.; Bin, B.H.; Koseki, H.; Hirano, T. The zinc transporter SLC39A14/ZIP14 controls G-protein coupled receptor-mediated signaling required for systemic growth. *PLoS ONE* **2011**, *6*, e18059. [CrossRef] [PubMed]

39. Gaither, L.A.; Eide, D.J. Functional expression of the human hZIP2 zinc transporter. *J. Biol. Chem.* **2000**, *275*, 5560–5564. [CrossRef] [PubMed]

40. Inoue, Y.; Hasegawa, S.; Ban, S.; Yamada, T.; Date, Y.; Mizutani, H.; Nakata, S.; Tanaka, M.; Hirashima, N. ZIP2 protein, a zinc transporter, is associated with keratinocyte differentiation. *J. Biol. Chem.* **2014**, *289*, 21451–21462. [CrossRef] [PubMed]

41. Kelleher, S.L.; Lönnerdal, B. Zip3 plays a major role in zinc uptake into mammary epithelial cells and is regulated by prolactin. *Am. J. Physiol. Cell Physiol.* **2005**, *288*, C1042–C1047. [CrossRef] [PubMed]

42. Dufner-Beattie, J.; Huang, Z.L.; Geiser, J.; Xu, W.; Andrews, G.K. Generation and characterization of mice lacking the zinc uptake transporter ZIP3. *Mol. Cell. Biol.* **2005**, *25*, 5607–5615. [CrossRef] [PubMed]

43. Franklin, R.B.; Zou, J.; Costello, L.C. The cytotoxic role of RREB1, ZIP3 zinc transporter, and zinc in human pancreatic adenocarcinoma. *Cancer Biol. Ther.* **2014**, *15*, 1431–1437. [CrossRef] [PubMed]

44. Dufner-Beattie, J.; Wang, F.; Kuo, Y.-M.; Gitschier, J.; Eide, D.; Andrews, G.K. The acrodermatitis enteropathica gene ZIP4 encodes a tissue-specific, zinc-regulated zinc transporter in mice. *J. Biol. Chem.* **2003**, *278*, 33474–33481. [CrossRef] [PubMed]

45. Kambe, T.; Andrews, G.K. Novel proteolytic processing of the ectodomain of the zinc transporter ZIP4 (SLC39A4) during zinc deficiency is inhibited by acrodermatitis enteropathica mutations. *Mol. Cell. Biol.* **2009**, *29*, 129–139. [CrossRef] [PubMed]

46. Wang, K.; Zhou, B.; Kuo, Y.-M.; Zemansky, J.; Gitschier, J. A novel member of a zinc transporter family is defective in acrodermatitis enteropathica. *Am. J. Hum. Genet.* **2002**, *71*, 66–73. [CrossRef] [PubMed]

47. Küry, S.; Kharfi, M.; Kamoun, R.; Taïeb, A.; Mallet, E.; Baudon, J.-J.; Glastre, C.; Michel, B.; Sebag, F.; Brooks, D.; et al. Mutation spectrum of human SLC39A4 in a panel of patients with acrodermatitis enteropathica. *Hum. Mutat.* **2003**, *22*, 337–338. [CrossRef] [PubMed]

48. Wang, X.; Zhou, B. Dietary zinc absorption: A play of Zips and ZnTs in the gut. *IUBMB Life* **2010**, *62*, 176–182. [CrossRef] [PubMed]

49. Feng, C.; Huang, X.; Cheng, X.; Wu, R.; Lu, F.; Jin, Z. Mutational screening of SLC39A5, LEPREL1 and LRPAP1 in a cohort of 187 high myopia patients. *Sci. Rep.* **2017**, *7*, 1120. [CrossRef] [PubMed]

50. Li, D.; Achkar, J.-P.; Haritunians, T.; Jacobs, J.P.; Hui, K.Y.; D'Amato, M.; Brand, S.; Radford-Smith, G.; Halfvarson, J.; Niess, J.-H.; et al. A Pleiotropic missense variant in SLC39A8 is associated with *Crohn's disease* and human gut microbiome composition. *Gastroenterology* **2016**, *151*, 724–732. [CrossRef] [PubMed]

51. Boycott, K.M.; Beaulieu, C.L.; Kernohan, K.D.; Gebril, O.H.; Mhanni, A.; Chudley, A.E.; Redl, D.; Qin, W.; Hampson, S.; Küry, S.; et al. Autosomal-recessive intellectual disability with cerebellar atrophy syndrome caused by mutation of the manganese and zinc transporter gene SLC39A8. *Am. J. Hum. Genet.* **2015**, *97*, 886–893. [CrossRef] [PubMed]

52. Tuschl, K.; Meyer, E.; Valdivia, L.E.; Zhao, N.; Dadswell, C.; Abdul-Sada, A.; Hung, C.Y.; Simpson, M.A.; Chong, W.K.; Jacques, T.S.; et al. Mutations in SLC39A14 disrupt manganese homeostasis and cause childhood-onset parkinsonism–dystonia. *Nat. Commun.* **2016**, *7*, 11601. [CrossRef] [PubMed]

53. Dufner-Beattie, J.; Kuo, Y.M.; Gitschier, J.; Andrews, G.K. The adaptive response to dietary zinc in mice involves the differential cellular localization and zinc regulation of the zinc transporters ZIP4 and ZIP5. *J. Biol. Chem.* **2004**, *279*, 49082–49090. [CrossRef] [PubMed]

54. Kong, B.Y.; Duncan, F.E.; Que, E.L.; Kim, A.M.; O'Halloran, T.V.; Woodruff, T.K. Maternally-derived zinc transporters ZIP6 and ZIP10 drive the mammalian oocyte-to-egg transition. *Mol. Hum. Reprod.* **2014**, *20*, 1077–1089. [CrossRef] [PubMed]

55. Tozlu, S.; Girault, I.; Vacher, S.; Vendrell, J.; Andrieu, C.; Spyratos, F.; Cohen, P.; Lidereau, R.; Bieche, I. Identification of novel genes that co-cluster with estrogen receptor alpha in breast tumor biopsy specimens, using a large-scale real-time reverse transcription-PCR approach. *Endocr. Relat. Cancer* **2006**, *13*, 1109–1120. [CrossRef] [PubMed]

56. Manning, D.L.; Daly, R.J.; Lord, P.G.; Kelly, K.F.; Green, C.D. Effects of oestrogen on the expression of a 4.4 kb mRNA in the ZR-75-1 human breast cancer cell line. *Mol. Cell. Endocrinol.* **1988**, *59*, 205–212. [CrossRef]

57. Brethour, D.; Mehrabian, M.; Williams, D.; Wang, X.; Ghodrati, F.; Ehsani, S.; Rubie, E.A.; Woodgett, J.R.; Sevalle, J.; Xi, Z.; et al. A ZIP6-ZIP10 heteromer controls NCAM1 phosphorylation and integration into focal adhesion complexes during epithelial-to-mesenchymal transition. *Sci. Rep.* **2017**, *7*, 40313. [CrossRef] [PubMed]

58. Taylor, K.M.; Muraina, I.A.; Brethour, D.; Schmitt-Ulms, G.; Nimmanon, T.; Ziliotto, S.; Kille, P.; Hogstrand, C. Zinc transporter ZIP10 forms a heteromer with ZIP6 which regulates embryonic development and cell migration. *Biochem. J.* **2016**, *473*, 2531–2544. [CrossRef] [PubMed]

59. Huang, L.; Kirschke, C.P.; Zhang, Y.; Yan, Y.Y. The ZIP7 gene (Slc39a7) encodes a zinc transporter involved in zinc homeostasis of the *Golgi apparatus*. *J. Biol. Chem.* **2005**, *280*, 15456–15463. [CrossRef] [PubMed]

60. Taylor, K.M.; Morgan, H.E.; Johnson, A.; Nicholson, R.I. Structure-function analysis of HKE4, a member of the new LIV-1 subfamily of zinc transporters. *Biochem. J.* **2004**, *377*, 131–139. [CrossRef] [PubMed]

61. Hogstrand, C.; Kille, P.; Nicholson, R.I.; Taylor, K.M. Zinc transporters and cancer: A potential role for ZIP7 as a hub for tyrosine kinase activation. *Trends Mol. Med.* **2009**, *15*, 101–111. [CrossRef] [PubMed]

62. Taylor, K.M.; Morgan, H.E.; Smart, K.; Zahari, N.M.; Pumford, S.; Ellis, I.O.; Robertson, J.F.R.; Nicholson, R.I. The emerging role of the LIV-1 subfamily of zinc transporters in breast cancer. *Mol. Med.* **2007**, *13*, 396–406. [CrossRef] [PubMed]

63. Taylor, K.M.; Hiscox, S.; Nicholson, R.I.; Hogstrand, C.; Kille, P. Protein kinase CK2 triggers cytosolic zinc signaling pathways by phosphorylation of zinc channel ZIP7. *Sci. Signal.* **2012**, *5*, ra11. [CrossRef] [PubMed]

64. Ellis, C.D.; Wang, F.; MacDiarmid, C.W.; Clark, S.; Lyons, T.; Eide, D.J. Zinc and the Msc2 zinc transporter protein are required for endoplasmic reticulum function. *J. Cell Biol.* **2004**, *166*, 325–335. [CrossRef] [PubMed]

65. Ishihara, K.; Yamazaki, T.; Ishida, Y.; Suzuki, T.; Oda, K.; Nagao, M.; Yamaguchi-Iwai, Y.; Kambe, T. Zinc transport complexes contribute to the homeostatic maintenance of secretory pathway function in vertebrate cells. *J. Biol. Chem.* **2006**, *281*, 17743–17750. [CrossRef] [PubMed]

66. Groth, C.; Sasamura, T.; Khanna, M.R.; Whitley, M.; Fortini, M.E. Protein trafficking abnormalities in *Drosophila* tissues with impaired activity of the ZIP7 zinc transporter *Catsup*. *Development* **2013**, *140*, 3018–3027. [CrossRef] [PubMed]

67. He, L.; Girijashanker, K.; Dalton, T.P.; Reed, J.; Li, H.; Soleimani, M.; Nebert, D.W. ZIP8, member of the solute-carrier-39 (SLC39) metal-transporter family: Characterization of transporter properties. *Mol. Pharmacol.* **2006**, *70*, 171–180. [CrossRef] [PubMed]

68.   Besecker, B.; Bao, S.; Bohacova, B.; Papp, A.; Sadee, W.; Knoell, D.L. The human zinc transporter SLC39A8 (Zip8) is critical in zinc-mediated cytoprotection in lung epithelia. *Am. J. Physiol. Lung Cell. Mol. Physiol.* **2008**, *294*, L1127–L1136. [CrossRef] [PubMed]

69.   Liu, M.J.; Bao, S.; Gálvez-Peralta, M.; Pyle, C.J.; Rudawsky, A.C.; Pavlovicz, R.E.; Killilea, D.W.; Li, C.; Nebert, D.W.; Wewers, M.D.; et al. ZIP8 regulates host defense through zinc-mediated inhibition of NF-κB. *Cell Rep.* **2013**, *3*, 386–400. [CrossRef] [PubMed]

70.   Ovesen, J.; Møller-Madsen, B.; Nielsen, P.T.; Christensen, P.H.; Simonsen, O.; Hoeck, H.C.; Laursen, M.B.; Thomsen, J.S. Differences in zinc status between patients with osteoarthritis and osteoporosis. *J. Trace Elem. Med. Biol.* **2009**, *23*, 1–8. [CrossRef] [PubMed]

71.   Lin, W.; Vann, D.R.; Doulias, P.-T.; Wang, T.; Landesberg, G.; Li, X.; Ricciotti, E.; Scalia, R.; He, M.; Hand, N.J.; et al. Hepatic metal ion transporter ZIP8 regulates manganese homeostasis and manganese-dependent enzyme activity. *J. Clin. Investig.* **2017**, *127*, 2407–2417. [CrossRef] [PubMed]

72.   Thomas, P.; Pang, Y.; Dong, J.; Berg, A.H. Identification and characterization of membrane androgen receptors in the ZIP9 zinc transporter subfamily: II. Role of human ZIP9 in testosterone-induced prostate and breast cancer cell apoptosis. *Endocrinology* **2014**, *155*, 4250–4265. [CrossRef] [PubMed]

73.   Lichten, L.A.; Ryu, M.-S.; Guo, L.; Embury, J.; Cousins, R.J. MTF-1-mediated repression of the zinc transporter Zip10 is alleviated by zinc restriction. *PLoS ONE* **2011**, *6*, e21526. [CrossRef] [PubMed]

74.   Pal, D.; Sharma, U.; Singh, S.K.; Prasad, R. Association between ZIP10 gene expression and tumor aggressiveness in renal cell carcinoma. *Gene* **2014**, *552*, 195–198. [CrossRef] [PubMed]

75.   Martin, A.B.; Aydemir, T.B.; Guthrie, G.J.; Samuelson, D.A.; Chang, S.-M.; Cousins, R.J. Gastric and colonic zinc transporter ZIP11 (Slc39a11) in mice responds to dietary zinc and exhibits nuclear localization. *J. Nutr.* **2013**, *143*, 1882–1888. [CrossRef] [PubMed]

76.   Kelleher, S.L.; Velasquez, V.; Croxford, T.P.; McCormick, N.H.; Lopez, V.; MacDavid, J. Mapping the zinc-transporting system in mammary cells: Molecular analysis reveals a phenotype-dependent zinc-transporting network during lactation. *J. Cell. Physiol.* **2012**, *227*, 1761–1770. [CrossRef] [PubMed]

77.   Wu, L.; Chaffee, K.G.; Parker, A.S.; Sicotte, H.; Petersen, G.M. Zinc transporter genes and urological cancers: Integrated analysis suggests a role for ZIP11 in bladder cancer. *Tumour Biol.* **2015**, *36*, 7431–7437. [CrossRef] [PubMed]

78.   Chowanadisai, W.; Graham, D.M.; Keen, C.L.; Rucker, R.B.; Messerli, M.A. Neurulation and neurite extension require the zinc transporter ZIP12 (slc39a12). *Proc. Natl. Acad. Sci. USA* **2013**, *110*, 9903–9908. [CrossRef] [PubMed]

79.   Scarr, E.; Udawela, M.; Greenough, M.A.; Neo, J.; Suk Seo, M.; Money, T.T.; Upadhyay, A.; Bush, A.I.; Everall, I.P.; Thomas, E.A.; et al. Increased cortical expression of the zinc transporter SLC39A12 suggests a breakdown in zinc cellular homeostasis as part of the pathophysiology of schizophrenia. *NPJ Schizophr.* **2016**, *2*, 16002. [CrossRef] [PubMed]

80.   Xiao, G.; Wan, Z.; Fan, Q.; Tang, X.; Zhou, B. The metal transporter ZIP13 supplies iron into the secretory pathway in *Drosophila melanogaster*. *Elife* **2014**, *3*, e03191. [CrossRef] [PubMed]

81.   Mandilaras, K.; Pathmanathan, T.; Missirlis, F. Iron absorption in *Drosophila melanogaster*. *Nutrients* **2013**, *5*, 1622–1647. [CrossRef] [PubMed]

82.   Bin, B.-H.; Hojyo, S.; Hosaka, T.; Bhin, J.; Kano, H.; Miyai, T.; Ikeda, M.; Kimura-Someya, T.; Shirouzu, M.; Cho, E.-G.; et al. Molecular pathogenesis of spondylocheirodysplastic Ehlers-Danlos syndrome caused by mutant ZIP13 proteins. *EMBO Mol. Med.* **2014**, *6*, 1028–1042. [CrossRef] [PubMed]

83.   Giunta, C.; Elçioglu, N.H.; Albrecht, B.; Eich, G.; Chambaz, C.; Janecke, A.R.; Yeowell, H.; Weis, M.; Eyre, D.R.; Kraenzlin, M.; et al. Spondylocheiro dysplastic form of the Ehlers-Danlos Syndrome-an autosomal-recessive entity caused by mutations in the zinc transporter gene SLC39A13. *Am. J. Hum. Genet.* **2008**, *82*, 1290–1305. [CrossRef] [PubMed]

84.   Fukunaka, A.; Fukada, T.; Bhin, J.; Suzuki, L.; Tsuzuki, T.; Takamine, Y.; Bin, B.; Yoshihara, T.; Ichinoseki-Sekine, N.; Naito, H.; et al. Zinc transporter ZIP13 suppresses beige adipocyte biogenesis and energy expenditure by regulating C/EBP-β expression. *PLoS Genet.* **2017**, *13*, e1006950. [CrossRef] [PubMed]

85.   Zhao, N.; Gao, J.; Enns, C.A.; Knutson, M.D. ZRT/IRT-like protein 14 (ZIP14) promotes the cellular assimilation of iron from transferrin. *J. Biol. Chem.* **2010**, *285*, 32141–32150. [CrossRef] [PubMed]

86. Taylor, K.M.; Morgan, H.E.; Johnson, A.; Nicholson, R.I. Structure-function analysis of a novel member of the LIV-1 subfamily of zinc transporters, ZIP14. *FEBS Lett.* **2005**, *579*, 427–432. [CrossRef] [PubMed]
87. Sveen, A.; Bakken, A.C.; Ågesen, T.H.; Lind, G.E.; Nesbakken, A.; Nordgård, O.; Brackmann, S.; Rognum, T.O.; Lothe, R.A.; Skotheim, R.I. The exon-level biomarker SLC39A14 has organ-confined cancer-specificity in colorectal cancer. *Int. J. Cancer* **2012**, *131*, 1479–1485. [CrossRef] [PubMed]
88. Franklin, R.B.; Levy, B.A.; Zou, J.; Hanna, N.; Desouki, M.M.; Bagasra, O.; Johnson, L.A.; Costello, L.C. ZIP14 zinc transporter downregulation and zinc depletion in the development and progression of hepatocellular cancer. *J. Gastrointest. Cancer* **2012**, *43*, 249–257. [CrossRef] [PubMed]
89. Zhang, L.; Liu, X.; Zhang, X.; Chen, R. Identification of important long non-coding RNAs and highly recurrent aberrant alternative splicing events in hepatocellular carcinoma through integrative analysis of multiple RNA-Seq datasets. *Mol. Genet. Genom.* **2016**, *291*, 1035–1051. [CrossRef] [PubMed]
90. Xu, X.-M.; Wang, C.-G.; Zhu, Y.-D.; Chen, W.-H.; Shao, S.-L.; Jiang, F.-N.; Liao, Q.-D. Decreased expression of SLC 39A14 is associated with tumor aggressiveness and biochemical recurrence of human prostate cancer. *Onco Targets Ther.* **2016**, *9*, 4197–4205. [PubMed]
91. Guthrie, G.J.; Aydemir, T.B.; Troche, C.; Martin, A.B.; Chang, S.-M.; Cousins, R.J. Influence of ZIP14 (slc39A14) on intestinal zinc processing and barrier function. *Am. J. Physiol. Gastrointest. Liver Physiol.* **2015**, *308*, G171–G178. [CrossRef] [PubMed]
92. Wessels, I.; Cousins, R.J. Zinc dyshomeostasis during polymicrobial sepsis in mice involves zinc transporter Zip14 and can be overcome by zinc supplementation. *Am. J. Physiol. Gastrointest. Liver Physiol.* **2015**, *309*, G768–G778. [CrossRef] [PubMed]
93. Xin, Y.; Gao, H.; Wang, J.; Qiang, Y.; Imam, M.U.; Li, Y.; Wang, J.; Zhang, R.; Zhang, H.; Yu, Y.; et al. Manganese transporter Slc39a14 deficiency revealed its key role in maintaining manganese homeostasis in mice. *Cell Discov.* **2017**, *3*, 17025. [CrossRef] [PubMed]
94. Jenkitkasemwong, S.; Wang, C.-Y.; Coffey, R.; Zhang, W.; Chan, A.; Biel, T.; Kim, J.-S.; Hojyo, S.; Fukada, T.; Knutson, M.D. SLC39A14 is required for the development of hepatocellular iron overload in murine models of hereditary hemochromatosis. *Cell Metab.* **2015**, *22*, 138–150. [CrossRef] [PubMed]
95. Pietrangelo, A. Hereditary hemochromatosis: Pathogenesis, diagnosis, and treatment. *Gastroenterology* **2010**, *139*, 393–408. [CrossRef] [PubMed]
96. Liuzzi, J.P.; Aydemir, F.; Nam, H.; Knutson, M.D.; Cousins, R.J. Zip14 (Slc39a14) mediates non-transferrin-bound iron uptake into cells. *Proc. Natl. Acad. Sci. USA* **2006**, *103*, 13612–13617. [CrossRef] [PubMed]
97. Zhao, F.; Edwards, R.; Dizon, D.; Afrasiabi, K.; Mastroianni, J.R.; Geyfman, M.; Ouellette, A.J.; Andersen, B.; Lipkin, S.M. Disruption of Paneth and goblet cell homeostasis and increased endoplasmic reticulum stress in Agr2-/- mice. *Dev. Biol.* **2010**, *338*, 270–279. [CrossRef] [PubMed]
98. Kaser, A.; Lee, A.-H.; Franke, A.; Glickman, J.N.; Zeissig, S.; Tilg, H.; Nieuwenhuis, E.E.S.; Higgins, D.E.; Schreiber, S.; Glimcher, L.H.; et al. XBP1 Links ER stress to intestinal inflammation and confers genetic risk for human inflammatory bowel disease. *Cell* **2008**, *134*, 743–756. [CrossRef] [PubMed]
99. Prasad, A.S. Discovery of human zinc deficiency and studies in an experimental human model. *Am. J. Clin. Nutr.* **1991**, *53*, 403–412. [PubMed]
100. Prasad, A.S. Zinc and immunity. *Mol. Cell. Biochem.* **1998**, *188*, 63–69. [CrossRef] [PubMed]
101. Shankar, A.H.; Prasad, A.S. Zinc and immune function: The biological basis of altered resistance to infection. *Am. J. Clin. Nutr.* **1998**, *68*, 447S–463S. [PubMed]
102. Hojyo, S.; Miyai, T.; Fukada, T. B-cell receptor strength and zinc signaling: Unraveling the role of zinc transporter ZIP10 in humoral immunity. *Recept. Clin. Investig.* **2015**, 2–5. [CrossRef]
103. Heltemes-Harris, L.M.; Farrar, M.A. The role of STAT5 in lymphocyte development and transformation. *Curr. Opin. Immunol.* **2012**, *24*, 146–152. [CrossRef] [PubMed]
104. Malin, S.; McManus, S.; Busslinger, M. STAT5 in B cell development and leukemia. *Curr. Opin. Immunol.* **2010**, *22*, 168–176. [CrossRef] [PubMed]
105. Kurosaki, T.; Shinohara, H.; Baba, Y. B Cell signaling and fate decision. *Annu. Rev. Immunol.* **2010**, *28*, 21–55. [CrossRef] [PubMed]
106. Kurosaki, T. Genetic analysis of B cell antigen receptor signaling. *Annu. Rev. Immunol.* **1999**, *17*, 555–592. [CrossRef] [PubMed]

107. Ogawa, Y.; Kawamura, T.; Shimada, S. Zinc and skin biology. *Arch. Biochem. Biophys.* **2016**, *611*, 113–119. [CrossRef] [PubMed]

108. Tichý, V.; Navrátilová, L.; Adámik, M.; Fojta, M.; Brázdová, M. Redox state of p63 and p73 core domains regulates sequence-specific DNA binding. *Biochem. Biophys. Res. Commun.* **2013**, *433*, 445–449. [CrossRef] [PubMed]

109. Bin, B.-H.; Hojyo, S.; Ryong Lee, T.; Fukada, T. Spondylocheirodysplastic Ehlers-Danlos syndrome (SCD-EDS) and the mutant zinc transporter ZIP13. *Rare Dis.* **2014**, e974982. [CrossRef] [PubMed]

110. Roth, J.A. Correlation between the biochemical pathways altered by mutated parkinson-related genes and chronic exposure to manganese. *Neurotoxicology* **2014**, *44*, 314–325. [CrossRef] [PubMed]

111. Salazar, J.; Mena, N.; Hunot, S.; Prigent, A.; Alvarez-Fischer, D.; Arredondo, M.; Duyckaerts, C.; Sazdovitch, V.; Zhao, L.; Garrick, L.M.; et al. Divalent metal transporter 1 (DMT1) contributes to neurodegeneration in animal models of *Parkinson's disease*. *Proc. Natl. Acad. Sci. USA* **2008**, *105*, 18578–18583. [CrossRef] [PubMed]

112. Mitchell, C.J.; Shawki, A.; Ganz, T.; Nemeth, E.; Mackenzie, B. Functional properties of human ferroportin, a cellular iron exporter reactive also with cobalt and zinc. *AJP Cell Physiol.* **2014**, *306*, C450–C459. [CrossRef] [PubMed]

113. Tuschl, K.; Clayton, P.T.; Gospe, S.M.; Gulab, S.; Ibrahim, S.; Singhi, P.; Aulakh, R.; Ribeiro, R.T.; Barsottini, O.G.; Zaki, M.S.; et al. Syndrome of hepatic cirrhosis, dystonia, polycythemia, and hypermanganesemia caused by mutations in SLC30A10, a manganese transporter in man. *Am. J. Hum. Genet.* **2016**, *99*, 521. [CrossRef] [PubMed]

114. Fujishiro, H.; Yoshida, M.; Nakano, Y.; Himeno, S. Interleukin-6 enhances manganese accumulation in SH-SY5Y cells: Implications of the up-regulation of ZIP14 and the down-regulation of ZnT10. *Metallomics* **2014**, *6*, 944–949. [CrossRef] [PubMed]

115. Aydemir, T.B.; Kim, M.-H.; Kim, J.; Colon-Perez, L.M.; Banan, G.; Mareci, T.H.; Febo, M.; Cousins, R.J. Metal transporter Zip14 (*Slc39a14*) deletion in mice increases manganese deposition and produces neurotoxic signatures and diminished motor activity. *J. Neurosci.* **2017**, *37*, 5996–6006. [CrossRef] [PubMed]

116. Fukada, T.; Kambe, T. *Zinc Signals in Cellular Functions and Disorders*, 1st ed.; Springer: Tokyo, Japan, 2014.

117. Fukada, T.; Hojyo, S.; Furuichi, T. Zinc signal: A new player in osteobiology. *J. Bone Miner. Metab.* **2013**, *31*, 129–135. [CrossRef] [PubMed]

International Journal of
*Molecular Sciences*

MDPI

*Review*

# The Zinc Sensing Receptor, ZnR/GPR39, in Health and Disease

Michal Hershfinkel [ID]

Department of Physiology and Cell Biology and The Zlotowski Center for Neuroscience, Faculty of Health Sciences, POB 653, Ben-Gurion Ave. Ben-Gurion University of the Negev, Beer Sheva 84105, Israel; hmichal@bgu.ac.il, Tel.: +972-8-6477318

Received: 7 January 2018; Accepted: 29 January 2018; Published: 1 February 2018

**Abstract:** A distinct G-protein coupled receptor that senses changes in extracellular $Zn^{2+}$, ZnR/GPR39, was found in cells from tissues in which $Zn^{2+}$ plays a physiological role. Most prominently, ZnR/GPR39 activity was described in prostate cancer, skin keratinocytes, and colon epithelial cells, where zinc is essential for cell growth, wound closure, and barrier formation. ZnR/GPR39 activity was also described in neurons that are postsynaptic to vesicular $Zn^{2+}$ release. Activation of ZnR/GPR39 triggers $G\alpha q$-dependent signaling and subsequent cellular pathways associated with cell growth and survival. Furthermore, ZnR/GPR39 was shown to regulate the activity of ion transport mechanisms that are essential for the physiological function of epithelial and neuronal cells. Thus, ZnR/GPR39 provides a unique target for therapeutically modifying the actions of zinc in a specific and selective manner.

**Keywords:** zinc; ZnR/GPR39; zinc signaling; neuron; keratinocyte; epithelium; intestine; colon; bone

## 1. Introduction

The symptoms of zinc deficiency are particularly prominent in the digestive, immune, nervous, endocrine, and integumentary systems [1–5]. In many cases dietary zinc supplementation can ameliorate the symptoms and indeed zinc supplementation is widely used to treat diarrhea, the common cold, and skin conditions. The mechanisms underlying the roles of zinc have been revealed in the last two decades, but there is still a lot to learn about the pathways and regulation of zinc ions ($Zn^{2+}$). Initially, $Zn^{2+}$ was identified as a structural element and cofactor in enzymes [6,7] and transcription factors [8–10]. It is estimated that about 3000 proteins contain $Zn^{2+}$ binding sites, and interaction with $Zn^{2+}$ regulates or modulates the activity of these proteins, thereby affecting numerous cellular processes [11]. Cellular $Zn^{2+}$ is associated with these proteins with a very high affinity and is considered a tightly bound pool of $Zn^{2+}$ [10,12]. The labile $Zn^{2+}$ pool in cells includes proteins that interact with $Zn^{2+}$ via histidines, cysteines, or glutamate/aspartate residues; most prominent are the metallothioneiens (MTs) $Zn^{2+}$ binding proteins [13]. This is a dynamic pool that releases $Zn^{2+}$ upon redox signaling and oxidative or nitrosative stress, and contributes to cellular signaling [14–18]. In addition, cytosolic $Zn^{2+}$ rise, likely mediated by $Zn^{2+}$ transporters on the endoplasmic reticulum (ER), was monitored in mast cells following activation of the immunoglobulin receptor [19,20]. Subsequent studies determined that $Zn^{2+}$ transporters found on various cellular organelles induce changes in cytosolic or organellar $Zn^{2+}$ and thereby modulate cellular signaling [21–26]. Indeed, $Zn^{2+}$ transport from the ER, Golgi, or mitochondria plays an important role in the function of mammary gland or prostate epithelial cells and other secretory cells [27–29]. Similar release of $Zn^{2+}$, from the ER, during cardiac function regulates $Ca^{2+}$ leakage from the ER in these cells [30,31]. These studies established $Zn^{2+}$ as a second messenger that is released following diverse stimuli and triggers the regulation of kinases or phosphatases as well as protein expression [20,32].

Cellular $Zn^{2+}$ is buffered by interaction with proteins and formation of complexes to rapidly reduce levels of $Zn^{2+}$ to the picomolar range [17,33]. Importantly, transient changes in extracellular levels of $Zn^{2+}$ can also occur following release of $Zn^{2+}$-containing vesicles. Such vesicular $Zn^{2+}$ is found in neurons, epithelial Paneth cells of the intestine or the salivary gland, as well as in pancreatic β-cells [34]. The vesicular $Zn^{2+}$ can be released during normal activity of the cells; for example, $Zn^{2+}$ is released into the synapse during neuronal activity or is secreted from β-cells or mammary epithelial cells [35–40]. Release of $Zn^{2+}$ from cells can also occur following cellular injury and cell death, which liberates $Zn^{2+}$ from the numerous $Zn^{2+}$-binding proteins or cellular organelles [41]. Extracellular $Zn^{2+}$ can interact with specific binding sites on numerous proteins and regulate their activity. For example, extracellular $Zn^{2+}$ allosterically modulates numerous neuronal receptors, i.e., N-methyl-D-aspartate (NMDA), γ-Aminobutyric acid (GABA), or glycine receptors, thereby modulating the excitatory and inhibitory responses [42–46]. In epithelial cells, extracellular $Zn^{2+}$ regulates the activity of purinergic receptors and the store-operated $Ca^{2+}$ (SOC), representing an important link between $Zn^{2+}$ and intracellular $Ca^{2+}$ [47–49]. Application of $Zn^{2+}$ was also suggested to upregulate the phosphatidylinositol-4,5-bisphosphate 3 (PI3) kinase/AKT pathway [50] or mitogen-activated protein kinases (MAPKs) [51], both essential to cell survival and proliferation.

## 2. Identification of a $Zn^{2+}$-Sensing Receptor, ZnR/GPR39

In addition to the large numbers of $Zn^{2+}$ homeostatic proteins described above, a distinct target for extracellular $Zn^{2+}$ is the plasma membrane G-protein coupled receptor that is sensitive to $Zn^{2+}$, ZnR/GPR39 [52–54]. G-protein coupled receptors are a large family of seven-transmembrane proteins that mediate cellular signaling in response to a diverse array of extracellular stimuli [55]. The endogenous $Zn^{2+}$, released during physiological activity, acts as a first messenger and triggers intracellular $Ca^{2+}$ signaling via the specific Gαq-coupled receptor ZnR/GPR39 [34,56]. Activity of ZnR/GPR39 in tissues relevant to $Zn^{2+}$ signaling has been identified in neurons, colon epithelial cells (colonocytes), skin epidermal cells (keratinocytes), pancreatic cells, prostate cancer cells, salivary gland cells, and in bones [57–61]. In neurons, stimulation of the mossy fibers triggers ZnR/GPR39-dependent $Ca^{2+}$ rises in postsynaptic CA3 (Cornu Ammonis 3) neurons [62] that are diminished in the presence of a non-permeable $Zn^{2+}$ chelator, or in the absence of the $Zn^{2+}$ transporter-3 (ZnT3), which is responsible for synaptic $Zn^{2+}$ accumulation. Similar ZnR/GPR39 responses were observed in postsynaptic neurons of the auditory brainstem nucleus, the dorsal cochlear nucleus [63]. Importantly, ZnR/GPR39 activity was shown to enhance neuronal inhibitory tone, and zinc deficiency is associated with epilepsy and seizures, suggesting the significant physiological role of ZnR/GPR39 [53,64–68]. Luminal application of $Zn^{2+}$ to colon epithelial cells, colonocytes, was sufficient to activate the plasma membrane ZnR/GPR39 [69], which is highly expressed in this tissue [70,71]. In colonocytes, ZnR/GPR39 activated cellular pathways that are strongly associated with cell growth, MAP, and PI3 kinases. The prominent role of zinc supplementation in digestive system function, taste disorders, and salivary secretion suggests that ZnR/GPR39 may play an important role in the physiological functions of this system. A specific role for zinc in wound healing and the strong link between its deficiency and skin lesions suggested that ZnR/GPR39 may mediate cell proliferation and wound healing, thereby contributing to skin health. A recent study also describes ZnR/GPR39 expression in the oviduct, where it colocalized with a higher concentration of $Zn^{2+}$ but its activity has not been studied [72]. While a link to $Zn^{2+}$ physiology is still not clear, ZnR/GPR39 was also associated with adipocyte and myoblast proliferation and differentiation [73,74]. Activation of ZnR/GPR39 was triggered by transient changes in extracellular $Zn^{2+}$. While exogenous application of $Zn^{2+}$ may trigger ZnR/GPR39 activation, the endogenous sources of vesicular $Zn^{2+}$ may be the physiological trigger of ZnR/GPR39 activation, i.e., $Zn^{2+}$ released from neuronal vesicles, salivary gland vesicles, pancreatic enzymes, or Paneth cells in the intestinal epithelium [35–40,75]. In addition, extracellular $Zn^{2+}$ levels may transiently change following efflux mediated by $Zn^{2+}$ transporters, such as ZnT6 [76], or following injury and cell death [41].

### 3. ZnR/GPR39-Dependent Signaling

Intracellular $Ca^{2+}$ signaling triggered by extracellular $Zn^{2+}$ was the first functional identification of a distinct $Zn^{2+}$ sensing receptor, named ZnR [77]. Use of pharmacological inhibitors of $G\alpha q$ [78,79], inositol 1,4,5-trisphosphate (IP3) receptor and the phospholipase C (PLC), indicated that a $Zn^{2+}$-dependent $Ca^{2+}$ rise is mediated by activation of a $G\alpha q$-coupled receptor, such that the $Ca^{2+}$ is released from thapsigargin-sensitive stores following activation of the $IP_3$ receptor [52,57] (see Figure 1). Importantly, the $Zn^{2+}$-dependent signaling was mediated by changes in extracellular, and not intracellular, levels of this ion, as expected from a G-coupled receptor [52,57]. The search for the protein that mediates $Zn^{2+}$-dependent signaling focused on members of the $G\alpha q$ family of receptors, their possible isoforms, or interactions between these receptors that may affect the affinity towards $Zn^{2+}$; the main candidate in this family was the $Ca^{2+}$-sensing receptor (CaSR). Most G-protein coupled receptors are activated by peptides and not cations, but a CaSR was already identified and its physiological significance to cellular signaling was established [80,81]. The similarity of the ligands and the signaling pathway activated by the CaSR and the putative ZnR suggested that these may be isoforms of the same receptor. Surprisingly, $Zn^{2+}$ turned up in a screen of serum for the agonist of GPR39, which was an orphan receptor until then [82], subsequent studies confirmed that ZnR and GPR39 are one receptor, termed ZnR/GPR39. Despite their ligand similarity, CaSR and GPR39 are not members of the same subfamily of G-protein coupled receptors. The GPR39 is a member of ghrelin receptor family A, while CaSR is a member of family C of the G-protein coupled receptors [83]. It is important to note that ZnR/GPR39 is not activated by extracellular $Ca^{2+}$, nor is the CaSR activated by $Zn^{2+}$ [52,84]. Nevertheless, the affinity of ZnR/GPR39 to $Zn^{2+}$ is modulated by $Ca^{2+}$, as the $K_{0.5}$ of ZnR/GPR39 in salivary gland cells was ~55 $\mu$M in the presence of $Ca^{2+}$ and only ~36 $\mu$M in its absence [58]. This may be mediated by a direct effect of CaSR on ZnR/GPR39 conformation or its membrane expression or by a direct effect of $Ca^{2+}$ on the $Zn^{2+}$-binding site. Indeed, ZnR/GPR39 and the CaSR have been shown to directly interact in an exogenous overexpression system and may thereby modulate cation-dependent signaling in many systems where they are both expressed [84]. Importantly, the orphan receptor GPR39 was initially suggested to mediate signaling triggered by the obesity-related peptide obestatin [85]. These results were not reproduced by other laboratories and a study using serum identified $Zn^{2+}$ as the endogenous ligand of GPR39 [82]. Using silencing and overexpression, it has been shown that the endogenous $Zn^{2+}$-dependent signaling is mediated by GPR39, which is highly selective for $Zn^{2+}$ and is not activated by $Mn^{2+}$, $Cu^{2+}$, or $Fe^{2+}$ [52,53]. The affinity of ZnR/GPR39 to $Zn^{2+}$ was physiologically adapted to the relevant tissues. For example, $Zn^{2+}$ concentration in the digestive system lumen may reach hundreds of $\mu$M [86–88] and the colonocytic ZnR/GPR39 has an EC50 (half maximal effective concentration) of 80 $\mu$M [52,57]. Physiological relevance was further established when $Zn^{2+}$ release from Caco-2 colonocytes was sufficient to induce ZnR/GPR39-dependent cell growth and tight junction formation [69,89]. In addition, in a cholera toxin model of diarrhea or a dextran sodium sulfate model of colitis, ZnR/GPR39-dependent pathways were not activated following dietary $Zn^{2+}$ depletion [90,91]. Similarly, in the prostate, where there are high concentrations of $Zn^{2+}$/citrate complex and transient release of this ion is likely to occur following cell death or changes in pH, ZnR/GPR39 is adapted to the relevant concentrations, which range from 10 to 200 $\mu$M [59]. In contrast, in keratinocytes ZnR/GPR39 EC50 to $Zn^{2+}$ is in the nanomolar range, likely because this tissue contains much lower concentrations of labile $Zn^{2+}$ [41]. Most importantly, the ZnR/GPR39 is triggered during keratinocytic injury, as shown using a scratch assay [41]. In addition, the neuronal ZnR/GPR39 has an affinity that is adapted to the release of $Zn^{2+}$ from the synaptic mossy fiber terminals, and indeed very mild activation of these fibers induces sufficient $Zn^{2+}$ levels to trigger postsynaptic ZnR/GPR39 signaling [62,92]. The differences in the affinity of the ZnR/GPR39 may result from its interaction with other, physiologically relevant G-protein coupled receptors in the tissues, as has been established for many receptors of this family [93].

Since $Zn^{2+}$ can interact with numerous intracellular and extracellular proteins, application of this ion to study the effects of ZnR/GPR39 may yield confusing results and distinct agonists or antagonists would be of importance. Using various screening methods, agonists for ZnR/GPR39 have been suggested but very few were successfully tested in endogenous tissues. A recent study identified several compounds that may interact with ZnR/GPR39 and were shown to affect gastric function in wild-type but not *GPR39* knockout mice, yet these compounds only potentiated the response of the ZnR/GPR39 to $Zn^{2+}$ itself [94]. The use of molecular approaches to modulate expression of ZnR/GPR39, together with pharmacological inhibition of its signaling pathway, is therefore still important to study the effects of ZnR/GPR39. Indeed, the first description of the role of ZnR/GPR39 was established using a knockout mouse, which exhibited accelerated gastric emptying and increased body weight and fat composition [70]. This phenotype strengthened the link between the receptor and the well-known effects of $Zn^{2+}$ on the gastrointestinal system. Future studies using knockout mice required challenging the mice to trigger a phenotypic distinction from the wild-type mice, suggesting that ZnR/GPR39 has a role in stress conditions. Finally, overexpression of ZnR/GPR39 in exogenous systems resulted in signaling that exhibited constitutive activity or was suggested to trigger $G\alpha s$ or $G\alpha 12/13$ signaling and CRE- or SRE-dependent gene expression [83], but the physiological significance of these pathways is yet to be determined.

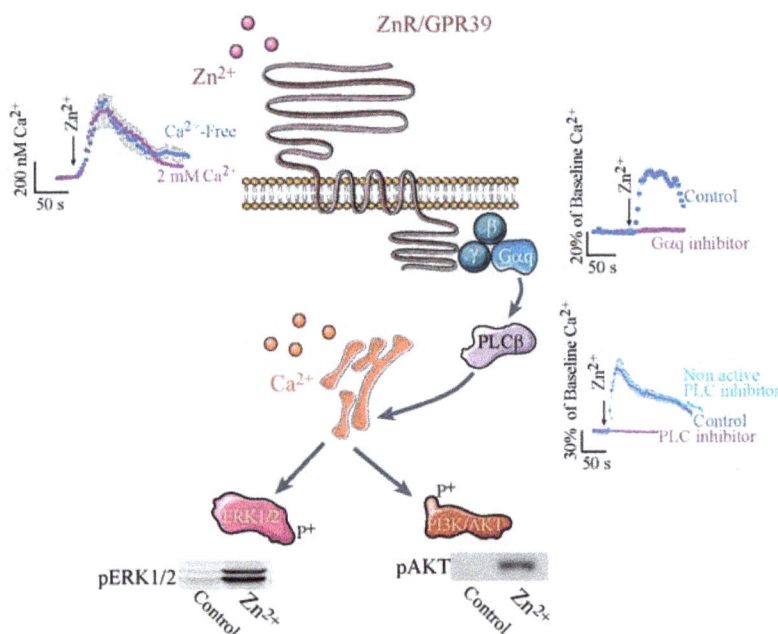

**Figure 1.** Schematic representations of common $Zn^{2+}$ sensing receptor, ZnR/GPR39, signaling in epithelial cells. Extracellular signal–regulated kinases, ERK; Phosphatidylinositol-4,5-bisphosphate 3 (PI3) kinase/AKT, PI3K/AKT; Phospholipase C, PLC.

Activation of the $G\alpha q$ is triggering PLCβ activation and subsequent $Ca^{2+}$ release from thapsigargin-sensitive ER stores. Insets show the Fura-2 fluorescent signals in cells expressing ZnR/GPR39 following application of $Zn^{2+}$. The top left inset shows the calibrated level of $Ca^{2+}$ change, monitored with Fura-2, obtained in the presence or absence of extracellular $Ca^{2+}$; the right upper inset shows the % change of $Ca^{2+}$ levels, relative to baseline Fura-2 fluorescence, in the presence or absence of the $G\alpha q$ inhibitor (YM-254890); and the right bottom panel shows the % change of $Ca^{2+}$ levels in the presence of the PLC inhibitor (U73122 active form, or U73343 inactive form). Subsequent to the $Ca^{2+}$ signal ERK1/2 (extracellular regulated kinase) or AKT phosphorylation is monitored (shown in blots in the lower panels), indicating activation of the MAPK or PI3K pathways, respectively. (The figure was composed using Servier Medical Art templates (http://smart.servier.com/)).

Subsequent to the $Ca^{2+}$ rise, ZnR/GPR39-triggers activation of the ERK/MAPK and AKT/PI3K pathways [57,84] that are essential for cell survival and proliferation [95]. ZnR/GPR39 activation in keratinocytes, colonocytes, and prostate cancer cells was shown to upregulate ERK and AKT phosphorylation and thereby cell growth. Activation of the $Zn^{2+}$-dependent $Ca^{2+}$ response was first shown to activate ERK1/2 phosphorylation, which was attenuated by functional de-sensitization of ZnR/GPR39, critical for protecting cells from excessive activation of the signaling [84]. In androgen-insensitive prostate cancer cell lines, ZnR/GPR39 activation by $Zn^{2+}$ triggers PI3K pathway upregulation, which is reflected by increased expression and phosphorylation of AKT [84], associated with more malignant phenotypes of carcinomas [96–98]. Butyrate is a short-chain fatty acid found to affect colon epithelial cell growth and carcinogenesis [99–102]. In the colonocytic cell line, butyrate-induced apoptosis was attenuated by ZnR/GPR39-dependent activation of MAPK and PI3K pathways that increased expression of the pro-survival protein clusterin [69]. Moreover, enhanced cell proliferation was monitored using BrdU in colon tissue from ZnR/GPR39 expressing mice, but not in *GPR39* knockout mice, during recovery from treatment with the toxin dextran sodium sulfate [90]. Under normal conditions BrdU staining in knockout mice lacking ZnR/GPR39 did not show differences from the wild-type tissue, suggesting that the baseline proliferation is intact, in agreement with the mild phenotype of these mice. The requirement for enhanced proliferation following the injury is the process that is impaired in the absence of ZnR/GPR39. As such, a role for ZnR/GPR39 may also underlie the healing effects of $Zn^{2+}$ on gastric ulcers [103]. Topical application of zinc-containing ointments to enhance wound healing and re-epithelialization of the skin is well established [104–107]. Indeed ZnR/GPR39 activation in keratinocytes was shown to trigger MAPK phosphorylation and increased rate of scratch closure, suggesting that the receptor may mediate the effects of $Zn^{2+}$ [41]. Finally, pre-adipocyte proliferation and differentiation are also induced following AKT activation, associated with ZnR/GPR39 expression [73,108]. In neurons, ZnR/GPR39 and subsequent $Ca^{2+}$ release are essential for activation of MAPK by $Zn^{2+}$ [92,109]. Such activation of the MAPK pathway by metabotropic signaling mediates changes in synaptic plasticity [110,111]. Finally, activation of ZnR/GPR39 in a salivary gland ductal cell line was shown to induce ATP release that mediated metabotropic signaling via the purinergic system in neighboring smooth muscle cells [58]. Thus ZnR/GPR39 has paracrine effects on neighboring cells, which may provide an important mechanism by which $Zn^{2+}$ can affect physiological processes in tissues where not all cells express ZnR/GPR39 itself.

$Zn^{2+}$, in contrast to most ligands of G-protein coupled receptors, is not rapidly degraded and a desensitization mechanism to protect cells from excessive $Ca^{2+}$ signals is important for the regulation of ZnR/GPR39 signaling. Indeed, profound and prolonged desensitization [112] is monitored following exposure to subtoxic concentrations of $Zn^{2+}$ [57,59,92]. The desensitization of ZnR/GPR39 by prolonged $Zn^{2+}$ treatment induces internalization and possible degradation of the receptor, and profound loss of ZnR/GPR39 signaling is sustained for several hours. As such, $Zn^{2+}$-induced desensitization was also used to specifically identify the roles of $Zn^{2+}$ via ZnR/GPR39. For example, following ZnR/GPR39 desensitization the $Zn^{2+}$-dependent ERK1/2 phosphorylation was diminished in prostate cancer cells [59]. The pathways that lead to ZnR/GPR39 desensitization are not fully understood. Recruitment of β-arrestin following ZnR/GPR39 with an allosteric modulator in the presence of $Zn^{2+}$ did not induce desensitization but inhibition of Rho kinase blocked this process [113].

$Zn^{2+}$ binding to ZnR/GPR39 occurs via two histidine residues, His17 and His19 [114], and an aspartate residue, Asp313. The pH sensitivity of these residues matched the regulation of ZnR/GPR39 response by extracellular pH. The ZnR/GPR39-dependent $Ca^{2+}$ response and subsequent phosphorylation of MAP or PI3 kinase is completely abolished at pH 6.5 [41,109,115]. Hence, ZnR/GPR39 activity is regulated by physiologically relevant changes in extracellular $Zn^{2+}$ or pH [115]. Thus, ZnR/GPR39 may be the mediator for many of the well-established, health-promoting functions of $Zn^{2+}$ [116]. In contrast, local pH changes during inflammatory bowel disease may attenuate ZnR/GPR39-dependent cell proliferation in the digestive system and may contribute to epithelial erosion and barrier breakdown [117].

## 4. ZnR/GPR39 Regulation of Physiological Functions

### 4.1. ZnR/GPR39 Regulates Ion Transport Mechanisms

Downstream to activation of ZnR/GPR39, it has been shown that transport of $Na^+$, $K^+$, and $Cl^-$ are regulated. The movement of these ions is essential for the physiological functions of epithelial cells and neurons.

The ubiquitously expressed $Na^+/H^+$ exchanger (NHE) is upregulated following cytoplasmic acidification, to induce recovery of intracellular pH [118]. Activation of ZnR/GPR39 upregulates NHE activity in colonocytes, keratinocytes, and neurons [41,57,69,89,109], thereby providing a $Zn^{2+}$-dependent homeostatic mechanism. Colonocytes are constantly exposed to cellular acidification, for example by short-chain fatty acid penetration [119], which can be recovered by NHE activity. Indeed, activation of ZnR/GPR39 in colonocytes and native colon tissues induced activation of NHE, downstream to the $Ca^{2+}$ signaling and ERK1/2 activation, which enhanced the recovery of the colonocytic pH [57,69]. Thus, ZnR/GPR39 plays a role in pH homeostasis that is essential for colonocytes' survival. Importantly, $Na^+$-dependent $H^+$ export can lower the extracellular pH. In keratinocytes, ZnR/GPR39 upregulation of NHE activity was also mediated via the same signaling pathway [41]. The extracellular acidification triggered by ZnR/GPR39-dependent activation of NHE may be required for migration of cells during wound healing or for the formation of an effective permeability barrier [120,121]. Intracellular acid loading in neurons affects neuronal excitability and results from metabolic $H^+$ generation during repetitive firing [122]. Neuronal ZnR/GPR39 activation following release of $Zn^{2+}$, concomitant with the neurotransmitter, resulted in increased NHE activity, thus relieving the metabolic acidification [109]. However, acidification of neuronal surfaces, by upregulating NHE activity, may contribute to tissue acidosis during ischemic neuronal injury [123]. Interestingly, ZnR/GPR39 itself is inactive at acidic pH [109], suggesting a homeostatic mechanism that can attenuate ZnR/GPR39 activation of NHE and excessive tissue acidification.

The $K^+/Cl^-$ cotransporters (KCC) family is responsible for mediating $Cl^-$ efflux and thereby maintaining cell volume, as well as transepithelial ion transport and neuronal excitability [124]. These transporters are highly regulated via their phosphorylation and changes in surface expression [125,126]. In neurons, KCC2 is crucial for mediating $Cl^-$ efflux and thereby rendering the $GABA_A$ and glycine receptors inhibitory [127–130]. Activation of ZnR/GPR39 results in enhanced $K^+$-dependent $Cl^-$ transport, which is mediated by KCC2 [62,131]. This $Zn^{2+}$-dependent upregulation is abolished in the absence of ZnR/GPR39, or its downstream $Ca^{2+}$ and MAPK activation. Moreover, $G\alpha q$-dependent signaling triggered by ZnR/GPR39 enhances KCC2 surface expression and thereby upregulates KCC2-dependent $Cl^-$ transport [62]. Similar upregulation of $K^+$-dependent $Cl^-$ transport was also monitored following ZnR/GPR39 activation in colonocytes [91]. Loss of $Cl^-$ and $Na^+$ into the colon lumen, via CFTR (cystic fibrosis transmembrane conductance regulator) upregulation for example, produces the driving force for water loss, thereby inducing diarrhea [132]. Yet, $Cl^-$ absorption pathways are not fully identified. Activation of ZnR/GPR39 in native colon epithelial tissue or in colonocytic cell lines resulted in activation of KCC1, which was mitogen activated kinase (MAPK)-dependent [91]. Moreover, KCC1 expression was shown to be basolateral, thereby providing a pathway for modulation of $Cl^-$ absorption in the colon.

### 4.2. ZnR/GPR39 Regulates Tight Junction Formation

Formation of epithelial barriers strongly depends on expression of junctional proteins, such as E-cadherin of the adherens junctions and zonula occludens-1 (ZO-1) or occludin of the tight junctions. This physical barrier is essential for the function of all epithelia and is particularly important in regions exposed to pathogens, such as the digestive tract. A role for $Zn^{2+}$ in modulating colon epithelial tight junctions was previously described, but the prolonged application in that study may have resulted in changes of intracellular $Zn^{2+}$ and not only activation of ZnR/GPR39 [133,134]. Using siRNA silencing of ZnR/GPR39 in Caco-2 colonocytic cell line revealed that ZnR/GPR39 was essential for $Zn^{2+}$-dependent upregulation of tight junction formation, thus establishing that ZnR/GPR39 has a specific role in enhancing tight junctional complexes and epithelial barrier function [89]. It was further established that these colonocytic cells release $Zn^{2+}$ in a manner that activates the ZnR/GPR39-dependent formation of the barrier, since a chelator of extracellular $Zn^{2+}$ attenuated tight junction formation. Colon from ZnR/GPR39 deficient mice exhibited diminished expression level for the tight junction protein occludin, further revealing an important role of ZnR/GPR39 in barrier formation in vivo [90]. This loss of tight junctions may underlie some of the immune system effects associated with $Zn^{2+}$ deficiency: as the permeation of pathogens is easier, inflammation may be prevalent during $Zn^{2+}$ deficiency. A recent study showed that $Zn^{2+}$ enhanced the expression of protein kinase C ζ (PKCζ), which was associated with ZnR/GPR39 levels, and linked to tight junction formation during *Salmonella enterica serovar Typhimurium* infection [135].

## 5. A Role for ZnR/GPR39 in Disease

### 5.1. ZnR/GPR39 in Wound Healing

Perhaps the oldest known use of zinc as a treatment is in dermal ointments for enhancing wound healing [104–107]. Zinc has been associated with proliferating tissues and is indeed accumulated in the skin [136]. Zinc transporters, i.e., ZIP4 (Zrt-Irt-like protein) or ZIP7, knockdown or mutations in these proteins also reveal an important role for these $Zn^{2+}$ homeostatic proteins in skin formation during development [137]. Activation of ZnR/GPR39 in primary keratinocytes and in HaCaT cells suggested that $Zn^{2+}$ may trigger this receptor signaling and may be the missing link between topical application of zinc and wound healing [41,52]. Indeed, the pro-proliferation/migration pathways were activated by ZnR/GPR39: ERK1/2 phosphorylation was increased via ZnR/GPR39-dependent activation of PKC and PI3K. In a scratch assay model, silencing of ZnR/GPR39 expression or activity inhibited the $Zn^{2+}$-dependent increased rate of scratch closure [41]. One of the suggested benefits of zinc application was associated with anti-inflammatory effects. The ZnR/GPR39-dependent activation of NHE in keratinocytes induces acidification of the extracellular region. Such acidification is essential for reducing barrier permeability in the skin [120]. Hence NHE activation and the subsequent acidification by the ZnR/GPR39 may also exert an anti-inflammatory effect. Finally, if paracrine release of ATP following ZnR/GPR39 activation [58] also occurs in keratinocytes, it suggests another mechanism to increase the proliferation of neighboring fibroblasts that do not express ZnR/GPR39. Activation of cellular signaling by ZnR/GPR39 may affect numerous pathways and $Zn^{2+}$ binding proteins. As such, a role for MG53, a $Zn^{2+}$ binding protein, has been associated with myoblasts' cell membrane recovery following permeation of $Zn^{2+}$ into the cells [138]. The ZnR/GPR39 has also been described in myogenic processes, but the role of $Zn^{2+}$ in this aspect has not been addressed [74], hence future studies on the role of ZnR/GPR39 in muscle cell recovery would be of interest.

### 5.2. Diarrhea and Inflammatory Bowel Diseases

Prominent roles of $Zn^{2+}$ include attenuation of diarrhea and amelioration of symptoms of inflammatory ulcerative disease, such as Crohn's disease and colitis [139–143]. Initial breakdown of tight junctions is considered a trigger to recurrence of inflammatory bowel diseases. The ZnR/GPR39-dependent enhancement of junctional complex proteins ZO-1 and occludin [69,89] suggested that ZnR/GPR39 may be involved in ameliorating symptoms of inflammatory bowel diseases. Indeed, ZnR/GPR39 deficient mice showed increased susceptibility to dextran sodium sulfate (DSS) model of colitis [90]. Even more profound was the effect of ZnR/GPR39 during a recovery phase. ZnR/GPR39 expression was essential for rapid recovery of the epithelial layer, via increased proliferation and crypt formation, and formation of the physical barrier, via increased expression of occludin. The benefit of $Zn^{2+}$ treatment in inflammatory bowel disease is controversial. The results described here suggest that during bouts of the inflammatory state the epithelial erosion and loss of ZnR/GPR39 on the epithelial barrier may render $Zn^{2+}$ inefficient, yet if provided during remission $Zn^{2+}$, via ZnR/GPR39, may extend this period. In fact, ZnR/GPR39 expression in the epithelial cells may serve as a therapeutic target that can be specifically activated to extend the remission periods.

Maintenance of osmotic gradients, for proper water movement, is mediated by ion transporters found on the epithelial cells [144]. In diarrhea, impaired transporters function results in excessive loss of $Na^+$ and $Cl^-$ into the lumen and subsequent water loss. The $Zn^{2+}$ and ZnR/GPR39 upregulation of $Na^+/H^+$ exchanger activity [57,69] can serve to enhance uptake of $Na^+$ from the lumen. Indeed many previous studies showed that the colonocytic apical NHE3 upregulation enhances $Na^+$ absorption and thereby reduces water loss and diarrhea [144–146]. In addition, activation of a basolateral KCC1 by ZnR/GPR39 increases absorption of $Cl^-$, which is also essential to reducing fluid loss. Cholera toxin infection, a common cause of diarrhea, induced significantly worse diarrhea in *GPR39* knockout mice, lacking ZnR/GPR39 signaling, compared to WT mice [91]. Thus, ZnR/GPR39 activation can reduce fluid loss during the disease, but reduced luminal $Zn^{2+}$, which may be a dietary or disease-mediated condition, may diminish the protective effect of this pathway. While $Zn^{2+}$ is suggested by the World Health Organization (WHO) as an important supplement to treat diarrhea [142,147], ZnR/GPR39 is a novel and specific target that may be more effectively targeted.

## 5.3. Epilepsy

Several studies linked the loss of synaptic $Zn^{2+}$ or $Zn^{2+}$ deficiency with increased incidence of seizures [148–151]. Despite a well-known role for $Zn^{2+}$ in modulating numerous excitatory and inhibitory post synaptic targets, how synaptically released $Zn^{2+}$ can affect epileptogenesis was not clear. Nevertheless, the major phenotype of the *ZnT3* knockout mice, lacking synaptic $Zn^{2+}$, is enhanced sensitivity to kainate-induced or febrile hyperthermia induced seizures [152,153]. This indicated that synaptic $Zn^{2+}$ itself does have a role in epilepsy. The results showing regulation of $Cl^-$ transport by ZnR/GPR39 activation of KCC2, taken together with the prominent role of loss of KCC2 function in increasing seizure susceptibility [128,154,155], suggested that ZnR/GPR39 may play a role in epilepsy via this pathway. Indeed, *GPR39* knockout animals, lacking ZnR/GPR39 signaling, exhibit enhanced susceptibility to kainate-induced seizures, with significantly higher behavioral seizure severity scores and more seizures over longer periods of time compared to wild-type controls [156]. Kainate-induced upregulation of KCC2 activity is dependent on $Zn^{2+}$, which is released by the increased firing under these enhanced excitability conditions. Moreover, ZnR/GPR39 signaling via the G$\alpha$q and subsequent MAPK pathway are required for increased KCC2 activity and thereby inhibitory tone. Thus the homeostatic role of ZnR/GPR39, activated by $Zn^{2+}$ co-released with glutamate, is essential during excessive firing to reduce excitatory activity via enhancing GABAergic responses. In contrast, loss of this signaling in the absence of synaptic $Zn^{2+}$ or ZnR/GPR39 may result in epileptogenesis [53]. A similar effect on increasing inhibitory neuronal signaling is monitored in the dorsal cochlear neurons, where ZnR/GRP39 activation enhances endocannabinoid release and reduces excitatory glutamate release [63]. In addition, enhanced excitability and thereby seizure activity has been associated with neuronal acidification and loss of $Na^+/H^+$ exchanger (NHE) activity [157,158]. Thus ZnR/GPR39-dependent upregulation of NHE activity, which was monitored in primary neurons, may also link the receptor to reduced seizures [109]. In Alzheimer's disease, A$\beta$ oligomers interact with $Zn^{2+}$ [159,160], thus lowering levels of labile $Zn^{2+}$. Indeed, in the presence of A$\beta$, the ZnR/GPR39-dependent $Ca^{2+}$ responses in primary neurons were significantly reduced and resulted in much lower MAPK activation [161]. This decrease in ZnR/GPR39-dependent signaling, reducing the homeostatic activation of KCC2, may serve as a link to the increased incidence of seizure found in Alzheimer's disease patients compared to the general population.

## 5.4. Depression

Zinc deficiency is associated with neurological and psychiatric disorders [162]; however, it is not yet clear if the decrease in $Zn^{2+}$ results from aberrant intake, especially in depression, when appetite is lost and general uptake of nutrients is low, or is a cause of the disorder. Several studies reported a role for ZnR/GPR39 in depression, based on apparent changes in the expression level of this receptor following $Zn^{2+}$-deficiency that were correlated with behavioral changes, also in suicide victims [163,164]. Changes in ZnR/GPR39 expression were also shown following treatment with monoaminergic inhibitors, such as used to treat depression, thus suggesting a link between the receptor and this disease. Surprisingly, despite the extensive use of antibodies against ZnR/GPR39 in this study, the antibodies were not verified in *GPR39* knockout mice [165]. A role for ZnR/GPR39 in the regulation of the CREB/BDNF/TrkB (cyclic AMP response element binding protein/brain-derived neurotrophic factor/tyrosine receptor kinase B) pathway, and thereby in depression, has also been postulated, though it is not clear at present how Gαq signaling activates this pathway or whether these effects are lost in ZnR/GPR39 deficient mice [166,167].

## 5.5. Insulin Secretion

Pancreatic β-cells contain vesicular $Zn^{2+}$ that is released together with insulin [168]. Several studies have highlighted a role for $Zn^{2+}$ in the regulation of β-cell function and glucagon release [169–171]. The $Zn^{2+}$ transporter ZnT8 is responsible for transporting $Zn^{2+}$ into the insulin vesicles, and a mutation in this transporter of an Arg replacing Trp325 is associated with increased risk of developing type 2 diabetes [172]. Thus a role for ZnR/GPR39 in this tissue may have important physiological implications in the regulation of the $Zn^{2+}$ releasing β-cells or neighboring cells within the islets of Langerhans. Knockout of ZnR/GPR39 does not immediately produce a phenotype under baseline conditions, and the knockout mice show normal insulin secretion. However, when fed a sucrose-rich diet, older mice show increased glucose levels and decreased insulin compared to the wild type [173]. Similarly, higher glucose levels were monitored in *GPR39* knockout mice fed a high-fat diet [174]. In agreement, overexpression of ZnR/GPR39 in β-cells resulted in protection from streptozotocin-induced diabetes [175]. A recent study showed ZnR/GPR39 expression and $Zn^{2+}$-dependent $Ca^{2+}$ release in association with $Zn^{2+}$-dependent insulin secretion [176]. Yet how ZnR/GPR39 activity regulates insulin secretion and whether this is an autocrine effect of endogenous $Zn^{2+}$ released from the β-cells is still poorly understood.

## 5.6. Defects in Bone Composition

Zinc is accumulated in bone and plays a role in the dynamic maintenance of the structure of bones. Supplementation with dietary zinc enhances the strength of bones, but an underlying mechanism is not available. While several zinc transporters of the ZIP family have been associated with skeletal function [137], a role for ZnR/GPR39 was not described. Using *GPR39* knockout mice, a recent study indicates that this receptor is important for osteoblast differentiation [61]. Hence, mice lacking ZnR/GPR39 showed impaired bone composition with decreased collagen content, likely involving ADAMTS metalloproteinase, which regulates collagen processing [61]. Most importantly, ZnR/GPR39 deficient osteoblasts showed lower *ZIP13* expression, linking ZnR/GPR39 and $Zn^{2+}$ transporters for the first time. Future studies aiming to determine how ZnR/GPR39 modulates $Zn^{2+}$ transporters' activity or expression can provide a more complete picture of the network of zinc homeostasis and its physiological implications.

## 5.7. ZnR/GPR39 in Cancer

Increased cell proliferation and migration, as well as the activation of MAPK and AKT, suggest a possible role for ZnR/GPR39 in cancer. Activation of ZnR/GPR39 signaling was monitored in androgen-independent, but not androgen-dependent, prostate cancer cells [59]. Extracellular $Zn^{2+}$

via activation of ZnR/GPR39 in the prostate cancer cell line PC-3 enhances expression of S100A4 [84], a protein that is thought to enhance metastatic prostate cell proliferation and angiogenesis [177]. Other studies that associated ZnR/GPR39 expression in epithelial cells with cancer did not employ changes in extracellular or dietary $Zn^{2+}$ to specifically study whether signaling pathway activation or $Zn^{2+}$-dependent processes are affected in the cancer cells. These studies nevertheless indicate the importance of this receptor as a therapeutic target for cancer treatment. As such, GPR39 was overexpressed in primary human esophageal squamous cell carcinomas and its silencing reduced the tumorigenicity of these cells [178]. A recent study suggested that GPR39 expression is modulated in gastric adenocarcinoma [179], yet this study applied a previously incorrectly suggested ligand of GPR39 and not $Zn^{2+}$ [180]. Interestingly, a link between the ZnR/GPR39 and mRNA levels of the $Zn^{2+}$ transporter *ZIP13* was recently shown in bone [61], but whether ZnR/GPR39 regulates other members of the ZIP family of $Zn^{2+}$ transporters is unknown. Such a link between ZnR/GPR39 and ZIP transporters may further link the receptor to tumorigenesis. For example, ZIP6 and ZIP7 overexpression in breast cancer has been previously shown [181,182], and ZIP4 has recently been associated with ovarian stem cell growth and carcinoma [183]. Future studies aiming to specifically test the role of ZnR/GPR39 in cancer tissue and the link to ZIP transporters expression are of major interest and can provide a novel target for therapeutic tools.

## 6. Conclusions

ZnR/GPR39 is an important regulator of $Zn^{2+}$-dependent signaling, functional in numerous epithelial cells, bone cells, and neurons—all tissues associated with $Zn^{2+}$ homeostasis. Transient changes in extracellular $Zn^{2+}$ occur during physiological activity and are sufficient to activate ZnR/GPR39. While dietary or serum zinc itself has been suggested to affect the physiological function or pathological conditions in these tissues, these changes in zinc concentration do not directly reflect local or cellular changes in the concentrations of the ionic $Zn^{2+}$. In addition, $Zn^{2+}$ interacts with a multitude of intracellular or extracellular proteins and modulates their activity, as described in the introduction; therefore, changes in $Zn^{2+}$ concentration may affect many proteins and cellular functions and not just ZnR/GPR39 activity. Thus, this micronutrient is a poor therapeutic compound with inconsistent effects. However, elucidation of ZnR/GPR39 as a regulator of $Zn^{2+}$-dependent cellular signaling can offer a novel handle to effective therapeutic approaches that will depend on ZnR/GPR39 agonists. Of note, ZnR/GPR39 is a member of the G-protein coupled receptor family, which is currently considered a major candidate for targeted therapies [184,185]. Finally, what regulates the activity of $Zn^{2+}$ transporters is only partially understood; for example, it was previously shown that intracellular $Zn^{2+}$ activation of metal-responsive elements regulates ZnT expression or that phosphorylation of ZIP regulates their expression [24,25]. In this context, a possible link between ZnR/GPR39 and the transporters may be a key to understanding $Zn^{2+}$ homeostasis and is an important aim for future studies. Thus ZnR/GPR39 may serve as a specific and efficacious handle to modulate $Zn^{2+}$ homeostatic proteins and signaling, thereby ameliorating physiological processes to enhance recovery.

**Acknowledgments:** Michal Hershfinkel was supported by Israeli Science Foundation grant #891/14 and Bi-National US Israel Science Foundation grant (2011126).

**Conflicts of Interest:** The author declares no conflict of interest.

## References

1. Roohani, N.; Hurrell, R.; Kelishadi, R.; Schulin, R. Zinc and its importance for human health: An integrative review. *J. Res. Med. Sci.* **2013**, *18*, 144–157. [PubMed]
2. Prasad, A.S. Zinc in human health: Effect of zinc on immune cells. *Mol. Med.* **2008**, *14*, 353–357. [CrossRef] [PubMed]
3. Sandstead, H.H.; Frederickson, C.J.; Penland, J.G. History of zinc as related to brain function. *J. Nutr.* **2000**, *130*, 496S–502S. [CrossRef] [PubMed]

4.  Kelleher, S.L.; McCormick, N.H.; Velasquez, V.; Lopez, V. Zinc in specialized secretory tissues: Roles in the pancreas, prostate, and mammary gland. *Adv. Nutr.* **2011**, *2*, 101–111. [CrossRef] [PubMed]
5.  Kambe, T.; Weaver, B.P.; Andrews, G.K. The genetics of essential metal homeostasis during development. *Genesis* **2008**, *46*, 214–228. [CrossRef] [PubMed]
6.  Vallee, B.L. Zinc and carbonic anhydrase content of red cells in normals and in pernicious anemia. *J. Clin. Investig.* **1948**, *27*, 559. [PubMed]
7.  Vallee, B.L.; Altschule, M.D. Zinc in the mammalian organism, with particular reference to carbonic anhydrase. *Physiol. Rev.* **1949**, *29*, 370–388. [CrossRef] [PubMed]
8.  Maret, W. Zinc biochemistry, physiology, and homeostasis—Recent insights and current trends. *BioMetals* **2001**, *14*, 187–190. [CrossRef]
9.  Vallee, B.L.; Falchuk, K.H. The biochemical basis of zinc physiology. *Physiol. Rev.* **1993**, *73*, 79–118. [CrossRef] [PubMed]
10. Maret, W. Zinc biochemistry: From a single zinc enzyme to a key element of life. *Adv. Nutr.* **2013**, *4*, 82–91. [CrossRef] [PubMed]
11. Maret, W. Zinc in cellular regulation: The nature and significance of "zinc signals". *Int. J. Mol. Sci.* **2017**, *18*, 2285. [CrossRef] [PubMed]
12. Bellomo, E.; Massarotti, A.; Hogstrand, C.; Maret, W. Zinc ions modulate protein tyrosine phosphatase 1B activity. *Metallomics* **2014**, *6*, 1229–1239. [CrossRef] [PubMed]
13. Krezel, A.; Maret, W. The functions of metamorphic metallothioneins in zinc and copper metabolism. *Int. J. Mol. Sci.* **2017**, *18*, 1237. [CrossRef] [PubMed]
14. Sekler, I.; Sensi, S.L.; Hershfinkel, M.; Silverman, W.F. Mechanism and regulation of cellular zinc transport. *Mol. Med.* **2007**, *13*, 337–343. [CrossRef] [PubMed]
15. Liuzzi, J.P.; Cousins, R.J. Mammalian zinc transporters. *Annu. Rev. Nutr.* **2004**, *24*, 151–172. [CrossRef] [PubMed]
16. Eide, D.J. Zinc transporters and the cellular trafficking of zinc. *Biochim. Biophys. Acta* **2006**, *1763*, 711–722. [CrossRef] [PubMed]
17. Krezel, A.; Maret, W. Zinc-buffering capacity of a eukaryotic cell at physiological pZn. *J. Biol. Inorg. Chem.* **2006**, *11*, 1049–1062. [CrossRef] [PubMed]
18. Aizenman, E.; Mastroberardino, P.G. Metals and neurodegeneration. *Neurobiol. Dis.* **2015**, *81*, 1–3. [CrossRef] [PubMed]
19. Yamasaki, S.; Hasegawa, A.; Hojyo, S.; Ohashi, W.; Fukada, T.; Nishida, K.; Hirano, T. A novel role of the l-type calcium channel α1D subunit as a gatekeeper for intracellular zinc signaling: Zinc wave. *PLoS ONE* **2012**, *7*, e39654. [CrossRef] [PubMed]
20. Yamasaki, S.; Sakata-Sogawa, K.; Hasegawa, A.; Suzuki, T.; Kabu, K.; Sato, E.; Kurosaki, T.; Yamashita, S.; Tokunaga, M.; Nishida, K.; et al. Zinc is a novel intracellular second messenger. *J. Cell Biol.* **2007**, *177*, 637–645. [CrossRef] [PubMed]
21. Hojyo, S.; Fukada, T.; Shimoda, S.; Ohashi, W.; Bin, B.H.; Koseki, H.; Hirano, T. The zinc transporter SLC39A14/ZIP14 controls G-protein coupled receptor-mediated signaling required for systemic growth. *PLoS ONE* **2011**, *6*, e18059. [CrossRef] [PubMed]
22. Aydemir, T.B.; Sitren, H.S.; Cousins, R.J. The zinc transporter Zip14 influences c-Met phosphorylation and hepatocyte proliferation during liver regeneration in mice. *Gastroenterology* **2012**, *142*, 1536–1546.e5. [CrossRef] [PubMed]
23. Aydemir, T.B.; Liuzzi, J.P.; McClellan, S.; Cousins, R.J. Zinc transporter ZIP8 (SLC39A8) and zinc influence IFN-γ expression in activated human T cells. *J. Leukoc. Biol.* **2009**, *86*, 337–348. [CrossRef] [PubMed]
24. Nimmanon, T.; Ziliotto, S.; Morris, S.; Flanagan, L.; Taylor, K.M. Phosphorylation of zinc channel ZIP7 drives MAPK, PI3K and mTOR growth and proliferation signalling. *Metallomics* **2017**, *9*, 471–481. [CrossRef] [PubMed]
25. Taylor, K.M.; Hiscox, S.; Nicholson, R.I.; Hogstrand, C.; Kille, P. Protein kinase CK2 triggers cytosolic zinc signaling pathways by phosphorylation of zinc channel ZIP7. *Sci. Signal* **2012**, *5*, ra11. [CrossRef] [PubMed]
26. Taylor, K.M.; Vichova, P.; Jordan, N.; Hiscox, S.; Hendley, R.; Nicholson, R.I. ZIP7-mediated intracellular zinc transport contributes to aberrant growth factor signaling in antihormone-resistant breast cancer cells. *Endocrinology* **2008**, *149*, 4912–4920. [CrossRef] [PubMed]

27. Hessels, A.M.; Taylor, K.M.; Merkx, M. Monitoring cytosolic and ER $Zn^{2+}$ in stimulated breast cancer cells using genetically encoded FRET sensors. *Metallomics* **2016**, *8*, 211–217. [CrossRef] [PubMed]

28. Hara, T.; Takeda, T.A.; Takagishi, T.; Fukue, K.; Kambe, T.; Fukada, T. Physiological roles of zinc transporters: Molecular and genetic importance in zinc homeostasis. *J. Physiol. Sci.* **2017**, *67*, 283–301. [CrossRef] [PubMed]

29. Jiang, D.; Sullivan, P.G.; Sensi, S.L.; Steward, O.; Weiss, J.H. $Zn^{2+}$ induces permeability transition pore opening and release of pro-apoptotic peptides from neuronal mitochondria. *J. Biol. Chem.* **2001**, *276*, 47524–47529. [CrossRef] [PubMed]

30. Reilly-O'Donnell, B.; Robertson, G.B.; Karumbi, A.; McIntyre, C.; Bal, W.; Nishi, M.; Takeshima, H.; Stewart, A.J.; Pitt, S.J. Dysregulated $Zn^{2+}$ homeostasis impairs cardiac type-2 ryanodine receptor and mitsugumin 23 functions, leading to sarcoplasmic reticulum $Ca^{2+}$ leakage. *J. Biol. Chem.* **2017**, *292*, 13361–13373. [CrossRef] [PubMed]

31. Woodier, J.; Rainbow, R.D.; Stewart, A.J.; Pitt, S.J. Intracellular zinc modulates cardiac ryanodine receptor-mediated calcium release. *J. Biol. Chem.* **2015**, *290*, 17599–17610. [CrossRef] [PubMed]

32. Maret, W. Zinc in the biosciences. *Metallomics* **2014**, *6*, 1174. [CrossRef] [PubMed]

33. Kocyla, A.; Adamczyk, J.; Krezel, A. Interdependence of free zinc changes and protein complex assembly—Insights into zinc signal regulation. *Metallomics* **2017**, *10*, 120–131. [CrossRef] [PubMed]

34. Hershfinkel, M. Zinc, a dynamic signaling molecule. In *Molecular Biology of Metal Homeostasis and Detoxification*; Tamas, M., Martinoia, E., Eds.; Springer: Berlin/Heidelberg, Germany, 2006; Volume 14, pp. 131–153.

35. Frederickson, C.J.; Rampy, B.A.; Reamy Rampy, S.; Howell, G.A. Distribution of histochemically reactive zinc in the forebrain of the rat. *J. Chem. Neuroanat.* **1992**, *5*, 521–530, ISSN: 0891-0618. [CrossRef]

36. Frederickson, C.J.; Perez-Clausell, J.; Danscher, G. Zinc-containing 7S-NGF complex. Evidence from zinc histochemistry for localization in salivary secretory granules. *J. Histochem. Cytochem.* **1987**, *35*, 579–583. [CrossRef] [PubMed]

37. Frederickson, C.J.; Danscher, G. Zinc-containing neurons in hippocampus and related CNS structures. *Prog. Brain Res.* **1990**, *83*, 71–84. [PubMed]

38. Danscher, G.; Stoltenberg, M. Zinc-enriched neurons. *J. Neurochem.* **2003**, *85*, 10. [CrossRef]

39. Ishii, K.; Sato, M.; Akita, M.; Tomita, H. Localization of zinc in the rat submandibular gland and the effect of its deficiency on salivary secretion. *Ann. Otol. Rhinol. Laryngol.* **1999**, *108*, 300–308. [CrossRef] [PubMed]

40. McCormick, N.; Velasquez, V.; Finney, L.; Vogt, S.; Kelleher, S.L. X-ray fluorescence microscopy reveals accumulation and secretion of discrete intracellular zinc pools in the lactating mouse mammary gland. *PLoS ONE* **2010**, *5*, e11078. [CrossRef] [PubMed]

41. Sharir, H.; Zinger, A.; Nevo, A.; Sekler, I.; Hershfinkel, M. Zinc released from injured cells is acting via the $Zn^{2+}$-sensing receptor, ZnR, to trigger signaling leading to epithelial repair. *J. Biol. Chem.* **2010**, *285*, 26097–26106. [CrossRef] [PubMed]

42. Hosie, A.M.; Dunne, E.L.; Harvey, R.J.; Smart, T.G. Zinc-mediated inhibition of GABA$_A$ receptors: Discrete binding sites underlie subtype specificity. *Nat. Neurosci.* **2003**, *6*, 362–369. [CrossRef] [PubMed]

43. Han, Y.; Wu, S.M. Modulation of glycine receptors in retinal ganglion cells by zinc. *Proc. Natl. Acad. Sci. USA* **1999**, *96*, 3234–3238. [CrossRef] [PubMed]

44. Lynch, J.W.; Jacques, P.; Pierce, K.D.; Schofield, P.R. Zinc potentiation of the glycine receptor chloride channel is mediated by allosteric pathways. *J. Neurochem.* **1998**, *71*, 2159–2168. [CrossRef] [PubMed]

45. Paoletti, P.; Ascher, P.; Neyton, J. High-affinity zinc inhibition of nmda NR1-NR2A receptors. *J. Neurosci.* **1997**, *17*, 5711–5725. [PubMed]

46. Herin, G.A.; Aizenman, E. Amino terminal domain regulation of NMDA receptor function. *Eur. J. Pharmacol.* **2004**, *500*, 101–111. [CrossRef] [PubMed]

47. Gore, A.; Moran, A.; Hershfinkel, M.; Sekler, I. Inhibitory mechanism of store-operated $Ca^{2+}$ channels by zinc. *J. Biol. Chem.* **2004**, *279*, 11106–11111. [CrossRef] [PubMed]

48. Wildman, S.S.; King, B.F.; Burnstock, G. Modulatory activity of extracellular $h^+$ and $Zn^{2+}$ on ATP-responses at $rP2X_1$ and $rP2X_3$ receptors. *Br. J. Pharmacol.* **1999**, *128*, 486–492. [CrossRef] [PubMed]

49. Acuna-Castillo, C.; Morales, B.; Huidobro-Toro, J.P. Zinc and copper modulate differentially the $P2X_4$ receptor. *J. Neurochem.* **2000**, *74*, 1529–1537. [CrossRef] [PubMed]

50. Kim, S.; Jung, Y.; Kim, D.; Koh, H.; Chung, J. Extracellular zinc activates p70 S6 kinase through the phosphatidylinositol 3-kinase signaling pathway. *J. Biol. Chem.* **2000**, *275*, 25979–25984. [CrossRef] [PubMed]

51. Oh, S.Y.; Park, K.S.; Kim, J.A.; Choi, K.Y. Differential modulation of zinc-stimulated p21(Cip/WAF1) and cyclin D1 induction by inhibition of PI3 kinase in HT-29 colorectal cancer cells. *Exp. Mol. Med.* **2002**, *34*, 27–31. [CrossRef] [PubMed]

52. Hershfinkel, M.; Moran, A.; Grossman, N.; Sekler, I. A zinc-sensing receptor triggers the release of intracellular $Ca^{2+}$ and regulates ion transport. *Proc. Natl. Acad. Sci. USA* **2001**, *98*, 11749–11754. [CrossRef] [PubMed]

53. Sunuwar, L.; Gilad, D.; Hershfinkel, M. The zinc sensing receptor, ZnR/GPR39, in health and disease. *Front. Biosci.* **2017**, *22*, 1469–1492.

54. Hershfinkel, M. The zinc-sensing receptor, ZnR/GPR39: Signaling and significance. In *Zinc Signals in Cellular Functions and Disorders*; Fukada, T., Kambe, T., Eds.; Springer: Tokyo, Japan, 2014; pp. 11–134.

55. Hilger, D.; Masureel, M.; Kobilka, B.K. Structure and dynamics of GPCR signaling complexes. *Nat. Struct. Mol. Biol.* **2018**, *25*, 4–12. [CrossRef] [PubMed]

56. Fukada, T.; Yamasaki, S.; Nishida, K.; Murakami, M.; Hirano, T. Zinc homeostasis and signaling in health and diseases: Zinc signaling. *J. Biol. Inorg. Chem.* **2011**, *16*, 1123–1134. [CrossRef] [PubMed]

57. Azriel-Tamir, H.; Sharir, H.; Schwartz, B.; Hershfinkel, M. Extracellular zinc triggers ERK-dependent activation of $Na^+/H^+$ exchange in colonocytes mediated by the zinc-sensing receptor. *J. Biol. Chem.* **2004**, *279*, 51804–51816. [CrossRef] [PubMed]

58. Sharir, H.; Hershfinkel, M. The extracellular zinc-sensing receptor mediates intercellular communication by inducing ATP release. *Biochem. Biophys. Res. Commun.* **2005**, *332*, 845–852. [CrossRef] [PubMed]

59. Dubi, N.; Gheber, L.; Fishman, D.; Sekler, I.; Hershfinkel, M. Extracellular zinc and zinc-citrate, acting through a putative zinc-sensing receptor, regulate growth and survival of prostate cancer cells. *Carcinogenesis* **2008**, *29*, 1692–1700. [CrossRef] [PubMed]

60. Holst, B.; Egerod, K.L.; Jin, C.; Petersen, P.S.; Ostergaard, M.V.; Hald, J.; Sprinkel, A.M.; Storling, J.; Mandrup-Poulsen, T.; Holst, J.J.; et al. G protein-coupled receptor 39 deficiency is associated with pancreatic islet dysfunction. *Endocrinology* **2009**, *150*, 2577–2585. [CrossRef] [PubMed]

61. Jovanovic, M.; Schmidt, F.; Guterman-Ram, G.; Khayyeri, H.; Jähn, K.; Hiram-Bab, S.; Orenbuch, A.; Katchkovsky, S.; Aflalo, A.; Isaksson, H.; et al. Perturbed bone composition and integrity with disorganized osteoblast function in zinc receptor/GPR39 deficient mice. *FASEB J.* **2018**. [CrossRef] [PubMed]

62. Chorin, E.; Vinograd, O.; Fleidervish, I.; Gilad, D.; Herrmann, S.; Sekler, I.; Aizenman, E.; Hershfinkel, M. Upregulation of KCC2 activity by zinc-mediated neurotransmission via the mZnR/GPR39 receptor. *J. Neurosci.* **2011**, *31*, 12916–12926. [CrossRef] [PubMed]

63. Perez-Rosello, T.; Anderson, C.T.; Schopfer, F.J.; Zhao, Y.; Gilad, D.; Salvatore, S.R.; Freeman, B.A.; Hershfinkel, M.; Aizenman, E.; Tzounopoulos, T. Synaptic $Zn^{2+}$ inhibits neurotransmitter release by promoting endocannabinoid synthesis. *J. Neurosci.* **2013**, *33*, 9259–9272. [CrossRef] [PubMed]

64. Reid, C.A.; Hildebrand, M.S.; Mullen, S.A.; Hildebrand, J.M.; Berkovic, S.F.; Petrou, S. Synaptic $Zn^{2+}$ and febrile seizure susceptibility. *Br. J. Pharmacol.* **2017**, *174*, 119–125. [CrossRef] [PubMed]

65. Takeda, A.; Iida, M.; Ando, M.; Nakamura, M.; Tamano, H.; Oku, N. Enhanced susceptibility to spontaneous seizures of noda epileptic rats by loss of synaptic $Zn^{2+}$. *PLoS ONE* **2013**, *8*, e71372. [CrossRef] [PubMed]

66. Nasehi, M.M.; Sakhaei, R.; Moosazadeh, M.; Aliramzany, M. Comparison of serum zinc levels among children with simple febrile seizure and control group: A systematic review. *Iran. J. Child. Neurol.* **2015**, *9*, 17–24. [PubMed]

67. Ganesh, R.; Janakiraman, L. Serum zinc levels in children with simple febrile seizure. *Clin. Pediatr. (Phila)* **2008**, *47*, 164–166. [CrossRef] [PubMed]

68. Elsas, S.M.; Hazany, S.; Gregory, W.L.; Mody, I. Hippocampal zinc infusion delays the development of afterdischarges and seizures in a kindling model of epilepsy. *Epilepsia* **2009**, *50*, 870–879. [CrossRef] [PubMed]

69. Cohen, L.; Azriel-Tamir, H.; Arotsker, N.; Sekler, I.; Hershfinkel, M. Zinc sensing receptor signaling, mediated by GPR39, reduces butyrate-induced cell death in HT29 colonocytes via upregulation of clusterin. *PLoS ONE* **2012**, *7*, e35482.

70. Moechars, D.; Depoortere, I.; Moreaux, B.; de Smet, B.; Goris, I.; Hoskens, L.; Daneels, G.; Kass, S.; Ver Donck, L.; Peeters, T.; et al. Altered gastrointestinal and metabolic function in the GPR39-obestatin receptor-knockout mouse. *Gastroenterology* **2006**, *131*, 1131–1141. [CrossRef] [PubMed]

71. Depoortere, I. Gi functions of GPR39: Novel biology. *Curr. Opin. Pharmacol.* **2012**, *12*, 647–652. [CrossRef] [PubMed]

72. Qiao, J.; Zhao, H.; Zhang, Y.; Peng, H.; Chen, Q.; Zhang, H.; Zheng, X.; Jin, Y.; Ni, H.; Duan, E.; et al. GPR39 is region-specifically expressed in mouse oviduct correlating with the $Zn^{2+}$ distribution. *Theriogenology* **2017**, *88*, 98–105. [CrossRef] [PubMed]

73. Dong, X.; Tang, S.; Zhang, W.; Gao, W.; Chen, Y. GPR39 activates proliferation and differentiation of porcine intramuscular preadipocytes through targeting the PI3K/AKT cell signaling pathway. *J. Recept. Signal Transduct. Res.* **2016**, *36*, 130–138. [CrossRef] [PubMed]

74. Yang, Q.; Li, Y.; Zhang, X.; Chen, D. Zac1/GPR39 phosphorylating CaMK-II contributes to the distinct roles of Pax3 and Pax7 in myogenic progression. *Biochim. Biophys. Acta* **2018**, *1864*, 407–419. [CrossRef] [PubMed]

75. Gopalsamy, G.L.; Alpers, D.H.; Binder, H.J.; Tran, C.D.; Ramakrishna, B.S.; Brown, I.; Manary, M.; Mortimer, E.; Young, G.P. The relevance of the colon to zinc nutrition. *Nutrients* **2015**, *7*, 572–583. [CrossRef] [PubMed]

76. Yu, Y.Y.; Kirschke, C.P.; Huang, L. Immunohistochemical analysis of ZnT1, 4, 5, 6, and 7 in the mouse gastrointestinal tract. *J. Histochem. Cytochem.* **2007**, *55*, 223–234. [CrossRef] [PubMed]

77. Hershfinkel, M.; Silverman, W.F.; Sekler, I. The zinc sensing receptor, a link between zinc and cell signaling. *Mol. Med.* **2007**, *13*, 331–336. [CrossRef] [PubMed]

78. Takasaki, J.; Saito, T.; Taniguchi, M.; Kawasaki, T.; Moritani, Y.; Hayashi, K.; Kobori, M. A novel $G\alpha_{q/11}$-selective inhibitor. *J. Biol. Chem.* **2004**, *279*, 47438–47445. [CrossRef] [PubMed]

79. Taniguchi, M.; Suzumura, K.; Nagai, K.; Kawasaki, T.; Takasaki, J.; Sekiguchi, M.; Moritani, Y.; Saito, T.; Hayashi, K.; Fujita, S.; et al. YM-254890 analogues, novel cyclic depsipeptides with $G\alpha_{q/11}$ inhibitory activity from *Chromobacterium* sp. QS3666. *Bioorg. Med. Chem.* **2004**, *12*, 3125–3133. [CrossRef] [PubMed]

80. Brown, E.M.; Gamba, G.; Riccardi, D.; Lombardi, M.; Butters, R.; Kifor, O.; Sun, A.; Hediger, M.A.; Lytton, J.; Hebert, S.C. Cloning and characterization of an extracellular $Ca^{2+}$-sensing receptor from bovine parathyroid. *Nature* **1993**, *366*, 575–580. [CrossRef] [PubMed]

81. Pearce, S.H.; Brown, E.M. Disorders of calcium ion sensing. *J. Clin. Endocrinol. Metab.* **1996**, *81*, 2030–2035. [PubMed]

82. Yasuda, S.; Miyazaki, T.; Munechika, K.; Yamashita, M.; Ikeda, Y.; Kamizono, A. Isolation of $Zn^{2+}$ as an endogenous agonist of GPR39 from fetal bovine serum. *J. Recept. Signal Transduct. Res.* **2007**, *27*, 235–246. [CrossRef] [PubMed]

83. Popovics, P.; Stewart, A.J. GPR39: A $Zn^{2+}$-activated G protein-coupled receptor that regulates pancreatic, gastrointestinal and neuronal functions. *Cell. Mol. Life Sci.* **2011**, *68*, 85–95. [CrossRef] [PubMed]

84. Asraf, H.; Salomon, S.; Nevo, A.; Sekler, I.; Mayer, D.; Hershfinkel, M. The ZnR/GPR39 interacts with the CaSR to enhance signaling in prostate and salivary epithelia. *J. Cell. Physiol.* **2013**, *229*, 868–877. [CrossRef] [PubMed]

85. Zhang, J.V.; Ren, P.G.; Avsian-Kretchmer, O.; Luo, C.W.; Rauch, R.; Klein, C.; Hsueh, A.J. Obestatin, a peptide encoded by the ghrelin gene, opposes ghrelin's effects on food intake. *Science* **2005**, *310*, 996–999. [CrossRef] [PubMed]

86. Sandstrom, B. Consideration in estimates of requirements and critical intake of zinc. Adaption, availability and interactions. *Analyst* **1995**, *120*, 913–915. [CrossRef] [PubMed]

87. Knudsen, E.; Jensen, M.; Solgaard, P.; Sorensen, S.S.; Sandstrom, B. Zinc absorption estimated by fecal monitoring of zinc stable isotopes validated by comparison with whole-body retention of zinc radioisotopes in humans. *J. Nutr.* **1995**, *125*, 1274–1282. [PubMed]

88. Sandstrom, B.; Cederblad, A.; Kivisto, B.; Stenquist, B.; Andersson, H. Retention of zinc and calcium from the human colon. *Am. J. Clin. Nutr.* **1986**, *44*, 501–504. [CrossRef] [PubMed]

89. Cohen, L.; Sekler, I.; Hershfinkel, M. The zinc sensing receptor, ZnR/GPR39, controls proliferation and differentiation of colonocytes and thereby tight junction formation in the colon. *Cell Death Dis.* **2014**, *5*, e1307. [CrossRef] [PubMed]

90. Sunuwar, L.; Medini, M.; Cohen, L.; Sekler, I.; Hershfinkel, M. The zinc sensing receptor, ZnR/GPR39, triggers metabotropic calcium signalling in colonocytes and regulates occludin recovery in experimental colitis. *Philos. Trans. R Soc. Lond. B Biol. Sci.* **2016**, *371*. [CrossRef]

91. Sunuwar, L.; Asraf, H.; Donowitz, M.; Sekler, I.; Hershfinkel, M. The $Zn^{2+}$-sensing receptor, ZnR/GPR39, upregulates colonocytic $Cl^-$ absorption, via basolateral KCC1, and reduces fluid loss. *Biochim. Biophys. Acta* **2017**, *1863*, 947–960. [CrossRef] [PubMed]

92. Besser, L.; Chorin, E.; Sekler, I.; Silverman, W.F.; Atkin, S.; Russell, J.T.; Hershfinkel, M. Synaptically released zinc triggers metabotropic signaling via a zinc-sensing receptor in the hippocampus. *J. Neurosci.* **2009**, *29*, 2890–2901. [CrossRef] [PubMed]

93. Albizu, L.; Balestre, M.N.; Breton, C.; Pin, J.P.; Manning, M.; Mouillac, B.; Barberis, C.; Durroux, T. Probing the existence of g protein-coupled receptor dimers by positive and negative ligand-dependent cooperative binding. *Mol. Pharmacol.* **2006**, *70*, 1783–1791. [CrossRef] [PubMed]

94. Frimurer, T.M.; Mende, F.; Graae, A.S.; Engelstoft, M.S.; Egerod, K.L.; Nygaard, R.; Gerlach, L.O.; Hansen, J.B.; Schwartz, T.W.; Holst, B. Model-based discovery of synthetic agonists for the $Zn^{2+}$-sensing G-protein-coupled receptor 39 (GPR39) reveals novel biological functions. *J. Med. Chem.* **2017**, *60*, 886–898. [CrossRef] [PubMed]

95. Chappell, W.H.; Steelman, L.S.; Long, J.M.; Kempf, R.C.; Abrams, S.L.; Franklin, R.A.; Basecke, J.; Stivala, F.; Donia, M.; Fagone, P.; et al. Ras/raf/MEK/ERK and PI3K/PTEN/AKT/MTOR inhibitors: Rationale and importance to inhibiting these pathways in human health. *Oncotarget* **2011**, *2*, 135–164. [CrossRef] [PubMed]

96. Hasson, S.P.; Rubinek, T.; Ryvo, L.; Wolf, I. Endocrine resistance in breast cancer: Focus on the phosphatidylinositol 3-kinase/AKT/mammalian target of rapamycin signaling pathway. *Breast Care* **2013**, *8*, 248–255. [CrossRef] [PubMed]

97. Miller, T.W.; Rexer, B.N.; Garrett, J.T.; Arteaga, C.L. Mutations in the phosphatidylinositol 3-kinase pathway: Role in tumor progression and therapeutic implications in breast cancer. *Breast Cancer Res.* **2011**, *13*, 224. [CrossRef] [PubMed]

98. Fassnacht, M.; Weismann, D.; Ebert, S.; Adam, P.; Zink, M.; Beuschlein, F.; Hahner, S.; Allolio, B. AKT is highly phosphorylated in pheochromocytomas but not in benign adrenocortical tumors. *J. Clin. Endocrinol. Metab.* **2005**, *90*, 4366–4370. [CrossRef] [PubMed]

99. Zhang, Y.; Zhou, L.; Bao, Y.L.; Wu, Y.; Yu, C.L.; Huang, Y.X.; Sun, Y.; Zheng, L.H.; Li, Y.X. Butyrate induces cell apoptosis through activation of JNK MAP kinase pathway in human colon cancer RKO cells. *Chem. Biol. Interact.* **2010**, *185*, 174–181. [CrossRef] [PubMed]

100. Yu, D.C.; Waby, J.S.; Chirakkal, H.; Staton, C.A.; Corfe, B.M. Butyrate suppresses expression of neuropilin I in colorectal cell lines through inhibition of Sp1 transactivation. *Mol. Cancer* **2010**, *9*, 276. [CrossRef] [PubMed]

101. Scharlau, D.; Borowicki, A.; Habermann, N.; Hofmann, T.; Klenow, S.; Miene, C.; Munjal, U.; Stein, K.; Glei, M. Mechanisms of primary cancer prevention by butyrate and other products formed during gut flora-mediated fermentation of dietary fibre. *Mutat. Res.* **2009**, *682*, 39–53. [CrossRef] [PubMed]

102. Bordonaro, M.; Lazarova, D.L.; Sartorelli, A.C. Butyrate and wnt signaling: A possible solution to the puzzle of dietary fiber and colon cancer risk? *Cell Cycle* **2008**, *7*, 1178–1183. [CrossRef] [PubMed]

103. Opoka, W.; Adamek, D.; Plonka, M.; Reczynski, W.; Bas, B.; Drozdowicz, D.; Jagielski, P.; Sliwowski, Z.; Adamski, P.; Brzozowski, T. Importance of luminal and mucosal zinc in the mechanism of experimental gastric ulcer healing. *J. Physiol. Pharmacol.* **2010**, *61*, 581–591. [PubMed]

104. Barceloux, D.G. Zinc. *J. Toxicol. Clin. Toxicol.* **1999**, *37*, 279–292. [CrossRef] [PubMed]

105. Lansdown, A.B.; Mirastschijski, U.; Stubbs, N.; Scanlon, E.; Agren, M.S. Zinc in wound healing: Theoretical, experimental, and clinical aspects. *Wound Repair Regen.* **2007**, *15*, 2–16. [CrossRef] [PubMed]

106. Lansdown, A.B. Zinc in the healing wound. *Lancet* **1996**, *347*, 706–707. [CrossRef]

107. Schwartz, J.R.; Marsh, R.G.; Draelos, Z.D. Zinc and skin health: Overview of physiology and pharmacology. *Dermatol. Surg.* **2005**, *31*, 837–847. [CrossRef] [PubMed]

108. Tang, S.; Dong, X.; Zhang, W. Obestatin changes proliferation, differentiation and apoptosis of porcine preadipocytes. *Ann. Endocrinol.* **2014**, *75*, 1–9. [CrossRef] [PubMed]

109. Ganay, T.; Asraf, H.; Aizenman, E.; Bogdanovic, M.; Sekler, I.; Hershfinkel, M. Regulation of neuronal pH by the metabotropic zinc receptor mZnR/GPR39. *J. Neurochem.* **2015**, *135*, 897–907. [CrossRef] [PubMed]

110. Volk, L.J.; Daly, C.A.; Huber, K.M. Differential roles for group 1 mGluR subtypes in induction and expression of chemically induced hippocampal long-term depression. *J. Neurophysiol.* **2006**, *95*, 2427–2438. [CrossRef] [PubMed]

111. Wang, J.Q.; Fibuch, E.E.; Mao, L. Regulation of mitogen-activated protein kinases by glutamate receptors. *J. Neurochem.* **2007**, *100*, 1–11. [CrossRef] [PubMed]

112. Mohan, M.L.; Vasudevan, N.T.; Gupta, M.K.; Martelli, E.E.; Naga Prasad, S.V. G-protein coupled receptor resensitization-appreciating the balancing act of receptor function. *Curr. Mol. Pharmacol.* **2012**, *5*, 350–361. [CrossRef]

113. Shimizu, Y.; Koyama, R.; Kawamoto, T. Rho kinase-dependent desensitization of GPR39; a unique mechanism of gpcr downregulation. *Biochem. Pharmacol.* **2017**, *140*, 105–114. [CrossRef] [PubMed]

114. Storjohann, L.; Holst, B.; Schwartz, T.W. Molecular mechanism of $Zn^{2+}$ agonism in the extracellular domain of GPR39. *FEBS Lett.* **2008**, *582*, 2583–2588. [CrossRef] [PubMed]

115. Cohen, L.; Asraf, H.; Sekler, I.; Hershfinkel, M. Extracellular ph regulates zinc signaling via an ASP residue of the zinc-sensing receptor (ZnR/GPR39). *J. Biol. Chem.* **2012**, *287*, 33339–33350. [CrossRef] [PubMed]

116. MacDonald, R.S. The role of zinc in growth and cell proliferation. *J. Nutr.* **2000**, *130*, 1500S–1508S. [CrossRef] [PubMed]

117. Nugent, S.G.; Kumar, D.; Rampton, D.S.; Evans, D.F. Intestinal luminal ph in inflammatory bowel disease: Possible determinants and implications for therapy with aminosalicylates and other drugs. *Gut* **2001**, *48*, 571–577. [CrossRef] [PubMed]

118. Orlowski, J.; Grinstein, S. $Na^+/H^+$ exchangers. *Compr. Physiol.* **2011**, *1*, 2083–2100. [PubMed]

119. Vaneckova, I.; Vylitova-Pletichova, M.; Beskid, S.; Zicha, J.; Pacha, J. Intracellular pH regulation in colonocytes of rat proximal colon. *Biochim. Biophys. Acta* **2001**, *1536*, 103–115. [CrossRef]

120. Hachem, J.P.; Behne, M.; Aronchik, I.; Demerjian, M.; Feingold, K.R.; Elias, P.M.; Mauro, T.M. Extracellular ph controls NHE1 expression in epidermis and keratinocytes: Implications for barrier repair. *J. Investig. Dermatol.* **2005**, *125*, 790–797. [CrossRef] [PubMed]

121. Stock, C.; Cardone, R.A.; Busco, G.; Krahling, H.; Schwab, A.; Reshkin, S.J. Protons extruded by NHE1: Digestive or glue? *Eur. J. Cell. Biol.* **2008**, *87*, 591–599. [CrossRef] [PubMed]

122. Kaila, K.; Panula, P.; Karhunen, T.; Heinonen, E. Fall in intracellular pH mediated by gabaa receptors in cultured rat astrocytes. *Neurosci. Lett.* **1991**, *126*, 9–12. [CrossRef]

123. Manhas, N.; Shi, Y.; Taunton, J.; Sun, D. $P90^{RSK}$ activation contributes to cerebral ischemic damage via phosphorylation of $Na^+/H^+$ exchanger isoform 1. *J. Neurochem.* **2010**, *114*, 1476–1486. [CrossRef] [PubMed]

124. Arroyo, J.P.; Kahle, K.T.; Gamba, G. The SLC12 family of electroneutral cation-coupled chloride cotransporters. *Mol. Aspects Med.* **2013**, *34*, 288–298. [CrossRef] [PubMed]

125. Lee, H.H.; Jurd, R.; Moss, S.J. Tyrosine phosphorylation regulates the membrane trafficking of the potassium chloride co-transporter KCC2. *Mol. Cell. Neurosci.* **2010**, *45*, 173–179. [CrossRef] [PubMed]

126. Wake, H.; Watanabe, M.; Moorhouse, A.J.; Kanematsu, T.; Horibe, S.; Matsukawa, N.; Asai, K.; Ojika, K.; Hirata, M.; Nabekura, J. Early changes in KCC2 phosphorylation in response to neuronal stress result in functional downregulation. *J. Neurosci.* **2007**, *27*, 1642–1650. [CrossRef] [PubMed]

127. Farrant, M.; Kaila, K. The cellular, molecular and ionic basis of $GABA_A$ receptor signalling. *Prog. Brain Res.* **2007**, *160*, 59–87. [PubMed]

128. Viitanen, T.; Ruusuvuori, E.; Kaila, K.; Voipio, J. The $K^+$-$Cl^-$ cotransporter KCC2 promotes gabaergic excitation in the mature rat hippocampus. *J. Physiol.* **2010**, *588*, 1527–1540. [CrossRef] [PubMed]

129. Lee, H.; Chen, C.X.; Liu, Y.J.; Aizenman, E.; Kandler, K. KCC2 expression in immature rat cortical neurons is sufficient to switch the polarity of gaba responses. *Eur J. Neurosci.* **2005**, *21*, 2593–2599. [CrossRef] [PubMed]

130. Lu, J.; Karadsheh, M.; Delpire, E. Developmental regulation of the neuronal-specific isoform of K-Cl cotransporter KCC2 in postnatal rat brains. *J. Neurobiol.* **1999**, *39*, 558–568. [CrossRef]

131. Saadi, R.A.; He, K.; Hartnett, K.A.; Kandler, K.; Hershfinkel, M.; Aizenman, E. Snare-dependent upregulation of potassium chloride co-transporter 2 activity after metabotropic zinc receptor activation in rat cortical neurons in vitro. *Neuroscience* **2012**, *210*, 38–46. [CrossRef] [PubMed]

132. Thiagarajah, J.R.; Donowitz, M.; Verkman, A.S. Secretory diarrhoea: Mechanisms and emerging therapies. *Nat. Rev. Gastroenterol. Hepatol.* **2015**, *12*, 446–457. [CrossRef] [PubMed]

133. Wang, X.; Valenzano, M.C.; Mercado, J.M.; Zurbach, E.P.; Mullin, J.M. Zinc supplementation modifies tight junctions and alters barrier function of CACO-2 human intestinal epithelial layers. *Dig. Dis. Sci.* **2013**, *58*, 77–87. [CrossRef] [PubMed]

134. Finamore, A.; Massimi, M.; Conti Devirgiliis, L.; Mengheri, E. Zinc deficiency induces membrane barrier damage and increases neutrophil transmigration in Caco-2 cells. *J. Nutr.* **2008**, *138*, 1664–1670. [CrossRef] [PubMed]

135. Shao, Y.X.; Lei, Z.; Wolf, P.G.; Gao, Y.; Guo, Y.M.; Zhang, B.K. Zinc supplementation, via GPR39, upregulates PKCζ to protect intestinal barrier integrity in Caco-2 cells challenged by salmonella enterica serovar typhimurium. *J. Nutr.* **2017**, *147*, 1282–1289. [CrossRef] [PubMed]
136. Nitzan, Y.B.; Sekler, I.; Silverman, W.F. Histochemical and histofluorescence tracing of chelatable zinc in the developing mouse. *J. Histochem. Cytochem.* **2004**, *52*, 529–539. [CrossRef] [PubMed]
137. Takagishi, T.; Hara, T.; Fukada, T. Recent advances in the role of SLC39A/ZIP zinc transporters in vivo. *Int. J. Mol. Sci.* **2017**, *18*, 2708. [CrossRef] [PubMed]
138. Cai, C.; Lin, P.; Zhu, H.; Ko, J.K.; Hwang, M.; Tan, T.; Pan, Z.; Korichneva, I.; Ma, J. Zinc binding to MG53 protein facilitates repair of injury to cell membranes. *J. Biol. Chem.* **2015**, *290*, 13830–13839. [CrossRef] [PubMed]
139. Krasovec, M.; Frenk, E. Acrodermatitis enteropathica secondary to crohn's disease. *Dermatology* **1996**, *193*, 361–363. [CrossRef] [PubMed]
140. Sturniolo, G.C.; Fries, W.; Mazzon, E.; Di Leo, V.; Barollo, M.; D'Inca, R. Effect of zinc supplementation on intestinal permeability in experimental colitis. *J. Lab. Clin. Med.* **2002**, *139*, 311–315. [CrossRef] [PubMed]
141. Luk, H.H.; Ko, J.K.; Fung, H.S.; Cho, C.H. Delineation of the protective action of zinc sulfate on ulcerative colitis in rats. *Eur. J. Pharmacol.* **2002**, *443*, 197–204. [CrossRef]
142. Walker, C.L.; Black, R.E. Zinc for the treatment of diarrhoea: Effect on diarrhoea morbidity, mortality and incidence of future episodes. *Int. J. Epidemiol.* **2010**, *39*, i63–i69. [CrossRef] [PubMed]
143. Alam, D.S.; Yunus, M.; El Arifeen, S.; Chowdury, H.R.; Larson, C.P.; Sack, D.A.; Baqui, A.H.; Black, R.E. Zinc treatment for 5 or 10 days is equally efficacious in preventing diarrhea in the subsequent 3 months among bangladeshi children. *J. Nutr.* **2011**, *141*, 312–315. [CrossRef] [PubMed]
144. Singh, V.; Yang, J.; Chen, T.E.; Zachos, N.C.; Kovbasnjuk, O.; Verkman, A.S.; Donowitz, M. Translating molecular physiology of intestinal transport into pharmacologic treatment of diarrhea: Stimulation of Na$^+$ absorption. *Clin. Gastroenterol. Hepatol.* **2014**, *12*, 27–31. [CrossRef] [PubMed]
145. Girardi, A.C.; Di Sole, F. Deciphering the mechanisms of the Na$^+$/H$^+$ exchanger-3 regulation in organ dysfunction. *Am. J. Physiol. Cell Physiol.* **2012**, *302*, C1569–C1587. [CrossRef] [PubMed]
146. Thiagarajah, J.R.; Ko, E.A.; Tradtrantip, L.; Donowitz, M.; Verkman, A.S. Discovery and development of antisecretory drugs for treating diarrheal diseases. *Clin. Gastroenterol. Hepatol.* **2014**, *12*, 204–209. [CrossRef] [PubMed]
147. Passariello, A.; Terrin, G.; De Marco, G.; Cecere, G.; Ruotolo, S.; Marino, A.; Cosenza, L.; Tardi, M.; Nocerino, R.; Berni Canani, R. Efficacy of a new hypotonic oral rehydration solution containing zinc and prebiotics in the treatment of childhood acute diarrhea: A randomized controlled trial. *J. Pediatr.* **2011**, *158*, 288–292.e281. [CrossRef] [PubMed]
148. Hildebrand, M.S.; Phillips, A.M.; Mullen, S.A.; Adlard, P.A.; Hardies, K.; Damiano, J.A.; Wimmer, V.; Bellows, S.T.; McMahon, J.M.; Burgess, R.; et al. Loss of synaptic Zn$^{2+}$ transporter function increases risk of febrile seizures. *Sci. Rep.* **2015**, *5*, 17816. [CrossRef] [PubMed]
149. Farahani, H.N.; Ashthiani, A.R.; Masihi, M.S. Study on serum zinc and selenium levels in epileptic patients. *Neurosciences* **2013**, *18*, 138–142. [PubMed]
150. Mitsuya, K.; Nitta, N.; Suzuki, F. Persistent zinc depletion in the mossy fiber terminals in the intrahippocampal kainate mouse model of mesial temporal lobe epilepsy. *Epilepsia* **2009**, *50*, 1979–1990. [CrossRef] [PubMed]
151. Qian, J.; Xu, K.; Yoo, J.; Chen, T.T.; Andrews, G.; Noebels, J.L. Knockout of Zn transporters ZIP-1 and ZIP-3 attenuates seizure-induced ca1 neurodegeneration. *J. Neurosci.* **2011**, *31*, 97–104. [CrossRef] [PubMed]
152. Cole, T.B.; Robbins, C.A.; Wenzel, H.J.; Schwartzkroin, P.A.; Palmiter, R.D. Seizures and neuronal damage in mice lacking vesicular zinc. *Epilepsy Res.* **2000**, *39*, 153–169. [CrossRef]
153. McAllister, B.B.; Dyck, R.H. Zinc transporter 3 (ZnT3) and vesicular zinc in central nervous system function. *Neurosci. Biobehav. Rev.* **2017**, *80*, 329–350. [CrossRef] [PubMed]
154. Zhu, L.; Lovinger, D.; Delpire, E. Cortical neurons lacking KCC2 expression show impaired regulation of intracellular chloride. *J. Neurophysiol.* **2005**, *93*, 1557–1568. [CrossRef] [PubMed]
155. Woo, N.S.; Lu, J.; England, R.; McClellan, R.; Dufour, S.; Mount, D.B.; Deutch, A.Y.; Lovinger, D.M.; Delpire, E. Hyperexcitability and epilepsy associated with disruption of the mouse neuronal-specific K-Cl cotransporter gene. *Hippocampus* **2002**, *12*, 258–268. [CrossRef] [PubMed]

156. Gilad, D.; Shorer, S.; Ketzef, M.; Friedman, A.; Sekler, I.; Aizenman, E.; Hershfinkel, M. Homeostatic regulation of KCC2 activity by the zinc receptor mZnR/GPR39 during seizures. *Neurobiol. Dis.* **2015**, *81*, 4–13. [CrossRef] [PubMed]

157. Gu, X.Q.; Yao, H.; Haddad, G.G. Increased neuronal excitability and seizures in the $Na^+/H^+$ exchanger null mutant mouse. *Am. J. Physiol. Cell Physiol.* **2001**, *281*, C496–C503. [CrossRef] [PubMed]

158. Xia, Y.; Zhao, P.; Xue, J.; Gu, X.Q.; Sun, X.; Yao, H.; Haddad, G.G. $Na^+$ channel expression and neuronal function in the $Na^+/H^+$ exchanger 1 null mutant mouse. *J. Neurophysiol.* **2003**, *89*, 229–236. [CrossRef] [PubMed]

159. Huang, X.; Atwood, C.S.; Moir, R.D.; Hartshorn, M.A.; Vonsattel, J.P.; Tanzi, R.E.; Bush, A.I. Zinc-induced alzheimer's Aβ1-40 aggregation is mediated by conformational factors. *J. Biol. Chem.* **1997**, *272*, 26464–26470. [CrossRef] [PubMed]

160. Takeda, A.; Tamano, H.; Tempaku, M.; Sasaki, M.; Uematsu, C.; Sato, S.; Kanazawa, H.; Datki, Z.L.; Adlard, P.A.; Bush, A.I. Extracellular $Zn^{2+}$ is essential for amyloid β1-42-induced cognitive decline in the normal brain and its rescue. *J. Neurosci.* **2017**, *37*, 7253–7262. [CrossRef] [PubMed]

161. Abramovitch-Dahan, C.; Asraf, H.; Bogdanovic, M.; Sekler, I.; Bush, A.I.; Hershfinkel, M. Amyloid β attenuates metabotropic zinc sensing receptor, mZnR/GPR39, dependent $Ca^{2+}$, ERK1/2 and clusterin signaling in neurons. *J. Neurochem.* **2016**, *139*, 221–233. [CrossRef] [PubMed]

162. Petrilli, M.A.; Kranz, T.M.; Kleinhaus, K.; Joe, P.; Getz, M.; Johnson, P.; Chao, M.V.; Malaspina, D. The emerging role for zinc in depression and psychosis. *Front. Pharmacol.* **2017**, *8*, 414. [CrossRef] [PubMed]

163. Mlyniec, K.; Doboszewska, U.; Szewczyk, B.; Sowa-Kucma, M.; Misztak, P.; Piekoszewski, W.; Trela, F.; Ostachowicz, B.; Nowak, G. The involvement of the GPR39-$Zn^{2+}$-sensing receptor in the pathophysiology of depression. Studies in rodent models and suicide victims. *Neuropharmacology* **2014**, *79*, 290–297. [CrossRef] [PubMed]

164. Mlyniec, K.; Nowak, G. GPR39 up-regulation after selective antidepressants. *Neurochem. Int.* **2013**, *62*, 936–939. [CrossRef] [PubMed]

165. Mlyniec, K.; Gawel, M.; Librowski, T.; Reczynski, W.; Bystrowska, B.; Holst, B. Investigation of the GPR39 zinc receptor following inhibition of monoaminergic neurotransmission and potentialization of glutamatergic neurotransmission. *Brain Res. Bull.* **2015**, *115*, 23–29. [CrossRef] [PubMed]

166. Cichy, A.; Sowa-Kucma, M.; Legutko, B.; Pomierny-Chamiolo, L.; Siwek, A.; Piotrowska, A.; Szewczyk, B.; Poleszak, E.; Pilc, A.; Nowak, G. Zinc-induced adaptive changes in NMDA/glutamatergic and serotonergic receptors. *Pharmacol. Rep.* **2009**, *61*, 1184–1191. [CrossRef]

167. Mlyniec, K.; Nowak, G. Up-regulation of the GPR39 $Zn^{2+}$-sensing receptor and CREB/BDNF/TrkB pathway after chronic but not acute antidepressant treatment in the frontal cortex of zinc-deficient mice. *Pharmacol. Rep.* **2015**, *67*, 1135–1140. [CrossRef] [PubMed]

168. Qian, W.J.; Gee, K.R.; Kennedy, R.T. Imaging of $Zn^{2+}$ release from pancreatic β-cells at the level of single exocytotic events. *Anal. Chem.* **2003**, *75*, 3468–3475. [CrossRef] [PubMed]

169. Bloc, A.; Cens, T.; Cruz, H.; Dunant, Y. Zinc-induced changes in ionic currents of clonal rat pancreatic β-cells: Activation of ATP-sensitive $K^+$ channels. *J. Physiol. (Lond.)* **2000**, *529*, 723–734. [CrossRef]

170. Ferrer, R.; Soria, B.; Dawson, C.M.; Atwater, I.; Rojas, E. Effects of $Zn^{2+}$ on glucose-induced electrical activity and insulin release from mouse pancreatic islets. *Am. J. Physiol.* **1984**, *246*, C520–C527. [CrossRef] [PubMed]

171. Franklin, I.; Gromada, J.; Gjinovci, A.; Theander, S.; Wollheim, C.B. Beta-cell secretory products activate α-cell ATP-dependent potassium channels to inhibit glucagon release. *Diabetes* **2005**, *54*, 1808–1815. [CrossRef] [PubMed]

172. Maret, W. Zinc in pancreatic islet biology, insulin sensitivity, and diabetes. *Prev. Nutr. Food Sci.* **2017**, *22*, 1–8. [CrossRef] [PubMed]

173. Tremblay, F.; Richard, A.M.; Will, S.; Syed, J.; Stedman, N.; Perreault, M.; Gimeno, R.E. Disruption of G protein-coupled receptor 39 impairs insulin secretion in vivo. *Endocrinology* **2009**, *150*, 2586–2595. [CrossRef] [PubMed]

174. Verhulst, P.J.; Lintermans, A.; Janssen, S.; Loeckx, D.; Himmelreich, U.; Buyse, J.; Tack, J.; Depoortere, I. GPR39, a receptor of the ghrelin receptor family, plays a role in the regulation of glucose homeostasis in a mouse model of early onset diet-induced obesity. *J. Neuroendocrinol.* **2011**, *23*, 490–500. [CrossRef] [PubMed]

175. Egerod, K.L.; Jin, C.; Petersen, P.S.; Wierup, N.; Sundler, F.; Holst, B.; Schwartz, T.W. Beta-cell specific overexpression of GPR39 protects against streptozotocin-induced hyperglycemia. *Int. J. Endocrinol.* **2011**, *2011*, 401258. [CrossRef] [PubMed]

176. Moran, B.M.; Abdel-Wahab, Y.H.; Vasu, S.; Flatt, P.R.; McKillop, A.M. GPR39 receptors and actions of trace metals on pancreatic β cell function and glucose homoeostasis. *Acta Diabetol.* **2016**, *53*, 279–293. [CrossRef] [PubMed]

177. Boye, K.; Maelandsmo, G.M. S100A4 and metastasis: A small actor playing many roles. *Am. J. Pathol.* **2010**, *176*, 528–535. [CrossRef] [PubMed]

178. Xie, F.; Liu, H.; Zhu, Y.H.; Qin, Y.R.; Dai, Y.; Zeng, T.; Chen, L.; Nie, C.; Tang, H.; Li, Y.; et al. Overexpression of GPR39 contributes to malignant development of human esophageal squamous cell carcinoma. *BMC Cancer* **2011**, *11*, 86. [CrossRef] [PubMed]

179. Alen, B.O.; Leal-Lopez, S.; Alen, M.O.; Viano, P.; Garcia-Castro, V.; Mosteiro, C.S.; Beiras, A.; Casanueva, F.F.; Gallego, R.; Garcia-Caballero, T.; et al. The role of the obestatin/GPR39 system in human gastric adenocarcinomas. *Oncotarget* **2016**, *7*, 5957–5971. [CrossRef] [PubMed]

180. Holst, B.; Egerod, K.L.; Schild, E.; Vickers, S.P.; Cheetham, S.; Gerlach, L.O.; Storjohann, L.; Stidsen, C.E.; Jones, R.; Beck-Sickinger, A.G.; et al. GPR39 signaling is stimulated by zinc ions but not by obestatin. *Endocrinology* **2007**, *148*, 13–20. [CrossRef] [PubMed]

181. Taylor, K.M. Liv-1 breast cancer protein belongs to new family of histidine-rich membrane proteins with potential to control intracellular $Zn^{2+}$ homeostasis. *IUBMB Life* **2000**, *49*, 249–253. [CrossRef] [PubMed]

182. Taylor, K.M. A distinct role in breast cancer for two liv-1 family zinc transporters. *Biochem. Soc. Trans.* **2008**, *36*, 1247–1251. [CrossRef] [PubMed]

183. Fan, Q.; Cai, Q.; Li, P.; Wang, W.; Wang, J.; Gerry, E.; Wang, T.L.; Shih, I.M.; Nephew, K.P.; Xu, Y. The novel ZIP4 regulation and its role in ovarian cancer. *Oncotarget* **2017**, *8*, 90090–90107. [CrossRef] [PubMed]

184. Wootten, D.; Christopoulos, A.; Sexton, P.M. Emerging paradigms in GPCR allostery: Implications for drug discovery. *Nat. Rev. Drug Discov.* **2013**, *12*, 630–644. [CrossRef] [PubMed]

185. Custodi, C.; Nuti, R.; Oprea, T.I.; Macchiarulo, A. Fitting the complexity of GPCRs modulation into simple hypotheses of ligand design. *J. Mol. Graph. Model.* **2012**, *38*, 70–81. [CrossRef] [PubMed]

International Journal of
*Molecular Sciences*

MDPI

*Review*

# Metallothioneins: Emerging Modulators in Immunity and Infection

**Kavitha Subramanian Vignesh \*** (ID) **and George S. Deepe Jr.** (ID)

Division of Infectious Diseases, College of Medicine, University of Cincinnati, Cincinnati, OH 45267, USA; george.deepe@uc.edu
\* Correspondence: Kavitha.subramanian@uc.edu; Tel.: +1-513-558-4717

Received: 29 September 2017; Accepted: 17 October 2017; Published: 23 October 2017

**Abstract:** Metallothioneins (MTs) are a family of metal-binding proteins virtually expressed in all organisms including prokaryotes, lower eukaryotes, invertebrates and mammals. These proteins regulate homeostasis of zinc (Zn) and copper (Cu), mitigate heavy metal poisoning, and alleviate superoxide stress. In recent years, MTs have emerged as an important, yet largely underappreciated, component of the immune system. Innate and adaptive immune cells regulate MTs in response to stress stimuli, cytokine signals and microbial challenge. Modulation of MTs in these cells in turn regulates metal ion release, transport and distribution, cellular redox status, enzyme function and cell signaling. While it is well established that the host strictly regulates availability of metal ions during microbial pathogenesis, we are only recently beginning to unravel the interplay between metal-regulatory pathways and immunological defenses. In this perspective, investigation of mechanisms that leverage the potential of MTs to orchestrate inflammatory responses and antimicrobial defenses has gained momentum. The purpose of this review, therefore, is to illumine the role of MTs in immune regulation. We discuss the mechanisms of MT induction and signaling in immune cells and explore the therapeutic potential of the MT-Zn axis in bolstering immune defenses against pathogens.

**Keywords:** Metallothioneins; zinc; cytokines; signaling; infection; antimicrobial defenses; metals; nutritional immunity

---

## 1. Introduction

Regulation of metal homeostasis is crucial in biological and cellular processes such as development and functions of organs, optimal enzyme activity, intracellular signaling and cell to cell communication [1]. Metallothioneins (MTs) are low molecular weight, cysteine-rich proteins that physiologically bind Zn and Cu in cells, but also sequester heavy metals such as cadmium (Cd) and mercury (Hg). They are induced by a variety of physiological and xenobiotic stimuli, buffer Zn and Cu ions, mitigate oxidative damage and protect against heavy metal intoxication [2]. In light of their ability to regulate metal ion availability, distribution and transport in cells, MTs have emerged as prominent players in maintaining overall organism fitness.

Development of mice genetically deficient in MT1/2 (MT-null) and MT3 isoforms has facilitated analysis of their functions in the resting state and in disease. In these models, in an unperturbed environment, MT deficiency does not result in apparent developmental, reproductive or age-related defects suggesting that MTs may be dispensable for normal development [3–5]. However, challenging MT-nullmice with common environmental stressors such as heavy metals, microbes or oxidative stress profoundly impacts fundamental processes such as DNA repair mechanisms, cell viability and inflammatory processes that rely on metal ion homeostasis and redox regulation for optimal functions [4–7]. For example, in mice and humans, Cd exposure induces MTs that sequester the metal to mitigate heavy metal poisoning. MT-null mice, however, exhibit decreased tolerance to Cd, and exposure results in nephrotoxicity and liver damage [3,4,8]. MT dysregulation is also observed under a wide spectrum of diseases including cancer, atherosclerosis, metabolic disease, autoimmunity, and infections [9–14]. Thus, it is clear that MTs respond to and are modulated in disease settings. However, a vast underlying gap in knowledge exists about the inducers and regulators of their complex functions in immunological responses. Deciphering how MTs shape the fate of development, dynamics and resolution of an immune response will be a crucial step in identifying novel therapeutic targets in pathways regulated by MTs. On the one hand, Zn and Cu have long been known to be involved in development and function of the innate and adaptive arms of our immune system [15] and on the other hand, numerous studies have reported MT regulation in the context of immunity [16]. However, our understanding of how this metalloprotein executes metal modulation in immune cells and molecular cues that drive these functions is fairly recent. The focus herein, stems from such recent insights into the fundamental role of mammalian MTs in immune regulation, with an emphasis on their ability to leverage host-pathogen interactions. We summarize the fundamental aspects of MT function and its role in Zn homeostasis (reviewed in greater detail elsewhere [2,17–20]) prior to exploring the interplay between MTs and immune responses.

## 2. The Metallothionein Family: Master Zinc Regulators

MTs are low molecular weight (6–7 kD), highly conserved, cysteine (Cys)-rich proteins that bind metals through thiol-clusters [17]. The MT protein family constitutes four isoforms (MT1–4) in mice and several isoforms with subtypes/variants in humans (MT1A, MT1B, MT1E, MT1F, MT1G1, MT1G2, MT1H, MT1HL1, MT1M, MT1X, MT2A, MT3, and MT4, and the pseudogenes *MT1DP*, *MT1JP*, *MT1L*, *MT2P1*, *MT1CP*, *MT1LP*, *MT1XP1*, *MT1P3*, *MT1P1* and *MTL3P*) [21–24].

In the early 1970s, a role for MTs in sequestering heavy metals such as Cd and Hg pinpointed their effect in alleviating xenobiotic stress [25]. Long since their discovery, the physiological functions of MTs remained unknown. Why did prokaryotes, lower eukaryotes and complex organisms evolve to express highly conserved thiol-rich proteins with an apparent crucial role in moderating heavy metal poisoning? The enigma was gradually dispelled when Zn homeostasis surfaced as being essential to all biological processes. About 10% of the mammalian genome encodes Zn binding proteins that regulate a diverse spectrum of biological functions [26]. Zn, unlike Cu and iron (Fe), is redox-inert, supporting evolutionary conservation of Zn binding sites in a large number of metalloproteins [27]. Undoubtedly, Zn availability is strictly regulated by MTs and Zn transporters [28,29]. MTs are master Zn regulators that sense intracellular cues and modulate Zn through sequestration, mobilization or release. The Zn binding constant of thionein was thought to be the highest ($>3 \times 10^{13}$/M) in biological systems, but, more recently, MTs have been shown to bind Zn ions sequentially with graded affinity and exist in metamorphic states [30,31]. The possibility of random occupancy of metal binding sites on MTs, followed by rearrangement of ions to a thermodynamically stable state has also been proposed [32]. Zn excess induces MTs, whereas Zn deficiency causes release of the metal from MTs, in effect scaling the intracellular Zn pool in response to cellular redox and energy state [33,34]. Intriguingly, Zn handling by MT1 and MT2 is distinct from that of MT3 [35]. The former sequesters Zn and readily releases only 1 Zn ion; the metal-thiolate cluster of MT3, however, assumes an "open conformation" to readily release Zn [35–37]. Moreover, Zn binding to MT3 is non-cooperative, suggesting that excess Zn may not stimulate saturation of all metal-ion binding sites in MT3 [38]. Thus, MT1/2 and MT3 share common ground in Zn regulation, but also exert discrete functions in scaling the intracellular Zn quota. The literature on MT4 function is scarce. Zn coordination by MT4 results in weaker folding of the protein compared to that of MT1; it has been suggested that MT4 may function as a Cu-thionein [39]. Of note, our knowledge of MT structure and function is gathered from an amalgamation of studies performed in solution, ex vivo, in vitro as well as in vivo. Thus, it is essential to recognize that the biology of MTs is highly complex; their behavior under different biochemical and cellular environments is likely heavily influenced by the nature of the stimulus, metal composition and redox environment.

Upstream of these events, the Zn-sensing metal-response element-binding transcription factor-1 (MTF-1) regulates MT expression to maintain precision in the size of the intracellular free Zn pool [40,41]. MTs mobilize Zn into the nucleus, mitochondria, Golgi apparatus, lysosomes, endoplasmic reticulum, cytosol and, possibly, zincosomes [42–47]. How MTs achieve this feat in intracellular compartments with diverse Zn demands is not clear. Their amino acid sequence lacks signals that dictate localization to specific organelles. The 3' untranslated region of MT1 mRNA signals transcript localization to the perinuclear region, arming the ability of MT1 protein to gain entry into the nucleus [48]. Another mechanism explicating MT targeting is protein-protein interaction. While it is well established that peptides and small molecules such as glutathione, ATP and GTP interact with MT, other interacting partner proteins have also been identified [27,33,34]. MT3 interacts with proteins involved in heat shock response, secretion, signaling pathways, metabolic enzymes and chaperones [49]. Such associations enable MT targeting to the extracellular milieu. Indeed, MTs have been detected outside cells [50]; whether they are actively secreted or passively released as a result of compromised membrane integrity is unclear. Mounting evidence points to an active involvement of MTs in modulating extracellular cues [51–53]. Nonetheless, in the field of MT biology, several unknowns remain. Do cells export MTs in their apo-form or as MT-Zn/Cu complexes? How do cells sense the need to tune extracellular Zn availability? Stretching beyond metal-ion buffering, what functions do MTs execute extracellularly? The rapid response of MTs to changing redox potential and Zn demands justifies their presence in this environment. Outside cells, MTs may participate in regulating chemotaxis, signaling, cell–cell communication, and mitigating oxidative damage [50,54]. A detailed understanding of MT functions will open new arenas for exploring their therapeutic potential in a variety of inflammatory disease conditions including Alzheimer's, coronary heart disease, arthritis, obesity, cancer and infections.

## 3. Immunity: Do Metallothioneins Take Center Stage?

Metal homeostasis, particularly Zn regulation, is essential for the development and adequate functioning of the innate and adaptive arms of immunity. The adverse impact of Zn deficiency on antibody production, cytokine production, chemotaxis, cell signaling, proliferation and functions of B, T helper (Th) and natural killer (NK) cells is well established [55–59]. Aberrant Zn regulation caused by Zn deficiency increases susceptibility to bacterial, viral and fungal infections, whereas Zn excess can exert toxic effects on immune cells [57,60]. MTs calibrate Zn availability; it is therefore conceivable that MTs are important regulators of immune cell function and promote immunological fitness. However, the underlying evidence supporting this hypothesis is still in its infancy.

Our current understanding of the role of MTs in immunity is fueled by studies on MT1 and MT2. In the mouse thymus, a primary lymphoid organ, MT expression peaks prior to thymic growth and wanes during thymic involution [61]. These changes parallel the impact of Zn on thymic mass, wherein Zn deficiency promotes thymic involution during aging [62]. The specific MT isoforms altered during thymic development and involution are undetermined, but the Zn-inducible nature of the MT response may result from dynamic changes in MT1 and MT2 expression [61,62]. The MT3 isoform, that is largely associated with neuronal functions, is also detected in the thymus [62]. However, MT3 is not Zn-inducible and data clearly elucidating the expression and functional roles of MT3 in the thymus are lacking. Overall, our understanding of how MTs and Zn control immunological functions in the thymus is limited and requires further investigation.

Thymic spatio-temporal regulation of MTs over the lifespan of an animal may impact T cell development and maturation in this organ. Thymic epithelial cells secrete thymulin, a Zn-dependent hormone [63]. MT expression in the thymic epithelial cells correlates with that of thymulin in humans and its expression is enhanced in thymomas [64,65]. Whether MTs regulate thymulin expression or vice versa is not known, but it may be postulated that in thymic epithelial cells, MTs deliver Zn to thymulin, whose function critically depends on the availability of this ion. Aberrant MT regulation may therefore impact downstream processes controlled by thymulin secretion such as T cell selection, differentiation and lymphocyte function. Interestingly, thymic abnormalities in MT-null mice have not been reported, albeit, this is a poorly studied field. An essential process in T cell selection is the presentation of self and non-self antigens by thymic non-lymphoid cells and professional antigen presenting cells (APCs) such as dendritic cells (DCs). Zinc influences the expression of major histocompatibility complex class (MHC)II on the surface of DCs [66,67]. A rise in intracellular Zn diminishes MHCII, while Zn chelation elevates it [66,67]. The occurrence of this phenomenon in DCs residing within the thymus remains unknown, but stimulates the proposition that MT-Zn sequestration may calibrate MHCII levels on DCs in the thymus, ultimately influencing thymic T cell selection.

The strategic placement of lymph nodes enables immunological surveillance and is the center for cross-talk between innate and adaptive immunity. Expression of different MT isoforms, their regulation and functional aspects in myeloid and lymphoid populations in the lymph nodes have not been characterized. Lymph-node associated MT expression is emerging as a prognostic marker in disease diagnosis, especially in patients with tumors. For example, MTs exhibit significant elevation in sentinel lymph node biopsies obtained from breast cancer and melanoma patients [68,69]. This modulation is prognostic and signals disease progression. It is noteworthy that the lymph node harbors a highly dynamic environment as circulating immune cells, signaling molecules and foreign agents continually drain in and out of these nodes through the lymphatic system. Thus, the myriad factors that condition the lymph node milieu may potentially influence MT regulation and function in macrophages, DCs, CD4$^+$ Th cells, cytotoxic CD8$^+$ cells and NK cells that enter and leave the lymph nodes. It may be conceived that MTs regulate Zn metabolism, proliferative, apoptotic, oxidative and nitrosative responses of these cells. However, as is evident, our understanding of MT regulation in lymph node biology is very limited and we may be far from uncovering the interplay between MT regulation and immune cell function in these tissues.

Only a handful of studies have shed light on MTs in the bone marrow, spleen and Peyer's Patches; our knowledge of the functional significance of MTs in these organs is extremely scarce. Hematopoietic stem cells in the bone marrow produce progenitors that differentiate into cells of the myeloid and lymphoid compartments [70]. Although data directly linking MTs to the immunological functions of the bone marrow are lacking, several lines of evidence suggest that MT-Zn homeostasis may play a significant role: (i) Dietary Zn deficiency or chronic Zn exposure results in precursor-B and -T cell apoptosis in the bone marrow [71,72]. MT1 expression in the bone marrow of rats is modulated by dietary Zn status [73]. The absence of MTs may perturb bone marrow Zn homeostasis, unless compensatory Zn transporters and other metalloproteins replace the loss; (ii) Several Zn-dependent transcription factors dictate terminal differentiation of precursor cells in the bone marrow. Early growth response-1 (Egr-1), a Zn-dependent transcription factor promotes monocyte differentiation to macrophages [74,75]. In contrast, another Zn-finger transcription factor, growth factor independent-1 (Gfi-1), antagonizes monocyte/macrophage lineage commitment and promotes neutrophil differentiation [75]. A number of studies support a role for Zn transfer from MTs to other metalloproteins including Zn-dependent transcription factors [76–78]. As such, metalation of proteins residing in the Golgi and endoplasmic reticulum (ER) must occur in these organelles [46,79]. The nucleus is another site where Zn availability must be tightly regulated to mediate Zn binding/release by gene-inducer and repressor molecules and prevent oxidative DNA damage [80,81]. MTs are detected and/or regulate the functions of these organelles [42,43,82]. The finding that MT resides in the nucleus raises questions about its contribution to nuclear functions such as gene regulation and provides clues to its crucial influence in regulating the size of nuclear Zn pool. (iii) Mt-null mice contain fewer CD4$^+$, CD8$^+$ T and B cells in blood [83]. The ubiquitous expression of MT1 and MT2 and control of proliferative, apoptotic, oxidative and chemotactic responses imply that MTs potentially shape the bone marrow immunological milieu.

The spleen is a major site for antibody production, erythrocyte clearance and filtration of pathogens. The splenic composition of CD4, CD8 and B cells in MT-null mice was found to be similar to that of WT animals under unperturbed conditions [84]. Contradictory findings have suggested that these MTs are involved in regulating splenic cellularity and function. One study reported elevated number of lymphoid cells and a 19% increase in splenic weight under MT1/2 deficiency. The number of circulating lymphocytes and splenic B cells were reduced in these mice, but number and proportion of splenic CD4[+] and CD8[+] T cells were marginally elevated [83]. Differences in age, gender and/or Zn concentration in the mouse diet used in the two studies may have led to discrepancies in the findings. Nonetheless, substantial evidence points to the role of MTs in regulating splenic T and B cell biology. MT-null splenic T cells fail to proliferate robustly upon stimulation with an antigen-independent T cell mitogen, concanavalin A or α-CD3. These splenocytes produce substantially lower interleukin (IL)-2 and the proliferative defect of purified T cells is reversed by addition of this cytokine [84]. Treatment of splenocytes with purified MT induces a hyper-proliferative response that is subdued in the presence of a reducing agent. Interestingly, while apo-MT, Zn-MT and Cd-MT comparably trigger proliferation, addition of Cu-MT inhibits this response [53], suggesting that MTs impart disparate effects pertinent to the metal bound to thionein. Accessibility to thiol groups on MTs is crucial in driving the lymphoproliferative response. Post activation, rapid T cell expansion is accompanied by superoxide burst [85]. The MT promoter possesses an antioxidant response element (ARE) that triggers MT expression in response to reactive oxygen species (ROS) [86]. An MT-ROS feedback loop may be operative in T cells, wherein basal MT levels mitigate oxidative damage, while ROS triggers the expression of MTs through ARE that feeds back to control the intracellular redox environment. Such a mechanism may operate to shield T cells from oxidative disruption of intracellular processes and improve T cell viability (Figure 1). Concentration, transport mechanisms and regulation of metals in the spleen await thorough investigation, but clearly, erythrocyte infiltration and hemoglobin metabolism in this organ place a demand on strictly regulating Fe homeostasis [87]. Although MT does not directly mobilize Fe, dietary Fe intake and MT1 expression share a reciprocal relationship in blood cells in rats, perhaps indicating changes in cellular Zn homeostasis or redox state [88,89]. Excess Fe interferes with Zn availability to proteins and vice versa, suggesting that, in addition to mechanisms that control Fe homeostasis, an MT-Zn axis may well be functional in sizing the Zn pool in this organ [90,91].

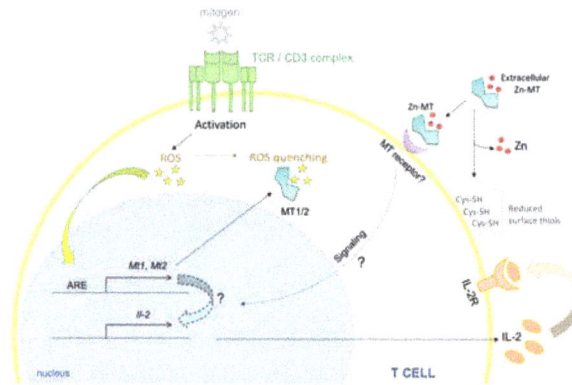

**Figure 1.** MTs (Metallothioneins) promote T cell survival and hyperproliferation. Activation of T cells with a T cell-specific mitogen generates ROS. MT1 and MT2 are induced by ROS produced during T cell activation, possibly by engagement of the antioxidant response elements (ARE) on the MT promoter. MTs quench ROS by oxidation of Cys residues, thereby preventing oxidative damage to T cells. MTs are required to produce optimal IL-2 that through autocrine and paracrine signaling, supports T cell survival. Exogenously added Zn-MT associates with the T cell membrane, possibly by binding to a yet unidentified MT receptor. Extracellular or membrane bound MTs reduce surface thiols (Cys-SH) on T cells. Mitigation of oxidative stress, reduction of surface thiols and an IL-2 response converge to promote T cell survival and hyper-proliferation by MTs. Red dots, Zn ions; all solid lines and arrows are known links; all dotted lines and arrows are predicted links.

Contrary its role in promoting T cell proliferation, an immunosuppressive effect of MTs on humoral immune responses has been described. The first evidence of modulation of humoral immune responses by MTs came from in vivo studies on ovalbumin (OVA) injected mice treated with Zn-MT or Cd-MT. MT injection suppresses serum anti-OVA specific IgG responses [52]. Moreover, neutralization of MT1 and MT2 boosts IgG responses, indicating that MTs are a negative modulator of T-cell antigen-dependent humoral immunity. Complementing these data, MT-null mice mount a pronounced IgG and IgM humoral response to OVA injection, 58% greater than that observed in wild type mice [83]. Perhaps, the splenic environment of these mice increases the propensity of B cell differentiation to plasma cells. The precise mechanism(s) that escalate plasma cell differentiation is enigmatic. One postulate is that MTs regulate nuclear factor-kappa B (NF-κB) activation that hastens B cell differentiation into plasma cells. Although this hypothesis has not been specifically explored in B cells, splenocytes from naïve MT-null mice exhibit increased NF-κB p50 subunit activation. Additionally, stimulation of these splenocytes with phorbol myristate acetate (PMA) and a calcium (Ca) ionophore increases p50 and p65 subunit activation compared to wild type splenocytes [83].

How does the function of a metal regulatory protein intersect with the globally significant transcriptional regulator, NF-κB? At least two hypotheses may be projected: (i) Redox control by MTs calibrates NF-κB activation; and (ii) Intracellular Zn buffering by MTs modulates the NF-κB activation pathway. These postulates are centered on the premise that ROS produced by mitochondria and phagosomes have signaling functions that may be stimulatory or inhibitory in the NF-κB pathway. Mitochondrial ROS and NADPH oxidase-derived phagosomal ROS enhance IκB degradation in the cytoplasm, resulting in increased NF-κB activity [92]. Oxygen intermediates also inactivate protein tyrosine phosphatases and dual specificity phosphatases, in turn, sustaining kinase activation [92]. While these mechanisms prolong NF-κB activation, MTs could subdue this response by quenching ROS through oxidation of Cys residues and Zn release from MTs. Zn itself is redox inert, but release of the metal upon oxidation tunes the intracellular redox state [27,93]. The redox regulating capacity of MTs is approximately 50 times greater than that of the major antioxidant, glutathione (GSH) [94]. Thus, when MTs are lacking, elevated ROS likely signals activation of NF-κB and its downstream pathways—resulting in enhanced humoral responses. On the flip side, Zn import by Zn importer protein (ZIP)8 directly modulates NF-κB function by inhibiting IκB kinase (IκK) [95]. This mechanism results in increased IκB mediated degradation of NF-κB. Whether MTs contribute to this process by altering Zn availability is not known. One may predict that heightened labile Zn in immune cells lacking MTs subdues NF-κB signaling. In fibroblasts, MT deficiency results in reduced levels of p65 [96]. These studies suggest that the impact of MTs on NF-κB signaling may vary by cell type. Whatever be the case, how MTs manipulate signaling to dictate immunological fate clearly deserves attention.

Of note, primary and secondary lymphoid organs exhibit remarkable architectural organization that regulate the development and function of immune cells [97–99]. Determining MT distribution and localization in these organs may reveal how MT expression patterns interweave with the thymic, lymph node and splenic architecture, providing clues to the mechanisms by which MTs guide immune cell programming, metabolism, maturation and lineage commitment.

## 4. Metallothionein Induction and Signaling in Immunity

Immunological surveillance places a unique "rapid adaptation" demand on cells of the innate and adaptive immune system. For example, unlike a hepatocyte that resides in the hepatic tissue, circulating immune cells constantly traffic in and out of tissues, through the lymphatic system and into primary and secondary lymphoid organs [100,101]. This implies that, in addition to preparation for an immune response, these cells must adapt to an ever-changing extracellular biological milieu that varies in oxygen tension and concentrations of cytokines, chemokines, hormones, metabolites, other small molecules and metals. Zn concentrations can exhibit large variations between circulation and within tissues—as much as 9 μg/cc (9 μg/g) in whole-blood to 520 μg/g in prostrate tissue [101,102]. It is not surprising then, that extracellular cues and immune signals modulate MTs in response to stress and a rapidly changing Zn environment. Cytokines such as tumor necrosis factor (TNF)α, IL-1α, IL-6 and interferon (IFN)γ modulate MTs and Zn metabolism in non-immunological organs such as the hepatic tissue [103]. However, do these mediators signal changes in MTs within immune cells? An integrative analysis of all immunological modulators of MTs in immune cells is lacking, but mounting evidence indicates that cytokines have a profound impact on MT gene regulation and functions in both myeloid and lymphoid compartments [44,104,105].

In response to the pro-inflammatory or M1 cytokine, granulocyte macrophage colony stimulating factor (GM-CSF), macrophages upregulate MT1 and MT2, with the latter exhibiting heightened changes. While both MT1 and MT2 are ubiquitously expressed and respond to Zn, MT2 induction is pronounced in response to GM-CSF [44]. Why MT2 responds more strongly to cytokine stimulation is elusive, but one possible explanation is that basal MT1 expression in macrophages is already high, perhaps near saturation. Further studies are needed to pinpoint the specific roles of MT1 versus MT2 in tackling redox changes and Zn metabolism in macrophages. In MT-null peritoneal macrophages, lipopolysaccharide (LPS) fails to strongly induce TNFα, suggesting that MT1 and MT2 are required for a macrophage pro-inflammatory response [106]. Moreover, MT-null macrophages manifest gross defects in antigen-presentation, expression of MHCII and co-stimulatory CD80 and CD86 molecules and cytokine production [107].

Interestingly, the isoform of MT augmented in immune cells is pertinent to the nature of the stimulating signal. The anti-inflammatory M2 cytokine, IL-4, unlike GM-CSF, strongly induces MT3, but not MT1 and MT2 in macrophages [104]. The distinct Zn demand and defense functions of pro-inflammatory vs. anti-inflammatory macrophages may determine preferential expression of specific MTs. Indeed, MT1 and MT2 in M1 macrophages sequester the Zn pool, whereas MT3 in M2 macrophages serves to expand this fraction [44,104] (Figure 2). An immunological role for the MT4 isoform remains to be uncovered.

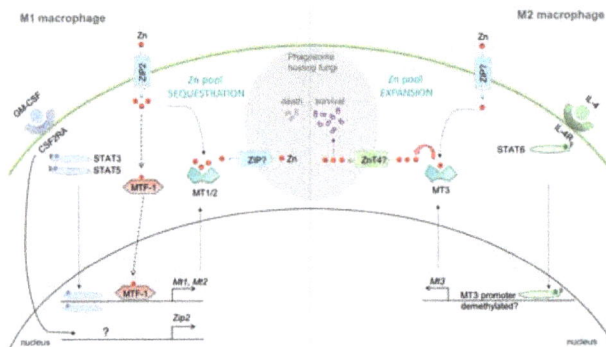

**Figure 2.** MTs shape the macrophage Zn pool and antifungal defenses. M1 macrophage (left side of center dotted line): GM-CSF binds to its receptor, CSF2RA, to trigger STAT3 and STAT5 activation. GM-CSF also upregulates Zn import via ZIP2. MT1 and MT2 induction is STAT3, STAT5 and ZIP2 dependent. While STAT3/5 can directly induce MTs by binding to the MT promoter, it is possible that ZIP2 mediated Zn import activates the transcription factor, MTF-1 to induce MT1 and MT2. In fungi infected macrophages, these MTs sequester Zn from *H. capsulatum* contained in phagosomes, possibly via Zn import into the cytosol, starving the pathogen of this essential element. M2 macrophage (right side of center dotted line): IL-4 binds to its receptor, IL-4RA to activate STAT6 mediated MT3 response. The *MT3* gene is not Zn inducible; IL-4 may prompt demethylation of the MT3 promoter to enable transcriptional activation by STAT6. M2 macrophages acquire Zn from extracellular sources that is rendered labile by MT3 and transported to the phagosome, possibly via ZnT4. This expanded labile Zn pool is exploited by *H. capsulatum* for survival within the macrophage. Red dots, Zn ions; all solid lines and arrows are known links; all dotted lines and arrows are predicted links.

DCs bridge innate and adaptive immunity through professional antigen presentation and expression of costimulatory molecules. The many flavors of DCs execute very distinct functions, in effect, shaping the fate of adaptive T cell immunity [108]. For example, inflammatory DCs promote effector T cell responses, whereas tolerogenic DCs induce fork head box P3 (FoxP3) expressing regulatory T cells (Tregs) that suppress inflammation [109,110]. Mounting evidence suggests that DCs modulate MTs in response to molecular cues, and that they may, in fact, regulate their phenotype and function. LPS upregulates MT2A in mature and activated human DCs [111], but downregulates MT1 in mouse bone marrow derived DCs [112]. Murine DCs express MT1 in response to thermal stress, dexamethasone or $ZnCl_2$ [112,113]. One may decipher that MT induction by stress modulates Zn distribution and regulates the intracellular redox environment in DCs. MTs (particularly MT1) promotes the development of tolerogenic DCs. These DCs express MT1 on their surface; blocking MT1 suppresses their tolerogenic potential, subduing naïve T cell differentiation into FoxP3 expressing Tregs [112]. This finding is intriguing because MT1 lacks a hydrophobic leader peptide sequence signaling surface expression or secretion. Whether DCs position MTs on the surface through interaction with other membrane proteins is not known. How MT expression on the DC surface confers a tolerogenic advantage also remains enigmatic. We and others have demonstrated that Zn handling in DCs is associated with modulation of DC phenotype [66,67]. DCs stimulated with LPS decrease their intracellular labile Zn pool leading to heightened surface MHCII expression. Overexpression of the Zn importer ZIP6 reverses this phenomenon [67]. Exposure of DCs to Zn salts also dampens surface MHCII, reduces pro-inflammatory cytokines such as IL-1β, IL-6 and IL-12 and promotes emergence of a tolerogenic signature characterized by increased programmed death ligand (PDL)1, PDL2 and enzyme indoleamine 2,3 dioxygenase expression [66]. Based on this premise, one may hypothesize that Zn, possibly via import through ZIP6, induces MTs in DCs that in turn regulate cellular processes associated with transport of MHCII containing vesicles. By localizing on the surface, MTs could gain access to the extracellular Zn pool, sequester it and subsequently interfere with Zn-dependent DC-T cell interactions. Indeed, superantigen presenting DCs rapidly trigger an ionic Zn signal that localizes to the subsynaptic region of the TCR activation complex in T cells [114]. Is it possible that MTs "soak up" Zn in extracellular spaces to stall TCR signaling? By scaling the size of the extracellular Zn pool, MTs potentially regulate T cell activation and proliferation in response to antigen presentation. Exogenously added MT binds to the membrane of purified $CD4^+$ T cells [115]. Thus, it is also possible that DC-MT directly interacts with the T cell plasma membrane (Figure 3). In this case, MT may be envisioned as a co-stimulatory molecule on the surface of DCs. Perhaps, the protein reduces thiols on the surface of T cells that serve dual function: (i) mitigate oxidative damage to T cells; and (ii) calibrate ROS mediated T cell activation [115]. How such an interaction favors $FoxP3^+$ Treg differentiation entails further investigation. Zn inhibits the histone deacetylase Sirt1, an enzyme that degrades FoxP3 in T cells [116]. Thus, on a paradoxical note, if MTs on the surface of DCs acted as Zn donors to T cells, the inhibitory effect of Zn on Sirt1 may in turn sustain FoxP3 expression.

MT expression within T cells regulates their activation, proliferation and differentiation potential. MT1 and MT2 manifest a late induction response in naïve T cells stimulated with IL-27 [105,117]. This cytokine stimulates generation of type-1 regulatory T (Tr)1 cells that drive immunosuppression via IL-10. In the absence of MT1 and MT2, Tr1 generation by IL-27 is greatly augmented, indicating that MTs thwart Tr1 differentiation [105]. In support of these data, exogenously added MT1/2 severely impair the emergence of IL-10 expressing Tr1 cells [118]. MTs interfere with signal transducer and activator of transcription (STAT)1 and STAT3 phosphorylation that are crucial for Tr1 generation [105]. The mechanism possibly involves modulation of Zn homeostasis by MTs. Numerous studies point to a role for Zn signaling in kinase and phosphatase functions [58,119]. In this view, one postulate is that by sequestering Zn, MT1 and MT2 sustain protein tyrosine phosphatase 1B (PTP1B) activity (an enzyme inhibited by Zn) that in turn dephosphorylates STATs (Figure 3). Indeed, MT-null Tr1 cells exhibit hyperphosphorylation of STAT1 and STAT3 and a pronounced IL-10 response [105]. T cells may orchestrate such a negative feedback mechanism through temporal control of MT expression.

This idea is supported by at least two observations: (i) a late MT1 and MT2 response (past 48 h) is observed in Tr1 differentiating cells [105]; and (ii) early burst in the T cell–intracellular labile Zn pool is MT independent [117], and is likely governed by ZIP6 mediated Zn import. Contrary to its role in STAT inhibition, MT-driven Zn mobilization supports p38 MAPK signaling in Tr1 cells [117,120]. These data suggest that the MT-Zn pool plays distinct roles during early and late stages of Tr1 cell differentiation. It is noteworthy that MT1 and MT2 are also highly upregulated in the Th17 subset generated by IL-6 and TGFβ stimulation, but not in Th1 and Th2 cells [105]. In Th17 cells however, a clear role for MTs remains to be determined as MT-null Th17 cells develop normally and produce IL-17 [105]. Whether MTs dictate transcriptional programming, lineage commitment, regulate T cell plasticity and skew the Treg-Th17 balance remain open areas of investigation.

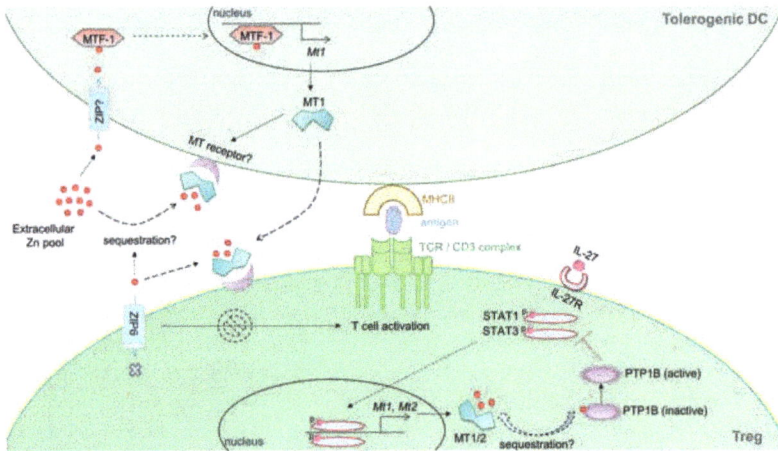

**Figure 3.** MT-Zn axis in tolerogenic DC–Treg interactions. DCs exposed to a pool of extracellular Zn develop a tolerogenic phenotype. Zn imported into DCs triggers MT-1 expression, possibly in an MTF-1 dependent manner. MT1 is expressed on the membrane of tolerogenic DCs and may also be actively secreted and bind to the T cell membrane. Upon activation, T cells import Zn via ZIP6 that mobilizes to the DC-T cell synapse. Zn sequestration by extracellular MTs may hinder this process, potentially stalling T cell activation. However, MT expression within T cells suppresses Treg development. IL-27 induces Tregs (in this case, IL-10 producing Tr1 cells) via STAT1 and STAT3 signaling. These STATs induce MT1 and MT2 that negatively feedback to inhibit STAT activation. MTs plausibly sequester Zn from PTP1B, an enzyme that is inhibited by Zn. Active PTP1B mediates dephosphorylation of STAT1 and STAT3 ultimately attenuating Treg development. Red dots, Zn ions; all solid lines and arrows are known links; all dotted lines and arrows are predicted links; red T bar depicts inhibition of STAT1/3 by PTP1B (active).

The immunomodulatory role of MTs is further highlighted by its impact on CD8+ T cell responses. Exposure of cytotoxic T lymphocytes (CTLs) to MTs diminishes surface MHCI and CD8 expression and impacts their ability to proliferate and mount cytotoxic responses against allogeneic target cells [121]. Zn deficiency reduces the proportion of CTLs in humans [122]. Moreover, in mice fed a Zn deficient diet, the ability of CTLs to kill tumor cells is impaired in vivo [123]. Taken together, our understanding of Zn-empowerment of CTL functions is vague, but the following postulates may be proposed: (i) MTs sequester Zn from the extracellular space, yielding a Zn-deficient environment; (ii) Zn deficiency alters membrane mobility, thereby affecting pore formation by perforins and exocytosis of cytolytic mediators by CTLs; and (iii) MTs quench ROS, in effect, impeding oxidative damage caused by CTLs. These postulates may well be interlinked, given the intimate relationship between MTs, Zn buffering and redox regulation. Although the data supporting MT mediated suppression of CTL functions are inclined towards an important role for Zn, the ability of MTs to donate Zn or modulate Cu homeostasis during CTL responses cannot be ruled out. MTs could additionally manipulate antigen presentation and interfere with surface receptors by formation of disulfide bridges with Cys residues, subsequently barring CTL-target cell interactions. Pending further research, these possibilities offer an opportunity to utilize MTs as therapeutic targets in disease conditions wherein CTL activity is crucial. For example, MT neutralization may benefit target cell killing by CTLs resulting in improved tumor outcomes or enhanced viral clearance.

Aside from its direct impact on immune cell behavior, MTs may act as chemoattractants to regulate cellular infiltration [54]. The arrangement of Cys residues in the MT molecule is reminiscent of its chemotactic properties; these motifs are also identified in chemokines. Clustal alignment of the MT protein sequence with that of C-C motif ligand (CCL)17 (both are encoded by genes on chromosome 8), reveals similarities in their Cys motifs, suggesting that MTs have a chemotactic attribute [54]. In vitro, exposure of Jurkat T cells to MTs results in F-actin reorganization and migration of the cells towards an MT gradient [54]. This phenomenon may be potentiated by direct MT and F-actin interactions [124]. In vivo evidence demonstrating MTs' chemotactic potential is lacking, but, if true, why should the immune system utilize MT as a chemoattractant? This thought is intriguing, particularly because the immune system possesses a highly-organized array of chemokines and corresponding receptors that cater to cell infiltration demands. Perhaps, extracellular release of MTs is recognized as an "early alarm" by the immune system, a premise, supported by the observation that a variety of immunological stressors rapidly and strongly induce MTs. In light of their chemotactic potential, important questions emerge. How does extracellular MT communicate with immune cells? As indicated previously, MTs bind the surface of T and B cells [53]. Surface and nuclear receptors for MTs have been identified in astrocytes [125], however, there are no known receptors for MTs on immune cells. The chemotactic property of MTs can be blocked by cholera or pertussis toxins that also block G-protein coupled receptor signaling [54]. Thus, it is reasonable to propose that that MTs share chemokine binding sites on chemokine receptors. The specific chemokine receptors involved, the impact of metal occupancy on MTs on receptor binding and the downstream signals relayed by MTs in immune cells need to be dissected.

Thus, it is clear that MT regulation is intricately tied to immunological responsiveness. As discussed earlier, the classical inducer of MT1 and MT2 is the Zn responsive transcription factor, MTF-1 [41]. Nitric oxide (NO) produced by immune cells causes Zn release from MTs that activates MTF-1-dependent transcription of genes involved in Zn regulatory pathways, including MT1 and MT2 [126]. However, studies on MT regulation in MTF-1 deficient immune cells will be required to decipher whether MTF-1 is absolutely necessary for MT gene regulation. In addition to metal response elements (MRE), the MT promoter harbors sequences that respond to STATs, hypoxia, redox status and hormones, implying that immune cells are armored with additional means of MT regulation. TLR4 stimulation induces IL-6 dependent MT1 expression via STAT1 binding to the MT1 promoter [127]. Likewise, GM-CSF driven activation of STAT3 and STAT5 elevates MT1 and MT2 in macrophages. This response can be abrogated by RNA interference or pharmacological inhibition of STAT3 and STAT5 signaling. In these cells however, MT1 and MT2 expression is also governed by Zn import via ZIP2 [44]. STAT binding sites have been identified on ZIP genes [128]. Whether STAT3 and STAT5 drive ZIP2 expression in macrophages is not known. It may be proposed that these STATs provide the initial impetus to elevate *ZIP2* as well as *MT1* and *MT2* genes and Zn import by the former sustains expression of the latter via an MTF-1 dependent mechanism. In essence, there is active STAT3/5–Zn–MT crosstalk that shapes the "Zn sequestering" phenotype of GM-CSF activated M1 macrophages [44]. In M2-IL-4 polarized macrophages, STAT6 specifically elevates MT3, but not MT1 and MT2 expression [104]. Thus, through distinct signaling pathways, immune cells calibrate levels of different MT isoforms, in effect, shaping the intracellular Zn environment and redox status. Based on this premise, it is conceivable that STATs function as "Zn modulators", whereby STAT signaling establishes the communication link between cytokines and Zn regulatory pathways in immune cells. Clearly, STAT activation plays a role in MT regulation (Figure 2), but the findings raise an intriguing question: Are MTs armored with the ability to modulate cytokine signaling? As discussed earlier, at least in the context of T cells, MT1 and MT2 subdue STAT1 and STAT3 activation during Tr1 differentiation [105] (Figure 3). The undisputed significance of Zn ions in enzyme function and cell signaling is reminiscent of the potential of MTs in controlling downstream effects of cytokine stimulation in immune cells.

Oxidative stress is intricately tied to innate and adaptive immunological functions. Rapidly dividing T cells heighten ROS production and phagocytes such as macrophages and neutrophils produce ROS to combat bacterial and fungal infections [85,129–132]. Thus, immune cells must mount antioxidant responses that selectively minimize damage to the host. Given its major antioxidant functions in addition to that of glutathione, MTs are rapidly induced by oxidative stress [93]. This response, is at least in part, driven by intracellular changes in Zn released from oxidized Cys and His residues on proteins that in turn modulate MTF-1 transcriptional activity. One contributor to this phenomenon is MT itself; the protein liberates Zn as a consequence of Cys oxidation [19,93]. Thus, MTs orchestrate a "self-amplifying response", whereby mitigating oxidative stress activates a Zn-driven positive feedback loop to further propel MT synthesis. However, under oxidative stress, cells may not solely rely on Zn status to dictate MT expression. This proposition may be especially true in immune responses, wherein Zn changes promote key defense functions such as DC maturation and antimicrobial responses of macrophages and neutrophils [44,66,133,134]. The proximal promoters of *MT1* and *MT2* genes contain the ARE consensus sequence GTGACnnnGC. In the presence of $H_2O_2$, the basic helix-loop-helix-leucine zipper upstream stimulatory factor family (USF) protein recognizes a USF binding site that overlaps with ARE on the MT1 promoter to trigger transcription [135]. The roles that USF and ARE play in driving MTs during oxidative stress in immune cells is unknown. Perhaps, in an effort to avert oxidative disruption of crucial host immunological processes, transcriptional activation via both, MRE and ARE synergize to maximize MT production. Mouse *MT1* and *MT2* genes as well as the human *MT2* gene possess enhancer glucorticoid response elements (GREs) [136]. Glucocorticoids bind the glucocorticoid receptor (GR) to transcriptionally regulate genes involved in suppressing inflammation. HeLa cells respond to the synthetic glucocorticoid, dexamethasone, by inducing MT1 and MT2 that is further elevated by Zn supplementation [136]. In the context of immunity, these steroids have been widely used to treat autoimmunity and inflammation, primarily due to their ability to exert immunosuppression by influencing macrophage polarization, DC tolerogenicity, skewing Th1/Th2 responses and promoting IL-10 production by Tregs [137]. The significance of MTs in the suppressive impact of glucocorticoids remains unraveled. Do MTs promote or inhibit glucocorticoid mediated immunosuppression? Perhaps, the answer to this question lays in determining temporal—acute versus late phase—MT responses driven by GREs, MREs and AREs. The effect of dexamethasone parallels that of IL-27 and Vitamin D3 in differentiation of IL-10 generating Tr1 cells; these factors also promote MT1 and MT2 gene expression [105,117]. At least in the case of IL-27, MTs are late response genes and thwart emergence of suppressive immunity [105]. These data and ancillary evidence from studies on UVB radiation reveal that absence of MT1 and MT2 enhances immunosuppression [138], implying that MTs incline immune response progression towards "inflammation sustenance". However, considering that MTs suppress B cell responses, the impact of MTs on humoral versus adaptive responses is discrete [83]. Deciphering the complex nature of MT regulation demands thorough investigation of how MT gene regulation, redox status, intracellular Zn levels intersect to influence signaling mechanisms and shape the outcome of inflammatory processes.

Evidently, a majority of the focus of MTs in immunity has rested on the functional significance of MT1 and MT2. This bias primarily stems from restricted association of MT3 with the CNS and a few other organs such as kidney and pancreas as opposed to ubiquitous MT1 and MT2 expression [139–141]; the former is also non-responsive to factors such as Zn and Cd that commonly induce MT1 and MT2. Epigenetic control has surfaced as another layer that regulates MTs; such a mechanism may especially be true for MT3 induction. This process has not been specifically investigated in different immune cells, but several lines of evidence point to a role for epigenetic machinery in influencing MT3 expression.

Hypermethylation of CpG islands is observed upstream of the MT3 transcription start site in breast cancer epithelial cells, myeloid leukemic cells and esophageal adenocarcinomas and is associated with poor disease outcome [142–144]. Chromatin compaction in the DNA region where *MT3* gene resides possibly explicates the highly restricted nature of MT3 transcription. The finding that MT3 is a cytokine-inducible gene in macrophages suggests that polarizing stimuli manipulate the epigenetic structure of this gene, rendering it accessible to signaling mediators such as STAT6 [104,145]. MT1 may also be subject to such epigenetic control, whereby its induction is associated with extensive promoter demethylation [146]. Epigenetic regulation is a critical component of immune cell development and functions [147]. It is likely that immunological stimuli influence the epigenetic structure of MT promoters, adding another layer of complexity to the control of MTs during an immune response. However, the epigenetic status of MT promoters in different immune cell types and the impact of immune mediators on epigenetic control of MT expression is unknown. Nevertheless, a multitude of factors including the nature of the immunological stimulus, signaling environment, Zn availability and oxidative stress dictate the isoform of MT expressed and intracellular level of each isoform in immune cells. Notably, MT regulation in immune responses is frequently reported, but there exists a vast gap in our knowledge of the factors that regulate their expression and the functional significance of these proteins. For example, do immune cells employ multiple signaling platforms to modulate MT levels? How do immune cells distinguish the need to preferentially trigger MT1 and MT2 versus MT3 and what benefit does this confer? Are MTs involved in regulating fundamental differentiation and lineage commitment processes of macrophages, DCs and T cell subtypes? Finally, does MT4 play a role in immunity? Addressing these questions will necessitate analysis of MT functions using models that exhibit conditional deficiency of the MT isoforms in specific myeloid and lymphoid compartments. MT regulation is context dependent, is influenced by the stimulus in question and controlled at multiple levels of gene regulation, translation and protein turnover. It is crucial to factor these in, especially when data about MT functions are extrapolated from one immunological setting to another.

## 5. Metallothioneins Respond to Microbial Stress

A primary function of immune surveillance is to screen microbial invaders and rapidly respond by activating myriad immune pathways that converge to mount effective antimicrobial defenses. Microbial challenge is thus, a potent "stressor" that immune cells must be prepared to cope with or alleviate to maintain the hosts' integrity during host-pathogen interactions. On the innate end, nutritional immunity and ROS are two principal defense mechanisms deployed by immune cells to curtail microbial growth and dissemination [132,148,149]. It is not surprising then, that numerous studies have reported dramatic upregulation of MTs, particularly MT1 and MT2 isoforms during bacterial, viral and fungal infections [44,103,150,151]. A classic example is hepatic MT1 and MT2 elevation during endotoxin (LPS) challenge. This is an acute phase response associated with rapid reduction in plasma Zn and ~29% increase in hepatic Zn. Production of IL-1$\alpha$, IL-1$\beta$, TNF$\alpha$ and IL-6 precede this response and each of these cytokines independently trigger LPS driven MT transcription [96]. Additionally, in children with sepsis, MT elevation is a predictor of survival [152]. While precise pathways that link the cytokine response to MT induction remain elusive, hyperglucagonemia is a feature of endotoxin challenge that regulates MTs. In this view, a role for metabolic alterations and elevated blood glucose in modulating MT levels upon endotoxin challenge may be proposed. Indeed, several immune mediators impose unique metabolic demands on glucose and fat utilization to cope with inflammatory stress [153,154]. Another notable acute phase response during endotoxemia is the rapid elevation in ZIP14 that imports Zn into hepatocytes [155]. MT elevation likely signifies sequestration and retention of the newly imported Zn, as mice lacking MT1 and MT2 exhibit hepatic Zn loss and concomitant rise in plasma Zn [103]. This peculiar phenomenon confers, at least, dual benefit to the host: first, it restricts Zn access to pathogens in the extracellular environment; and, second, Zn sequestration potentially favors chemotactic migration of immune cells towards the site of inflammation. The former surfaces as an arm of nutritional immunity, wherein the immune system attempts to starve microbial invaders by withholding nutrients essential to biological processes. Localized Zn redistribution by MTs may shape the course of immune response by escalating DC maturation and skewing T cell responses towards that of an inflammatory phenotype. MT-dependent hypozincemia has similarly been reported in other gram-negative infection models including *Salmonella typhimurium* and the live vaccine strain of *Francisella tularensis* [156], implying a broadly applicable role of MTs in orchestrating hypozincemic responses upon bacterial challenge. The MT response is transient; withdrawal of the endotoxin stimulus results in rapid degradation of accumulated MT justifying the short half-life of Zn-thionein complex [154]. The fate of liberated Zn, however, is obscure. One obvious consequence appears to be redistribution and restoration of Zn homeostasis after infection has subsided. Beyond these, is there a role for Zn in suppressing pro-inflammatory responses and promoting tissue repair? If yes, do MTs partake in this process? Zn has a well-recognized role in wound healing and inflammation. Experimental models of Zn deficiency and Zn deficient individuals exhibit markedly delayed tissue repair [157,158]. Moreover, recent revelations of the role of MTs in wound-healing M2 macrophages, discussed in the following paragraphs, illuminate this possibility [104].

Perhaps, the impact of MTs on immune responses to infection is highly complex, depending on the cell type, regulation dynamics, degree of induction and the pathogen in question. On the one hand, in mice infected with the gram-negative bacterium *Helicobacter pylori*, MT1 and MT2 do not contribute to bacterial clearance but confer protection against erosive lesions in the gastric epithelium [159]. In this model, a lack of MTs aggravates infiltration of inflammatory cells, enhances macrophage inflammatory protein (MIP)-1α and monocyte chemoattractant protein (MCP)-1 expression and augments NF-κB DNA binding activity in the stomach tissue. How Zn buffering and antioxidant mechanisms of MTs converge to modulate the inflammatory process to yield subdued ulceration awaits further experimentation. On the other hand, genetic deficiency of MT1 and MT2 or increased MT gene dose, accelerate clearance of the gram-positive bacterium, *Listeria monocytogenes* in mice [160]. Thus, swaying MT expression in either direction favorably promotes antibacterial defenses. This observation may be explicated by disrupted redox control, as prolonged MT deficiency or MT overdose interferes with alleviation of oxidative stress. Indeed, the splenic lymphocyte population exhibits increased oxidized surface thiols in MT-null mice and augmented ROS in MT overexpressing mice [160].

Complementing the aforementioned studies on bacteria, a plethora of studies have revealed MT modulation in viral infections. Influenza A/PR8 strain that causes upper respiratory tract and lung infections strongly elevates MT1 and MT2 in the lung and liver [86]. Infection triggers engagement of MREs, ARE, GRE and STAT3 binding sites on the MT promoters, perhaps, signifying an effort to rapidly maximize MT production to curtail oxidative injury caused by heightened inflammation [86]. This hypothesis is further supported by an inverse correlation between hepatic MT expression and hepatitis C virus (HCV) pathogenesis and that MTs improve response of chronic HCV patients to IFNα therapy [161,162]. In an experimental model of human coxsackievirus B type 3 infection, MT1 and MT2 are augmented in the liver, kidney and spleens of mice, facilitating Cu and Zn redistribution across different organs [163]. Metal ion redistribution is frequently reported but how such a process benefits the host is enigmatic [164]. Viruses must explicitly supply all of their metal ion demands through the host metal pool; thus, it is reasonable to predict that changes in MTs and Zn or Cu redistribution are "virus-induced" phenomena. However, from the hosts' perspective, can such a mechanism deter viral binding and dissemination? Studies on respiratory syncytial virus reveal an inhibitory role for Zn but not Ca, magnesium (Mg) and manganese (Mn) in viral replication [165]. Zn-ejecting compounds inhibit the activity of the transcription anti-termination cofactor protein, M2-1 that is crucial for viral RNA-dependent RNA polymerase function [166]. Zn also directly inhibits binding of common cold causing rhinoviruses to intracellular adhesion molecule-1 (ICAM1) receptor in the nasal epithelium, stalling viral entry into the host [167]. However, to directly obstruct viral entry, MTs must be present extracellularly. Whether cells release small amounts of MTs extracellularly at barrier interfaces in preparation for antiviral responses is unknown. Upon sensing viral insult, it is possible that some of the upregulated MT protein is released from cells to curtail further viral invasion. Nonetheless, a number of questions will need to be addressed. What signals MT release and how does the host decide how much should be released during infection? What is the fate of extracellular MTs after infection has subsided? Do viruses encapsulate host MTs as they are released from cells? The latter, if true, may be exploited as a mechanism of piracy by viral invaders that depletes the host of intracellular MT stores and manipulates MT levels in the extracellular space. An increased focus on Zn binding proteins and Zn ionophores in the development of antivirals suggests that harnessing MTs and their Zn modulating potential may prove beneficial to the host [168,169]. However, a thorough understanding of how MTs manipulate the course of an antiviral response will be necessary to pinpoint precise molecular processes involved in MT-mediated host-protection.

Immune responses against microbial challenge modulates host-MTs, but very little is known about the contribution and functions of immune-cell derived MTs in this context. The pulmonary fungal pathogen *Histoplasma capsulatum* establishes a replicative niche within macrophages prior to pro-inflammatory activation by cytokines such as IFNγ or GM-CSF [44,170]. The latter purposes as a "Zn reprogramming signal" that induces MT1 and MT2 to deplete intracellular labile Zn that would otherwise be readily accessible to *H. capsulatum* containing phagosomes (Figure 2). As indicated earlier, GM-CSF also triggers Zn import via ZIP2 in macrophages. This finding appears paradoxical, but the Zn flux probably serves to meet an increased demand of Zn-dependent processes that arm the macrophage defense machinery. In the host-pathogen combat, the Zn pool could be envisioned as "drifting" towards the former; one apparent advantage of this mechanism is that the host is guarded against oxidative disruption of critical cellular processes, while rendering the pathogen in a Zn-starved state [171]. A macrophage superoxide-boost is interwoven into the umbrella of MT-driven effects that ultimately cripples fungal replication [44,172]. Parallel to this phenomenon, is the finding that MT1 and MT2 detain intracellular labile Zn from *S. typhimurium* within macrophages [173]. Genetic deficiency of MT1 and MT2 elevates labile Zn and impairs oxidative and nitrosative macrophage defenses, a situation, that *S. typhimurium* exploits to acquire Zn and colonize the macrophage. Even in a Zn-restricted state, *S. typhimurium* secures the metal ion via the ZnuABC Zn importer [174]. On the flip side, macrophages also employ "Zn stress" to poison invading microbes. Human macrophages hosting *Mycobacterium tuberculosis* rapidly deploy Zn into the phagosome to intoxicate the pathogen. A similar mechanism is functional in *E. coli* eradication by macrophages [150]. Contrary to ZIP2 mediated Zn import in *H. capsulatum* infected macrophages, the source of this Zn pool is intracellular [44,150]. In these cells, MTF-1 localizes to the nucleus and various MT transcripts (*MT1H*, *MT1M*, *MT1X*, and *MT2A*) are elevated. However, whether MTs contribute to the Zn-burst is unclear. It has been proposed that ROS liberates Zn from MT-Zn complexes; the metal ion is channeled into the phagosome by yet unknown transport mechanism(s) to drive mycobacterial poisoning [150]. It is conceivable that cytosolic Zn is redirected by means of Zn transporters, but what signals the entry of this ion into vacuoles housing mycobacteria? Do MTs localize within late-endosomes and lysosomes, or do they bind to the phagosomal membrane and interact with transporters to guide entry of Zn into these vacuoles? Are extracellular Zn-MT complexes and mycobacteria "co-engulfed" as a result of phagocytosis? Regardless of the mechanism in action, it is reasonable to postulate that MT1 and MT2 arm phagocyte defenses, particularly that of macrophages. Pertinent to the pathogen in context, the precise molecular cues that instruct Zn poisoning as opposed to Zn sequestration by MTs remain enigmatic, but likely represents a carefully elected immunological decision that caters to incline Zn concentrations beyond the narrow "tolerance" window that succumbs pathogens to Zn-starvation versus Zn-overload (Figure 2).

From the pathogens' standpoint, survival within the host demands subversion strategies. In the case of mycobacteria, alleviation of Zn poisoning is achieved by activation of P1-type ATPases [150]. Interestingly, contrary to the hosts' Zn-restriction defense strategy reported for *S. typhimurium*, macrophages also attempt a Zn-overload defense mechanism to eliminate this pathogen [175].

The involvement of MTs in this situation has not been deciphered, but the pathogen evades this response by virtue of Salmonella pathogenicity islands, expression of the bacterial Zn exporter, ZntA and yet unidentified mechanisms [175]. Thus, in their struggle for survival, pathogens launch counter-strategies that rely on Zn regulation to subvert host defenses. Given the highly conserved nature of MTs from single-celled to multicellular organisms, that pathogen-derived MTs are determinants of virulence within the host is an interesting conjecture to be explored. The finding that the fungal pathogen, *Cryptococcus neoformans* produces Cu-binding MTs to neutralize toxic copper-overload imposed by host macrophages is a classic premise demonstrating that MTs may also belong to the pathogen-defense armor [176].

Over half a century of research has emphasized the roles of MT1 and MT2 in immune responses. Until recently, virtually nothing was known about the contribution of MT3 in host defenses. Existing data on the regulatory and functional attributes of MT3 suggests that from an immunity point of view, the impact of MT1 and MT2 versus MT3 on host immunity is unparalleled. In M2 macrophages polarized with IL-4 or IL-13, MT3 is elevated and enriches the labile-Zn pool [104]. These type-2 cytokines promote parasite clearance, wound healing and tissue repair, but exaggerated IL-4 and IL-13 production worsen allergic inflammation and render the host susceptible to invasion by intracellular pathogens [177,178]. M2 macrophages harbor a favorable niche for the survival of *H. capsulatum*. Although the regulation of transferrin and arginase in these cells imply roles for iron metabolism and subdued NO defenses [179,180], emerging evidence pinpoints to the significance of Zn homeostasis by MT3 in suppressing macrophage defenses. M2 macrophages source the metal ion from the extracellular environment, but instead of sequestering it, render it readily accessible to the pathogen. This "expansion" effect on the Zn-pool is MT3-dependent and is at least partially driven by cathepsins that trigger Zn release from the protein. The pathogen ultimately gains access to the host-Zn resource, possibly via the transporter ZnT4, to thrive within phagosomes of M2 macrophages [104]. MT1 and MT2 act in stark contrast in M1 macrophages invaded by *H. capsulatum*; in these cells, Zn that once belonged to the pathogen befits the macrophage Zn-pool [44,104] (Figure 2). It is intriguing that, rather than being poisoned, this fungal pathogen exploits Zn-release tactics of MT3 to its advantage. The findings seed a fundamental query: Why did M2 phagocytes evolve to increase the labile-Zn pool? The existence of such a Zn modulatory machinery in M2 macrophages is reminiscent of the importance of Zn in anti-parasitic defenses [181]. Nonetheless, much remains to be unraveled about the involvement of the MT3-Zn axis in immune processes regulated by IL-4 and IL-13 signaling.

## 6. Concluding Remarks

The evolutionary web of MTs has casted its presence on virtually every life form—prokaryotes to lower eukaryotes, invertebrates to higher vertebrates including mammals. Within its lifespan, a living cell must adequately tap the potential of highly dynamic redox chemistries and metal ion environments to conduct various biological processes. Evidently then, through duplication and functional segregation, a number of MT isoforms emerged to support metal-ion buffering, maintaining homeostasis and protecting the host from oxidative assault. The complexity of MT evolution is readily demonstrated by differences in the MT gene clusters, copy number and functional characteristics between different species within the same genera [182]. The mouse has four MT isoforms, but numerous isoforms and subtypes/variants contribute to the heterogeneity of this family in humans; that newer proteins may line the MT queue pending annotation and functional characterization may not be a surprise. The question is: Will the MT family evolve further, and why? From a mammalian-immunity viewpoint, innate and adaptive barriers are constantly confronted with tremendous evolutionary pressure from environmental cues such as xenobionts including pathogens that rapidly manipulate their genomes and virulence strategies to sustain infection [183]. While direct evidence that immunological pressure imposed divergence of the MT family remains to be discovered, the >10 trillion microbes that have harmlessly populated the human gut and skin for eons suggests that such an evolutionary demand may have aroused within the host itself. Indeed, MT expression in the gut, known to be guided largely by dietary Zn status, maintains integrity of the gut epithelium, an important barrier that guards against dissemination of opportunistic pathogens [184,185]. On a captivating contrary note, is it possible that, evolutionarily, MTs exerted selective pressure to shape the human microbiome? The fundamental differences in metal-ion requirements, susceptibility to sequestration or intoxication and ability to withstand superoxide damage between microbes provide a framework to test this hypothesis. Nonetheless, investigating this possibility demands a rigorous understanding of microbial metal-ion homeostasis and how immune responses employ MTs to lay "beneficial" microflora within the host. Perhaps, the functions of MTs are intricately interwoven into the complex attributes of innate and adaptive cells; thus, in the world of immunity, we may have only scraped the surface of vast perturbations that underlie a dysregulated MT response. With careful consideration to how MTs globally impact the immune response, unearthing the potential of MT as an "antimicrobial protein" will take the next leap forward in battling the rising menace of antibiotic resistance and emerging virulent pathogens. Such a proposition must be supported by an unparalleled understanding of interactions of MTs with microbes and immune cells, half-life, fate of the protein and significance of its apo- and metal-bound forms in bolstering immunological defenses. Finally, gaining deeper insight into the immunoregulatory role of MTs holds promise in levitating this class of proteins in the therapeutic ladder targeting infections and the myriad diseases that engage host immunity.

**Acknowledgments:** This work was supported by NIH grant AI106269 to George S. Deepe, Jr. and American Heart Association 15POST25700182, 2015 to Kavitha Subramanian Vignesh.

**Author Contributions:** Kavitha Subramanian Vignesh wrote the manuscript and George S. Deepe, Jr. reviewed it.

**Conflicts of Interest:** The authors declare no conflict of interest.

## Abbreviations

| | |
|---|---|
| APC | antigen presenting cell |
| ARE | antioxidant response element |
| Ca | calcium |
| CCL | C-C motif ligand |
| Cd | cadmium |
| CTL | cytotoxic T lymphocyte |
| Cu | copper |
| Cys | cysteine |
| DC | dendritic cell |
| Egr-1 | early growth response-1 |
| ER | endoplasmic reticulum |
| Fe | iron |
| FoxP3 | Fork head box P3 |
| Gfi-1 | growth factor independent-1 |
| GM-CSF | granulocyte macrophage-colony stimulating factor |
| GRE | glucocorticoid response element |
| GSH | glutathione (reduced) |
| HCV | hepatitis C virus |
| Hg | mercury |
| ICAM-1 | intracellular adhesion molecule-1 |
| IFN | interferon |
| IL- | interleukin |
| I$\kappa$B | inhibitor of $\kappa$B |
| I$\kappa\kappa$ | I$\kappa$B kinase |
| LPS | lipopolysaccharide |
| MCP | monocyte chemotactic protein |
| Mg | magnesium |
| MHC | major histocompatibility class |

| MIP | macrophage inflammatory protein |
| Mn | manganese |
| MRE | metal response element |
| MT | Metallothionein |
| MT-null | MT1/2 deficient |
| MTF-1 | metal-response element-binding transcription factor-1 |
| NF-κB | nuclear factor-kappa B |
| NK | natural killer |
| NO | nitric oxide |
| PMA | phorbol myristate acetate |
| PTP | protein tyrosine phosphatase |
| ROS | reactive oxygen species |
| STAT | signal transducer and activator of transcription |
| Th | T helper cell |
| TNF | tumor necrosis factor, |
| Tr1 | Type-1 regulatory T cell |
| Treg | regulatory T cell |
| USF | upstream stimulatory factor |
| ZIP | zinc importer |
| Zn | zinc |

## References

1. Nelson, N. Metal ion transporters and homeostasis. *EMBO J.* **1999**, *18*, 4361–4371. [CrossRef] [PubMed]
2. Coyle, P.; Philcox, J.C.; Carey, L.C.; Rofe, A.M. Metallothionein: The multipurpose protein. *Cell. Mol. Life Sci.* **2002**, *59*, 627–647. [CrossRef] [PubMed]
3. Klaassen, C.D.; Liu, J. Metallothionein transgenic and knock-out mouse models in the study of cadmium toxicity. *J. Toxicol. Sci.* **1998**, *23* (Suppl. 2), 97–102. [CrossRef] [PubMed]
4. Masters, B.A.; Kelly, E.J.; Quaife, C.J.; Brinster, R.L.; Palmiter, R.D. Targeted disruption of metallothionein i and ii genes increases sensitivity to cadmium. *Proc. Natl. Acad. Sci. USA* **1994**, *91*, 584–588. [CrossRef] [PubMed]
5. Davis, S.R.; Cousins, R.J. Metallothionein expression in animals: A physiological perspective on function. *J. Nutr.* **2000**, *130*, 1085–1088. [PubMed]
6. Davis, S.R.; McMahon, R.J.; Cousins, R.J. Metallothionein knockout and transgenic mice exhibit altered intestinal processing of zinc with uniform zinc-dependent zinc transporter-1 expression. *J. Nutr.* **1998**, *128*, 825–831. [PubMed]
7. Liu, J.; Liu, Y.; Hartley, D.; Klaassen, C.D.; Shehin-Johnson, S.E.; Lucas, A.; Cohen, S.D. Metallothionein-i/ii knockout mice are sensitive to acetaminophen-induced hepatotoxicity. *J. Pharmacol. Exp. Ther.* **1999**, *289*, 580–586. [PubMed]
8. Liu, Y.; Liu, J.; Habeebu, S.S.; Klaassen, C.D. Metallothionein protects against the nephrotoxicity produced by chronic cdmt exposure. *Toxicol. Sci.* **1999**, *50*, 221–227. [CrossRef] [PubMed]
9. Eckschlager, T.; Adam, V.; Hrabeta, J.; Figova, K.; Kizek, R. Metallothioneins and cancer. *Curr. Protein Pept. Sci.* **2009**, *10*, 360–375. [CrossRef] [PubMed]
10. Gobel, H.; van der Wal, A.C.; Teeling, P.; van der Loos, C.M.; Becker, A.E. Metallothionein in human atherosclerotic lesions: A scavenger mechanism for reactive oxygen species in the plaque? *Virchows Arch.* **2000**, *437*, 528–533. [CrossRef] [PubMed]
11. Sato, M.; Kawakami, T.; Kadota, Y.; Mori, M.; Suzuki, S. Obesity and metallothionein. *Curr. Pharm. Biotechnol.* **2013**, *14*, 432–440. [CrossRef] [PubMed]
12. Espejo, C.; Carrasco, J.; Hidalgo, J.; Penkowa, M.; Garcia, A.; Sáez-Torres, I.; Martínez-Cáceres, E.M. Differential expression of metallothioneins in the cns of mice with experimental autoimmune encephalomyelitis. *Neuroscience* **2001**, *105*, 1055–1065. [CrossRef]
13. Pedersen, M.O.; Jensen, R.; Pedersen, D.S.; Skjolding, A.D.; Hempel, C.; Maretty, L.; Penkowa, M. Metallothionein-i+ii in neuroprotection. *Biofactors* **2009**, *35*, 315–325. [CrossRef] [PubMed]

14. Subramanian Vignesh, K.; Deepe, G.S., Jr. Immunological orchestration of zinc homeostasis: The battle between host mechanisms and pathogen defenses. *Arch. Biochem. Biophys.* **2016**, *611*, 66–78. [CrossRef] [PubMed]

15. Sherman, A.R. Zinc, copper, and iron nutriture and immunity. *J. Nutr.* **1992**, *122*, 604–609. [PubMed]

16. Dziegiel, P.; Pula, B.; Kobierzycki, C.; Stasiolek, M.; Podhorska-Okolow, M. Metallothioneins and immune function. In *Metallothioneins in Normal and Cancer Cells*; Springer International Publishing: Cham, Switzerland, 2016; pp. 65–77.

17. Bert, L.V. The function of metallothionein. *Neurochem. Int.* **1995**, *27*, 23–33.

18. Blindauer, C.A.; Leszczyszyn, O.I. Metallothioneins: Unparalleled diversity in structures and functions for metal ion homeostasis and more. *Nat. Prod. Rep.* **2010**, *27*, 720–741. [CrossRef] [PubMed]

19. Maret, W. Redox biochemistry of mammalian metallothioneins. *J. Biol. Inorg. Chem.* **2011**, *16*, 1079–1086. [CrossRef] [PubMed]

20. Vasak, M. Advances in metallothionein structure and functions. *J. Trace Elem. Med. Biol.* **2005**, *19*, 13–17. [CrossRef] [PubMed]

21. Mehus, A.A.; Muhonen, W.W.; Garrett, S.H.; Somji, S.; Sens, D.A.; Shabb, J.B. Quantitation of human metallothionein isoforms: A family of small, highly conserved, cysteine-rich proteins. *Mol. Cell. Proteom.* **2014**, *13*, 1020–1033. [CrossRef] [PubMed]

22. National Library of Medicine (US); National Center for Biotechnology Information. *Gene (Metallothionein) and "Homo Sapiens" [Porgn:__txid9606]*; National Center for Biotechnology Information: Bethesda, MD, USA, 2004. Available online: https://www.ncbi.nlm.nih.gov/gene/ (accessed on 12 October 2017).

23. Quaife, C.J.; Findley, S.D.; Erickson, J.C.; Froelick, G.J.; Kelly, E.J.; Zambrowicz, B.P.; Palmiter, R.D. Induction of a new metallothionein isoform (MT-IV) occurs during differentiation of stratified squamous epithelia. *Biochemistry* **1994**, *33*, 7250–7259. [CrossRef] [PubMed]

24. West, A.K.; Stallings, R.; Hildebrand, C.E.; Chiu, R.; Karin, M.; Richards, R.I. Human metallothionein genes: Structure of the functional locus at 16q13. *Genomics* **1990**, *8*, 513–518. [CrossRef]

25. Piotrowski, J.K.; Trojanowska, B.; Sapota, A. Binding of cadmium and mercury by metallothionein in the kidneys and liver of rats following repeated administration. *Arch. Toxicol.* **1974**, *32*, 351–360. [CrossRef] [PubMed]

26. Andreini, C.; Banci, L.; Bertini, I.; Rosato, A. Counting the zinc-proteins encoded in the human genome. *J. Proteome Res.* **2006**, *5*, 196–201. [CrossRef] [PubMed]

27. Maret, W. The function of zinc metallothionein: A link between cellular zinc and redox state. *J. Nutr.* **2000**, *130*, 1455S–1458S. [PubMed]

28. Eide, D.J. Zinc transporters and the cellular trafficking of zinc. *Biochim. Biophys. Acta* **2006**, *1763*, 711–722. [CrossRef] [PubMed]

29. Cousins, R.J.; Liuzzi, J.P.; Lichten, L.A. Mammalian zinc transport, trafficking, and signals. *J. Biol. Chem.* **2006**, *281*, 24085–24089. [CrossRef] [PubMed]

30. Maret, W.; Vallee, B.L. Thiolate ligands in metallothionein confer redox activity on zinc clusters. *Proc. Natl. Acad. Sci. USA* **1998**, *95*, 3478–3482. [CrossRef] [PubMed]

31. Krezel, A.; Maret, W. The functions of metamorphic metallothioneins in zinc and copper metabolism. *Int. J. Mol. Sci.* **2017**, *18*, 1237. [CrossRef] [PubMed]

32. Rigby Duncan, K.E.; Stillman, M.J. Metal-dependent protein folding: Metallation of metallothionein. *J. Inorg. Biochem.* **2006**, *100*, 2101–2107. [CrossRef] [PubMed]

33. Jiang, L.J.; Maret, W.; Vallee, B.L. The ATP-metallothionein complex. *Proc. Natl. Acad. Sci. USA* **1998**, *95*, 9146–9149. [CrossRef] [PubMed]

34. Maret, W. Oxidative metal release from metallothionein via zinc-thiol/disulfide interchange. *Proc. Natl. Acad. Sci. USA* **1994**, *91*, 237–241. [CrossRef] [PubMed]

35. Ding, Z.C.; Ni, F.Y.; Huang, Z.X. Neuronal growth-inhibitory factor (metallothionein-3): Structure-function relationships. *FEBS J.* **2010**, *277*, 2912–2920. [CrossRef] [PubMed]

36. Wang, H.; Zhang, Q.; Cai, B.; Li, H.; Sze, K.H.; Huang, Z.X.; Wu, H.M.; Sun, H. Solution structure and dynamics of human metallothionein-3 (MT-3). *FEBS Lett.* **2006**, *580*, 795–800. [CrossRef] [PubMed]

37. Jacob, C.; Maret, W.; Vallee, B.L. Control of zinc transfer between thionein, metallothionein, and zinc proteins. *Proc. Natl. Acad. Sci. USA* **1998**, *95*, 3489–3494. [CrossRef] [PubMed]

38. Palumaa, P.; Eriste, E.; Njunkova, O.; Pokras, L.; Jornvall, H.; Sillard, R. Brain-specific metallothionein-3 has higher metal-binding capacity than ubiquitous metallothioneins and binds metals noncooperatively. *Biochemistry* **2002**, *41*, 6158–6163. [CrossRef] [PubMed]

39. Tio, L.; Villarreal, L.; Atrian, S.; Capdevila, M. Functional differentiation in the mammalian metallothionein gene family: Metal binding features of mouse MT4 and comparison with its paralog MT1. *J. Biol. Chem.* **2004**, *279*, 24403–24413. [CrossRef] [PubMed]

40. Andrews, G.K. Cellular zinc sensors: MTF-1 regulation of gene expression. *Biometals* **2001**, *14*, 223–237. [CrossRef] [PubMed]

41. Gunther, V.; Lindert, U.; Schaffner, W. The taste of heavy metals: Gene regulation by MTF-1. *Biochim. Biophys. Acta* **2012**, *1823*, 1416–1425. [CrossRef] [PubMed]

42. Cherian, M.G.; Apostolova, M.D. Nuclear localization of metallothionein during cell proliferation and differentiation. *Cell. Mol. Biol.* **2000**, *46*, 347–356. [PubMed]

43. Ye, B.; Maret, W.; Vallee, B.L. Zinc metallothionein imported into liver mitochondria modulates respiration. *Proc. Natl. Acad. Sci. USA* **2001**, *98*, 2317–2322. [CrossRef] [PubMed]

44. Subramanian Vignesh, K.; Landero Figueroa, J.A.; Porollo, A.; Caruso, J.A.; Deepe, G.S., Jr. Granulocyte macrophage-colony stimulating factor induced zn sequestration enhances macrophage superoxide and limits intracellular pathogen survival. *Immunity* **2013**, *39*, 697–710. [CrossRef] [PubMed]

45. Lee, S.J.; Koh, J.Y. Roles of zinc and metallothionein-3 in oxidative stress-induced lysosomal dysfunction, cell death, and autophagy in neurons and astrocytes. *Mol. Brain* **2010**, *3*, 30. [CrossRef] [PubMed]

46. Qin, Y.; Dittmer, P.J.; Park, J.G.; Jansen, K.B.; Palmer, A.E. Measuring steady-state and dynamic endoplasmic reticulum and golgi $Zn^{2+}$ with genetically encoded sensors. *Proc. Natl. Acad. Sci. USA* **2011**, *108*, 7351–7356. [CrossRef] [PubMed]

47. Wellenreuther, G.; Cianci, M.; Tucoulou, R.; Meyer-Klaucke, W.; Haase, H. The ligand environment of zinc stored in vesicles. *Biochem. Biophys. Res. Commun.* **2009**, *380*, 198–203. [CrossRef] [PubMed]

48. Levadoux, M.; Mahon, C.; Beattie, J.H.; Wallace, H.M.; Hesketh, J.E. Nuclear import of metallothionein requires its mrna to be associated with the perinuclear cytoskeleton. *J. Biol. Chem.* **1999**, *274*, 34961–34966. [CrossRef] [PubMed]

49. El Ghazi, I.; Martin, B.L.; Armitage, I.M. New proteins found interacting with brain metallothionein-3 are linked to secretion. *Int. J. Alzheimers's Dis.* **2010**, *2011*, 208634. [CrossRef] [PubMed]

50. Lynes, M.A.; Zaffuto, K.; Unfricht, D.W.; Marusov, G.; Samson, J.S.; Yin, X. The physiological roles of extracellular metallothionein. *Exp. Biol. Med. (Maywood)* **2006**, *231*, 1548–1554. [CrossRef] [PubMed]

51. Youn, J.; Borghesi, L.A.; Olson, E.A.; Lynes, M.A. Immunomodulatory activities of extracellular metallothionein. II. Effects on macrophage functions. *J. Toxicol. Environ. Health* **1995**, *45*, 397–413. [CrossRef] [PubMed]

52. Lynes, M.A.; Borghesi, L.A.; Youn, J.; Olson, E.A. Immunomodulatory activities of extracellular metallothionein. I. Metallothionein effects on antibody production. *Toxicology* **1993**, *85*, 161–177. [CrossRef]

53. Lynes, M.A.; Garvey, J.S.; Lawrence, D.A. Extracellular metallothionein effects on lymphocyte activities. *Mol. Immunol.* **1990**, *27*, 211–219. [CrossRef]

54. Yin, X.; Knecht, D.A.; Lynes, M.A. Metallothionein mediates leukocyte chemotaxis. *BMC Immunol.* **2005**, *6*, 21. [CrossRef] [PubMed]

55. Fraker, P.J.; Gershwin, M.E.; Good, R.A.; Prasad, A. Interrelationships between zinc and immune function. *Fed. Proc.* **1986**, *45*, 1474–1479. [PubMed]

56. Prasad, A.S. Effects of zinc deficiency on TH1 and TH2 cytokine shifts. *J. Infect. Dis.* **2000**, *182* (Suppl. 1), S62–S68. [CrossRef] [PubMed]

57. Shankar, A.H.; Prasad, A.S. Zinc and immune function: The biological basis of altered resistance to infection. *Am. J. Clin. Nutr.* **1998**, *68*, 447S–463S. [PubMed]

58. Haase, H.; Rink, L. Zinc signals and immune function. *Biofactors* **2014**, *40*, 27–40. [CrossRef] [PubMed]

59. Rink, L.; Gabriel, P. Zinc and the immune system. *Proc. Nutr. Soc.* **2000**, *59*, 541–552. [CrossRef] [PubMed]

60. Chandra, R. Excessive intake of zinc impairs immune responses. *JAMA* **1984**, *252*, 1443–1446. [CrossRef] [PubMed]

61. Olafson, R.W. Thymus metallothionein: Regulation of zinc-thionein in the aging mouse. *Can. J. Biochem. Cell Biol.* **1985**, *63*, 91–95. [CrossRef] [PubMed]

62. Mocchegiani, E.; Giacconi, R.; Cipriano, C.; Muti, E.; Gasparini, N.; Malavolta, M. Are zinc-bound metallothionein isoforms (i + ii and iii) involved in impaired thymulin production and thymic involution during ageing? *Immun. Ageing* **2004**, *1*, 5. [CrossRef] [PubMed]

63. Bach, J.F.; Dardenne, M. Thymulin, a zinc-dependent hormone. *Med. Oncol. Tumor Pharmacother.* **1989**, *6*, 25–29. [PubMed]

64. Savino, W.; Huang, P.C.; Corrigan, A.; Berrih, S.; Dardenne, M. Thymic hormone-containing cells. V. Immunohistological detection of metallothionein within the cells bearing thymulin (a zinc-containing hormone) in human and mouse thymuses. *J. Histochem. Cytochem.* **1984**, *32*, 942–946. [CrossRef] [PubMed]

65. Kuo, T.; Lo, S.K. Immunohistochemical metallothionein expression in thymoma: Correlation with histological types and cellular origin. *Histopathology* **1997**, *30*, 243–248. [CrossRef] [PubMed]

66. George, M.M.; Subramanian Vignesh, K.; Landero Figueroa, J.A.; Caruso, J.A.; Deepe, G.S. Zinc induces dendritic cell tolerogenic phenotype and skews regulatory T cell–TH17 balance. *J. Immunol.* **2016**. [CrossRef] [PubMed]

67. Kitamura, H.; Morikawa, H.; Kamon, H.; Iguchi, M.; Hojyo, S.; Fukada, T.; Yamashita, S.; Kaisho, T.; Akira, S.; Murakami, M.; et al. Toll-like receptor-mediated regulation of zinc homeostasis influences dendritic cell function. *Nat. Immunol.* **2006**, *7*, 971–977. [CrossRef] [PubMed]

68. Weinlich, G.; Topar, G.; Eisendle, K.; Fritsch, P.O.; Zelger, B. Comparison of metallothionein-overexpression with sentinel lymph node biopsy as prognostic factors in melanoma. *J. Eur. Acad. Dermatol. Venereol.* **2007**, *21*, 669–677. [CrossRef] [PubMed]

69. Haerslev, T.; Jacobsen, G.K.; Zedeler, K. The prognostic significance of immunohistochemically detectable metallothionein in primary breast carcinomas. *APMIS* **1995**, *103*, 279–285. [CrossRef] [PubMed]

70. Kondo, M. Lymphoid and myeloid lineage commitment in multipotent hematopoietic progenitors. *Immunol. Rev.* **2010**, *238*, 37–46. [CrossRef] [PubMed]

71. King, L.E.; Osati-Ashtiani, F.; Fraker, P.J. Depletion of cells of the b lineage in the bone marrow of zinc-deficient mice. *Immunology* **1995**, *85*, 69–73. [PubMed]

72. Fraker, P.J.; King, L.E. Reprogramming of the immune system during zinc deficiency. *Annu. Rev. Nutr.* **2004**, *24*, 277–298. [CrossRef] [PubMed]

73. Huber, K.L.; Cousins, R.J. Metallothionein expression in rat bone marrow is dependent on dietary zinc but not dependent on interleukin-1 or interleukin-6. *J. Nutr.* **1993**, *123*, 642–648. [PubMed]

74. Krishnaraju, K.; Nguyen, H.Q.; Liebermann, D.A.; Hoffman, B. The zinc finger transcription factor EGR-1 potentiates macrophage differentiation of hematopoietic cells. *Mol. Cell. Biol.* **1995**, *15*, 5499–5507. [CrossRef] [PubMed]

75. Hock, H.; Hamblen, M.J.; Rooke, H.M.; Traver, D.; Bronson, R.T.; Cameron, S.; Orkin, S.H. Intrinsic requirement for zinc finger transcription factor GFI-1 in neutrophil differentiation. *Immunity* **2003**, *18*, 109–120. [CrossRef]

76. Feng, W.; Cai, J.; Pierce, W.M.; Franklin, R.B.; Maret, W.; Benz, F.W.; Kang, Y.J. Metallothionein transfers zinc to mitochondrial aconitase through a direct interaction in mouse hearts. *Biochem. Biophys. Res. Commun.* **2005**, *332*, 853–858. [CrossRef] [PubMed]

77. Jiang, L.J.; Maret, W.; Vallee, B.L. The glutathione redox couple modulates zinc transfer from metallothionein to zinc-depleted sorbitol dehydrogenase. *Proc. Natl. Acad. Sci. USA* **1998**, *95*, 3483–3488. [CrossRef] [PubMed]

78. Roesijadi, G.; Bogumil, R.; Vasak, M.; Kagi, J.H. Modulation of DNA binding of a tramtrack zinc finger peptide by the metallothionein-thionein conjugate pair. *J. Biol. Chem.* **1998**, *273*, 17425–17432. [CrossRef] [PubMed]

79. McCormick, N.H.; Kelleher, S.L. Znt4 provides zinc to zinc-dependent proteins in the trans-golgi network critical for cell function and zn export in mammary epithelial cells. *Am. J. Physiol. Cell Physiol.* **2012**, *303*, C291–C297. [CrossRef] [PubMed]

80. Cousins, R.J. A role of zinc in the regulation of gene expression. *Proc. Nutr. Soc.* **1998**, *57*, 307–311. [CrossRef] [PubMed]

81. Song, Y.; Leonard, S.W.; Traber, M.G.; Ho, E. Zinc deficiency affects DNA damage, oxidative stress, antioxidant defenses, and DNA repair in rats. *J. Nutr.* **2009**, *139*, 1626–1631. [CrossRef] [PubMed]

82. Sato, M.; Suzuki, S. Endoplasmic reticulum stress and metallothionein. *Yakugaku Zasshi* **2007**, *127*, 703–708. [CrossRef] [PubMed]

83. Crowthers, K.C.; Kline, V.; Giardina, C.; Lynes, M.A. Augmented humoral immune function in metallothionein-null mice. *Toxicol. Appl. Pharmacol.* **2000**, *166*, 161–172. [CrossRef] [PubMed]

84. Mita, M.; Imura, N.; Kumazawa, Y.; Himeno, S. Suppressed proliferative response of spleen T cells from metallothionein null mice. *Microbiol. Immunol.* **2002**, *46*, 101–107. [CrossRef] [PubMed]

85. Jackson, S.H.; Devadas, S.; Kwon, J.; Pinto, L.A.; Williams, M.S. T cells express a phagocyte-type nadph oxidase that is activated after t cell receptor stimulation. *Nat. Immunol.* **2004**, *5*, 818–827. [CrossRef] [PubMed]

86. Ghoshal, K.; Majumder, S.; Zhu, Q.; Hunzeker, J.; Datta, J.; Shah, M.; Sheridan, J.F.; Jacob, S.T. Influenza virus infection induces metallothionein gene expression in the mouse liver and lung by overlapping but distinct molecular mechanisms. *Mol. Cell. Biol.* **2001**, *21*, 8301–8317. [CrossRef] [PubMed]

87. Ganz, T. Macrophages and systemic iron homeostasis. *J. Innate Immun.* **2012**, *4*, 446–453. [CrossRef] [PubMed]

88. Kojima, N.; Young, C.R.; Bates, G.W. Failure of metallothionein to bind iron or act as an iron mobilizing agent. *Biochim. Biophys. Acta* **1982**, *716*, 273–275. [CrossRef]

89. Robertson, A.; Morrison, J.N.; Wood, A.M.; Bremner, I. Effects of iron deficiency on metallothionein-I concentrations in blood and tissues of rats. *J. Nutr.* **1989**, *119*, 439–445. [PubMed]

90. Lonnerdal, B. Dietary factors influencing zinc absorption. *J. Nutr.* **2000**, *130*, 1378S–1383S. [PubMed]

91. Rossander-Hulten, L.; Brune, M.; Sandstrom, B.; Lonnerdal, B.; Hallberg, L. Competitive inhibition of iron absorption by manganese and zinc in humans. *Am. J. Clin. Nutr.* **1991**, *54*, 152–156. [PubMed]

92. Morgan, M.J.; Liu, Z.G. Crosstalk of reactive oxygen species and NF-κB signaling. *Cell Res.* **2011**, *21*, 103–115. [CrossRef] [PubMed]

93. Ruttkay-Nedecky, B.; Nejdl, L.; Gumulec, J.; Zitka, O.; Masarik, M.; Eckschlager, T.; Stiborova, M.; Adam, V.; Kizek, R. The role of metallothionein in oxidative stress. *Int. J. Mol. Sci.* **2013**, *14*, 6044–6066. [CrossRef] [PubMed]

94. Miura, T.; Muraoka, S.; Ogiso, T. Antioxidant activity of metallothionein compared with reduced glutathione. *Life Sci.* **1997**, *60*, PL 301–PL 309. [CrossRef]

95. Liu, M.J.; Bao, S.; Galvez-Peralta, M.; Pyle, C.J.; Rudawsky, A.C.; Pavlovicz, R.E.; Killilea, D.W.; Li, C.; Nebert, D.W.; Wewers, M.D.; et al. Zip8 regulates host defense through zinc-mediated inhibition of NF-κB. *Cell Rep.* **2013**, *3*, 386–400. [CrossRef] [PubMed]

96. Butcher, H.L.; Kennette, W.A.; Collins, O.; Zalups, R.K.; Koropatnick, J. Metallothionein mediates the level and activity of nuclear factor kappa b in murine fibroblasts. *J. Pharmacol. Exp. Ther.* **2004**, *310*, 589–598. [CrossRef] [PubMed]

97. Friedberg, S.H.; Weissman, I.L. Lymphoid tissue architecture. II. Ontogeny of peripheral T and B cells in mice: Evidence against peyer's patches as the site of generation of B cells. *J. Immunol.* **1974**, *113*, 1477–1492. [PubMed]

98. Gutman, G.A.; Weissman, I.L. Lymphoid tissue architecture. Experimental analysis of the origin and distribution of T-cells and B-cells. *Immunology* **1972**, *23*, 465–479. [PubMed]

99. Weissman, I.L.; Gutman, G.A.; Friedberg, S.H.; Jerabek, L. Lymphoid tissue architecture. III. Germinal centers, T cells, and thymus-dependent vs. thymus-independent antigens. *Adv. Exp. Med. Biol.* **1976**, *66*, 229–237. [PubMed]

100. Kataru, R.P.; Lee, Y.G.; Koh, G.Y. Interactions of immune cells and lymphatic vessels. *Adv. Anat. Embryol. Cell Biol.* **2014**, *214*, 107–118. [PubMed]

101. Vallee, B.L.; Gibson, J.G., II. The zinc content of normal human whole blood, plasma, leucocytes, and erythrocytes. *J. Biol. Chem.* **1948**, *176*, 445–457. [PubMed]

102. Schrodt, G.R.; Hall, T.; Whitmore, W.F., Jr. The concentration of zinc in diseased human prostate glands. *Cancer* **1964**, *17*, 1555–1566. [CrossRef]

103. De, S.K.; McMaster, M.T.; Andrews, G.K. Endotoxin induction of murine metallothionein gene expression. *J. Biol. Chem.* **1990**, *265*, 15267–15274. [PubMed]

104. Subramanian Vignesh, K.; Landero Figueroa, J.A.; Porollo, A.; Divanovic, S.; Caruso, J.A.; Deepe, G.S., Jr. Interleukin-4 induces metallothionein 3- and SLC30A4-dependent increase in intracellular $Zn^{2+}$ that promotes pathogen persistence in macrophages. *Cell Rep.* **2016**, *16*, 1–15. [CrossRef] [PubMed]

105. Wu, C.; Pot, C.; Apetoh, L.; Thalhamer, T.; Zhu, B.; Murugaiyan, G.; Xiao, S.; Lee, Y.; Rangachari, M.; Yosef, N.; et al. Metallothioneins negatively regulate IL-27-induced type 1 regulatory T-cell differentiation. *Proc. Natl. Acad. Sci. USA* **2013**, *110*, 7802–7807. [CrossRef] [PubMed]

106. Kanekiyo, M.; Itoh, N.; Kawasaki, A.; Matsuyama, A.; Matsuda, K.; Nakanishi, T.; Tanaka, K. Metallothionein modulates lipopolysaccharide-stimulated tumour necrosis factor expression in mouse peritoneal macrophages. *Biochem. J.* **2002**, *361*, 363–369. [CrossRef] [PubMed]

107. Sugiura, T.; Kuroda, E.; Yamashita, U. Dysfunction of macrophages in metallothionein-knock out mice. *J. UOEH* **2004**, *26*, 193–205. [CrossRef] [PubMed]

108. Shortman, K.; Liu, Y.J. Mouse and human dendritic cell subtypes. *Nat. Rev. Immunol.* **2002**, *2*, 151–161. [CrossRef] [PubMed]

109. Maldonado, R.A.; von Andrian, U.H. How tolerogenic dendritic cells induce regulatory T cells. *Adv. Immunol.* **2010**, *108*, 111–165. [PubMed]

110. Reis e Sousa, C. Dendritic cells in a mature age. *Nat. Rev. Immunol.* **2006**, *6*, 476–483. [CrossRef] [PubMed]

111. Hashimoto, S.I.; Suzuki, T.; Nagai, S.; Yamashita, T.; Toyoda, N.; Matsushima, K. Identification of genes specifically expressed in human activated and mature dendritic cells through serial analysis of gene expression. *Blood* **2000**, *96*, 2206–2214. [PubMed]

112. Spiering, R.; Wagenaar-Hilbers, J.; Huijgen, V.; van der Zee, R.; van Kooten, P.J.; van Eden, W.; Broere, F. Membrane-bound metallothionein 1 of murine dendritic cells promotes the expansion of regulatory T cells in vitro. *Toxicol. Sci.* **2014**, *138*, 69–75. [CrossRef] [PubMed]

113. Spiering, R.; van der Zee, R.; Wagenaar, J.; Kapetis, D.; Zolezzi, F.; van Eden, W.; Broere, F. Tolerogenic dendritic cells that inhibit autoimmune arthritis can be induced by a combination of carvacrol and thermal stress. *PLoS ONE* **2012**, *7*, e46336. [CrossRef] [PubMed]

114. Yu, M.; Lee, W.W.; Tomar, D.; Pryshchep, S.; Czesnikiewicz-Guzik, M.; Lamar, D.L.; Li, G.; Singh, K.; Tian, L.; Weyand, C.M.; et al. Regulation of T cell receptor signaling by activation-induced zinc influx. *J. Exp. Med.* **2011**, *208*, 775–785. [CrossRef] [PubMed]

115. Borghesi, L.A.; Youn, J.; Olson, E.A.; Lynes, M.A. Interactions of metallothionein with murine lymphocytes: Plasma membrane binding and proliferation. *Toxicology* **1996**, *108*, 129–140. [CrossRef]

116. Rosenkranz, E.; Metz, C.H.; Maywald, M.; Hilgers, R.D.; Wessels, I.; Senff, T.; Haase, H.; Jager, M.; Ott, M.; Aspinall, R.; et al. Zinc supplementation induces regulatory T cells by inhibition of SIRT-1 deacetylase in mixed lymphocyte cultures. *Mol. Nutr. Food Res.* **2015**. [CrossRef] [PubMed]

117. Rice, J.M.; Zweifach, A.; Lynes, M.A. Metallothionein regulates intracellular zinc signaling during CD4(+) T cell activation. *BMC Immunol.* **2016**, *17*, 13. [CrossRef] [PubMed]

118. Huh, S.; Lee, K.; Yun, H.S.; Paik, D.J.; Kim, J.M.; Youn, J. Functions of metallothionein generating interleukin-10-producing regulatory CD4$^+$ t cells potentiate suppression of collagen-induced arthritis. *J. Microbiol. Biotechnol.* **2007**, *17*, 348–358. [PubMed]

119. Haase, H.; Rink, L. Functional significance of zinc-related signaling pathways in immune cells. *Annu. Rev. Nutr.* **2009**, *29*, 133–152. [CrossRef] [PubMed]

120. Brockmann, L.; Gagliani, N.; Steglich, B.; Giannou, A.D.; Kempski, J.; Pelczar, P.; Geffken, M.; Mfarrej, B.; Huber, F.; Herkel, J.; et al. IL-10 receptor signaling is essential for TR1 cell function in vivo. *J. Immunol.* **2017**, *198*, 1130–1141. [CrossRef] [PubMed]

121. Youn, J.; Lynes, M.A. Metallothionein-induced suppression of cytotoxic T lymphocyte function: An important immunoregulatory control. *Toxicol. Sci.* **1999**, *52*, 199–208. [CrossRef] [PubMed]

122. Beck, F.W.; Prasad, A.S.; Kaplan, J.; Fitzgerald, J.T.; Brewer, G.J. Changes in cytokine production and T cell subpopulations in experimentally induced zinc-deficient humans. *Am. J. Physiol.* **1997**, *272*, E1002–E1007. [PubMed]

123. Fernandes, G.; Nair, M.; Onoe, K.; Tanaka, T.; Floyd, R.; Good, R.A. Impairment of cell-mediated immunity functions by dietary zinc deficiency in mice. *Proc. Natl. Acad. Sci. USA* **1979**, *76*, 457–461. [CrossRef] [PubMed]

124. Lee, S.J.; Cho, K.S.; Kim, H.N.; Kim, H.J.; Koh, J.Y. Role of zinc metallothionein-3 (ZNMT3) in epidermal growth factor (EGF)-induced C-ABL protein activation and actin polymerization in cultured astrocytes. *J. Biol. Chem.* **2011**, *286*, 40847–40856. [CrossRef] [PubMed]

125. El Refaey, H.; Ebadi, M.; Kuszynski, C.A.; Sweeney, J.; Hamada, F.M.; Hamed, A. Identification of metallothionein receptors in human astrocytes. *Neurosci. Lett.* **1997**, *231*, 131–134. [CrossRef]

126. Stitt, M.S.; Wasserloos, K.J.; Tang, X.; Liu, X.; Pitt, B.R.; St Croix, C.M. Nitric oxide-induced nuclear translocation of the metal responsive transcription factor, MTF-1 is mediated by zinc release from metallothionein. *Vasc. Pharmacol.* **2006**, *44*, 149–155. [CrossRef] [PubMed]

127. Lee, D.K.; Carrasco, J.; Hidalgo, J.; Andrews, G.K. Identification of a signal transducer and activator of transcription (STAT) binding site in the mouse metallothionein-i promoter involved in interleukin-6-induced gene expression. *Biochem. J.* **1999**, *337 Pt 1*, 59–65. [CrossRef] [PubMed]
128. Yamashita, S.; Miyagi, C.; Fukada, T.; Kagara, N.; Che, Y.S.; Hirano, T. Zinc transporter livi controls epithelial-mesenchymal transition in zebrafish gastrula organizer. *Nature* **2004**, *429*, 298–302. [CrossRef] [PubMed]
129. Minakami, R.; Sumimotoa, H. Phagocytosis-coupled activation of the superoxide-producing phagocyte oxidase, a member of the nadph oxidase (NOX) family. *Int. J. Hematol.* **2006**, *84*, 193–198. [CrossRef] [PubMed]
130. Munoz-Fernandez, M.A.; Fernandez, M.A.; Fresno, M. Activation of human macrophages for the killing of intracellular trypanosoma cruzi by TNF-α and IFN-γ through a nitric oxide-dependent mechanism. *Immunol. Lett.* **1992**, *33*, 35–40. [CrossRef]
131. Wolf, J.E.; Massof, S.E. In vivo activation of macrophage oxidative burst activity by cytokines and amphotericin B. *Infect. Immun.* **1990**, *58*, 1296–1300. [PubMed]
132. Segal, A.W. How superoxide production by neutrophil leukocytes kills microbes. *Novartis Found. Symp.* **2006**, *279*, 92–98. [PubMed]
133. Corbin, B.D.; Seeley, E.H.; Raab, A.; Feldmann, J.; Miller, M.R.; Torres, V.J.; Anderson, K.L.; Dattilo, B.M.; Dunman, P.M.; Gerads, R.; et al. Metal chelation and inhibition of bacterial growth in tissue abscesses. *Science* **2008**, *319*, 962–965. [CrossRef] [PubMed]
134. Hasan, R.; Rink, L.; Haase, H. Zinc signals in neutrophil granulocytes are required for the formation of neutrophil extracellular traps. *Innate Immun.* **2013**, *19*, 253–264. [CrossRef] [PubMed]
135. Andrews, G.K. Regulation of metallothionein gene expression by oxidative stress and metal ions. *Biochem. Pharmacol.* **2000**, *59*, 95–104. [CrossRef]
136. Kelly, E.J.; Sandgren, E.P.; Brinster, R.L.; Palmiter, R.D. A pair of adjacent glucocorticoid response elements regulate expression of two mouse metallothionein genes. *Proc. Natl. Acad. Sci. USA* **1997**, *94*, 10045–10050. [CrossRef] [PubMed]
137. Franchimont, D. Overview of the actions of glucocorticoids on the immune response: A good model to characterize new pathways of immunosuppression for new treatment strategies. *Ann. N. Y. Acad. Sci.* **2004**, *1024*, 124–137. [CrossRef] [PubMed]
138. Reeve, V.E.; Nishimura, N.; Bosnic, M.; Michalska, A.E.; Choo, K.H. Lack of metallothionein-I and -II exacerbates the immunosuppressive effect of ultraviolet B radiation and cis-urocanic acid in mice. *Immunology* **2000**, *100*, 399–404. [CrossRef] [PubMed]
139. Palmiter, R.D.; Findley, S.D.; Whitmore, T.E.; Durnam, D.M. Mt-iii, a brain-specific member of the metallothionein gene family. *Proc. Natl. Acad. Sci. USA* **1992**, *89*, 6333–6337. [CrossRef] [PubMed]
140. Hozumi, I.; Suzuki, J.S.; Kanazawa, H.; Hara, A.; Saio, M.; Inuzuka, T.; Miyairi, S.; Naganuma, A.; Tohyama, C. Metallothionein-3 is expressed in the brain and various peripheral organs of the rat. *Neurosci. Lett.* **2008**, *438*, 54–58. [CrossRef] [PubMed]
141. Slusser, A.; Zheng, Y.; Zhou, X.D.; Somji, S.; Sens, D.A.; Sens, M.A.; Garrett, S.H. Metallothionein isoform 3 expression in human skin, related cancers and human skin derived cell cultures. *Toxicol. Lett.* **2014**, *232*, 141–148. [CrossRef] [PubMed]
142. Tao, Y.F.; Xu, L.X.; Lu, J.; Cao, L.; Li, Z.H.; Hu, S.Y.; Wang, N.N.; Du, X.J.; Sun, L.C.; Zhao, W.L.; et al. Metallothionein III (MT3) is a putative tumor suppressor gene that is frequently inactivated in pediatric acute myeloid leukemia by promoter hypermethylation. *J. Transl. Med.* **2014**, *12*, 182. [CrossRef] [PubMed]
143. Somji, S.; Garrett, S.H.; Zhou, X.D.; Zheng, Y.; Sens, D.A.; Sens, M.A. Absence of metallothionein 3 expression in breast cancer is a rare, but favorable marker of outcome that is under epigenetic control. *Toxicol. Environ. Chem.* **2010**, *92*, 1673–1695. [CrossRef] [PubMed]
144. Peng, D.; Hu, T.L.; Jiang, A.; Washington, M.K.; Moskaluk, C.A.; Schneider-Stock, R.; El-Rifai, W. Location-specific epigenetic regulation of the metallothionein 3 gene in esophageal adenocarcinomas. *PLoS ONE* **2011**, *6*, e22009. [CrossRef] [PubMed]
145. Haq, F.; Mahoney, M.; Koropatnick, J. Signaling events for metallothionein induction. *Mutat. Res.* **2003**, *533*, 211–226. [CrossRef] [PubMed]

146. Majumder, S.; Kutay, H.; Datta, J.; Summers, D.; Jacob, S.T.; Ghoshal, K. Epigenetic regulation of metallothionein-I gene expression: Differential regulation of methylated and unmethylated promoters by DNA methyltransferases and methyl cpg binding proteins. *J. Cell. Biochem.* **2006**, *97*, 1300–1316. [CrossRef] [PubMed]

147. Busslinger, M.; Tarakhovsky, A. Epigenetic control of immunity. *Cold Spring Harb. Perspect. Biol.* **2014**, *6*. [CrossRef]

148. Hood, M.I.; Skaar, E.P. Nutritional immunity: Transition metals at the pathogen-host interface. *Nat. Rev. Microbiol.* **2012**, *10*, 525–537. [CrossRef] [PubMed]

149. Rada, B.K.; Geiszt, M.; Kaldi, K.; Timar, C.; Ligeti, E. Dual role of phagocytic nadph oxidase in bacterial killing. *Blood* **2004**, *104*, 2947–2953. [CrossRef] [PubMed]

150. Botella, H.; Peyron, P.; Levillain, F.; Poincloux, R.; Poquet, Y.; Brandli, I.; Wang, C.; Tailleux, L.; Tilleul, S.; Charriere, G.M.; et al. Mycobacterial P(1)-type atpases mediate resistance to zinc poisoning in human macrophages. *Cell Host Microbe* **2011**, *10*, 248–259. [CrossRef] [PubMed]

151. Nagamine, T.; Suzuki, K.; Kondo, T.; Nakazato, K.; Kakizaki, S.; Takagi, H.; Nakajima, K. Interferon-$\alpha$-induced changes in metallothionein expression in liver biopsies from patients with chronic hepatitis C. *Can. J. Gastroenterol.* **2005**, *19*, 481–486. [CrossRef] [PubMed]

152. Wong, H.R.; Shanley, T.P.; Sakthivel, B.; Cvijanovich, N.; Lin, R.; Allen, G.L.; Thomas, N.J.; Doctor, A.; Kalyanaraman, M.; Tofil, N.M.; et al. Genome-level expression profiles in pediatric septic shock indicate a role for altered zinc homeostasis in poor outcome. *Physiol. Genom.* **2007**, *30*, 146–155. [CrossRef] [PubMed]

153. Kominsky, D.J.; Campbell, E.L.; Colgan, S.P. Metabolic shifts in immunity and inflammation. *J. Immunol.* **2010**, *184*, 4062–4068. [CrossRef] [PubMed]

154. Sobocinski, P.Z.; Canterbury, W.J., Jr. Hepatic metallothionein induction in inflammation. *Ann. N. Y. Acad. Sci.* **1982**, *389*, 354–367. [CrossRef] [PubMed]

155. Aydemir, T.B.; Chang, S.M.; Guthrie, G.J.; Maki, A.B.; Ryu, M.S.; Karabiyik, A.; Cousins, R.J. Zinc transporter ZIP14 functions in hepatic zinc, iron and glucose homeostasis during the innate immune response (endotoxemia). *PLoS ONE* **2012**, *7*, e48679.

156. Sobocinski, P.Z.; Canterbury, W.J., Jr.; Mapes, C.A.; Dinterman, R.E. Involvement of hepatic metallothioneins in hypozincemia associated with bacterial infection. *Am. J. Physiol.* **1978**, *234*, E399–E406. [PubMed]

157. Knoell, D.L.; Julian, M.W.; Bao, S.; Besecker, B.; Macre, J.E.; Leikauf, G.D.; DiSilvestro, R.A.; Crouser, E.D. Zinc deficiency increases organ damage and mortality in a murine model of polymicrobial sepsis. *Crit. Care Med.* **2009**, *37*, 1380–1388. [CrossRef] [PubMed]

158. Kogan, S.; Sood, A.; Garnick, M.S. Zinc and wound healing: A review of zinc physiology and clinical applications. *Wounds* **2017**, *29*, 102–106. [PubMed]

159. Mita, M.; Satoh, M.; Shimada, A.; Okajima, M.; Azuma, S.; Suzuki, J.S.; Sakabe, K.; Hara, S.; Himeno, S. Metallothionein is a crucial protective factor against helicobacter pylori-induced gastric erosive lesions in a mouse model. *Am. J. Physiol. Gastrointest. Liver Physiol.* **2008**, *294*, G877–G884. [CrossRef] [PubMed]

160. Emeny, R.T.; Marusov, G.; Lawrence, D.A.; Pederson-Lane, J.; Yin, X.; Lynes, M.A. Manipulations of metallothionein gene dose accelerate the response to *Listeria monocytogenes*. *Chem. Biol. Interact.* **2009**, *181*, 243–253. [CrossRef] [PubMed]

161. Carrera, G.; Paternain, J.L.; Carrere, N.; Folch, J.; Courtade-Saidi, M.; Orfila, C.; Vinel, J.P.; Alric, L.; Pipy, B. Hepatic metallothionein in patients with chronic hepatitis C: Relationship with severity of liver disease and response to treatment. *Am. J. Gastroenterol.* **2003**, *98*, 1142–1149. [PubMed]

162. O'Connor, K.S.; Parnell, G.; Patrick, E.; Ahlenstiel, G.; Suppiah, V.; van der Poorten, D.; Read, S.A.; Leung, R.; Douglas, M.W.; Yang, J.Y.; et al. Hepatic metallothionein expression in chronic hepatitis C virus infection is IFNL3 genotype-dependent. *Genes Immun.* **2014**, *15*, 88–94. [CrossRef] [PubMed]

163. Ilback, N.G.; Glynn, A.W.; Wikberg, L.; Netzel, E.; Lindh, U. Metallothionein is induced and trace element balance changed in target organs of a common viral infection. *Toxicology* **2004**, *199*, 241–250. [CrossRef] [PubMed]

164. Lazarczyk, M.; Favre, M. Role of $Zn^{2+}$ ions in host-virus interactions. *J. Virol.* **2008**, *82*, 11486–11494. [CrossRef] [PubMed]

165. Suara, R.O.; Crowe, J.E., Jr. Effect of zinc salts on respiratory syncytial virus replication. *Antimicrob. Agents Chemother.* **2004**, *48*, 783–790. [CrossRef] [PubMed]

166. Cancellieri, M.; Bassetto, M.; Widjaja, I.; van Kuppeveld, F.; de Haan, C.A.; Brancale, A. In silico structure-based design and synthesis of novel anti-rsv compounds. *Antivir. Res.* **2015**, *122*, 46–50. [CrossRef] [PubMed]

167. Hulisz, D. Efficacy of zinc against common cold viruses: An overview. *J. Am. Pharm. Assoc.* **2004**, *44*, 594–603. [CrossRef]

168. Krenn, B.M.; Gaudernak, E.; Holzer, B.; Lanke, K.; Van Kuppeveld, F.J.; Seipelt, J. Antiviral activity of the zinc ionophores pyrithione and hinokitiol against picornavirus infections. *J. Virol.* **2009**, *83*, 58–64. [CrossRef] [PubMed]

169. Sera, T. Inhibition of virus DNA replication by artificial zinc finger proteins. *J. Virol.* **2005**, *79*, 2614–2619. [CrossRef] [PubMed]

170. Allendoerfer, R.; Deepe, G.S., Jr. Intrapulmonary response to *Histoplasma capsulatum* in gamma interferon knockout mice. *Infect. Immun.* **1997**, *65*, 2564–2569. [PubMed]

171. Haase, H. An element of life: Competition for zinc in host-pathogen interaction. *Immunity* **2013**, *39*, 623–624. [CrossRef] [PubMed]

172. Subramanian Vignesh, K.; Landero Figueroa, J.A.; Porollo, A.; Caruso, J.A.; Deepe, G.S., Jr. Zinc sequestration: Arming phagocyte defense against fungal attack. *PLoS Pathog.* **2013**, *9*, e1003815.

173. Wu, A.; Tymoszuk, P.; Haschka, D.; Heeke, S.; Dichtl, S.; Petzer, V.; Seifert, M.; Hilbe, R.; Sopper, S.; Talasz, H.; et al. Salmonella utilizes zinc to subvert anti-microbial host defense of macrophages via modulation of nf-kappab signaling. *Infect. Immun.* **2017**. [CrossRef] [PubMed]

174. Liu, J.Z.; Jellbauer, S.; Poe, A.J.; Ton, V.; Pesciaroli, M.; Kehl-Fie, T.E.; Restrepo, N.A.; Hosking, M.P.; Edwards, R.A.; Battistoni, A.; et al. Zinc sequestration by the neutrophil protein calprotectin enhances *salmonella* growth in the inflamed gut. *Cell Host Microbe* **2012**, *11*, 227–239. [CrossRef] [PubMed]

175. Kapetanovic, R.; Bokil, N.J.; Achard, M.E.; Ong, C.Y.; Peters, K.M.; Stocks, C.J.; Phan, M.D.; Monteleone, M.; Schroder, K.; Irvine, K.M.; et al. Salmonella employs multiple mechanisms to subvert the tlr-inducible zinc-mediated antimicrobial response of human macrophages. *FASEB J.* **2016**, *30*, 1901–1912. [CrossRef] [PubMed]

176. Ding, C.; Festa, R.A.; Chen, Y.L.; Espart, A.; Palacios, O.; Espin, J.; Capdevila, M.; Atrian, S.; Heitman, J.; Thiele, D.J. *Cryptococcus neoformans* copper detoxification machinery is critical for fungal virulence. *Cell Host Microbe* **2013**, *13*, 265–276. [CrossRef] [PubMed]

177. Szymczak, W.A.; Deepe, G.S., Jr. The CCL7-CCL2-CCR2 axis regulates IL-4 production in lungs and fungal immunity. *J. Immunol.* **2009**, *183*, 1964–1974. [CrossRef] [PubMed]

178. Verma, A.; Kroetz, D.N.; Tweedle, J.L.; Deepe, G.S., Jr. Type II cytokines impair host defense against an intracellular fungal pathogen by amplifying macrophage generation of IL-33. *Mucosal. Immunol.* **2015**, *8*, 380–389. [CrossRef] [PubMed]

179. Gordon, S. Alternative activation of macrophages. *Nat. Rev. Immunol.* **2003**, *3*, 23–35. [CrossRef] [PubMed]

180. Mosser, D.M. The many faces of macrophage activation. *J. Leukoc. Biol.* **2003**, *73*, 209–212. [CrossRef] [PubMed]

181. Scott, M.E.; Koski, K.G. Zinc deficiency impairs immune responses against parasitic nematode infections at intestinal and systemic sites. *J. Nutr.* **2000**, *130*, 1412S–1420S. [PubMed]

182. Ragusa, M.A.; Nicosia, A.; Costa, S.; Cuttitta, A.; Gianguzza, F. Metallothionein gene family in the sea urchin paracentrotus lividus: Gene structure, differential expression and phylogenetic analysis. *Int. J. Mol. Sci.* **2017**, *18*, 812. [CrossRef] [PubMed]

183. Flajnik, M.F.; Kasahara, M. Origin and evolution of the adaptive immune system: Genetic events and selective pressures. *Nat. Rev. Genet.* **2010**, *11*, 47–59. [CrossRef] [PubMed]

184. Tran, C.D.; Butler, R.N.; Philcox, J.C.; Rofe, A.M.; Howarth, G.S.; Coyle, P. Regional distribution of metallothionein and zinc in the mouse gut: Comparison with metallothionien-null mice. *Biol. Trace Elem. Res.* **1998**, *63*, 239–251. [CrossRef] [PubMed]

185. Waeytens, A.; de Vos, M.; Laukens, D. Evidence for a potential role of metallothioneins in inflammatory bowel diseases. *Mediat. Inflamm.* **2009**, *2009*, 729172. [CrossRef] [PubMed]

International Journal of
*Molecular Sciences*

MDPI

*Review*

# The Impact of Synaptic Zn$^{2+}$ Dynamics on Cognition and Its Decline

**Atsushi Takeda \*** [ID] **and Hanuna Tamano**

Department of Neurophysiology, School of Pharmaceutical Sciences, University of Shizuoka, 52-1 Yada,
Suruga-ku, Shizuoka 422-8526, Japan; tamano@u-shizuoka-ken.ac.jp
\* Correspondence: takedaa@u-shizuoka-ken.ac.jp; Tel.: +81-54-264-5733; Fax: +81-54-264-5909

Received: 22 September 2017; Accepted: 9 November 2017; Published: 14 November 2017

**Abstract:** The basal levels of extracellular Zn$^{2+}$ are in the range of low nanomolar concentrations and less attention has been paid to Zn$^{2+}$, compared to Ca$^{2+}$, for synaptic activity. However, extracellular Zn$^{2+}$ is necessary for synaptic activity. The basal levels of extracellular zinc are age-dependently increased in the rat hippocampus, implying that the basal levels of extracellular Zn$^{2+}$ are also increased age-dependently and that extracellular Zn$^{2+}$ dynamics are linked with age-related cognitive function and dysfunction. In the hippocampus, the influx of extracellular Zn$^{2+}$ into postsynaptic neurons, which is often linked with Zn$^{2+}$ release from neuron terminals, is critical for cognitive activity via long-term potentiation (LTP). In contrast, the excess influx of extracellular Zn$^{2+}$ into postsynaptic neurons induces cognitive decline. Interestingly, the excess influx of extracellular Zn$^{2+}$ more readily occurs in aged dentate granule cells and intracellular Zn$^{2+}$-buffering, which is assessed with ZnAF-2DA, is weakened in the aged dentate granule cells. Characteristics (easiness) of extracellular Zn$^{2+}$ influx seem to be linked with the weakened intracellular Zn$^{2+}$-buffering in the aged dentate gyrus. This paper deals with the impact of synaptic Zn$^{2+}$ signaling on cognition and its decline in comparison with synaptic Ca$^{2+}$ signaling.

**Keywords:** Zn$^{2+}$ signaling; hippocampus; memory; Ca$^{2+}$ signaling; perforant pathway; dentate granule cell

---

## 1. Introduction

Cognitive activity has been closely linked to strengthening and weakening synaptic connections between neurons that is synaptic plasticity such as long-term potentiation (LTP) and long-term depression (LTD). The hippocampal formation, which spans the posterior-to-anterior extent of the base of the temporal lobes, plays a key role in learning, memory, and recognition of novelty [1]. In its transverse axis, the hippocampal formation consists of the entorhinal cortex, the dentate gyrus, the CA3 and the CA1 subfields, and the subiculum. The entorhinal cortex functions as the gateway into the hippocampal formation [2]. The entorhinal cortex layer II projects to dentate granule cells via the perforant pathway, and dentate granule cells project to CA3 pyramidal cells via the mossy fibers. CA3 pyramidal cells interconnect with other CA3 neurons and project to CA1 pyramidal cells via the Shaffer collaterals. Finally, CA1 pyramidal cells connect to the subiculum. The entorhinal cortex layer II also projects to CA3 pyramidal cells, and the entorhinal cortex layer III also projects to CA1 pyramidal cells and the subiculum [3].

In the process of changes in synaptic structure for memory formation, glutamatergic neurons play a key role in the main neural circuit of the hippocampal formation. Research on synaptic plasticity opens a window for the molecular mechanisms of memory. Changes in both presynaptic and postsynaptic strength have been implicated in the mechanisms of LTP and LTD, and attention has been paid to changes in postsynaptic glutamate receptor density [4]. Intracellular Ca$^{2+}$ signaling via the network of signaling molecules controls the glutamate receptor density and induces synaptic signals, as is observed in learning situations [5], followed by cognitive performance and memory.

For *N*-methyl-D-aspartate (NMDA)-receptor-dependent plasticity, the influx of extracellular $Ca^{2+}$ into postsynaptic neurons through NMDA receptors plays a key role [6]. However, glutamate receptor activation by excess of extracellular glutamate, which is known as glutamate excitotoxicity [7,8], leads to a final common pathway for neuronal death and is linked with pathophysiological processes of neurological disorders [9,10]. CA1 pyramidal cells are the most vulnerable to neurodegeneration in the hippocampus after stroke/ischemia [11–13]. A well-known fact is that extracellular $Ca^{2+}$ influx into postsynaptic neurons, in addition to $Ca^{2+}$ release from the calcium stores results in neuronal death.

On the other hand, extracellular glutamate signaling also induces cellular transients in $Zn^{2+}$ concentration, i.e., intracellular $Zn^{2+}$ signaling, which is required for synaptic plasticity [14,15] and may have crosstalk to intracellular $Ca^{2+}$ signaling via calcium channels [16,17]. On the basis of the subsequent evidence that glutamate-induced neuronal death is due to extracellular $Zn^{2+}$ influx into postsynaptic neurons, which is dynamically linked with $Zn^{2+}$ release from zincergic neurons, a subclass of glutamatergic neurons that concentrate zinc in the presynaptic vesicles [18–22], this paper deals with the impact of synaptic $Zn^{2+}$ signaling on cognitive function and dysfunction in comparison with synaptic $Ca^{2+}$ signaling. While the hydrated $Ca^{2+}$ ion is the major species in intracellular $Ca^{2+}$ signaling, this is not the case in intracellular $Zn^{2+}$ signaling because the $Zn^{2+}$ ion has much higher affinities for donors of ligands [23]. The $Zn^{2+}$ ion is different from the $Ca^{2+}$ and $Mg^{2+}$ ions because it forms much stronger complexes with water and various anions and ligands. These characteristics are important for its synaptic functions.

## 2. Physiology of Brain $Zn^{2+}$

Divalent cations such as $Ca^{2+}$ and $Mg^{2+}$ are involved in synaptic neurotransmission [24]. Among divalent cations, $Ca^{2+}$ concentration is the highest in brain parenchyma cells and is approximately 1.2 mM in the cerebrospinal fluid (CSF) and brain extracellular fluid in the adult rats (Figure 1) [25]. Approximately 2 mM $Ca^{2+}$ is added to artificial cerebrospinal fluid (ACSF) based on essentiality of intracellular $Ca^{2+}$ signaling in neurons and glial cells [26,27]. However, excess influx of extracellular $Ca^{2+}$ into neurons is linked with the pathophysiological process of neurodegeneration [28–30].

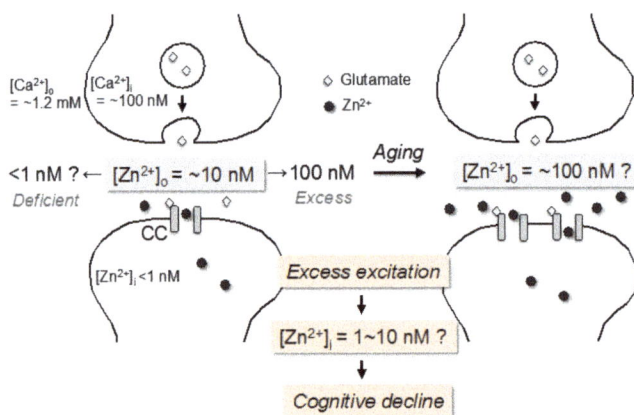

**Figure 1.** $Zn^{2+}$-mediated cognitive decline via rapid influx of extracellular $Zn^{2+}$. Estimated basal concentration of extracellular $Zn^{2+}$ is ~10 nM in the adult brain. When the basal concentration of extracellular $Zn^{2+}$ reaches 100 nM in the adult brain, it induces cognitive decline. In contrast, even if the basal concentration of extracellular $Zn^{2+}$ reaches 100 nM in the aged brain, it does not induce cognitive decline, suggesting that the basal concentration of extracellular $Zn^{2+}$ is ~100 nM in the aged brain. Extracellular $Zn^{2+}$-mediated cognitive decline is induced by glutameric synapse excitation, in which intracellular $Zn^{2+}$ may reach 1~10 nM. The synapses are non-zincergic. CC: $Ca^{2+}$-permeable channels.

Zinc concentration in the CSF is in the range of 150–380 nM [31–33]. It is estimated that the basal (static) concentration of extracellular $Zn^{2+}$ is approximately 10 nM in the brain of the adult rats (Figure 1) [34]. A small part of extracellular zinc is free ion ($Zn^{2+}$) in the brain under the basal condition. To research synaptic function, much less attention has been paid to the essentiality of $Zn^{2+}$ in brain extracellular fluid. ACSF, i.e., brain extracellular medium, without $Zn^{2+}$ has been used for in vitro and in vivo experiments. It is likely that not only neuronal excitation but also LTP is modified in brain slices immersed in ACSF without $Zn^{2+}$, in which original neurophysiology might be modified [35,36]. Clarifying the action of extracellular $Zn^{2+}$ in the range of physiological concentrations is important to precisely understand synaptic function. Furthermore, such clarification is also important to understand the bidirectional action of $Zn^{2+}$ under physiological and pathological conditions. It is recognized that low nanomolar concentrations of $Zn^{2+}$ are more physiological than micromolar concentrations of $Zn^{2+}$, which are widely used and often neurotoxic.

Spontaneous presynaptic activity assessed with FM4-64, an indicator of presynaptic activity (exocytosis), in the stratum lucidum where mossy fibers are contained is significantly suppressed in brain slices from young rats immersed in ACSF containing 10 nM $Zn^{2+}$, but not in ACSF containing 10 nM $Cu^{2+}$ or 10 nM $Fe^{3+}$, indicating that hippocampal presynaptic activity is enhanced in brain slices prepared with ACSF without $Zn^{2+}$ [36]. Suh et al. [37] report that acute brain slice preparations are poorly suitable to research the role of endogenous $Zn^{2+}$ released from zincergic neurons. Vesicular $Zn^{2+}$ levels are decreased in the process of slice preparation, and in vitro $Zn^{2+}$ release is reduced to approximately 25% of in vivo $Zn^{2+}$ release. While physiological concentration of extracellular $Zn^{2+}$ is low nanomolar in young rat brain, it may be elevated along with aging (Figure 1), based on the age-related increase in extracellular zinc concentration in the hippocampus [38].

## 3. Impact of Synaptic $Zn^{2+}$ Dynamics on Cognition

Extracellular $Ca^{2+}$ concentration is not affected by neuronal excitation. In contrast, extracellular $Zn^{2+}$ concentration is dynamically increased by zincergic excitation, but not by non-zincergic excitation. In any case, extracellular dynamics of $Ca^{2+}$ and $Zn^{2+}$ is critically linked with their intracellular dynamics. The basal concentration of intracellular (cytosol) $Ca^{2+}$ is 10–100 nM, while that of intracellular $Zn^{2+}$ is extremely low and estimated to be less than 1 nM (Figure 1) [39,40]. While intracellular $Ca^{2+}$ serves as a signaling factor for plastic changes at synapses, intracellular $Zn^{2+}$ is increased for not only signaling for plastic changes during learning and cognitive activity but also plastic changes in synapse structure [41,42]. The optimal range of intracellular $Zn^{2+}$ increased during learning and cognitive activity, which is dynamically linked with $Zn^{2+}$ release at zincergic synapses, remains to be clarified. Even at non-zincergic synapses, postsynaptic intracellular $Zn^{2+}$ may reach ~1 nM and the increase originates in internal stores/proteins unlike the neurotoxic increase via extracellular $Zn^{2+}$ influx as described below.

LTP at zincergic mossy fiber-CA3 pyramidal cell synapses is induced by the presynaptic mechanism, in which glutamate release is persistently increased. Mossy fiber LTP induction critically depends on the rise in presynaptic $Ca^{2+}$ [43–45], which activates the calcium-calmodulin-sensitive adenyl cyclase I [46]. $Zn^{2+}$ released from mossy fibers is immediately retaken up into presynaptic terminals through $Ca^{2+}$ channels and activates a Src family kinase, which promotes tropomyosin-related kinase B (TrkB) activation. The activation leads to the phosphorylation and activation of phospholipase C$\gamma$1, followed by calcium-calmodulin-sensitive adenyl cyclase I activation. $Zn^{2+}$ increases presynaptic glutamate release, while it inhibits postsynaptic mechanism of mossy fiber LTP via $Zn^{2+}$ influx [47,48].

LTP at the Schaffer collateral/commissural-CA1 pyramidal cell synapses depends on the postsynaptic activation of NMDA receptors [49]. NMDA receptor activation increases postsynaptic $Ca^{2+}$ concentration, which leads to LTP and LTD. NMDA receptors consist of multiple subclasses [50] and the subtypes have different sensitivities to $Zn^{2+}$, an endogenous blocker [51–54]. ZnAF-2DA is a useful tool to evaluate the direct involvement of $Zn^{2+}$ in cognitive function. ZnAF-2DA, a membrane-permeable $Zn^{2+}$ indicator, is taken up into neurons through the plasma membrane and is hydrolyzed by esterase in the cytosol, resulting in the production of ZnAF-2, which cannot permeate the plasma membrane [55,56]. When ZnAF-2DA is locally injected into the hippocampal CA1, intracellular ZnAF-2 is detected only in the injected area in the CA1 and can block cellular transients in $Zn^{2+}$ concentration ($K_d$, 2.7 nM for $Zn^{2+}$). The concurrent evaluations of in vivo LTP and learning behavior in separated experiments using ZnAF-2DA answer whether the in vivo LTP via intracellular $Zn^{2+}$ signaling is linked with learning behavior. The influx of extracellular $Zn^{2+}$ into CA1 pyramidal cells, which is linked with $Zn^{2+}$ release form the zincergic Schaffer collateral, is required for object recognition memory via in vivo Schaffer collateral LTP [57]. Glutamatergic input to CA1 pyramidal cells via the medial perforant pathway (the temporoammonic pathway) from the entorhinal cortex facilitates memory consolidation [58] and is required for temporal association memory [59] and spatial working memory [60]. Although the medial perforant pathway from the entorhinal cortex, is non-zincergic [61], intracellular $Zn^{2+}$ signaling, which originates in internal stores/proteins, is required for LTP at medial perforant pathway-CA1 pyramidal cell synapses [62]. It is likely that intracellular $Zn^{2+}$ signaling in CA1 pyramidal cells is also involved in cognitive function via in vivo perforant pathway LTP.

The lateral and medial entorhinal cortices are connected with the dentate gyrus. The lateral and the medial perforant pathways, which originate in the lateral and the medial entorhinal cortices, respectively, comprise physiologically distinct inputs to the dentate gyrus. The lateral perforant pathway transmits nonspatial information, while the medial perforant pathway transmits spatial information [63]. In regard to LTP at medial perforant pathway-dentate granule cell synapses, calmodulin-dependent protein kinase II $\alpha$ ($\alpha$-CaMKII)/brain-derived neurotrophic factor (BDNF) signaling pathway plays a key role for LTP induction. Zinc deficiency-induced cognitive and synaptic impairments are linked with disruption of $\alpha$-CaMKII/BDNF signaling pathway [64]. In dentate granule cells, intracellular $Zn^{2+}$ signaling originates in internal stores/proteins and is necessary for object and space recognition memory via medial perforant pathway LTP [65,66]. In postsynaptic neurons innervated by non-zincergic medial perforant pathway, glutamate receptor activation triggers off $Zn^{2+}$ release from internal stores/proteins that remain to be clarified.

Hippocampal neurogenesis always produces dentate granule cells, in which NMDA receptor-dependent synaptic plasticity is involved in learning and memory [67,68]. $Zn^{2+}$ is concentrated in the dentate gyrus of the hippocampus [69] and is required for neurogenesis process [70]. In human neuronal precursor cells, zinc deficiency induces apoptosis via mitochondrial p53- and caspase-dependent pathways [71], suggesting that dynamic $Zn^{2+}$ transport to neuronal precursor cells is critical for learning and memory via hippocampal neurogenesis [72].

## 4. Impact of Synaptic $Zn^{2+}$ Dynamics on Cognitive Decline

Aging has progressive pathophysiological features and is linked with altered cell metabolism, damaged nucleic acid, oxidative stress, and deposition of abnormal forms of proteins. Aging also is characterized by cognitive decline, neuronal loss, and vulnerability to neurological disorders [73] and may be often related with altered $Zn^{2+}$ homeostasis in the brain [74,75]. Hippocampal zinc concentration is decreased in aging, which decreases zinc transporter-3 (ZnT3) protein. ZnT3 controls synaptic vesicular $Zn^{2+}$ levels. $Zn^{2+}$ release from zincergic neuron terminals, which dynamically modifies the basal concentration of extracellular $Zn^{2+}$, is decreased in aging [74], while the basal concentration of extracellular $Zn^{2+}$ may be increased [38], probably as a compensatory mechanism. A negative modulation of extracellular glutamate signaling by extracellular $Zn^{2+}$ may be involved in cognitive function.

Metal chaperones i.e., clioquinol and PBT2, prevent normal age-related cognitive decline [76,77], suggest that metal chaperones are effective for preventing $Zn^{2+}$-mediated cognitive decline that is observed in aging and disease. The hippocampus is vulnerable to $Zn^{2+}$ neurotoxicity [78] and the dentate gyrus is the most vulnerable to aging process [2,79]. Although the vulnerability to aging is poorly understood, it is possible that synaptic $Zn^{2+}$ dynamics is involved in the vulnerability. New granule cells are continuously produced in the subgranular zone of the dentate gyrus (Figure 2) and the decreased rate of hippocampal neurogenesis is involved in age-related cognitive decline [80]. Neurogenesis-related apoptosis, which seems to be increased along with aging, always occurs in the dentate gyrus. In the subgranular zone, the apoptosis locally increases extracellular $K^+$ and the increase is due to the efflux of intracellular $K^+$ (approximately 140 mM) by disruption of the plasma membrane. The increase in extracellular $K^+$ may excite granule cells and pyramidal basket cells, which exist nearby in the dentate gyrus, and disturbs intracellular dynamics of $Ca^{2+}$ and $Zn^{2+}$ (Figure 2). As a matter of fact, both memory acquisition via LTP induction and memory retention via LTP maintenance are impaired after local injection of high $K^+$ into the dentate gyrus [81–83] or the CA1 [84]. The impairments are due to an increase in intracellular $Zn^{2+}$, but not that in intracellular $Ca^{2+}$, because the impairments are rescued with CaEDTA, which forms membrane-impermeable ZnEDTA in the extracellular compartment and inhibits the influx of extracellular $Zn^{2+}$, but not that in extracellular $Ca^{2+}$ [82]. The evidence indicates $Zn^{2+}$-mediated cognitive decline via transient $Zn^{2+}$ accumulation in dentate granule cells (Figure 1) and CA1 pyramidal cells.

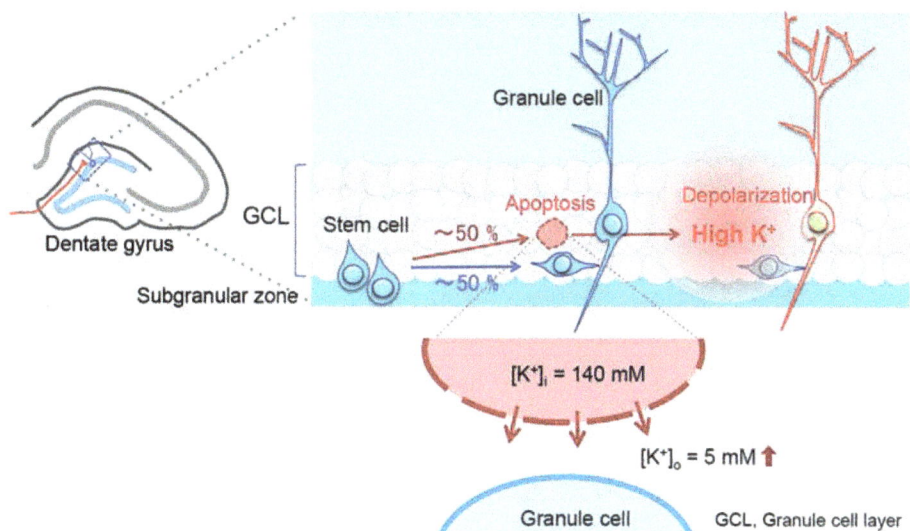

**Figure 2.** Neuronal depolarization via neurogenesis-related apoptosis. Neurogenesis-related apoptosis increases extracellular $K^+$ concentration ($[K^+]_o$), which is due to the efflux of intracellular $K^+$ as shown by red arrows, in the dentate granule cell layer and can lead dentate granule cells to depolarization, followed by extracellular $Zn^{2+}$ influx-mediated cognitive decline. The blue arrow shows the process of neurogenesis and red up-arrows show the process of apoptosis and efflux of intracellular $K^+$. $[K^+]_i$: intracellular $K^+$ concentration.

If the basal level of extracellular $Zn^{2+}$ is increased age-dependently in the hippocampus (Figure 1) [38], it is estimated that $Zn^{2+}$-mediated cognitive decline more readily occurs in the aged brain. High $K^+$-induced increase in intracellular $Zn^{2+}$ is facilitated in the aged dentate gyrus and leads to attenuating both LTP induction and maintained LTP at medial perforant pathway-dentate granule cell synapses of aged rats [38,83], suggesting that the influx of extracellular $Zn^{2+}$ into dentate granule cells more readily occurs in aged rats and is a cause of age-related cognitive decline via attenuation of LTP. It is likely that neurogenesis-related apoptosis is involved in $Zn^{2+}$-mediated cognitive decline.

*Int. J. Mol. Sci.* **2017**, *18*, 2411

GluR2-lacking calcium-permeable α-amino-3-hydroxy-5-methyl-4-isoxazolepropionate (AMPA) receptors are involved in $Zn^{2+}$-mediated neurodegeneration in the hippocampal CA1 and CA3 [18,85,86]. In the hippocampus, the levels of GluR1 and GluR2 mRNA are highest in the dentate gyrus and the GluR1/GluR2 mRNA ratios are elevated along with aging [87]. The findings suggest that $Zn^{2+}$ influx through $Ca^{2+}$-permeable AMPA receptors, which more readily occurs in the aged dentate gyrus, plays a key role for cognitive decline [38,83]. Intracellular $Zn^{2+}$ can reach approximately 10 nM via the rapid influx of extracellular $Zn^{2+}$ (Figure 1) [34]. Both increases in extracellular $Zn^{2+}$ and $Ca^{2+}$-permeable AMPA receptors contribute to $Zn^{2+}$-mediated cognitive decline in aging.

Although intracellular $Zn^{2+}$ level in the process of LTP maintenance is unknown, LTP maintenance at medial perforant pathway-dentate granule cell synapses is affected by chelation of intracellular $Zn^{2+}$ with intracellular ZnAF-2 [66] and the aged dentate gyrus is more susceptible to the chelating effect on LTP maintenance [83]. When ZnAF-2DA is used as an index of the capacity binding intracellular $Zn^{2+}$, interestingly, the capacity of intracellular ZnAF-2 for binding intracellular $Zn^{2+}$ is more rapidly lost in the aged dentate molecular layer where medial perforant pathway-dentate granule cell synapses are contained than in the young dentate molecular layer, suggesting that intracellular $Zn^{2+}$-buffering is weakened in the dentate gyrus along with aging (Figure 3) [83]. Characteristics (easiness) of extracellular $Zn^{2+}$ influx may be linked with weakened intracellular $Zn^{2+}$-buffering in the aged dentate gyrus [28]. Although the actual state of intracellular $Zn^{2+}$-buffering is poorly understood, $Ca^{2+}$-permeable channels, zinc transporters (ZIP and ZnT), zinc-binding proteins such as metallothioneins, and $Zn^{2+}$-containing internal stores are involved in the $Zn^{2+}$-buffering system.

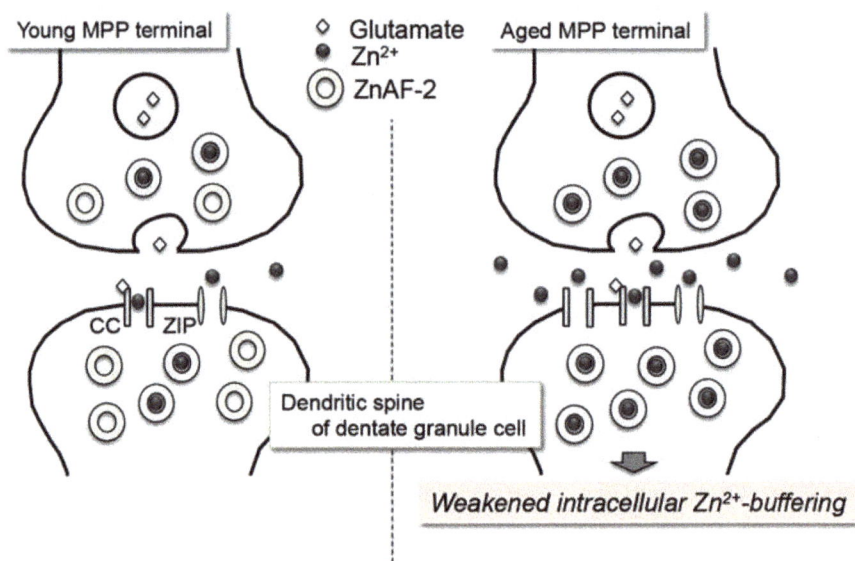

**Figure 3.** Is intracellular $Zn^{2+}$-buffering weakened in the dentate gyrus along with aging? In vivo intracellular $Zn^{2+}$-buffering is assessed in the dentate molecular layer where non-zincergic media performant pathway (MPP)-dentate granule cell synapses are contained. Intracellular ZnAF-2, an index of intracellular $Zn^{2+}$-buffering capacity, can bind $Zn^{2+}$ at young MPP synapse 2 h after ZnAF-2DA injection into the dentate molecular layer, but not aged MPP synapses. Capacity of intracellular ZnAF-2 for binding intracellular $Zn^{2+}$ is more rapidly lost in in aged dentate gyrus, probably due to easiness of extracellular $Zn^{2+}$ influx, suggesting a reduced capacity of intracellular $Zn^{2+}$-buffering in aged dentate gyrus. ZIP: Zrt/Irt-like proteins.

## 5. Perspectives

Vulnerability to $Ca^{2+}$ dysregulation has been observed in the process of brain aging [88–90]. It has been reported that $Ca^{2+}$ dysregulation is not ubiquitous. The mechanisms of dysregulation are observed in specific cell populations and areas in the brain. For example, L-type $Ca^{2+}$ channels is age-dependently increased in hippocampal pyramidal cells [91]. Age-dependent reduction in the NMDA receptor function is observed in the hippocampus and the frontal cortex [92], suggesting a compensatory mechanism to availability/restriction for intracellular $Ca^{2+}$ signaling. Intracellular $Ca^{2+}$-buffering, which is involved in cognitive function, is weakened during brain aging [89]. In contrast, intracellular $Zn^{2+}$-buffering is also dynamically involved in cognition and its decline. However, the $Zn^{2+}$-buffering system is more poorly understood than the $Ca^{2+}$-buffering system, and its clarification is required for understanding cognition and its decline.

**Author Contributions:** Atsushi Takeda wrote the manuscript and Haruna Tamano confirmed the contents.

**Conflicts of Interest:** The authors declare no conflict of interest.

## References

1. Knierim, J.J. The hippocampus. *Curr. Biol.* **2015**, *25*, R1116–R1121. [CrossRef] [PubMed]
2. Small, S.A.; Schobel, S.A.; Buxton, R.B.; Witter, M.P.; Barnes, C.A. A pathophysiological framework of hippocampal dysfunction in ageing and disease. *Nat. Rev. Neurosci.* **2011**, *12*, 585–601. [CrossRef] [PubMed]
3. Sasaki, T.; Leutgeb, S.; Leutgeb, J.K. Spatial and memory circuits in the medial entorhinal cortex. *Curr. Opin. Neurobiol.* **2015**, *32*, 16–23. [CrossRef] [PubMed]
4. Malenka, R.C.; Nicoll, R.A. Long-term potentiation—A decade of progress? *Science* **1999**, *285*, 1870–1874. [CrossRef] [PubMed]
5. Blitzer, R.D.; Iyengar, R.; Landau, E.M. Postsynaptic signaling networks: Cellular cogwheels underlying long-term plasticity. *Biol. Psychiatry* **2005**, *57*, 113–119. [CrossRef] [PubMed]
6. Nicoll, R.A.; Malenka, R.C. Contrasting properties of two forms of long-term potentiation in the hippocampus. *Nature* **1995**, *377*, 115–118. [CrossRef] [PubMed]
7. Danbolt, N.C. Glutamate uptake. *Prog. Neurobiol.* **2001**, *65*, 1–105. [CrossRef]
8. Dong, X.X.; Wang, Y.; Qin, Z.H. Molecular mechanisms of excitotoxicity and their relevance to pathogenesis of neurodegenerative diseases. *Acta Pharmacol. Sin.* **2009**, *30*, 379–387. [CrossRef] [PubMed]
9. Lai, T.W.; Zhang, S.; Wang, Y.T. Excitotoxicity and stroke: Identifying novel targets for neuroprotection. *Prog. Neurobiol.* **2014**, *115*, 157–188. [CrossRef] [PubMed]
10. Lewerenz, J.; Maher, P. Chronic glutamate toxicity in neurodegenerative diseases—What is the evidence? *Front. Neurosci.* **2015**, *9*, 469. [CrossRef] [PubMed]
11. Lo, E.H.; Dalkara, T.; Moskowitz, M.A. Mechanisms, challenges and opportunities in stroke. *Nat. Rev. Neurosci.* **2003**, *4*, 399–415. [CrossRef] [PubMed]
12. Malairaman, U.; Dandapani, K.; Katyal, A. Effect of Ca2EDTA on zinc mediated inflammation and neuronal apoptosis in hippocampus of an in vivo mouse model of hypobaric hypoxia. *PLoS ONE* **2014**, *9*, e110253. [CrossRef] [PubMed]
13. Medvedeva, Y.V.; Weiss, J.H. Intramitochondrial $Zn^{2+}$ accumulation via the $Ca^{2+}$ uniporter contributes to acute ischemic neurodegeneration. *Neurobiol. Dis.* **2014**, *68*, 137–144. [CrossRef] [PubMed]
14. Tamano, H.; Koike, Y.; Nakada, H.; Shakushi, Y.; Takeda, A. Significance of synaptic $Zn^{2+}$ signaling in zincergic and non-zincergic synapses in the hippocampus in cognition. *J. Trace Elem. Med. Biol.* **2016**, *38*, 93–98. [CrossRef] [PubMed]
15. Takeda, A.; Tamano, H. Significance of the degree of synaptic $Zn^{2+}$ signaling in cognition. *BioMetals* **2016**, *29*, 177–185. [CrossRef] [PubMed]
16. Takeda, A.; Tamano, H. Insight into zinc signaling from dietary zinc deficiency. *Brain Res. Rev.* **2009**, *62*, 33–34. [CrossRef] [PubMed]
17. Lavoie, N.; Jeyaraju, D.V.; Peralta, M.R., 3rd; Seress, L.; Pellegrini, L.; Tóth, K. Vesicular zinc regulates the $Ca^{2+}$ sensitivity of a subpopulation of presynaptic vesicles at hippocampal mossy fiber terminals. *J. Neurosci.* **2011**, *31*, 18251–18265. [CrossRef] [PubMed]

18. Colbourne, F.; Grooms, S.Y.; Zukin, R.S.; Buchan, A.M.; Bennett, M.V. Hypothermia rescues hippocampal CA1 neurons and attenuates down-regulation of the AMPA receptor GluR2 subunit after forebrain ischemia. *Proc. Natl. Acad. Sci. USA* **2003**, *100*, 2906–2910. [CrossRef] [PubMed]

19. Frederickson, C.J.; Koh, J.Y.; Bush, A.I. The neurobiology of zinc in health and disease. *Nat. Rev. Neurosci.* **2005**, *6*, 449–462. [CrossRef] [PubMed]

20. Noh, K.M.; Yokota, H.; Mashiko, T.; Castillo, P.E.; Zukin, R.S.; Bennett, M.V. Blockade of calcium-permeable AMPA receptors protects hippocampal neurons against global ischemia-induced death. *Proc. Natl. Acad. Sci. USA* **2005**, *102*, 12230–12235. [CrossRef] [PubMed]

21. Stork, C.J.; Li, Y.V. Rising zinc: A significant cause of ischemic neuronal death in the CA1 region of rat hippocampus. *J. Cereb. Blood Flow Metab.* **2009**, *29*, 1399–1408. [CrossRef] [PubMed]

22. Sensi, S.L.; Paoletti, P.; Bush, A.I.; Sekler, I. Zinc in the physiology and pathology of the CNS. *Nat. Rev. Neurosci.* **2009**, *10*, 780–791. [CrossRef] [PubMed]

23. Kreżel, A.; Maret, W. The biological inorganic chemistry of zinc ions. *Arch. Biochem. Biophys.* **2016**, *611*, 3–19. [CrossRef] [PubMed]

24. Kim, N.K.; Robinson, H.P. Effects of divalent cations on slow unblock of native NMDA receptors in mouse neocortical pyramidal neurons. *Eur. J. Neurosci.* **2011**, *34*, 199–212. [CrossRef] [PubMed]

25. Jones, H.C.; Keep, R.F. Brain fluid calcium concentration and response to acute hypercalcaemia during development in the rat. *J. Physiol.* **1988**, *402*, 579–593. [CrossRef] [PubMed]

26. Neves, G.; Cooke, S.F.; Bliss, T.V.P. Synaptic plasticity, memory and the hippocampus: A neural network approach to causality. *Nat. Rev. Neurosci.* **2008**, *9*, 65–67. [CrossRef] [PubMed]

27. Lisman, J.; Yasuda, R.; Raghavachari, S. Mechanisms of CaMKII action in long-term potentiation. *Nat. Rev. Neurosci.* **2012**, *13*, 169–182. [CrossRef] [PubMed]

28. Alberdi, E.; Sánchez-Gómez, M.V.; Cavaliere, F.; Pérez-Samartín, A.; Zugaza, J.L.; Trullas, R.; Domercq, M.; Matute, C. Amyloid β oligomers induce $Ca^{2+}$ dysregulation and neuronal death through activation of ionotropic glutamate receptors. *Cell Calcium* **2010**, *47*, 264–272. [CrossRef] [PubMed]

29. Bickler, P.E.; Warren, D.E.; Clark, J.P.; Gabatto, P.; Gregersen, M.; Brosnan, H. Anesthetic protection of neurons injured by hypothermia and rewarming: Roles of intracellular $Ca^{2+}$ and excitotoxicity. *Anesthesiology* **2012**, *117*, 280–292. [CrossRef] [PubMed]

30. Abushik, P.A.; Sibarov, D.A.; Eaton, M.J.; Skatchkov, S.N.; Antonov, S.M. Kainate-induced calcium overload of cortical neurons in vitro: Dependence on expression of AMPAR GluA2-subunit and down-regulation by subnanomolar ouabain. *Cell Calcium* **2013**, *54*, 95–104. [CrossRef] [PubMed]

31. Hershey, C.O.; Hershey, L.A.; Varnes, A.; Vibhakar, S.D.; Lavin, P.; Strain, W.H. Cerebrospinal fluid trace element content in dementia: Clinical, radiologic, and pathologic correlations. *Neurology* **1983**, *33*, 1350–1353. [CrossRef] [PubMed]

32. Gellein, K.; Skogholt, J.H.; Aaseth, J.; Thoresen, G.B.; Lierhagen, S.; Steinnes, E.; Syversen, T.; Flaten, T.P. Trace elements in cerebrospinal fluid and blood from patients with a rare progressive central and peripheral demyelinating disease. *J. Neurol. Sci.* **2008**, *266*, 70–78. [CrossRef] [PubMed]

33. Michalke, B.; Nischwitz, V. Review on metal speciation analysis in cerebrospinal fluid-current methods and results: A review. *Anal. Chim. Acta* **2010**, *682*, 23–36. [CrossRef] [PubMed]

34. Frederickson, C.J.; Giblin, L.J.; Krezel, A.; McAdoo, D.J.; Muelle, R.N.; Zeng, Y.; Balaji, R.V.; Masalha, R.; Thompson, R.B.; Fierke, C.A.; et al. Concentrations of extracellular free zinc (pZn)e in the central nervous system during simple anesthetization, ischemia and reperfusion. *Exp. Neurol.* **2006**, *198*, 285–293. [CrossRef] [PubMed]

35. Takeda, A.; Tamano, H. Significance of low nanomolar concentration of $Zn^{2+}$ in artificial cerebrospinal fluid. *Mol. Neurobiol.* **2017**, *54*, 2477–2482. [CrossRef] [PubMed]

36. Tamano, H.; Nishio, R.; Shakushi, Y.; Sasaki, M.; Koike, Y.; Osawa, M.; Takeda, A. In vitro and in vivo physiology of low nanomolar concentrations of $Zn^{2+}$ in artificial cerebrospinal fluid. *Sci. Rep.* **2017**, *7*, 42897. [CrossRef] [PubMed]

37. Suh, S.W.; Danscher, G.; Jensen, M.S.; Thompson, R.; Motamedi, M.; Frederickson, C.J. Release of synaptic zinc is substantially depressed by conventional brain slice preparations. *Brain Res.* **2000**, *879*, 7–12. [CrossRef]

38. Takeda, A.; Koike, Y.; Osaw, M.; Tamano, H. Characteristic of extracellular $Zn^{2+}$ influx in the middle-aged dentate gyrus and its involvement in attenuation of LTP. *Mol. Neurobiol.* **2017**. [CrossRef] [PubMed]

39. Sensi, S.L.; Canzoniero, L.M.T.; Yu, S.P.; Ying, H.S.; Koh, J.Y.; Kerchner, G.A.; Choi, D.W. Measurement of intracellular free zinc in living cortical neurons: Routes of entry. *J. Neurosci.* **1997**, *15*, 9554–9564.

40. Colvin, R.A.; Bush, A.I.; Volitakis, I.; Fontaine, C.P.; Thomas, D.; Kikuchi, K.; Holmes, W.R. Insights into $Zn^{2+}$ homeostasis in neurons from experimental and modeling studies. *Am. J. Physiol. Cell Physiol.* **2008**, *294*, C726–C742. [CrossRef] [PubMed]

41. Grabrucker, A.M.; Knight, M.J.; Proepper, C.; Bockmann, J.; Joubert, M.; Rowan, M.; Nienhaus, G.U.; Garner, C.C.; Bowie, J.U.; Kreutz, M.R.; et al. Concerted action of zinc and ProSAP/Shank in synaptogenesis and synapse maturation. *EMBO J.* **2011**, *30*, 569–581. [CrossRef] [PubMed]

42. Perrin, L.; Roudeau, S.; Carmona, A.; Domart, F.; Petersen, J.D.; Bohic, S.; Yang, Y.; Cloetens, P.; Ortega, R. Zinc and Copper Effects on Stability of Tubulin and Actin Networks in Dendrites and Spines of Hippocampal Neurons. *ACS Chem. Neurosci.* **2017**, *8*, 1490–1499. [CrossRef] [PubMed]

43. Castillo, P.E.; Weisskopf, M.G.; Nicoll, R.A. The role of $Ca^{2+}$ channels in hippocampal mossy fiber synaptic transmission and long-term potentiation. *Neuron* **1994**, *12*, 261–269. [CrossRef]

44. Tong, G.; Malenka, R.C.; Nicoll, R.A. Long-term potentiation in cultures of single hippocampal granule cells: A presynaptic form of plasticity. *Neuron* **1996**, *16*, 1147–1157.

45. Breustedt, J.; Vogt, K.E.; Miller, R.J.; Nicoll, R.A.; Schmitz, D. $\alpha$1E-containing $Ca^{2+}$ channels are involved in synaptic plasticity. *Proc. Natl. Acad. Sci. USA* **2003**, *100*, 12450–12455. [CrossRef] [PubMed]

46. Wang, H.; Storm, D.R. Calmodulin-regulated adenylyl cyclases: Cross-talk and plasticity in the central nervous system. *Mol. Pharmacol.* **2003**, *63*, 463–468. [CrossRef] [PubMed]

47. Huang, Y.Z.; Pan, E.; Xiong, Z.Q.; McNamara, J.O. Zinc-mediated transactivation of TrkB potentiates the hippocampal mossy fiber-CA3 pyramid synapse. *Neuron* **2008**, *57*, 546–558. [CrossRef] [PubMed]

48. Pan, E.; Zhang, X.A.; Huang, Z.; Krezel, A.; Zhao, M.; Tinberg, C.E.; Lippard, S.J.; McNamara, J.O. Vesicular zinc promotes presynaptic and inhibits postsynaptic long-term potentiation of mossy fiber-CA3 synapse. *Neuron* **2011**, *71*, 1116–1126. [CrossRef] [PubMed]

49. Malenka, R.C.; Bear, M.F. LTP and LTD: An embarrassment of riches. *Neuron* **2004**, *44*, 5–21. [CrossRef] [PubMed]

50. Cull-Candy, S.G.; Leszkiewicz, D.N. Role of distinct NMDA receptor subtypes at central synapses. *Sci. STKE* **2004**, *255*, re16. [CrossRef] [PubMed]

51. Westbrook, G.L.; Mayer, M.L. Micromolar concentrations of $Zn^{2+}$ antagonize NMDA and GABA responses of hippocampal neurons. *Nature* **1987**, *328*, 640–643. [CrossRef] [PubMed]

52. Chen, N.; Moshaver, A.; Raymond, L.A. Differential sensitivity of recombinant *N*-methyl-D-aspartate receptor subtypes to zinc inhibition. *Mol. Pharmacol.* **1997**, *51*, 1015–1023. [PubMed]

53. Paoletti, P.; Ascher, P.; Neyton, J. High-affinity zinc inhibition of NMDA NR1-NR2A receptors. *J. Neurosci.* **1998**, *17*, 5711–5725.

54. Choi, Y.B.; Lipton, S.A. Identification and mechanism of action of two histidine residues underlying high-affinity $Zn^{2+}$ inhibition of the NMDA receptor. *Neuron* **1999**, *23*, 171–180. [CrossRef]

55. Hirano, T.; Kikuchi, K.; Urano, Y.; Nagano, T. Improvement and biological applications of fluorescent probes for zinc, ZnAFs. *J. Am. Chem. Soc.* **2002**, *124*, 6555–6562. [CrossRef] [PubMed]

56. Ueno, S.; Tsukamoto, M.; Hirano, T.; Kikuchi, K.; Yamada, M.K.; Nishiyama, N.; Nagano, T.; Matsuki, N.; Ikegaya, Y. Mossy fiber $Zn^{2+}$ spillover modulates heterosynaptic *N*-methyl-D-aspartate receptor activity in hippocampal CA3 circuits. *J. Cell Biol.* **2002**, *158*, 215–220. [CrossRef] [PubMed]

57. Takeda, A.; Suzuki, M.; Tempaku, M.; Ohashi, K.; Tamano, H. Influx of extracellular $Zn^{2+}$ into the hippocampal CA1 neurons is required for cognitive performance via long-term potentiation. *Neuroscience* **2015**, *304*, 209–216. [CrossRef] [PubMed]

58. Remondes, M.; Schuman, E.M. Role for a cortical input to hippocampal area CA1 in the consolidation of a long-term memory. *Nature* **2004**, *431*, 699–703. [CrossRef] [PubMed]

59. Suh, J.; Rivest, A.J.; Nakashiba, T.; Tominaga, T.; Tonegawa, S. Entorhinal cortex layer III input to the hippocampus is crucial for temporal association memory. *Science* **2011**, *334*, 1415–1420. [CrossRef] [PubMed]

60. Vago, D.R.; Kesner, R.P. Disruption of the direct perforant path input to the CA1 subregion of the dorsal hippocampus interferes with spatial working memory and novelty detection. *Behav. Brain Res.* **2008**, *189*, 273–283. [CrossRef] [PubMed]

61. Sindreu, C.B.; Varoqui, H.; Erickson, J.D.; Pérez-Clausell, J. Boutons containing vesicular zinc define a subpopulation of synapses with low AMPAR content in rat hippocampus. *Cereb. Cortex* **2003**, *13*, 823–829. [CrossRef] [PubMed]

62. Tamano, H.; Nishio, R.; Takeda, A. Involvement of intracellular $Zn^{2+}$ signaling in LTP at perforant pathway-CA1 pyramidal cell synapse. *Hippocampus* **2017**, *27*, 777–783. [CrossRef] [PubMed]

63. Gonzalez, J.; Morales, I.S.; Villarreal, D.M.; Derrick, B.E. Low-frequency stimulation induces long-term depression and slow onset long-term potentiation at perforant path-dentate gyrus synapses in vivo. *J. Neurophysiol.* **2014**, *111*, 1259–1273. [CrossRef] [PubMed]

64. Yu, X.; Ren, T.; Yu, X. Disruption of calmodulin-dependent protein kinase II $\alpha$/brain-derived neurotrophic factor ($\alpha$-CaMKII/BDNF) signalling is associated with zinc deficiency-induced impairments in cognitive and synaptic plasticity. *Br. J. Nutr.* **2013**, *110*, 2194–2200. [CrossRef] [PubMed]

65. Takeda, A.; Tamano, H.; Ogawa, T.; Takada, S.; Nakamura, M.; Fujii, H.; Ando, M. Intracellular $Zn^{2+}$ signaling in the dentate gyrus is required for object recognition memory. *Hippocampus* **2014**, *24*, 1404–1412. [CrossRef] [PubMed]

66. Tamano, H.; Minamino, T.; Fujii, H.; Takada, S.; Nakamura, M.; Ando, M.; Takeda, A. Blockade of intracellular $Zn^{2+}$ signaling in the dentate gyrus erases recognition memory via impairment of maintained LTP. *Hippocampus* **2015**, *25*, 952–962. [CrossRef] [PubMed]

67. Schmidt-Hieber, C.; Jonas, P.; Bischofberger, J. Enhanced synaptic plasticity in newly generated granule cells of the adult hippocampus. *Nature* **2004**, *429*, 184–187. [CrossRef] [PubMed]

68. Kheirbek, M.A.; Tannenholz, L.; Hen, R. NR2B-dependent plasticity of adult-born granule cells is necessary for context discrimination. *J. Neurosci.* **2012**, *32*, 8696–8702. [CrossRef] [PubMed]

69. Takeda, A.; Sawashita, J.; Okada, S. Biological half-lives of zinc and manganese in rat brain. *Brain Res.* **1995**, *695*, 53–58. [CrossRef]

70. Suh, S.W.; Won, S.J.; Hamby, A.M.; Yoo, B.H.; Fan, Y.; Sheline, C.T.; Tamano, H.; Takeda, A.; Liu, J. Decreased brain zinc availability reduces hippocampal neurogenesis in mice and rats. *J. Cereb. Blood Flow Metab.* **2009**, *29*, 1579–1588. [CrossRef] [PubMed]

71. Seth, R.; Corniola, R.S.; Gower-Winter, S.D.; Morgan, T.J., Jr.; Bishop, B.; Levenson, C.W. Zinc deficiency induces apoptosis via mitochondrial p53- and caspase-dependent pathways in human neuronal precursor cells. *J. Trace Elem. Med. Biol.* **2015**, *30*, 59–65. [CrossRef] [PubMed]

72. Cope, E.C.; Morris, D.R.; Gower-Winter, S.D.; Brownstein, N.C.; Levenson, C.W. Effect of zinc supplementation on neuronal precursor proliferation in the rat hippocampus after traumatic brain injury. *Exp. Neurol.* **2016**, *279*, 96–103. [CrossRef] [PubMed]

73. Mocchegiani, E.; Bertoni-Freddari, C.; Marcellini, F.; Malavolta, M. Brain, aging and neurodegeneration: Role of zinc ion availability. *Prog. Neurobiol.* **2005**, *75*, 367–390. [CrossRef] [PubMed]

74. Adlard, P.A.; Parncutt, J.M.; Finkelstein, D.I.; Bush, A.I. Cognitive loss in zinc transporter-3 knock-out mice: A phenocopy for the synaptic and memory deficits of Alzheimer's disease? *J. Neurosci.* **2010**, *30*, 1631–1636. [CrossRef] [PubMed]

75. Sindreu, C.; Palmiter, R.D.; Storm, D.R. Zinc transporter ZnT-3 regulates presynaptic Erk1/2 signaling and hippocampus-dependent memory. *Proc. Natl. Acad. Sci. USA* **2012**, *108*, 3366–3370. [CrossRef] [PubMed]

76. Adlard, P.A.; Sedjahtera, A.; Gunawan, L.; Bray, L.; Hare, D.; Lear, J.; Doble, P.; Bush, A.I.; Finkelstein, D.I.; Cherny, R.D. A novel approach to rapidly prevent age-related cognitive decline. *Aging Cell* **2014**, *13*, 351–359. [CrossRef] [PubMed]

77. Adlard, P.A.; Parncutt, J.; Lal, V.; James, S.; Hare, D.; Doble, P.; Finkelstein, D.I.; Bush, A.I. Metal chaperones prevent zinc-mediated cognitive decline. *Neurobiol. Dis.* **2015**, *81*, 196–202. [CrossRef] [PubMed]

78. Stork, C.J.; Li, Y.V. Intracellular zinc elevation measured with a "calcium-specific" indicator during ischemia and reperfusion in rat hippocampus: A question on calcium overload. *J. Neurosci.* **2006**, *26*, 10430–10437. [CrossRef] [PubMed]

79. Brickman, A.M.; Khan, U.A.; Provenzano, F.A.; Yeung, L.K.; Suzuki, W.; Schroeter, H.; Wall, M.; Sloan, R.P.; Small, S.A. Enhancing dentate gyrus function with dietary flavanols improves cognition in older adults. *Nat. Neurosci.* **2014**, *17*, 1798–1803. [CrossRef] [PubMed]

80. Jinno, S. Aging affects new cell production in the adult hippocampus: A quantitative anatomic review. *J. Chem. Neuroanat.* **2016**, *76*, 64–72. [CrossRef] [PubMed]

81. Suzuki, M.; Fujise, Y.; Tsuchiya, Y.; Tamano, H.; Takeda, A. Excess influx of $Zn^{2+}$ into dentate granule cells affects object recognition memory via attenuated LTP. *Neurochem. Int.* **2015**, *87*, 60–65. [CrossRef] [PubMed]

82. Takeda, A.; Tamano, H.; Hisatsune, M.; Murakami, T.; Nakada, H.; Fujii, H. Maintained LTP and memory are lost by $Zn^{2+}$ influx into dentate granule cells, but not $Ca^{2+}$ influx. *Mol. Neurobiol.* **2017**. [CrossRef] [PubMed]

83. Takeda, A.; Tamano, H.; Murakami, T.; Nakada, H.; Minamino, T.; Koike, Y. Weakened intracellular $Zn^{2+}$-buffering in the aged dentate gyrus and its involvement in erasure of maintained LTP. *Mol. Neurobiol.* **2017**. [CrossRef] [PubMed]

84. Takeda, A.; Takada, S.; Nakamura, M.; Suzuki, M.; Tamano, H.; Ando, M.; Oku, N. Transient increase in $Zn^{2+}$ in hippocampal CA1 pyramidal neurons causes reversible memory deficit. *PLoS ONE* **2011**, *6*, e28615. [CrossRef] [PubMed]

85. Liu, S.; Lau, L.; Wei, J.; Zhu, D.; Zou, S.; Sun, H.S.; Fu, Y.; Liu, F.; Lu, Y. Expression of Ca(2+)-permeable AMPA receptor channels primes cell death in transient forebrain ischemia. *Neuron* **2004**, *43*, 43–55. [CrossRef] [PubMed]

86. Weiss, J.H. Ca permeable AMPA channels in diseases of the nervous system. *Front. Mol. Neurosci.* **2011**, *4*, 42. [CrossRef] [PubMed]

87. Pagliusi, S.R.; Gerrard, P.; Abdallah, M.; Talabot, D.; Catsicas, S. Age-related changes in expression of AMPA-selective glutamate receptor subunits: Is calcium-permeability altered in hippocampal neurons? *Neuroscience* **1994**, *61*, 429–433. [CrossRef]

88. Foster, T.C. Calcium homeostasis and modulation of synaptic plasticity in the aged brain. *Aging Cell* **2007**, *6*, 319–325. [CrossRef] [PubMed]

89. Kumar, A.; Bodhinathan, K.; Foster, T.C. Susceptibility to calcium dysregulation during brain aging. *Front. Aging Neurosci.* **2009**, *1*, 2. [CrossRef] [PubMed]

90. Toescu, E.C.; Vreugdenhil, M. Calcium and normal brain ageing. *Cell Calcium* **2010**, *47*, 158–164. [CrossRef] [PubMed]

91. Thibault, O.; Landfield, P.W. Increase in single L-type calcium channels in hippocampal neurons during aging. *Science* **1996**, *272*, 1017–1020. [CrossRef] [PubMed]

92. Berridge, M.J. Dysregulation of neural calcium signaling in Alzheimer disease, bipolar disorder and schizophrenia. *Prion* **2013**, *7*, 2–13. [CrossRef] [PubMed]

International Journal of
*Molecular Sciences*

*Review*

# Zinc Signal in Brain Diseases

Stuart D. Portbury and Paul A. Adlard *

The Florey Institute of Neuroscience and Mental Health, The University of Melbourne, Melbourne, Victoria 3052, Australia; stuart.portbury@florey.edu.au
* Correspondence: paul.adlard@florey.edu.au; Tel.: +61-3-903-567-75; Fax: +61-3-903-531-03

Received: 27 October 2017; Accepted: 16 November 2017; Published: 23 November 2017

**Abstract:** The divalent cation zinc is an integral requirement for optimal cellular processes, whereby it contributes to the function of over 300 enzymes, regulates intracellular signal transduction, and contributes to efficient synaptic transmission in the central nervous system. Given the critical role of zinc in a breadth of cellular processes, its cellular distribution and local tissue level concentrations remain tightly regulated via a series of proteins, primarily including zinc transporter and zinc import proteins. A loss of function of these regulatory pathways, or dietary alterations that result in a change in zinc homeostasis in the brain, can all lead to a myriad of pathological conditions with both acute and chronic effects on function. This review aims to highlight the role of zinc signaling in the central nervous system, where it may precipitate or potentiate diverse issues such as age-related cognitive decline, depression, Alzheimer's disease or negative outcomes following brain injury.

**Keywords:** zinc; brain; neurodegeneration; cognition

## 1. Introduction

The essential trace ion zinc is a stable divalent cation that participates in numerous biological processes, and after iron, is the second most abundant trace element [1]. It is a known contributor to the functionality of over 2000 proteins [2], and as such plays a role in many diverse cellular mechanisms including cell division [3], DNA synthesis [4], protein synthesis [5], wound healing [6], immunity [7], and cognition [8]. Thus, zinc is an indispensable micronutrient for humans, the deficiency of which has been linked to numerous disorders [1]. Indeed, zinc deficiency is a recognized global public health concern in developing countries [9], and is also becoming a prevalent concern in the ageing population of developed countries [10]. Whilst supplementation of dietary zinc may be efficacious in the prevention of certain zinc deficiency related conditions, excess zinc can also induce adverse effects due to its role in numerous biochemical reactions in the human body [11]. Thus, the maintenance of zinc at natural homeostatic levels, usually attainable through a balanced diet, is both desirable and essential for optimal physiological function. In the central nervous system (CNS), maintenance of zinc homeostasis is critical for brain health, particularly as it pertains to cognition [12,13]. Moreover, altered zinc homeostasis is considered a contributing factor to the pathogenesis of multiple CNS diseases [14]. In this review, we will provide a brief overview of zinc regulation and will then focus on detailing a number of examples of the role of zinc in different brain diseases.

## 2. Zinc in the Brain

Compared to other organs in the human body, zinc concentration is highest in the brain where it is estimated at 150 $\mu$mol/L, representing a 10-fold increase over serum zinc [15]. Zinc occurs in the brain as a structural component of about 70% of proteins, contributing to the efficient performance of over 2000 transcription factors and over 300 enzymes. 10–15% of brain zinc occurs in a "free" or chelatable form, and is present at much lower concentrations (~500 nM) in brain extracellular fluids [15]. However, the concentration of zinc in synaptic vesicles of excitatory glutamatergic forebrain neurons has been shown to be >1 mmol/L [16], and as such has resulted in these cells commonly being referred to as "gluzinergic" neurons. Other cell populations in the brain containing zinc ions in pre-synaptic boutons have similarly been designated as "zinc-enriched" (ZEN) neurons [17].

Zinc homeostasis in the brain is tightly regulated, primarily via three families of proteins. They are the metallothioneins (MTs) involved in the regulation and maintenance of intracellular zinc homeostasis [12], the zinc- and iron-like regulatory proteins (ZIPs) responsible for zinc uptake from extracellular fluids into both neurons and glia [18], and the zinc transporters (ZnTs) which are associated with cellular zinc efflux [1,18]. Whilst historical data has suggested that a number of the ZIP and ZnT proteins were specific for zinc, there is evidence now emerging that suggests that many of these zinc regulatory proteins also regulate other metal ions and are found in a more diverse suite of cellular compartments than originally thought.

## 3. Metallothioneins

Approximately 5–15% of the cytosolic zinc pool is bound by metallothioneins, of which there are four isoforms, designated MT-1, MT-2, MT-3, and MT-4 [19]. Each metallothionein is composed of 61–68 amino acids, comprising 20–21 cysteines which can incorporate up to 7 atoms of zinc for storage, or for the function of a zinc acceptor or donor [1] (these proteins typically also bind copper, and can also bind metals such as cadmium, lead and others). MT-1, MT-2 and MT-3 are all synthesized in the CNS, however MT-1 and MT-2 are expressed in all tissues, and in the CNS are primarily expressed by astrocytes. Similarly, MT-3 is principally expressed in the CNS [19], whereas MT-4 is a minor metallothionein localized in stratified epithelial cells [20].

## 4. Zinc- and Iron-like Regulatory Proteins

There are 14 ZIP transporters currently known in humans [18], each of which contain eight transmembrane domains with extracellular N- and C-termini. Whilst zinc transport activity has been confirmed for ZIPs 1-8 and -14 [21], ZIP transporters do not transport exclusively zinc. They have been shown to regulate intracellular iron, copper, manganese and cadmium transport as well [1]. Most ZIP transporters are localized on plasma membranes where they function in the cytosolic replenishment of zinc from the extracellular space and within the lumen of intracellular components [22]. Cell surface localization and expression of ZIP transporters accordingly increase in zinc-deficient environments, and rapid internalization of the transporters has been observed in conditions of excess zinc [22]. In the CNS, ZIP1 and ZIP3 appear to be the major regulators of zinc uptake, however, ZIP1 expression is greater in the brain than ZIP3, and as such is believed to be the key facilitator of neuronal zinc uptake [22].

## 5. Zinc Transporter Proteins

Currently there are ten known human ZnT proteins, each of which function to regulate zinc primarily via efflux out of cells and intracellular compartments [23]. Whilst there is no precise structural information on ZnTs, based on the structure of the bacterial ZnT homolog YiiP [24], ZnTs are predicted to have six transmembrane domains with cytoplasmic amino- and carboxy-termini.

The first mammalian zinc transporter to be identified and characterized was ZnT1, and it was shown to be primarily localized in the plasma membrane where it functioned by exporting cytosolic zinc ions into the extracellular space [25]. As such, ZnT1 was predicted to protect cells from zinc influx during pathological conditions. Indeed, the mRNA for ZnT1 is upregulated in response to increases in cellular zinc levels in transient forebrain ischemia [26], and to high dietary zinc intake [27]. ZnT2 and ZnT3 both function to transport cytosolic zinc into the lumen of vesicular compartments [23]. ZnT3, which is enriched in the hippocampus and cortex, plays a significant role in modulating neurotransmission and plasticity in glutamatergic neurons [28]. ZnT3 achieves this via its role in loading zinc into pre-synaptic vesicles (whilst other proteins have been reported to perform similar functions, ZnT3 is the primary synaptic vesicular zinc transporter).

When zinc ions are then released into the synaptic cleft, which is co-incident with the vesicular release of glutamate [29], it aids in regulating the neuronal processes related to cognition and memory such as long-term potentiation (LTP) and long-term depression (LTD), by acting on *N*-methyl-D-aspartate (NMDA) receptors [30]. Zinc released into the synaptic cleft can also regulate α-amino-3-hydroxy-5-methyl-4-isoxazolepropionic acid receptors on post-synaptic cells, further regulating synaptic function and plasticity [31]. Other interactions with targets such as the zinc sensing receptor (ZnR/GPR39 [32]), the tyrosine kinase receptor TrkB [33], glutamate receptors [34] and p75 [35] also contribute to its role as a modulator of synaptic transmission and plasticity [36] and have led to zinc being labelled as an "atypical neurotransmitter" [37]. These data, together with other detailed studies examining the role of zinc in various aspects of learning and memory in the hippocampus [38–43], highlight the significance of zinc in cognition. Significantly, the ablation of ZnT3 has been demonstrated to result in age-dependent cognitive deficits in mice [44]. ZnT3 protein levels are decreased across normal ageing in humans, and further decreased in Alzheimer's disease postmortem brain tissue [44], together with significant reductions in ZnT3 mRNA levels in disease [45]. Furthermore, there is an emerging notion that single nucleotide polymorphisms in specific genes, such as that for ZnT3 (*SLC30A3*), may have an interaction with the nutrient status that impacts upon short- and long-term memory scores in the normal population [46]. Thus, changes in specific ZnT proteins, or indeed ZIPs or other alterations that alter the normal zinc regulatory apparatus, and that subsequently then precipitate the abnormal homeostasis of zinc within critical brain structures, are likely to alter cognition across both "healthy" and "pathological" ageing. ZnT4 is expressed in the endosomes/lysososmes, Golgi apparatus and trans-Golgi network (TGN) and cytoplasmic vesicles, in the brain, mammary glands and intestinal epithelial cells, where it is involved in vesicular secretory functions [1,23]. ZnT5 and ZnT6 uniquely form heterodimer complexes and function by delivering zinc into early secretory pathways, and both are located in the Golgi apparatus and TGN [23]. ZnT7 is expressed in the large and small intestine, and as such is involved in the absorption of dietary zinc, evidenced through poor growth and reduced adiposity in mice with a disruption in the ZnT7 gene [1,23]. ZnT8 is a pancreatic specific zinc transporter localized to the membrane of insulin storage granules [27]. ZnT9 has been classified as a zinc transporter, however, due to an absence of essential histidine residues, is believed to have no zinc transport functions [1]. ZnT10 has been localized to early/recycling endosomes or the Golgi apparatus, and data suggests that the primary function of ZnT10 is manganese transport. Indeed, patients with homozygous mutations in the ZnT10 gene show disturbances in cellular manganese homeostasis, not zinc perturbations [1].

In the brain, there is a low expression of ZnT2, 5, 7, and 8, however ZnT1, 3, 4, and 6 are highly expressed [23]. Whilst each of these proteins are essential in the modulation of intracellular cytosolic zinc in the brain, it should be noted that ZnT1 and ZnT3 are unique with regards to the other ZnTs in that ZnT1 is present on plasma membranes and ZnT3 is located on synaptic vesicle membranes, with the latter playing a significant role in cognition in aging and pathogenesis of disease.

## 6. Zinc Signal in Brain Diseases

There exists a wide-range of neurological diseases where zinc homeostasis is impacted and subsequently associated with the pathogenesis of the disorder. These include Alzheimer's disease (AD), amyotrophic lateral sclerosis (ALS), traumatic brain injury (TBI), depression, schizophrenia (SCZ), and Parkinson's disease (PD). The role of zinc in each of these disorders will be briefly discussed in the following sections.

### 6.1. Alzheimer's Disease

Alzheimer's disease is the most common progressive dementia affecting the elderly today. It is a multi-factorial disease of both genetic and non-genetic aetiology. The principal theory of AD pathogenesis has focused on the accumulation of amyloid-β (Aβ), a cleavage product of the much larger amyloid precursor protein (APP). Significant efforts have also focused on the role of abnormally phosphorylated forms of the microtubule associated protein tau in the onset and progression of AD.

The interest of zinc in AD pathogenesis is derived from the observation that zinc, above 300 nM concentration, can precipitate Aβ to result in its aggregation into senile plaques, one of the major pathological hallmarks of the disease [47]. The extracellular concentration of zinc during synaptic transmission rises to 300 μM, and as such it is possible that synaptic transmission could contribute to Aβ deposition in AD [48]. More recently, Deshpande and colleagues [49] also demonstrated that the zinc emanating from the glutamatergic synapse (following neurotransmission) was critical for the targeting of Aβ oligomers to the synapse (where the Aβ subsequently colocalised with the NMDA receptor subunit NR2B). Supporting this notion is the observation that β-amyloid deposits are pronounced in the neocortex, an area in which the highest zinc concentration occurs [48]. The amyloid plaques are also enriched in metals such as zinc [50,51], and Aβ itself is a metalloprotein containing binding sites for metals such as zinc [52]. Further evidence of zinc involvement in Aβ deposition in AD arose from the observation of a significantly reduced plaque load in the brains of Tg2576 transgenic mice (a common mouse model of AD that overexpresses mutant human APP) that were cross-bred with ZnT3 knockout mice, indicating that synaptic zinc does indeed contribute to amyloid deposition in the TG2576 mouse [53]. In contrast to this, plaque number and size have been shown to be enhanced by dietary zinc modulation (both by zinc supplementation, which also resulted in cognitive deficits [54,55], and severe zinc deficiency) in various transgenic mouse models of AD [56,57]. These seemingly paradoxical data may reflect the diverse roles played by zinc in the regulation of Aβ—as zinc has been shown to prevent the proteolytic degradation of Aβ by matrix metalloprotease 2 [58], and also to modulate the activity of α, γ and β-secretase [59] that processes APP to generate Aβ and various cleavage products. Together with the impact of zinc on other pathways, and indeed on other metals (such as copper) that are reported to be involved in the evolution and progression of AD-like neuropathology, it is clear that there is not a simple linear relationship between zinc and Aβ/AD. These observations are not definitive, however, with other studies suggesting a minor role for Aβ and zinc binding [60]. The apparent effect of zinc on modulating the toxicity of Aβ also appears to be concentration dependent, with low concentrations being protective [61]. Similarly, post mortem analysis of AD brain zinc concentrations was contradictory with several studies showing increased zinc levels in the AD brain [50,62], decreased levels in the AD brain [63,64], or no change at all [65].

There is an additional pathway through which zinc may be involved in the pathogenesis of AD, which involves its role in the development of the hyperphosphorylated tau protein and causing the polymerization and subsequent generation of neurofibrillary tangles (NFTs), the other major neuropathological feature of the AD brain. Zinc has been revealed to regulate phosphorylation of tau protein through the extracellular signal-regulated kinase pathway (MAP/ERK) [66]. Additionally, low micromolar zinc concentrations can cause the aggregation of human tau fragments [67–69], and in rat hippocampal slices synaptically released zinc has been demonstrated to promote the hyperphosphorylation of tau [70]. Recent studies have also demonstrated that zinc inactivates the major tau phosphatase, protein phosphatase 2A (PP2A), via an Src-dependent phosphorylation of PP2A to result in the hyperphosphorylation of tau [71]. Furthermore, zinc may bind to and directly mediate the toxicity of tau [72]. Adding validity to these observations is the fact that the use of zinc chelators or a blockade of synaptic zinc signaling abolishes zinc mediated tau hyperphosphorylation [70]. Cumulatively, these data support an interaction of zinc with the two key proteins, and their regulatory pathways, that are believed to drive disease pathogenesis in AD. This has been the subject of numerous reviews over the last two decades [48,73,74].

### 6.2. Amyotrophic Lateral Sclerosis

Amyotrophic lateral sclerosis (ALS) is a fatal neurodegenerative disease of the human motor system. It is both sporadic and familial in nature with around 5–10% of cases being familial and 90% sporadic. The most common cause of familial ALS is a mutation to the copper, zinc superoxide dismutase (*SOD1*) gene [75]. The mutation, which results in the SOD1 protein having a reduced affinity for zinc, leads to a toxic gain of function in motor neurons when zinc is missing from its active site [76]. In addition to the *SOD1* mutation, both ZnT3 and ZnT6 are downregulated in the spinal cords of patients with sporadic ALS, independent of the loss of motor neurons, suggesting that ZnTs may also have a role in disease pathogenesis [77]. A parallel study in *SOD1* (G93A) mutant transgenic mice, however, indicated that ZnT3 and ZnT6 protein levels were not altered in the mouse spinal cord before or after the onset of ALS symptoms when compared with controls [77]. In another study, it has also been shown that the levels of zinc are significantly higher in the cerebrospinal fluid (CSF) of patients with ALS [78], however, the precise mechanism underlying this elevation and the potential implications for this in disease are yet to be clarified.

### 6.3. Traumatic Brain Injury

Traumatic brain injury (TBI) is a disruption in the normal structure and/or function of the brain that can be attributed to a blow or jolt to the head, or by a penetrating head injury. TBI can destroy neurons via direct mechanical damage such as cellular membrane rupture and diffuse axonal injury (DAI), and indirectly through the ischemia [79]. The ischemia associated with TBI (and ischemic stroke) initiates a release of glutamate from pre-synaptic axon terminals after injury, leading to excitotoxicity and cell death of post-synaptic neurons [80]. Evidence also suggests that the synaptic release of zinc from presynaptic boutons can additionally cause injury and death to post synaptic neurons under excitotoxic conditions. Research findings in both ischemia [81] and status epilepticus [82], both of which occur as secondary complications in TBI [83,84], clearly demonstrate that zinc is translocated from zinc-containing presynaptic boutons to dying postsynaptic soma. In TBI, synaptically released zinc has been shown to contribute to neuronal injury [85], and occurs concomitantly with the glutamate release observed after head injury [86]. However, the role of zinc in the injured brain is not yet clearly defined, with multiple chelation [87,88] and supplementation [89,90] studies demonstrating both neuroprotective and neurotoxic roles for zinc in the pathogenesis of TBI. However, in a rat brain model of TBI, protective effects of zinc chelation were shown to be associated with the upregulation of neuroprotective genes in combination with decreased neuronal death, potentially indicating a toxic role of zinc in TBI [88].

### 6.4. Depression

Depression affects millions of individuals world-wide, and is comorbid with many neurodegenerative diseases [91]. The evidence for a role of zinc in depression has gained much traction over the last decade after the observation that depressed patients exhibited lower serum zinc levels than psychiatrically normal controls, and that zinc levels negatively correlated with the severity of depressive symptoms [92]. Moreover, other studies indicated that serum zinc levels may be normalized after successful antidepressant therapy [93,94]. One explanation as to the role of zinc in depression is its ability to regulate the NMDA receptor [95]. As noted previously, zinc can act as an antagonist to NMDA receptors, and studies of depression have shown that zinc and other antagonists of the NMDA receptor show antidepressant-like effects [96]. Rodent models of depression also support the notion that zinc and NMDA receptors are intimately involved in depression. A recent study by Szewczyk et al., indicated that zinc pre-treatment negated depressive features in the forced swim test, whereby it was observed that zinc treated rodents exhibit longer periods of escape behavior before immobility [97]. Moreover, other studies have demonstrated that zinc administration in rodents can reduce the number of NMDA receptor complexes, indicative of downregulation [98,99]. Other studies have examined the interaction of zinc in relation to the monoaminergic theory of depression, focusing on the serotonergic, noradrenergic and dopaminergic systems [100].

## 6.5. Schizophrenia

Schizophrenia (SCZ) is a long-term mental disorder typified by a breakdown in the association between thought, emotion, and behaviour, leading to faulty perception, inappropriate actions and moods, withdrawal from reality into fantasy, and delusion. It is a condition with both neurodegenerative and neurodevelopmental pathologies with potential causation from maternal zinc deficiency and genetic risk factors [101]. Supporting the notion of the contribution by maternal zinc deficiency is that prenatal zinc deficiency in rodent models produces decreased brain volume [102]. In human SCZ patients, there is a 30–50% reduction in brain zinc content demonstrated for early onset cases compared to control samples in postmortem brain tissue [103,104]. Whilst pre-natal zinc deficiency may not be solely causative of SCZ, interactions with other risk genes, and/or ongoing zinc deficiency following birth may be a contributing factor.

Recently, a genome-wide association study between SCZ and control patients revealed a nonsynonymous single nucleotide polymorphism (nsSNP) in the ZIP8 gene as a risk factor for SCZ [105]. Its direct effect upon zinc transport in SCZ, however, has yet to be elucidated. Additionally, allelic variants in the ZnT3 gene, *SLC30A3*, revealed an increased and gender-specific effect of allele on the risk of SCZ in females [106,107]. Similarly, another study demonstrated an increased cortical expression of the zinc transporter SLC39A12. The expression microarray study revealed messenger RNA (mRNA) for solute carrier family 39 (zinc transporter). Member 12 (*SLC39A12*) was higher in the dorsolateral prefrontal cortex from subjects with SCZ in comparison with controls [108]. These results suggest that a breakdown in zinc cellular homeostasis is likely a part of the pathophysiology of schizophrenia.

## 6.6. Parkinson's Disease

Parkinson's disease (PD) is a long-term degenerative disorder of the CNS that mainly affects the motor system. Symptoms develop slowly over time, the most obvious being shaking, rigidity, slowness of movement and difficulty with walking. Both thinking and behavioral problems can occur, and dementia is not uncommon at the advanced disease stage.

There is an observed clinical zinc deficiency in patients presenting with PD [109,110], and while scientific evidence of zinc supplementation in PD patients cohorts are scarce, other animal models of PD demonstrate the efficacy of zinc supplementation. A Drosophila melanogaster PD disease model in which the orthologue of the human Parkin gene was disrupted, was shown to have beneficial responses to zinc supplements [111]. Parkin mutant flies exhibit muscle abnormalities, locomotor defects, an inability to fly owing to the degeneration of indirect flight muscles, as well as a severely reduced lifespan, and as such mimic human PD symptoms. Zinc supplementation, however, ameliorated these deficits. In humans, early onset PD is associated with a mutation of *PARK9*, a lysosomal type 5 P-type ATPase, which has been shown to lead to a reduction of lysosomal zinc storage with an increase in cytosolic zinc and α-synuclein accumulation, a pathological hallmark of the disease [112,113]. Additionally, in PD patients there is an observed accumulation of zinc in the substantia nigra, caudate nucleus and lateral putamen areas associated with PD pathology [114]. Taken together, these data suggest that zinc homeostasis is distorted in PD, and as such may play a contributory role to the pathogenesis of the disease.

## 7. Conclusions

As noted in this review, irregularities in zinc homeostasis may represent a point of intersection for both the pathogenesis and the symptoms that characterize multiple neurodegenerative disorders, in addition to potentially also being involved in the ageing process and associated cognitive decline. If zinc is indeed critical to such a breadth of conditions, then understanding how and why zinc levels change across age and/or prior to or during disease is key to providing the insight necessary to harness the therapeutic potential of zinc (and conversely, to avoid any complications arising from potential zinc toxicity). For example, do zinc levels change because of dietary insufficiency (or perhaps some other aspect of diet or a disease state that alters the absorption of zinc)? Is it due to an age- or disease-related change in zinc importer/transporter levels or function that alters the distribution of zinc? Or perhaps some other change in a different metal or aspect of the metalloproteome that adversely impacts zinc? Is there just one or multiple zinc signaling pathways that are impaired? All these questions and many more require thorough interrogation in order to optimize a zinc-based targeted therapy (e.g., at what stage might zinc supplementation versus chelation be optimal in a given disease state; if it is supplementation, then is bulk dietary modulation sufficient, or is a more targeted pharmacological approach required?). Answers to these questions will ultimately need to be validated in a human clinical trial in order to gain the burden of proof necessary for the wide spread acceptance of zinc as a critical player, and therapeutic target, in disorders of the CNS.

**Acknowledgments:** The Florey Institute of Neuroscience and Mental Health acknowledge the strong support from the Victorian Government and in particular the funding from the Operational Infrastructure Support Grant.

**Conflicts of Interest:** The authors declare no conflict of interest.

## References

1. Kambe, T.; Tsuji, T.; Hashimoto, A.; Itsumura, N. The Physiological, Biochemical, and Molecular Roles of Zinc Transporters in Zinc Homeostasis and Metabolism. *Physiol. Rev.* **2015**, *95*, 749–784. [CrossRef] [PubMed]
2. Andreini, C.; Banci, L.; Bertini, I.; Rosato, A. Counting the zinc-proteins encoded in the human genome. *J. Proteome Res.* **2006**, *5*, 196–201. [CrossRef] [PubMed]
3. Wu, S.; Lao, X.Y.; Sun, T.T.; Ren, L.L.; Kong, X.; Wang, J.L.; Wang, Y.C.; Du, W.; Yu, Y.N.; Weng, Y.R.; et al. Knockdown of ZFX inhibits gastric cancer cell growth in vitro and in vivo via downregulating the ERK-MAPK pathway. *Cancer Lett.* **2013**, *337*, 293–300. [CrossRef] [PubMed]
4. Townsend, S.F.; Briggs, K.K.; Krebs, N.F.; Hambidge, K.M. Zinc supplementation selectively decreases fetal hepatocyte DNA synthesis and insulin-like growth factor II gene expression in primary culture. *Pediatr. Res.* **1994**, *35*, 404–408. [CrossRef] [PubMed]
5. Giugliano, R.; Millward, D.J. The effects of severe zinc deficiency on protein turnover in muscle and thymus. *Br. J. Nutr.* **1987**, *57*, 139–155. [CrossRef] [PubMed]
6. Rojas, A.I.; Phillips, T.J. Patients with chronic leg ulcers show diminished levels of vitamins A and E, carotenes, and zinc. *Dermatol. Surg.* **1999**, *25*, 601–604. [CrossRef] [PubMed]
7. Beck, F.W.; Prasad, A.S.; Kaplan, J.; Fitzgerald, J.T.; Brewer, G.J. Changes in cytokine production and T cell subpopulations in experimentally induced zinc-deficient humans. *Am. J. Physiol.* **1997**, *272*, E1002–E1007. [PubMed]
8. Takeda, A. Significance of $Zn^{(2+)}$ signaling in cognition: Insight from synaptic $Zn^{(2+)}$ dyshomeostasis. *J. Trace Elem. Med. Biol.* **2014**, *28*, 393–396. [CrossRef] [PubMed]
9. Wessells, K.R.; Brown, K.H. Estimating the global prevalence of zinc deficiency: Results based on zinc availability in national food supplies and the prevalence of stunting. *PLoS ONE* **2012**, *7*, e50568. [CrossRef] [PubMed]
10. Bertoni-Freddari, C.; Fattoretti, P.; Casoli, T.; Di Stefano, G.; Giorgetti, B.; Balietti, M. Brain aging: The zinc connection. *Exp. Gerontol.* **2008**, *43*, 389–393. [CrossRef] [PubMed]
11. Maret, W.; Sandstead, H.H. Zinc requirements and the risks and benefits of zinc supplementation. *J. Trace Elem. Med. Biol.* **2006**, *20*, 3–18. [CrossRef] [PubMed]
12. Mocchegiani, E.; Giacconi, R.; Cipriano, C.; Muzzioli, M.; Fattoretti, P.; Bertoni-Freddari, C.; Isani, G.; Zambenedetti, P.; Zatta, P. Zinc-bound metallothioneins as potential biological markers of ageing. *Brain Res. Bull.* **2001**, *55*, 147–153. [CrossRef]
13. Nuttall, J.R.; Oteiza, P.I. Zinc and the aging brain. *Genes Nutr.* **2014**, *9*, 379. [CrossRef] [PubMed]
14. Szewczyk, B. Zinc homeostasis and neurodegenerative disorders. *Front. Aging Neurosci.* **2013**, *5*, 33. [CrossRef] [PubMed]

15.   Mocchegiani, E.; Bertoni-Freddari, C.; Marcellini, F.; Malavolta, M. Brain, aging and neurodegeneration: Role of zinc ion availability. *Prog. Neurobiol.* **2005**, *75*, 367–390. [CrossRef] [PubMed]

16.   Frederickson, C.J.; Suh, S.W.; Silva, D.; Frederickson, C.J.; Thompson, R.B. Importance of zinc in the central nervous system: The zinc-containing neuron. *J. Nutr.* **2000**, *130*, 1471S–1483S. [PubMed]

17.   Wang, Z.Y.; Li, J.Y.; Danscher, G.; Dahlstrom, A. Localization of zinc-enriched neurons in the mouse peripheral sympathetic system. *Brain Res.* **2002**, *928*, 165–174. [CrossRef]

18.   Liuzzi, J.P.; Cousins, R.J. Mammalian zinc transporters. *Annu. Rev. Nutr.* **2004**, *24*, 151–172. [CrossRef] [PubMed]

19.   Manso, Y.; Adlard, P.A.; Carrasco, J.; Vasak, M.; Hidalgo, J. Metallothionein and brain inflammation. *J. Biol. Inorg. Chem.* **2011**, *16*, 1103–1113. [CrossRef] [PubMed]

20.   Thirumoorthy, N.; Sunder, A.S.; Kumar, K.T.M.; Kumar, M.S.; Ganesh, G.N.K.; Chatterjee, M. A Review of Metallothionein Isoforms and their Role in Pathophysiology. *World J. Surg. Oncol.* **2011**, *9*, 54. [CrossRef] [PubMed]

21.   Cousins, R.J.; Liuzzi, J.P.; Lichten, L.A. Mammalian zinc transport, trafficking, and signals. *J. Biol. Chem.* **2006**, *281*, 24085–24089. [CrossRef] [PubMed]

22.   Levenson, C.W.; Tassabehji, N.M. Role and Regulation of Copper and Zinc Transport Proteins in the Central Nervous System. In *Handbook of Neurochemistry and Molecular Neurobiology*; Abel, L., Reith, M.E.A., Eds.; Springer: New York, NY, USA, 2007; pp. 257–284.

23.   Hancock, S.M.; Bush, A.I.; Adlard, P.A. The clinical implications of impaired zinc signaling in the brain. In *Zinc Signals in Cellular Functions and Disorders*; Springer: Tokyo, Janpan, 2014.

24.   Coudray, N.; Valvo, S.; Hu, M.; Lasala, R.; Kim, C.; Vink, M.; Zhou, M.; Provasi, D.; Filizola, M.; Tao, J.; et al. Inward-facing conformation of the zinc transporter YiiP revealed by cryoelectron microscopy. *Proc. Natl. Acad. Sci. USA* **2013**, *110*, 2140–2145. [CrossRef] [PubMed]

25.   Palmiter, R.D.; Findley, S.D. Cloning and functional characterization of a mammalian zinc transporter that confers resistance to zinc. *EMBO J.* **1995**, *14*, 639–649. [PubMed]

26.   Tsuda, M.; Imaizumi, K.; Katayama, T.; Kitagawa, K.; Wanaka, A.; Tohyama, M.; Takagi, T. Expression of zinc transporter gene, ZnT-1, is induced after transient forebrain ischemia in the gerbil. *J. Neurosci.* **1997**, *17*, 6678–6684. [PubMed]

27.   Lichten, L.A.; Cousins, R.J. Mammalian zinc transporters: Nutritional and physiologic regulation. *Annu. Rev. Nutr.* **2009**, *29*, 153–176. [CrossRef] [PubMed]

28.   Cole, T.B.; Wenzel, H.J.; Kafer, K.E.; Schwartzkroin, P.A.; Palmiter, R.D. Elimination of zinc from synaptic vesicles in the intact mouse brain by disruption of the ZnT3 gene. *Proc. Natl. Acad. Sci. USA* **1999**, *96*, 1716–1721. [CrossRef] [PubMed]

29.   Palmiter, R.D.; Cole, T.B.; Quaife, C.J.; Findley, S.D. ZnT-3, a putative transporter of zinc into synaptic vesicles. *Proc. Natl. Acad. Sci. USA* **1996**, *93*, 14934–14939. [CrossRef] [PubMed]

30.   Takeda, A.; Tamano, H. Proposed glucocorticoid-mediated zinc signaling in the hippocampus. *Metallomics* **2012**, *4*, 614–618. [CrossRef] [PubMed]

31.   Fukada, T.; Yamasaki, S.; Nishida, K.; Murakami, M.; Hirano, T. Zinc homeostasis and signaling in health and diseases: Zinc signaling. *J. Biol. Inorg. Chem.* **2011**, *16*, 1123–1134. [CrossRef] [PubMed]

32.   Besser, L.; Chorin, E.; Sekler, I.; Silverman, W.F.; Atkin, S.; Russell, J.T.; Hershfinkel, M. Synaptically Released Zinc Triggers Metabotropic Signaling via a Zinc-Sensing Receptor in the Hippocampus. *J. Neurosci.* **2009**, *29*, 2890–2901. [CrossRef] [PubMed]

33.   Huang, Y.Z.; Pan, E.; Xiong, Z.Q.; McNamara, J.O. Zinc-mediated transactivation of TrkB potentiates the hippocampal mossy Fiber-CA3 pyramid synapse. *Neuron* **2008**, *57*, 546–558. [CrossRef] [PubMed]

34.   Paoletti, P.; Vergnano, A.M.; Barbour, B.; Casado, M. Zinc at Glutamatergic Synapses. *Neuroscience* **2009**, *158*, 126–136. [CrossRef] [PubMed]

35.   Lee, J.Y.; Kim, Y.J.; Kim, T.Y.; Koh, J.Y.; Kim, Y.H. Essential Role for Zinc-Triggered p75(NTR) Activation in Preconditioning Neuroprotection. *J. Neurosci.* **2008**, *28*, 10919–10927. [CrossRef] [PubMed]

36.   Frederickson, C.J.; Koh, J.Y.; Bush, A.I. The neurobiology of zinc in health and disease. *Nat. Rev. Neurosci.* **2005**, *6*, 449–462. [CrossRef] [PubMed]

37.   Haase, H.; Rink, L. Zinc Signaling. In *Zinc in Human Health*; IOS Press: Amsterdam, The Netherlands, 2011; pp. 94–117.

38.  Fujise, Y.; Kubota, M.; Suzuki, M.; Tamano, H.; Takeda, A. Blockade of intracellular $Zn^{2+}$ signaling in the basolateral amygdala affects object recognition memory via attenuation of dentate gyrus LTP. *Neurochem. Int.* **2017**, *108*, 1–6. [CrossRef] [PubMed]

39.  Takeda, A.; Koike, Y.; Osaw, M.; Tamano, H. Characteristic of Extracellular $Zn^{2+}$ Influx in the Middle-Aged Dentate Gyrus and Its Involvement in Attenuation of LTP. *Mol. Neurobiol.* **2017**, 1–11. [CrossRef] [PubMed]

40.  Takeda, A.; Suzuki, M.; Tempaku, M.; Ohashi, K.; Tamano, H. Influx of extracellular $Zn(^{2+})$ into the hippocampal CA1 neurons is required for cognitive performance via long-term potentiation. *Neuroscience* **2015**, *304*, 209–216. [CrossRef] [PubMed]

41.  Takeda, A.; Tamano, H.; Hisatsune, M.; Murakami, T.; Nakada, H.; Fujii, H. Maintained LTP and Memory Are Lost by $Zn^{2+}$ Influx into Dentate Granule Cells, but Not $Ca^{2+}$ Influx. *Mol. Neurobiol.* **2017**. [CrossRef] [PubMed]

42.  Takeda, A.; Tamano, H.; Ogawa, T.; Takada, S.; Nakamura, M.; Fujii, H.; Ando, M. Intracellular $Zn^{2+}$ Signaling in the Dentate Gyrus Is Required for Object Recognition Memory. *Hippocampus* **2014**, *24*, 1404–1412. [CrossRef] [PubMed]

43.  Tamano, H.; Minamino, T.; Fujii, H.; Takada, S.; Nakamura, M.; Ando, M.; Takeda, A. Blockade of intracellular $Zn^{2+}$ signaling in the dentate gyrus erases recognition memory via impairment of maintained LTP. *Hippocampus* **2015**, *25*, 952–962. [CrossRef] [PubMed]

44.  Adlard, P.A.; Parncutt, J.M.; Finkelstein, D.I.; Bush, A.I. Cognitive loss in zinc transporter-3 knock-out mice: A phenocopy for the synaptic and memory deficits of Alzheimer's disease? *J. Neurosci.* **2010**, *30*, 1631–1636. [CrossRef] [PubMed]

45.  Beyer, N.; Coulson, D.T.; Heggarty, S.; Ravid, R.; Irvine, G.B.; Hellemans, J.; Johnston, J.A. ZnT3 mRNA levels are reduced in Alzheimer's disease post-mortem brain. *Mol. Neurodegener.* **2009**, *4*, 53. [CrossRef] [PubMed]

46.  Da Rocha, T.J.; Blehm, C.J.; Bamberg, D.P.; Fonseca, T.L.R.; Tisser, L.A.; de Oliveira, A.A.; de Andrade, F.M.; Fiegenbaum, M. The effects of interactions between selenium and zinc serum concentration and SEP15 and *SLC30A3* gene polymorphisms on memory scores in a population of mature and elderly adults. *Genes Nutr.* **2014**, *9*, 377. [CrossRef] [PubMed]

47.  Bush, A.I.; Pettingell, W.H.; Multhaup, G.; d Paradis, M.; Vonsattel, J.P.; Gusella, J.F.; Beyreuther, K.; Masters, C.L.; Tanzi, R.E. Rapid induction of Alzheimer Aβ amyloid formation by zinc. *Science* **1994**, *265*, 1464–1467. [CrossRef] [PubMed]

48.  Bush, A.I.; Tanzi, R.E. The galvanization of β-amyloid in Alzheimer's disease. *Proc. Natl. Acad. Sci. USA* **2002**, *99*, 7317–7319. [CrossRef] [PubMed]

49.  Deshpande, A.; Kawai, H.; Metherate, R.; Glabe, C.G.; Busciglio, J. A role for synaptic zinc in activity-dependent Aβ oligomer formation and accumulation at excitatory synapses. *J. Neurosci.* **2009**, *29*, 4004–4015. [CrossRef] [PubMed]

50.  Lovell, M.A.; Robertson, J.D.; Teesdale, W.J.; Campbell, J.L.; Markesbery, W.R. Copper, iron and zinc in Alzheimer's disease senile plaques. *Neurol. Sci.* **1998**, *158*, 47–52. [CrossRef]

51.  Maynard, C.J.; Bush, A.I.; Masters, C.L.; Cappai, R.; Li, Q.X. Metals and amyloid-β in Alzheimer's disease. *Int. J. Exp. Pathol.* **2005**, *86*, 147–159. [CrossRef] [PubMed]

52.  Bush, A.I.; Multhaup, G.; Moir, R.D.; Williamson, T.G.; Small, D.H.; Rumble, B.; Pollwein, P.; Beyreuther, K.; Masters, C.L. A novel zinc(II) binding site modulates the function of the β A4 amyloid protein precursor of Alzheimer's disease. *J. Biol. Chem.* **1993**, *268*, 16109–16112. [PubMed]

53.  Lee, J.Y.; Cole, T.B.; Palmiter, R.D.; Suh, S.W.; Koh, J.Y. Contribution by synaptic zinc to the gender-disparate plaque formation in human Swedish mutant APP transgenic mice. *Proc. Natl. Acad. Sci. USA* **2002**, *99*, 7705–7710. [CrossRef] [PubMed]

54.  Linkous, D.H.; Adlard, P.A.; Wanschura, P.B.; Conko, K.M.; Flinn, J.M. The effects of enhanced zinc on spatial memory and plaque formation in transgenic mice. *J. Alzheimers Dis.* **2009**, *18*, 565–579. [CrossRef] [PubMed]

55.  Railey, A.M.; Groeber, C.M.; Flinn, J.M. The effect of metals on spatial memory in a transgenic mouse model of Alzheimer's disease. *J. Alzheimers Dis.* **2011**, *24*, 375–381. [PubMed]

56.  Stoltenberg, M.; Bush, A.I.; Bach, G.; Smidt, K.; Larsen, A.; Rungby, J.; Lund, S.; Doering, P.; Danscher, G. Amyloid plaques arise from zinc-enriched cortical layers in APP/PS1 transgenic mice and are paradoxically enlarged with dietary zinc deficiency. *Neuroscience* **2007**, *150*, 357–369. [CrossRef] [PubMed]

57. Wang, C.Y.; Wang, T.; Zheng, W.; Zhao, B.L.; Danscher, G.; Chen, Y.H.; Wang, Z.Y. Zinc overload enhances APP cleavage and Aβ deposition in the Alzheimer mouse brain. *PLoS ONE* **2010**, *5*, e15349. [CrossRef] [PubMed]

58. Crouch, P.J.; Savva, M.S.; Hung, L.W.; Donnelly, P.S.; Mot, A.I.; Parker, S.J.; Greenough, M.A.; Volitakis, I.; Adlard, P.A.; Cherny, R.A.; et al. The Alzheimer's therapeutic PBT2 promotes amyloid-β degradation and GSK3 phosphorylation via a metal chaperone activity. *J. Neurochem.* **2011**, *119*, 220–230. [CrossRef] [PubMed]

59. Capasso, M.; Jeng, J.M.; Malavolta, M.; Mocchegiani, E.; Sensi, S.L. Zinc dyshomeostasis: A key modulator of neuronal injury. *J. Alzheimers Dis.* **2005**, *8*, 93–108. [CrossRef] [PubMed]

60. Clements, A.; Allsop, D.; Walsh, D.M.; Williams, C.H. Aggregation and metal-binding properties of mutant forms of the amyloid Aβ peptide of Alzheimer's disease. *J. Neurochem.* **1996**, *66*, 740–747. [CrossRef] [PubMed]

61. Moreira, P.; Pereira, C.; Santos, M.S.; Oliveira, C. Effect of zinc ions on the cytotoxicity induced by the amyloid β-peptide. *Antioxid. Redox Signal.* **2000**, *2*, 317–325. [CrossRef] [PubMed]

62. Danscher, G.; Jensen, K.B.; Frederickson, C.J.; Kemp, K.; Andreasen, A.; Juhl, S.; Stoltenberg, M.; Ravid, R. Increased amount of zinc in the hippocampus and amygdala of Alzheimer's diseased brains: A proton-induced X-ray emission spectroscopic analysis of cryostat sections from autopsy material. *J. Neurosci. Methods* **1997**, *76*, 53–59. [CrossRef]

63. Andrasi, E.; Farkas, E.; Gawlik, D.; Rosick, U.; Bratter, P. Brain Iron and Zinc Contents of German Patients with Alzheimer Disease. *J. Alzheimers Dis.* **2000**, *2*, 17–26. [CrossRef] [PubMed]

64. Corrigan, F.M.; Reynolds, G.P.; Ward, N.I. Hippocampal tin, aluminum and zinc in Alzheimer's disease. *Biometals* **1993**, *6*, 149–154. [CrossRef] [PubMed]

65. Rulon, L.L.; Robertson, J.D.; Lovell, M.A.; Deibel, M.A.; Ehmann, W.D.; Markesber, W.R. Serum zinc levels and Alzheimer's disease. *Biol. Trace Elem. Res.* **2000**, *75*, 79–85. [CrossRef]

66. Kim, I.; Park, E.J.; Seo, J.; Ko, S.J.; Lee, J.; Kim, C.H. Zinc stimulates tau S214 phosphorylation by the activation of Raf/mitogen-activated protein kinase-kinase/extracellular signal-regulated kinase pathway. *Neuroreport* **2011**, *22*, 839–844. [CrossRef] [PubMed]

67. Mo, Z.Y.; Zhu, Y.Z.; Zhu, H.L.; Fan, J.B.; Chen, J.; Liang, Y. Low micromolar zinc accelerates the fibrillization of human tau via bridging of Cys-291 and Cys-322. *J. Biol. Chem.* **2009**, *284*, 34648–34657. [CrossRef] [PubMed]

68. An, W.L.; Bjorkdahl, C.; Liu, R.; Cowburn, R.F.; Winblad, B.; Pei, J.J. Mechanism of zinc-induced phosphorylation of p70 S6 kinase and glycogen synthase kinase 3β in SH-SY5Y neuroblastoma cells. *J. Neurochem.* **2005**, *92*, 1104–1115. [CrossRef] [PubMed]

69. Pei, J.J.; An, W.L.; Zhou, X.W.; Nishimura, T.; Norberg, J.; Benedikz, E.; Gotz, J.; Winblad, B. P70 S6 kinase mediates tau phosphorylation and synthesis. *FEBS Lett.* **2006**, *580*, 107–114. [CrossRef] [PubMed]

70. Sun, X.Y.; Wei, Y.P.; Xiong, Y.; Wang, X.C.; Xie, A.J.; Wang, X.L.; Yang, Y.; Wang, Q.; Lu, Y.M.; Liu, R.W.J. Synaptic released zinc promotes tau hyperphosphorylation by inhibition of protein phosphatase 2A (PP2A). *J. Biol. Chem.* **2012**, *287*, 11174–11182. [CrossRef] [PubMed]

71. Xiong, Y.; Jing, X.P.; Zhou, X.W.; Wang, X.L.; Yang, Y.; Sun, X.Y.; Qiu, M.; Cao, F.Y.; Lu, Y.M.; Liu, R.; et al. Zinc induces protein phosphatase 2A inactivation and tau hyperphosphorylation through Src dependent PP2A (tyrosine 307) phosphorylation. *Neurobiol. Aging* **2013**, *34*, 745–756. [CrossRef] [PubMed]

72. Huang, Y.; Wu, Z.; Cao, Y.; Lang, M.; Lu, B.; Zhou, B. Zinc binding directly regulates tau toxicity independent of tau hyperphosphorylation. *Cell Rep.* **2014**, *8*, 831–842. [CrossRef] [PubMed]

73. Adlard, P.A.; Bush, A.I. Metals and Alzheimer's disease. *J. Alzheimers Dis.* **2006**, *10*, 145–163. [CrossRef] [PubMed]

74. Bush, A.I.; Tanzi, R.E. The role of zinc in the cerebral deposition of Aβ amyloid in Alzheimer's disease. In *Research Advances in Alzheimer's Disease and Related Disorders*, 1st ed.; Khalid, I., James, M., Bengt, W., Henry, W., Eds.; Wiley: Ann Arbor, MI, USA, 1995; pp. 607–618.

75. Rosen, D.R.; Siddique, T.; Patterson, D.; Figlewicz, D.A.; Sapp, P.; Hentati, A.; Donaldson, D.; Goto, J.; O'Regan, J.P.; Deng, H.X.; et al. Mutations in Cu/Zn superoxide dismutase gene are associated with familial amyotrophic lateral sclerosis. *Nature* **1993**, *362*, 59–62. [CrossRef] [PubMed]

76. Roberts, B.R.; Tainer, J.A.; Getzoff, E.D.; Malencik, D.A.; Anderson, S.R.; Bomben, V.C.; Meyers, K.R.; Karplus, P.A.; Beckman, J.S. Structural characterization of zinc-deficient human superoxide dismutase and implications for ALS. *J. Mol. Biol.* **2007**, *373*, 877–890. [CrossRef] [PubMed]

77. Kaneko, M.; Noguchi, T.; Ikegami, S.; Sakurai, T.; Kakita, A.; Toyoshima, Y.; Kambe, T.; Yamada, M.; Inden, M.; Hara, H.; et al. Zinc transporters ZnT3 and ZnT6 are downregulated in the spinal cords of patients with sporadic amyotrophic lateral sclerosis. *J. Neurosci. Res.* **2015**, *93*, 370–379. [CrossRef] [PubMed]

78. Hozumi, I.; Hasegawa, T.; Honda, A.; Ozawa, K.; Hayashi, Y.; Hashimoto, K.; Yamada, M.; Koumura, A.; Sakurai, T.; Kimura, A.; et al. Patterns of levels of biological metals in CSF differ among neurodegenerative diseases. *J. Neurol. Sci.* **2011**, *303*, 95–99. [CrossRef] [PubMed]

79. Veenith, T.V.; Carter, E.L.; Geeraerts, T.; Grossac, J.; Newcombe, V.F.; Outtrim, J.; Gee, G.S.; Lupson, V.; Smith, R.; Aigbirhio, F.I.; et al. Pathophysiologic Mechanisms of Cerebral Ischemia and Diffusion Hypoxia in Traumatic Brain Injury. *JAMA Neurol.* **2016**, *73*, 542–550. [CrossRef] [PubMed]

80. Arundine, M.; Tymianski, M. Molecular mechanisms of glutamate-dependent neurodegeneration in ischemia and traumatic brain injury. *Cell. Mol. Life Sci.* **2004**, *61*, 657–668. [CrossRef] [PubMed]

81. Tonder, N.; Johansen, F.F.; Frederickson, C.J.; Zimmer, J.; Diemer, N.H. Possible role of zinc in the selective degeneration of dentate hilar neurons after cerebral ischemia in the adult rat. *Neurosci. Lett.* **1990**, *109*, 247–252. [CrossRef]

82. Frederickson, C.J.; Hernandez, M.D.; McGinty, J.F. Translocation of zinc may contribute to seizure-induced death of neurons. *Brain Res.* **1989**, *480*, 317–321. [CrossRef]

83. Coles, J.P. Regional ischemia after head injury. *Curr. Opin. Crit. Care* **2004**, *10*, 120–125. [CrossRef] [PubMed]

84. Peets, A.D.; Berthiaume, L.R.; Bagshaw, S.M.; Federico, P.; Doig, C.J.; Zygun, D.A. Prolonged refractory status epilepticus following acute traumatic brain injury: A case report of excellent neurological recovery. *Crit. Care* **2005**, *9*, R725–R728. [CrossRef] [PubMed]

85. Suh, S.W.; Chen, J.W.; Motamedi, M.; Bell, B.; Listiak, K.; Pons, N.F.; Danscher, G.; Frederickson, C.J. Evidence that synaptically-released zinc contributes to neuronal injury after traumatic brain injury. *Brain Res.* **2000**, *852*, 268–273. [CrossRef]

86. Faden, A.I.; Demediuk, P.; Panter, S.S.; Vink, R. The role of excitatory amino acids and NMDA receptors in traumatic brain injury. *Science* **1989**, *244*, 798–800. [CrossRef] [PubMed]

87. Doering, P.; Stoltenberg, M.; Penkowa, M.; Rungby, J.; Larsen, A.; Danscher, G. Chemical blocking of zinc ions in CNS increases neuronal damage following traumatic brain injury (TBI) in mice. *PLoS ONE* **2010**, *5*, e10131. [CrossRef] [PubMed]

88. Hellmich, H.L.; Frederickson, C.J.; DeWitt, D.S.; Saban, R.; Parsley, M.O.; Stephenson, R.; Velasco, M.; Uchida, T.; Shimamura, M.; Prough, D.S. Protective effects of zinc chelation in traumatic brain injury correlate with upregulation of neuroprotective genes in rat brain. *Neurosci. Lett.* **2004**, *355*, 221–225. [CrossRef] [PubMed]

89. Cope, E.C.; Morris, D.R.; Scrimgeour, A.G.; VanLandingham, J.W.; Levenson, C.W. Zinc supplementation provides behavioral resiliency in a rat model of traumatic brain injury. *Physiol. Behav.* **2011**, *104*, 942–947. [CrossRef] [PubMed]

90. Cope, E.C.; Morris, D.R.; Scrimgeour, A.G.; Levenson, C.W. Scrimgeour; Levenson, C.W. Use of Zinc as a Treatment for Traumatic Brain Injury in the Rat: Effects on Cognitive and Behavioral Outcomes. Neurorehabilit. *Neural Repair* **2012**, *26*, 907–913. [CrossRef] [PubMed]

91. Grabrucker, A.M.; Rowan, M.; Garner, C.C. Brain-Delivery of Zinc-Ions as Potential Treatment for Neurological Diseases: Mini Review. *Drug Deliv. Lett.* **2011**, *1*, 13–23. [PubMed]

92. Maes, M.; D'Haese, P.C.; Scharpe, S.; D'Hondt, P.; Cosyns, P.; de Broe, M.E. Hypozincemia in depression. *J. Affect. Disord.* **1994**, *31*, 135–140. [CrossRef]

93. Maes, M.; Vandoolaeghe, E.; Neels, H.; Demedts, P.; Wauters, A.; Meltzer, H.Y.; Altamura, C.; Desnyder, R. Lower serum zinc in major depression is a sensitive marker of treatment resistance and of the immune/inflammatory response in that illness. *Biol. Psychiatry* **1997**, *42*, 349–358. [CrossRef]

94. McLoughlin, I.J.; Hodge, J.S. Zinc in depressive disorder. *Acta Psychiatr. Scand.* **1990**, *82*, 451–453. [CrossRef] [PubMed]

95. Low, C.M.; Zheng, F.; Lyuboslavsky, P.; Traynelis, S.F. Molecular determinants of coordinated proton and zinc inhibition of N-methyl-D-aspartate NR1/NR2A receptors. *Proc. Natl. Acad. Sci. USA* **2000**, *97*, 11062–11067. [CrossRef] [PubMed]

96. Szewczyk, B.; Poleszak, E.; Sowa-Kucma, M.; Siwek, M.; Dudek, D.; Ryszewska-Pokrasniewicz, B.; Radziwon-Zaleska, M.; Opoka, W.; Czekaj, J.; Pilc, A.; et al. Antidepressant activity of zinc and magnesium in view of the current hypotheses of antidepressant action. *Pharmacol. Rep.* **2008**, *60*, 588–589. [PubMed]

97.  Szewczyk, B.; Poleszak, E.; Sowa-Kucma, M.; Wrobel, A.; Slotwinski, S.; Listos, J.; Wlaz, P.; Cichy, A.; Siwek, A.; Dybala, M.; et al. The involvement of NMDA and AMPA receptors in the mechanism of antidepressant-like action of zinc in the forced swim test. *Amino Acids* **2010**, *39*, 205–217. [CrossRef] [PubMed]

98.  Cichy, A.; Sowa-Kucma, M.; Legutko, B.; Pomierny-Chamiolo, L.; Siwek, A.; Piotrowska, A.; Szewczyk, B.; Poleszak, E.; Pilc, A.; Nowak, G. Zinc-induced adaptive changes in NMDA/glutamatergic and serotonergic receptors. *Pharmacol. Rep.* **2009**, *61*, 1184–1191. [CrossRef]

99.  Szewczyk, B.; Poleszak, E.; Wlaz, P.; Wrobel, A.; Blicharska, E.; Cichy, A.; Dybala, M.; Siwek, A.; Pomierny-Chamiolo, L.; Piotrowska, A.; et al. The involvement of serotonergic system in the antidepressant effect of zinc in the forced swim test. *Prog. Neuropsychopharmacol. Biol. Psychiatry* **2009**, *33*, 323–329. [CrossRef] [PubMed]

100. Doboszewska, U.; Wlaz, P.; Nowak, G.; Radziwon-Zaleska, M.; Cui, R.J.; Mlyniec, K. Zinc in the Monoaminergic Theory of Depression: Its Relationship to Neural Plasticity. *Neural Plast.* **2017**. [CrossRef] [PubMed]

101. Petrilli, M.A.; Kranz, T.M.; Kleinhaus, K.; Joe, P.; Getz, M.; Johnson, P.; Chao, M.V.; Malaspina, D. The Emerging Role for Zinc in Depression and Psychosis. *Front. Pharmacol.* **2017**, *8*, 414. [CrossRef] [PubMed]

102. Takeda, A.; Tamano, H. Insight into zinc signaling from dietary zinc deficiency. *Brain Res. Rev.* **2009**, *62*, 33–44. [CrossRef] [PubMed]

103. Kimura, K.; Kumura, J. Preliminary reports on the metabolismof trace elements in neuro psychiatric diseases. I. Zinc in schizophrenia. *Proc. Jpn. Acad. Sci.* **1965**, *41*, 943–947.

104. McLardy, T. Hippocampal zinc in chronic alcoholism and schizophrenia. *IRCS Med. Sci.* **1973**, *2*, 1010.

105. Carrera, N.; Arrojo, M.; Sanjuan, J.; Ramos-Rios, R.; Paz, E.; Suarez-Rama, J.J.; Paramo, M.; Agra, S.; Brenlla, J.; Martinez, S.; et al. Association study of nonsynonymous single nucleotide polymorphisms in schizophrenia. *Biol. Psychiatry* **2012**, *71*, 169–177. [CrossRef] [PubMed]

106. Perez-Becerril, C.; Morris, A.G.; Mortimer, A.; McKenna, P.J.; de Belleroche, J. Allelic variants in the zinc transporter-3 gene, SLC30A3, a candidate gene identified from gene expression studies, show gender-specific association with schizophrenia. *Eur. Psychiatry* **2014**, *29*, 172–178. [CrossRef] [PubMed]

107. Perez-Becerril, C.; Morris, A.G.; Mortimer, A.; McKenna, P.J.; de Belleroche, J. Common variants in the chromosome 2p23 region containing the *SLC30A3* (*ZnT3*) gene are associated with schizophrenia in female but not male individuals in a large collection of European samples. *Psychiatry Res.* **2016**, *246*, 335–340. [CrossRef] [PubMed]

108. Scarr, E.; Udawela, M.; Greenough, M.A.; Neo, J.; Suk Seo, M.; Money, T.T.; Upadhyay, A.; Bush, A.I.; Everall, I.P.; Thomas, E.A.; et al. Increased cortical expression of the zinc transporter SLC39A12 suggests a breakdown in zinc cellular homeostasis as part of the pathophysiology of schizophrenia. *NPJ Schizophr.* **2016**, *2*, 16002. [CrossRef] [PubMed]

109. Brewer, G.J.; Kanzer, S.H.; Zimmerman, E.A.; Molho, E.S.; Celmins, D.F.; Heckman, S.M.; Dick, R. Subclinical zinc deficiency in Alzheimer's disease and Parkinson's disease. *Am. J. Alzheimers Dis. Dement.* **2010**, *25*, 572–575. [CrossRef] [PubMed]

110. Forsleff, L.; Schauss, A.G.; Bier, I.D.; Stuart, S. Evidence of functional zinc deficiency in Parkinson's disease. *J. Altern. Complement. Med.* **1999**, *5*, 57–64. [CrossRef] [PubMed]

111. Saini, N.; Schaffner, W. Zinc supplement greatly improves the condition of parkin mutant Drosophila. *Biol. Chem.* **2010**, *391*, 513–518. [CrossRef] [PubMed]

112. Kong, S.M.; Chan, B.K.; Park, J.S.; Hill, K.J.; Aitken, J.B.; Cottle, L.; Farghaian, H.; Cole, A.R.; Lay, P.A.; Sue, C.M.; et al. Parkinson's disease-linked human PARK9/ATP13A2 maintains zinc homeostasis and promotes α-Synuclein externalization via exosomes. *Hum. Mol. Genet.* **2014**, *23*, 2816–2833. [CrossRef] [PubMed]

113. Tsunemi, T.; Krainc, D. Zn($^{2+}$) dyshomeostasis caused by loss of ATP13A2/PARK9 leads to lysosomal dysfunction and α-synuclein accumulation. *Hum. Mol. Genet.* **2014**, *23*, 27912801. [CrossRef] [PubMed]
114. Dexter, D.T.; Carayon, A.; Javoy-Agid, F.; Agid, Y.; Wells, F.R.; Daniel, S.E.; Lees, A.J.; Jenner, P.; Marsden, C.D. Alterations in the levels of iron, ferritin and other trace metals in Parkinson's disease and other neurodegenerative diseases affecting the basal ganglia. *Brain* **1991**, *114 Pt 4*, 1953–1975. [CrossRef] [PubMed]

International Journal of
*Molecular Sciences*

MDPI

*Review*

# Role of Zinc Homeostasis in the Pathogenesis of Diabetes and Obesity

**Ayako Fukunaka * and Yoshio Fujitani ***

Laboratory of Developmental Biology & Metabolism, Institute for Molecular & Cellular Regulation,
Gunma University, 3-39-15 Showa-machi, Maebashi, Gunma 371-8512, Japan
* Correspondence: fukunaka@gunma-u.ac.jp (A.F.); fujitani@gunma-u.ac.jp (Y.F.); Tel.: +81-27-220-8855 (Y.F.)

Received: 27 November 2017; Accepted: 2 February 2018; Published: 6 February 2018

**Abstract:** Zinc deficiency is a risk factor for obesity and diabetes. However, until recently, the underlying molecular mechanisms remained unclear. The breakthrough discovery that the common polymorphism in zinc transporter *SLC30A8*/ZnT8 may increase susceptibility to type 2 diabetes provided novel insights into the role of zinc in diabetes. Our group and others showed that altered ZnT8 function may be involved in the pathogenesis of type 2 diabetes, indicating that the precise control of zinc homeostasis is crucial for maintaining health and preventing various diseases, including lifestyle-associated diseases. Recently, the role of the zinc transporter ZIP13 in the regulation of beige adipocyte biogenesis was clarified, which indicated zinc homeostasis regulation as a possible therapeutic target for obesity and metabolic syndrome. Here we review advances in the role of zinc homeostasis in the pathophysiology of diabetes, and propose that inadequate zinc distribution may affect the onset of diabetes and metabolic diseases by regulating various critical biological events.

**Keywords:** zinc; zinc transporters; diabetes; obesity; ZnT8; ZIP13; pancreatic β cell; beige adipocyte; therapeutic target

## 1. Introduction

Type 2 diabetes is now a crucial health problem in many parts of the world. Type 2 diabetes mellitus (T2DM) is characterized by peripheral insulin resistance and pancreatic beta (β) cell dysfunction. The disease is thought to be caused by defects in insulin signaling or secretion, the activation of various stress pathways, and dysregulation of the central nervous system (CNS). It is well accepted that the most accurate predictor for developing T2DM is obesity. Therefore, much attention has also been paid to the contribution of nutrients and nutrient-sensing pathways in situations of chronic caloric excess. Most of the interest in the role of nutrients in diabetes is centered on macronutrients, such as carbohydrate and fat, but micronutrients, such as iron and zinc, are also closely associated with diabetes [1,2]. Whole-body level dysregulation of zinc is known to occur in both type 1 and type 2 diabetes. However, it remains unclear as to whether zinc deficiency causes the disease or is merely a consequence of the disease. A possible causal link between changes in zinc homeostasis and pancreatic β cell function was suggested in 2007 with the identification of an association between the risk of T2DM and polymorphisms in the *SLC30A8* gene, which encodes zinc transporter ZnT8 [3].

Several groups have been analyzing the roles of zinc homeostasis in the health and disease of endocrine organs, with particular focus on zinc transporter function. In this review, we will discuss the roles of zinc homeostasis in glucose metabolism, particularly in association with ZnT8. Furthermore, we will discuss the role of ZIP13 in beige adipocyte biogenesis and energy expenditure, which we have recently elucidated.

## 2. Zinc Homeostasis and Pancreatic β Cells

### 2.1. Insulin Biosynthesis in Pancreatic β Cells

Pancreatic β cells are known to contain very high concentrations of zinc compared with various other cells. In particular, insulin secretory granules have been shown to have the highest zinc content within β cells [4,5]. In vertebrates, three protein families have been shown to regulate cellular zinc homeostasis, namely, metallothioneins (MTs), zinc importers (ZIP, *SLC39A*), and zinc exporters (ZnT, *SLC30A*). MTs have been shown to bind zinc with low affinity, whereas zinc transporters mediate the compartmentalization of zinc into various organelles and vesicles for their storage, and to supply zinc to various proteins that require zinc for their function [6]. Nine ZnTs and 14 ZIP transporters have been identified to play important roles in whole body maintenance, as well as in zinc homeostasis at the cellular and subcellular levels. These transporters act either independently or coordinately, and in a cell-specific or tissue-specific manner [7,8]. ZnT8 plays a key role in the accumulation of zinc within insulin secretory granules [9]. Furthermore, zinc is essential for the appropriate synthesis of insulin, as well as its storage and structural stability [10].

Insulin comprises a hexamer of six insulin and two zinc molecules [11,12]. The mature insulin molecule comprises two polypeptide chains, namely, chains A and B. Initially, insulin mRNA is translated into an inactive preproinsulin molecule, which comprises two chains that are connected by a c-peptide, with the signal peptide at the N-terminus. Proinsulin is formed from preproinsulin by signal peptide cleavage in the endoplasmic reticulum (ER). Proinsulin then folds into the final three-dimensional structure, upon formation of the correct disulfide bonds. Subsequently, the proinsulin protein forms dimers via electrostatic interactions. Proinsulin hexamers are then formed by electrostatically-coupled proinsulin dimers and zinc binding to histidine residue 10 of the B chain (His B10) [13]. After entering the Golgi apparatus, insulin hexamer formation is completed upon the dissociation of c-peptide, mediated by prohormone convertase (PC) dissociation [14]. Insulin crystallization occurs under specific conditions in insulin secretory granules, in which both insulin and zinc exist in high concentrations and acidic pH is maintained [15,16]. This crystallized insulin can be observed as "dense core granules" by electron microscopy [17,18], and insulin crystals that are secreted from pancreatic β cells are believed to dissociate rapidly into monomers as they enter the bloodstream.

### 2.2. Zinc Supplementation in Diabetic Animals and Patients

Mice with zinc deficiency were found to have a decreased number of insulin granules in their pancreatic β cells [19], as well as impaired glucose-stimulated insulin secretion (GSIS) [20]. Since pancreatic β cells synthesize a large amount of ATP, this makes them prone to oxidative stress exposure, which can subsequently cause cellular damage [21]. As zinc is required for the actions of many antioxidative enzymes, including Cu-Zn-SOD (superoxide dismutase) [22] and catalase [23], a lack of zinc will lead to further damage of pancreatic β cells under oxidative stress, such as in T2DM.

Regarding humans, a prospective cohort study in the United States analyzed 82,000 women and demonstrated that low zinc intake results in a 17% increased risk of developing diabetes compared with those women taking sufficient amounts of zinc [24]. Recently, a study from China has reported a negative correlation between concentrations of plasma zinc and the onset of diabetes [25]. Interestingly, this report suggested that an interaction between SLC30A8 (ZnT8) dysfunction and decreased plasma zinc concentrations regulates glucose tolerance and diabetes. Furthermore, the authors suggested that a decrease in plasma zinc concentrations as well as ZnT8 function may coordinately increase the risk of diabetes. Although these data suggest that zinc supplementation prevents disruption of glucose homeostasis, particularly in people with zinc deficiency, prospective intervention studies should be performed to clarify the efficacy of zinc supplementation in preventing the onset of diabetes. On the other hand, excessive supplementation with zinc may have deleterious effects, as excessive zinc intake may cause an undesirable increase in HbA1c levels and high blood pressure [26].

A genome-wide association study (GWAS) demonstrated that a nonsynonymous single-nucleotide polymorphism, namely, rs13266634 in the *SLC30A8* gene, results in the replacement of tryptophan-325 to arginine, which modestly increases the risk of T2DM [3]. Furthermore, recent studies on human SNPs demonstrated the association of 12 rare loss-of-function ZnT8 mutants with a 65% decreased risk of T2DM [27]. Taken together, the data indicate that polymorphisms in the *SLC30A8* gene are associated with altered risk of T2DM.

## 3. ZnT8 Plays a Crucial Role in Glucose Homeostasis

### 3.1. Insulin Secretory Granule of ZnT8-KO Mice

ZnT8 is found in the plasma membrane of insulin secretory granules of pancreatic β cells, and is implicated in zinc transport into insulin secretory granules [9]. Several groups, including our own have aimed to clarify the role of ZnT8 in glucose homeostasis by establishing *SLC30A8*-deficient (ZnT8-KO) mice [18,28,29] (Supplemental Table S1). Each mouse model shows variation in certain phenotype traits, which are attributed to differences in deletion strategy, genetic background, and housing condition. Most of the ZnT8-KO mice have been reported to have mildly impaired glucose tolerance and there have been no reports of the improvement of glucose tolerance in the ZnT8-KO mice model [18,30–35] (Supplemental Table S1). Furthermore, hZnT8 transgenic mice showed mildly improved glucose tolerance [31], suggesting that the expression levels of ZnT8 determines the risk of T2DM in these mouse models. Electron microscopy analysis revealed that dense-core granules, which are a hallmark of crystallized insulin usually seen in normal β cells, were absent in the ZnT8-KO mouse β cells. In most of the ZnT8-KO mice, some of the granules appeared atypical granules possessing abnormal "rod-like" or empty cores, while some ZnT8-KO mice revealed the nearly complete loss of crystal containing granules [18,30].

### 3.2. Phenotypes of ZnT8-KO Mice Regarding Glucose Metabolism

As described above, glucose tolerance was mildly impaired in ZnT8-KO mice. This demonstrates that peripheral insulin levels in ZnT8-KO mice were reduced compared with control mice. Indeed, we and others have observed decreased insulin levels in ZnT8-KO mice, despite GSIS levels being unchanged or slightly increased in ZnT8-KO mouse islets [18,34] (Supplemental Table S1). To understand this discrepancy, we performed a pancreas perfusion experiment, and found that insulin secretion was still enhanced upon pancreas infusion in ZnT8-KO mice, further supporting that insulin secretion is increased in ZnT8-KO mice. However, pancreas-liver dual perfusion analysis demonstrated that in these mice, a large proportion of the secreted insulin is actually degraded during its passage through the liver, suggesting that ZnT8 regulates hepatic insulin clearance [18]. A set of in vivo and in vitro experiments demonstrated that ZnT8-mediated zinc inhibits hepatic insulin uptake by counteracting clathrin-mediated endocytosis of the insulin receptor. These results suggested that ZnT8 plays an important role in determining the amount of insulin that is delivered to the liver and other peripheral organs, and thus optimizes the effect of insulin on whole body glucose metabolism [18].

For the assessment of in vivo insulin clearance, the c-peptide/insulin ratio may be useful [36,37]. Consistently, although insulin secretion was increased in ZnT8-KO mice, peripheral insulin levels were lower and c-peptide/insulin ratios were increased [18]. Studies of c-peptide/insulin ratios and rates of insulin clearance in humans with the rs13266634 polymorphism also demonstrated results consistent with this idea [18]. Furthermore, the Eugene study showed that when human homozygous carriers of the *SLC30A8* risk allele are subjected to the intravenous glucose tolerance test, they demonstrate low peripheral insulin levels in the early phases [38]. These results indicate that SLC30A8/ZnT8 regulates hepatic insulin clearance, and importantly, the same mechanism appears to be conserved in humans (Figure 1) [18].

Clinically, the inhibition of hepatic insulin clearance seems likely to be a therapeutic target for diabetes. A previous study reported the therapeutic potential of small molecule inhibitors of insulin-degrading enzyme, which regulates insulin catabolism [39,40]. Our findings hence might provide novel insights into the molecular pathology of diabetes, which involves dysregulated insulin clearance from the liver, and hence may be a promising future therapeutic target for diabetes [18].

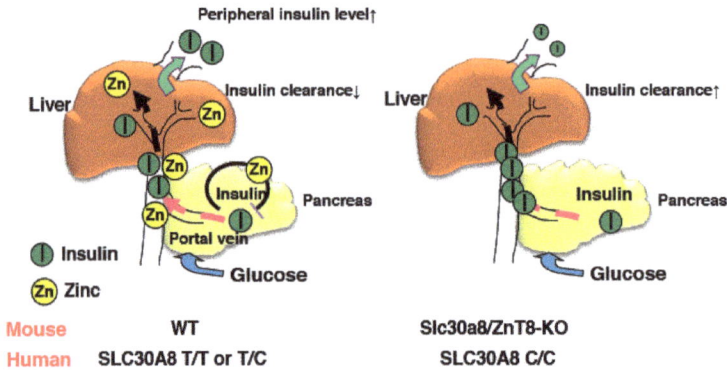

**Figure 1.** Schematic representation of insulin clearance in WT and ZnT8-KO mice. Zinc co-secreted with insulin suppresses insulin secretion from pancreatic β cells and inhibits hepatic insulin clearance in WT mice (**left**). In contrast, reduced zinc secretion results in enhanced insulin secretion from β cells in ZnT8-KO mice and hepatic insulin clearance is not suppressed (**right**). Thus, peripheral insulin levels in ZnT8-KO mice are maintained at lower levels than in WT mice.

### 3.3. Involvement of Other ZnT Transporters

Although there is a substantial decrease in total zinc levels in the islets of ZnT8-KO mice compared with wild-type mice, the phenotypes of ZnT8-KO mice regarding glucose metabolism were fairly modest. Several other ZnT isoforms were expressed at low levels in the pancreatic islets. Thus, functional compensation by other ZnT isoforms might reduce the effect of the ZnT8-KO phenotype. ZnT3 is a candidate ZnT transporter for this compensation. ZnT3 is known to play a role in the uptake of zinc in the synaptic vesicles of glutaminergic hippocampal neurons [41,42]. Considering that β cells and neurons share some similar characteristics, ZnT3 might be involved in the transport of zinc into insulin secretory vesicles. However, it is unclear whether ZnT3 is expressed in the islets of mice [43], and ZnT3-KO mice appear to undergo normal glucose metabolism [44], suggesting that ZnT3 is not involved in this process.

As zinc is required for the hexamerization of insulin and its conversion from proinsulin to insulin in the Golgi compartment, a sufficient amount of import of zinc to this compartment is also required. ZnT5 and ZnT7 are also reported to be expressed in β cells and to co-localize with the Golgi apparatus and secretory vesicles [43,45,46]. Thus, ZnT5 and ZnT7 transporters might be involved in these processes. A recent study analyzed this possibility by crossing ZnT7-KO mice with ZnT8-KO mice. However, whether ZnT7 has a redundant role of ZnT8 remains to be clarified because global ZnT7-KO mice displayed several defects in insulin-sensitive tissues outside of β cells, as described below, and because the report did not include data on ZnT8 single-knockout mice [47]. Further analyses are needed to identify the zinc transporters involved in each step from insulin processing to storage.

### 3.4. Zinc Transport Activity of ZnT8 Variants

One of the important unresolved issue regarding ZnT8 is whether the ZnT8 variant 325Arg(R) increases or decreases zinc transport activity. In one study, the fluorescent dye FluoZin-3 was used to monitor cytosolic zinc and the fluorescent dye Zinquin was used to monitor vacuolar zinc accumulation in MIN6 cells transiently expressing the Arg(R) or Trp(W) variants of hZnT8. Cells expressing the variant (W) showed significantly greater fluorescence of both dyes, and the authors concluded that this variant was a more active transporter of zinc [34]. In another study, HEK293 cells inducibly expressing hZnT8 variants were established, and both the R and W variants of hZnT8 were purified in the native state, and a reconstitution system was developed to measure zinc transport activities [48]. The authors found the R variant to be more active than the W variant, suggesting that the common high-risk R variant is hyperactive and thus may be a therapeutic target to reduce the risk of T2DM in the general population [48]. An additional report described the possible association between the diabetes risk allele in hZnT8 and the higher zinc concentration in human islets [49]. These results suggest that β cells with lower zinc levels may be protection from T2DM, whereas higher zinc in β cells may be associated with T2DM. Interestingly, these new findings appear to be consistent with the finding that rare loss-of-function mutations in *ZnT8* are associated with reduced T2DM risk in humans [27]. This suggests that the role of ZnT8 might be contradictory between humans and mice, as a loss-of-function of ZnT8 in humans decreases the risk for T2DM, whereas ZnT8-KO mice have impaired glucose tolerance [50]. In the course of evolution, the role of ZnT8 in glucose homeostasis has been altered. There are some factors to explain this discrepancy. Since synaptic ZnT3-mediated zinc contributes predominantly to amyloid deposition in human amyloid precursor protein (hAPP) mice [51,52], human islet amyloid polypeptide (hIAPP) might be able to explain this discrepancy. Compared to mouse IAPP, hIAPP can form toxic oligomers, which affect β cells by inducing apoptosis and amyloidogenesis in T2DM [53,54] (Figure 2A). A recent computational analysis showed that zinc concentration determines insulin oligomer equilibrium and that the hIAPP monomer preferentially binds to both the insulin monomer and dimer, compared to the formation of hIAPP homodimer. Therefore, regarding the loss of ZnT8 function, the zinc deficiency shifts the equilibrium of the insulin oligomers toward monomers and dimers, which isolate hIAPP monomers (nontoxic form) and prevent hIAPP from self-association and subsequent aggregation (toxic forms), thereby reducing the risk of T2DM (Figure 2B) [55]. The theory of altered hIAPP aggregation in β cells in response to altered ZnT8 function is a promising but still correlative hypothesis at this point. Therefore, this hypothesis can be validated by creating hIAPP transgenic (hIAPP-Tg) mice from ZnT8-KO mice, to investigate whether hIAPP cytotoxicity can be ameliorated by the deletion of ZnT8.

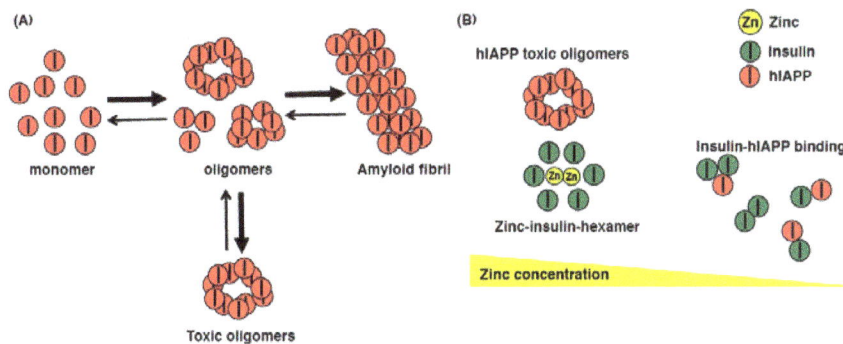

**Figure 2.** Schematic model of the relationship among hIAPP, insulin, and zinc. (**A**) hIAPP can easily form toxic oligomers that induce apoptosis and amyloidogenesis in β cells in T2DM patients; (**B**) When the zinc concentration in insulin secretory granules is high, zinc is used to form zinc-insulin-hexamer, and hIAPP can easily form toxic oligomers. On the other hand, when the zinc concentration is low in insulin secretory granules (such as insulin secretory granule in ZnT8-KO mice), insulin exists as monomer or dimer, which preferentially binds to hIAPP monomer and prevents hIAPP from self-associating and aggregating.

## 4. Zinc Distribution Affects Adipocyte Metabolism

### 4.1. Zinc Distribution in Obesity

Chronic low intake of zinc is associated with an increased risk of diabetes. Hence, zinc supplementation is expected to be an effective method for preventing metabolic syndrome and diabetes. A previous study analyzed the effect of zinc supplementation to prepubertal obese children on insulin resistance and metabolic syndrome. Zinc supplementation was suggested to be a useful and safe additional intervention treatment [56]. However, to our knowledge, the effectiveness of zinc supplementation for the treatment of obesity and diabetes has not been demonstrated in large-scale studies, particularly in adults. Therefore, it is important to establish the safety, efficacy, and effective dose of zinc supplementation in adults.

Nevertheless, zinc might be referred to as the insulin-mimetic, since zinc stimulates lipogenesis and glucose uptake in isolated adipocytes, and zinc ion acts as an insulin-mimetics through their direct effect on the insulin-signaling pathway [2]. The insulin-sensitizing effect of zinc has been attributed to the inhibition of the tyrosine phosphatase activity of protein tyrosine phosphatase 1B (PTP1B) (Figure 3) [57,58]. Zinc ion inactivates PTP1B by non-covalent binding to its cysteine residues which is crucial for the enzymatic activity and reactive oxygen species is also known to inactivate the enzyme in a similar manner. In fact, oxidative stress is upregulated in the most of the patients with diabetes and diabetic animals, further mechanistic analysis is needed to examine the risks and benefits of zinc supplementation as a mean for mitigating obesity and type 2 diabetes.

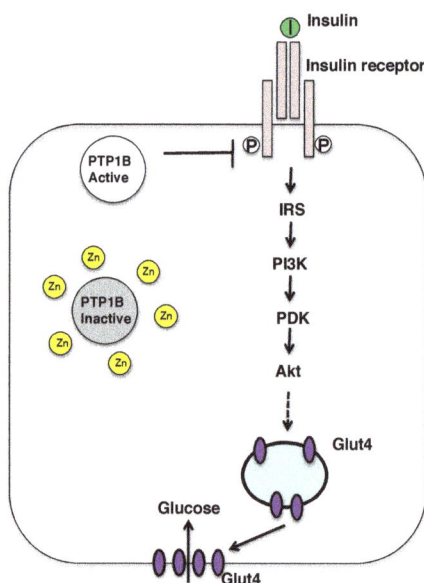

**Figure 3.** Insulin signaling pathway and insulin mimicking function of zinc ions. Insulin binds to the insulin receptor located in the plasma membrane in the peripheral tissues, such as liver and muscle. The insulin-signaling pathway is activated and the glucose transporter GLUT4 is translocated to the plasma membrane. Zinc might inhibit the activity of PTP1B, which activates the insulin-signaling pathway. PI3K, phosphatidylinositol-3-kinase; IRS, insulin receptor substrate; PKD, protein kinase D.

### 4.2. Association between Adipocyte Metabolism and Zinc Homeostasis

Obesity and its associated metabolic diseases develop when energy intake exceeds energy expenditure; this can be caused by decreased physical activity, the inability of the CNS to downregulate appetite, or the ingestion of high-calorie foods [59]. Adipose tissue is involved in energy storage and also functions as an endocrine organ to release free fatty acids (FFA) and adipokines, such as leptin, tumor necrosis factor-alpha (TNF-$\alpha$), interleukin-6 (IL-6), and adiponectin [60]. In addition, adipose tissue comprises numerous types of stromal cells, including preadipocytes, endothelial cells, immune cells, and fibroblasts. During the course of obesity, adipocyte cells and stromal cells in adipose tissue change in number and characteristics. Invasion of macrophages into the adipose tissue of obese individuals is associated with increased secretion of inflammatory adipocytokines, including TNF-$\alpha$ and IL-6, which leads to insulin resistance [61]. Macrophages have recently been reported to be involved in adipose tissue inflammation as well as in the regulation of adipose metabolism through the disrupted modulation of adipocytokines production [61].

Several reports have addressed the association between zinc transporters and adipose metabolism. For example, in the adipose tissue of *Zip14*-KO mice, hypertrophy together with enhanced proinflammatory signaling is observed through activation of nuclear factor-kappaB (NF-$\kappa$B) and the Janus-activating kinase 2 (JAK2)/signal transducer and activator of transcription 3 (STAT3) pathway, and this might contribute to obesity-induced insulin resistance [62]. Consistent with this idea, *Zip14* expression is significantly reduced in obese individuals compared with non-obese individuals, and is increased markedly following weight loss [63].

Adipose tissue plays a crucial role in controlling energy balance. It comprises white and brown adipocytes, which perform different functions. White adipocytes store excess energy, whereas brown adipocytes play a role in energy expenditure [59]. In mammals, brown adipose tissue (BAT) dissipates energy in the form of heat and acts as a defense mechanism against hypothermia. Brown adipocytes are unique in that they have a very large number of mitochondria and are also able to metabolize glucose and fats to produce heat rather than ATP. This thermogenic activity of brown adipocytes is mediated largely via the actions of uncoupling protein-1 (UCP1) [64]. Furthermore, two distinct types of thermogenic adipocytes have recently been identified, namely, the "classical brown adipocytes" and the "beige adipocytes". Beige adipocytes are induced in white adipose tissue (WAT), particularly in inguinal WAT upon various external factors, including exercise, chronic cold exposure, and bariatric surgery [65]. Identification and implementation of therapies based on beige fat require a detailed understanding of the differences in the developmental mechanisms and functions of white, brown, and beige adipocytes.

Many transcriptional regulators and transcription factors are used for differentiation into these various fat cell types, such as peroxisome proliferator-activated receptor gamma (PPAR$\gamma$) and the CCAAT/enhancer binding protein (C/EBP) family of transcription factors, respectively [66]. The induced differentiation of preadipocytes triggers DNA replication and reentry into the cell cycle (mitotic clonal expansion). Mitotic clonal expansion involves a transcription factor cascade, followed by the expression of adipocyte genes. A critical event is phosphorylation of C/EBP-$\beta$, which is a form of activated C/EBP-$\beta$, which then triggers PPAR$\gamma$ and C/EBP-$\alpha$, which in turn coordinately activate genes whose expression produces the adipocyte phenotype, such as *aP2* [67]. Several studies have analyzed the roles of zinc and the zinc transporter in the differentiation process of white adipocytes. Expression of the *Zip14* gene was found to be upregulated during early adipocyte differentiation [63,68]. During mitotic cell expansion, zinc and *MT* levels within cells are rapidly increased. This elevation is essential for the transition from G0/G1- to S-phase of the cell cycle [69] (Figure 4). ZnT7-KO mice show a mild zinc deficiency, with low body weight gain as well as body fat accumulation. The underlying mechanism of these characteristics in ZnT7-KO mice is that ZnT7 is likely to be involved in lipogenesis in adipocytes, rather than in the early adipocyte differentiation process, such as mitotic clonal expansion [70].

Most of the transcription factors that are known to direct cells toward a brown/beige adipocyte lineage instead of a white adipocyte lineage act via the core transcriptional machinery of adipogenesis. PR domain containing 16 (PRDM16), which is an essential transcriptional coregulators of brown/beige adipocyte differentiation [71], determines the brown/beige adipocyte lineage mainly via its interaction with various transcriptional factors, including PPAR$\gamma$, peroxisome proliferator-activated receptor $\gamma$ coactivator 1-$\alpha$ (PGC-1$\alpha$), C/EBP-$\beta$, and zinc finger protein 516 [65,72]. Among them, many zinc-containing transcriptional factors participate in brown/beige adipocyte differentiation and function.

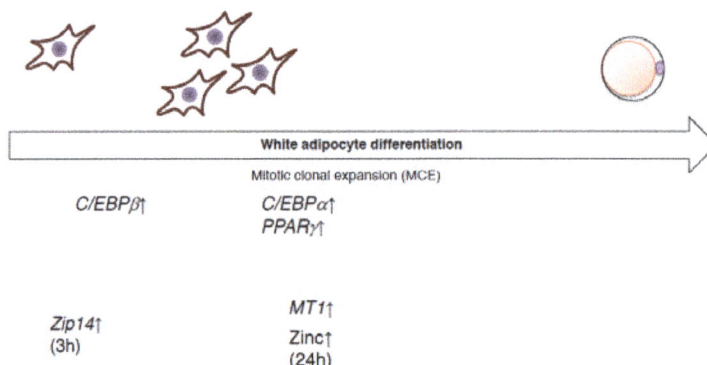

**Figure 4.** The expression of zinc transporters during white adipocyte differentiation. Preadipocytes are differentiated and trigger DNA replication and reentry into the cell cycle (mitotic clonal expansion, MCE). The expression of several genes related to zinc homeostasis is altered.

### 4.3. ZIP13 Regulates Beige Adipocyte Biogenesis and Energy Expenditure

As explained above, we have been investigating the roles of zinc homeostasis in the health and disease of endocrine organs by focusing on the biological functions of zinc transporters. *Zip13*-KO mice were reported to show impaired bone formation and growth retardation [73]. Importantly, ZIP13 is also plays a crucial role in connective tissue development in humans. Patients with a loss-of-function mutation in ZIP13 showed a similar phenotype to *Zip13*-KO mice and were diagnosed as having a novel type of Ehlers-Danlos syndrome. Interestingly, human patients of Ehlers-Danlos syndrome were reported to display lipoatrophy [73]. Therefore, we aimed to clarify the roles of ZIP13 in fat tissue. During our investigation, we found that *Zip13*-KO inguinal WAT had a high number of functional beige adipocytes [74]. Furthermore, *Zip13*-KO mice showed a significantly higher oxygen consumption rate than wild-type mice, although there were no differences in food intake, suggesting that *Zip13*-KO mice have a tendency to not gain weight. Consistent with this idea, *Zip13*-KO mice showed resistance to high fat diet-induced obesity [74].

Furthermore, both gain-of-function and loss-of-function experiments have demonstrated that the accumulation of C/EBP-β, which is involved in determining brown/beige adipocyte lineage in cooperation with the dominant transcriptional coregulator PRDM16, is crucial for the increased adipocyte browning resulting from the loss of ZIP13 [74], implying most likely that ZIP13-mediated zinc transport is required for the inhibition of adipocyte browning (Figure 5), and that ZIP13 may deliver zinc ions to specific molecular targets that control the function of C/EBP-β or/and other target proteins. Further analyses of *Zip13*-deficient cells will clarify the specific roles of ZIP13 in beige adipocyte biogenesis.

Importantly, the presence and activity of thermogenic beige adipocytes are associated with improved global metabolic fitness, such as improvements in insulin resistance and glucose homeostasis [65]. Indeed, *Zip13*-KO mice reportedly have improved glucose tolerance and insulin tolerance compared with control mice in addition to increased energy expenditure [74]. Furthermore, given that recent studies demonstrated that adult human brown adipocytes share biological characteristics with rodent beige adipocytes, rather than rodent brown adipocyte [75]. The results of our study may hence contribute to the establishment of novel treatments for obese diabetic patients via the manipulation of adipocyte identity.

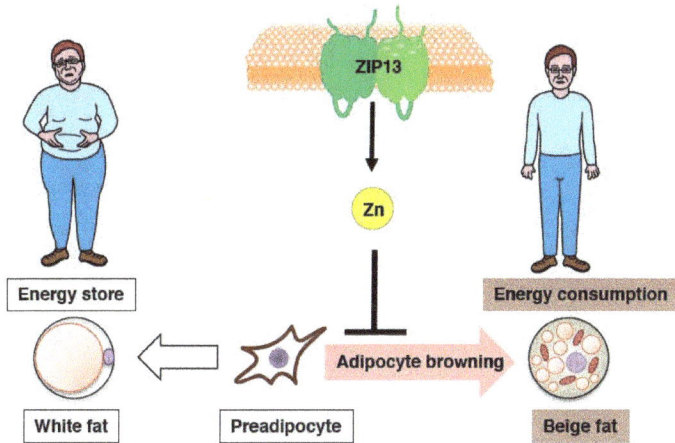

**Figure 5.** Zinc transporter ZIP13 inhibits adipocyte browning. Schematic model of the role of ZIP13 in adipocyte browning.

## 5. Involvement of Zinc in the Insulin Signaling Pathway

### 5.1. Zinc Transporters Affect Skeletal Muscle Insulin Signaling

Skeletal muscle plays a key role in insulin-stimulated glucose uptake in the postprandial state. Under insulin-resistant conditions, there is a decrease in insulin signaling via IRS-1, PI3K, and Akt, resulting in decreased translocation of the GLUT4 glucose transporter, to the plasma membrane, as well as decreased insulin-stimulated glucose transport into cells (Figure 3).

Skeletal muscle is the major reservoir for zinc, containing approximately 60% of the total whole-body zinc [76]. Zinc transporter SLC39A7 (ZIP7) has been reported to be involved in glycemic control within skeletal muscle. The knockdown of *Zip7* in the mouse skeletal muscle cell line C2C12 demonstrated that glucose metabolism is enhanced by ZIP7 via Akt phosphorylation, in which ZIP7-mediated zinc activates insulin receptor signaling via its binding to PTP1B [77]. A different study reported that ZnT7-KO mice demonstrate impaired glucose tolerance and insulin sensitivity, resulting from a decrease in the insulin signaling pathway, such as decreased levels of Akt phosphorylation in skeletal muscle and adipocytes, as described above, which thereby reduce glucose uptake [78]. Although many studies have been performed to analyze the insulin-mimetic actions of zinc [79], the specific zinc transporters that are responsible for initiating these signaling processes remain unclear. In particular, given that ZIP7 acts as the gatekeeper of zinc release from the Golgi apparatus [80] and that insulin-signaling cascades in skeletal muscle are affected by ZIP7-mediated zinc, it will be very important to analyze the metabolic phenotypes of skeletal muscle in tissue-specific *Zip7*-KO mice.

### 5.2. Zinc Homeostasis and Sarcopenia

Sarcopenia is the age-associated degenerative loss of skeletal muscle mass, quality, and strength. In older people, sarcopenia is often accompanied by diabetes [81]. Some of the mechanisms involved in the development of sarcopenia, such as insulin resistance, mitochondrial dysfunction, and chronic inflammation, are also thought to play roles in the pathogenesis of diabetes [82]. Recently, the zinc/zinc transporter and *MT* expression levels have been reported to be upregulated during skeletal muscle atrophy. For example, the blocking of MTs 1 and 2 has been shown to increase skeletal muscle mass and strength [83], suggesting the potential of MTs as therapeutic targets for sarcopenia. Furthermore, muscle zinc levels and *Zip14* expression levels are known to increase with age [84], although the underlying mechanisms and physiological meaning of these observations remain to be clarified.

Further analyses are required to understand the precise roles of zinc transporters in sarcopenia that occurs in association with diabetes.

## 6. Summary

Since human patients with *Zip13* deficiency have been reported to show lipoatrophy, we expected *Zip13*-KO mice to be resistant to insulin [73]. However, on the contrary, *Zip13*-KO mice showed improved insulin sensitivity owing to the acceleration of adipocyte browning. To our knowledge, our data are the first to show that the Golgi-to-cytoplasm transport of zinc by ZIP13 is necessary for the regulation of beige adipocyte metabolism and insulin sensitivity, because simple zinc supplementation cannot inhibit adipocyte browning. On the other hand, a study of ZnT8-KO mice also showed that the release of adequate amounts of zinc from β cells together with insulin is important for the regulation of insulin clearance, suggesting that the zinc transported by ZnT8 acts as a signaling molecule between organs and mediates pancreas-to-liver organ communication. ZnT7-KO mice demonstrate mild zinc deficiency accompanied with low body weights and very little accumulation of body fat. Dietary zinc supplementation in ZnT7-KO mice cannot rescue these phenotypes. The main tissue responsible for the phenotypes of ZnT7-KO mice might be adipose tissue, which results in impaired glucose tolerance and insulin sensitivity.

Considering the aforementioned phenotypes of zinc transporter KO mice, we would like to suggest a new perspective; that adequate local zinc delivery by zinc transporters are important, and their disruption leads to the pathogenesis of a variety of diseases. In particular, as sarcopenia is an important issue in aging societies, zinc transporters or factors associated with zinc homeostasis might be candidate biomarkers or therapeutic targets. Further analyses are required toward the development of zinc transporter-mediated therapies against both obesity and diabetes.

**Supplementary Materials:** Supplementary materials can be found at www.mdpi.com/xxx/s1.

**Acknowledgments:** We thank H. Akiko Popiel for helpful advice and criticism. This work was supported by grants from the Ministry of Education, Sports and Culture of Japan (to AF [16K09764, 26860700, and 24790936] and YF [25461364]), Fumi Yamamura Memorial Foundation for Female Natural Scientists, Suzuken Memorial Foundation, Banyu Life Science Foundation International, Front Runner of Future Diabetes Research, Astellas Foundation of Research on Metabolic Disorders, and Japan Diabetes Foundation (to Ayako Fukunaka).

**Author Contributions:** Ayako Fukunaka wrote the manuscript and Ayako Fukunaka and Yoshio Fujitani reviewed the manuscript.

**Conflicts of Interest:** The authors declare no conflict of interest.

## References

1.  Simcox, J.A.; McClain, D.A. Iron and diabetes risk. *Cell Metab.* **2013**, *17*, 329–341. [CrossRef] [PubMed]
2.  Maret, W. Zinc in Pancreatic Islet Biology, Insulin Sensitivity, and Diabetes. *Prev. Nutr. Food Sci.* **2017**, *22*, 1–8. [CrossRef] [PubMed]
3.  Sladek, R.; Rocheleau, G.; Rung, J.; Dina, C.; Shen, L.; Serre, D.; Boutin, P.; Vincent, D.; Belisle, A.; Hadjadj, S.; et al. A genome-wide association study identifies novel risk loci for type 2 diabetes. *Nature* **2007**, *445*, 881–885. [CrossRef] [PubMed]
4.  Foster, M.C.; Leapman, R.D.; Li, M.X.; Atwater, I. Elemental composition of secretory granules in pancreatic islets of Langerhans. *Biophys. J.* **1993**, *64*, 525–532. [CrossRef]
5.  Hutton, J.C.; Penn, E.J.; Peshavaria, M. Low-molecular-weight constituents of isolated insulin-secretory granules. Bivalent cations, adenine nucleotides and inorganic phosphate. *Biochem. J.* **1983**, *210*, 297–305. [CrossRef] [PubMed]
6.  Kimura, T.; Kambe, T. The Functions of Metallothionein and ZIP and ZnT Transporters: An Overview and Perspective. *Int. J. Mol. Sci.* **2016**, *17*, 336. [CrossRef] [PubMed]
7.  Hara, T.; Takeda, T.A.; Takagishi, T.; Fukue, K.; Kambe, T.; Fukada, T. Physiological roles of zinc transporters: Molecular and genetic importance in zinc homeostasis. *J. Physiol. Sci.* **2017**, *67*, 283–301. [CrossRef] [PubMed]

8.  Kambe, T.; Tsuji, T.; Hashimoto, A.; Itsumura, N. The Physiological, Biochemical, and Molecular Roles of Zinc Transporters in Zinc Homeostasis and Metabolism. *Physiol. Rev.* **2015**, *95*, 749–784. [CrossRef] [PubMed]
9.  Chimienti, F.; Devergnas, S.; Pattou, F.; Schuit, F.; Garcia-Cuenca, R.; Vandewalle, B.; Kerr-Conte, J.; Van Lommel, L.; Grunwald, D.; Favier, A.; et al. In vivo expression and functional characterization of the zinc transporter ZnT8 in glucose-induced insulin secretion. *J. Cell Sci.* **2006**, *119*, 4199–4206. [CrossRef] [PubMed]
10. Dodson, G.; Steiner, D. The role of assembly in insulin's biosynthesis. *Curr. Opin. Struct. Biol.* **1998**, *8*, 189–194. [CrossRef]
11. Dodson, E.J.; Dodson, G.G.; Hodgkin, D.C.; Reynolds, C.D. Structural relationships in the two-zinc insulin hexamer. *Can. J. Biochem.* **1979**, *57*, 469–479. [CrossRef] [PubMed]
12. Dunn, M.F. Zinc-ligand interactions modulate assembly and stability of the insulin hexamer—A review. *Biometals* **2005**, *18*, 295–303. [CrossRef] [PubMed]
13. Havu, N.; Lundgren, G.; Falkmer, S. Zinc and manganese contents of micro-dissected pancreatic islets of some rodents. A microchemical study in adult and newborn guinea pigs, rats, Chinese hamsters and spiny mice. *Acta Endocrinol.* **1977**, *86*, 570–577. [PubMed]
14. Steiner, D.F.; Rouille, Y.; Gong, Q.; Martin, S.; Carroll, R.; Chan, S.J. The role of prohormone convertases in insulin biosynthesis: Evidence for inherited defects in their action in man and experimental animals. *Diabetes Metab.* **1996**, *22*, 94–104. [PubMed]
15. Emdin, S.O.; Dodson, G.G.; Cutfield, J.M.; Cutfield, S.M. Role of zinc in insulin biosynthesis. Some possible zinc-insulin interactions in the pancreatic B-cell. *Diabetologia* **1980**, *19*, 174–182. [CrossRef] [PubMed]
16. Hutton, J.C. The insulin secretory granule. *Diabetologia* **1989**, *32*, 271–281. [CrossRef] [PubMed]
17. Hou, J.C.; Min, L.; Pessin, J.E. Insulin granule biogenesis, trafficking and exocytosis. *Vitam. Horm.* **2009**, *80*, 473–506. [PubMed]
18. Tamaki, M.; Fujitani, Y.; Hara, A.; Uchida, T.; Tamura, Y.; Takeno, K.; Kawaguchi, M.; Watanabe, T.; Ogihara, T.; Fukunaka, A.; et al. The diabetes-susceptible gene SLC30A8/ZnT8 regulates hepatic insulin clearance. *J. Clin. Investig.* **2013**, *123*, 4513–4524. [CrossRef] [PubMed]
19. Boquist, L.; Lernmark, A. Effects on the endocrine pancreas in Chinese hamsters fed zinc deficient diets. *Acta Pathol. Microbiol. Scand.* **1969**, *76*, 215–228. [CrossRef] [PubMed]
20. Huber, A.M.; Gershoff, S.N. Effect of zinc deficiency in rats on insulin release from the pancreas. *J. Nutr.* **1973**, *103*, 1739–1744. [CrossRef] [PubMed]
21. Donath, M.Y.; Ehses, J.A.; Maedler, K.; Schumann, D.M.; Ellingsgaard, H.; Eppler, E.; Reinecke, M. Mechanisms of beta-cell death in type 2 diabetes. *Diabetes* **2005**, *54* (Suppl. S2), S108–S113. [CrossRef] [PubMed]
22. Mysore, T.B.; Shinkel, T.A.; Collins, J.; Salvaris, E.J.; Fisicaro, N.; Murray-Segal, L.J.; Johnson, L.E.; Lepore, D.A.; Walters, S.N.; Stokes, R.; et al. Overexpression of glutathione peroxidase with two isoforms of superoxide dismutase protects mouse islets from oxidative injury and improves islet graft function. *Diabetes* **2005**, *54*, 2109–2116. [CrossRef] [PubMed]
23. Marklund, S.L.; Westman, N.G.; Lundgren, E.; Roos, G. Copper- and zinc-containing superoxide dismutase, manganese-containing superoxide dismutase, catalase, and glutathione peroxidase in normal and neoplastic human cell lines and normal human tissues. *Cancer Res.* **1982**, *42*, 1955–1961. [PubMed]
24. Sun, Q.; van Dam, R.M.; Willett, W.C.; Hu, F.B. Prospective study of zinc intake and risk of type 2 diabetes in women. *Diabetes Care* **2009**, *32*, 629–634. [CrossRef] [PubMed]
25. Shan, Z.; Bao, W.; Zhang, Y.; Rong, Y.; Wang, X.; Jin, Y.; Song, Y.; Yao, P.; Sun, C.; Hu, F.B.; et al. Interactions between zinc transporter-8 gene (*SLC30A8*) and plasma zinc concentrations for impaired glucose regulation and type 2 diabetes. *Diabetes* **2014**, *63*, 1796–1803. [CrossRef] [PubMed]
26. Miao, X.; Sun, W.; Fu, Y.; Miao, L.; Cai, L. Zinc homeostasis in the metabolic syndrome and diabetes. *Front. Med.* **2013**, *7*, 31–52. [CrossRef] [PubMed]
27. Flannick, J.; Thorleifsson, G.; Beer, N.L.; Jacobs, S.B.; Grarup, N.; Burtt, N.P.; Mahajan, A.; Fuchsberger, C.; Atzmon, G.; Benediktsson, R.; et al. Loss-of-function mutations in SLC30A8 protect against type 2 diabetes. *Nat. Genet.* **2014**, *46*, 357–363. [CrossRef] [PubMed]
28. Rutter, G.A.; Chabosseau, P.; Bellomo, E.A.; Maret, W.; Mitchell, R.K.; Hodson, D.J.; Solomou, A.; Hu, M. Intracellular zinc in insulin secretion and action: A determinant of diabetes risk? *Proc. Nutr. Soc.* **2016**, *75*, 61–72. [CrossRef] [PubMed]

29. Chabosseau, P.; Rutter, G.A. Zinc and diabetes. *Arch. Biochem. Biophys.* **2016**, *611*, 79–85. [CrossRef] [PubMed]
30. Wijesekara, N.; Dai, F.F.; Hardy, A.B.; Giglou, P.R.; Bhattacharjee, A.; Koshkin, V.; Chimienti, F.; Gaisano, H.Y.; Rutter, G.A.; Wheeler, M.B. Beta cell-specific Znt8 deletion in mice causes marked defects in insulin processing, crystallisation and secretion. *Diabetologia* **2010**, *53*, 1656–1668. [CrossRef] [PubMed]
31. Mitchell, R.K.; Hu, M.; Chabosseau, P.L.; Cane, M.C.; Meur, G.; Bellomo, E.A.; Carzaniga, R.; Collinson, L.M.; Li, W.H.; Hodson, D.J.; et al. Molecular Genetic Regulation of SLC30A8/ZnT8 Reveals a Positive Association With Glucose Tolerance. *Mol. Endocrinol.* **2016**, *30*, 77–91. [CrossRef] [PubMed]
32. Pound, L.D.; Sarkar, S.A.; Ustione, A.; Dadi, P.K.; Shadoan, M.K.; Lee, C.E.; Walters, J.A.; Shiota, M.; McGuinness, O.P.; Jacobson, D.A.; et al. The physiological effects of deleting the mouse *SLC30A8* gene encoding zinc transporter-8 are influenced by gender and genetic background. *PLoS ONE* **2012**, *7*, e40972. [CrossRef] [PubMed]
33. Pound, L.D.; Sarkar, S.A.; Benninger, R.K.; Wang, Y.; Suwanichkul, A.; Shadoan, M.K.; Printz, R.L.; Oeser, J.K.; Lee, C.E.; Piston, D.W.; et al. Deletion of the mouse *SLC30A8* gene encoding zinc transporter-8 results in impaired insulin secretion. *Biochem. J.* **2009**, *421*, 371–376. [CrossRef] [PubMed]
34. Nicolson, T.J.; Bellomo, E.A.; Wijesekara, N.; Loder, M.K.; Baldwin, J.M.; Gyulkhandanyan, A.V.; Koshkin, V.; Tarasov, A.I.; Carzaniga, R.; Kronenberger, K.; et al. Insulin storage and glucose homeostasis in mice null for the granule zinc transporter ZnT8 and studies of the type 2 diabetes-associated variants. *Diabetes* **2009**, *58*, 2070–2083. [CrossRef] [PubMed]
35. Lemaire, K.; Ravier, M.A.; Schraenen, A.; Creemers, J.W.; Van de Plas, R.; Granvik, M.; Van Lommel, L.; Waelkens, E.; Chimienti, F.; Rutter, G.A.; et al. Insulin crystallization depends on zinc transporter ZnT8 expression, but is not required for normal glucose homeostasis in mice. *Proc. Natl. Acad. Sci. USA* **2009**, *106*, 14872–14877. [CrossRef] [PubMed]
36. Meier, J.J.; Holst, J.J.; Schmidt, W.E.; Nauck, M.A. Reduction of hepatic insulin clearance after oral glucose ingestion is not mediated by glucagon-like peptide 1 or gastric inhibitory polypeptide in humans. *Am. J. Physiol. Endocrinol. Metab.* **2007**, *293*, E849–E856. [CrossRef] [PubMed]
37. Poy, M.N.; Yang, Y.; Rezaei, K.; Fernstrom, M.A.; Lee, A.D.; Kido, Y.; Erickson, S.K.; Najjar, S.M. CEACAM1 regulates insulin clearance in liver. *Nat. Genet.* **2002**, *30*, 270–276. [CrossRef] [PubMed]
38. Boesgaard, T.W.; Zilinskaite, J.; Vanttinen, M.; Laakso, M.; Jansson, P.A.; Hammarstedt, A.; Smith, U.; Stefan, N.; Fritsche, A.; Haring, H.; et al. The common SLC30A8 Arg325Trp variant is associated with reduced first-phase insulin release in 846 non-diabetic offspring of type 2 diabetes patients—The EUGENE2 study. *Diabetologia* **2008**, *51*, 816–820. [CrossRef] [PubMed]
39. Leissring, M.A.; Malito, E.; Hedouin, S.; Reinstatler, L.; Sahara, T.; Abdul-Hay, S.O.; Choudhry, S.; Maharvi, G.M.; Fauq, A.H.; Huzarska, M.; et al. Designed inhibitors of insulin-degrading enzyme regulate the catabolism and activity of insulin. *PLoS ONE* **2010**, *5*, e10504. [CrossRef] [PubMed]
40. Maianti, J.P.; McFedries, A.; Foda, Z.H.; Kleiner, R.E.; Du, X.Q.; Leissring, M.A.; Tang, W.J.; Charron, M.J.; Seeliger, M.A.; Saghatelian, A.; et al. Anti-diabetic activity of insulin-degrading enzyme inhibitors mediated by multiple hormones. *Nature* **2014**, *511*, 94–98. [CrossRef] [PubMed]
41. Cole, T.B.; Wenzel, H.J.; Kafer, K.E.; Schwartzkroin, P.A.; Palmiter, R.D. Elimination of zinc from synaptic vesicles in the intact mouse brain by disruption of the ZnT3 gene. *Proc. Natl. Acad. Sci. USA* **1999**, *96*, 1716–1721. [CrossRef] [PubMed]
42. Wenzel, H.J.; Cole, T.B.; Born, D.E.; Schwartzkroin, P.A.; Palmiter, R.D. Ultrastructural localization of zinc transporter-3 (ZnT-3) to synaptic vesicle membranes within mossy fiber boutons in the hippocampus of mouse and monkey. *Proc. Natl. Acad. Sci. USA* **1997**, *94*, 12676–12681. [CrossRef] [PubMed]
43. Bellomo, E.A.; Meur, G.; Rutter, G.A. Glucose regulates free cytosolic $Zn^{2+}$ concentration, Slc39 (ZiP), and metallothionein gene expression in primary pancreatic islet beta-cells. *J. Biol. Chem.* **2011**, *286*, 25778–25789. [CrossRef] [PubMed]
44. Smidt, K.; Jessen, N.; Petersen, A.B.; Larsen, A.; Magnusson, N.; Jeppesen, J.B.; Stoltenberg, M.; Culvenor, J.G.; Tsatsanis, A.; Brock, B.; et al. SLC30A3 responds to glucose- and zinc variations in beta-cells and is critical for insulin production and in vivo glucose-metabolism during beta-cell stress. *PLoS ONE* **2009**, *4*, e5684. [CrossRef] [PubMed]
45. Kambe, T.; Narita, H.; Yamaguchi-Iwai, Y.; Hirose, J.; Amano, T.; Sugiura, N.; Sasaki, R.; Mori, K.; Iwanaga, T.; Nagao, M. Cloning and characterization of a novel mammalian zinc transporter, zinc transporter 5, abundantly expressed in pancreatic beta cells. *J. Biol. Chem.* **2002**, *277*, 19049–19055. [CrossRef] [PubMed]

46. Huang, L.; Yan, M.; Kirschke, C.P. Over-expression of ZnT7 increases insulin synthesis and secretion in pancreatic beta-cells by promoting insulin gene transcription. *Exp. Cell Res.* **2010**, *316*, 2630–2643. [CrossRef] [PubMed]

47. Syring, K.E.; Boortz, K.A.; Oeser, J.K.; Ustione, A.; Platt, K.A.; Shadoan, M.K.; McGuinness, O.P.; Piston, D.W.; Powell, D.R.; O'Brien, R.M. Combined Deletion of *Slc30a7* and *SLC30A8* Unmasks a Critical Role for ZnT8 in Glucose-Stimulated Insulin Secretion. *Endocrinology* **2016**, *157*, 4534–4541. [CrossRef] [PubMed]

48. Merriman, C.; Huang, Q.; Rutter, G.A.; Fu, D. Lipid-tuned Zinc Transport Activity of Human ZnT8 Protein Correlates with Risk for Type-2 Diabetes. *J. Biol. Chem.* **2016**, *291*, 26950–26957. [CrossRef] [PubMed]

49. Wong, W.P.; Allen, N.B.; Meyers, M.S.; Link, E.O.; Zhang, X.; MacRenaris, K.W.; El Muayed, M. Exploring the Association Between Demographics, SLC30A8 Genotype, and Human Islet Content of Zinc, Cadmium, Copper, Iron, Manganese and Nickel. *Sci. Rep.* **2017**, *7*, 473. [CrossRef] [PubMed]

50. Rutter, G.A.; Chimienti, F. SLC30A8 mutations in type 2 diabetes. *Diabetologia* **2015**, *58*, 31–36. [CrossRef] [PubMed]

51. Lee, J.Y.; Cole, T.B.; Palmiter, R.D.; Suh, S.W.; Koh, J.Y. Contribution by synaptic zinc to the gender-disparate plaque formation in human Swedish mutant APP transgenic mice. *Proc. Natl. Acad. Sci. USA* **2002**, *99*, 7705–7710. [CrossRef] [PubMed]

52. Frederickson, C.J.; Koh, J.Y.; Bush, A.I. The neurobiology of zinc in health and disease. *Nat. Rev. Neurosci.* **2005**, *6*, 449–462. [CrossRef] [PubMed]

53. Costes, S.; Langen, R.; Gurlo, T.; Matveyenko, A.V.; Butler, P.C. beta-Cell failure in type 2 diabetes: A case of asking too much of too few? *Diabetes* **2013**, *62*, 327–335. [CrossRef] [PubMed]

54. Haataja, L.; Gurlo, T.; Huang, C.J.; Butler, P.C. Islet amyloid in type 2 diabetes, and the toxic oligomer hypothesis. *Endocr. Rev.* **2008**, *29*, 303–316. [CrossRef] [PubMed]

55. Nedumpully-Govindan, P.; Ding, F. Inhibition of IAPP aggregation by insulin depends on the insulin oligomeric state regulated by zinc ion concentration. *Sci. Rep.* **2015**, *5*, 8240. [CrossRef] [PubMed]

56. Hashemipour, M.; Kelishadi, R.; Shapouri, J.; Sarrafzadegan, N.; Amini, M.; Tavakoli, N.; Movahedian-Attar, A.; Mirmoghtadaee, P.; Poursafa, P. Effect of zinc supplementation on insulin resistance and components of the metabolic syndrome in prepubertal obese children. *Hormones* **2009**, *8*, 279–285. [CrossRef] [PubMed]

57. Haase, H.; Maret, W. Protein tyrosine phosphatases as targets of the combined insulinomimetic effects of zinc and oxidants. *Biometals* **2005**, *18*, 333–338. [CrossRef] [PubMed]

58. Maret, W. Zinc in Cellular Regulation: The Nature and Significance of "Zinc Signals". *Int. J. Mol. Sci.* **2017**, *18*. [CrossRef] [PubMed]

59. Cypess, A.M.; Kahn, C.R. Brown fat as a therapy for obesity and diabetes. *Curr. Opin. Endocrinol. Diabetes Obes.* **2010**, *17*, 143–149. [CrossRef] [PubMed]

60. Rosen, E.D.; Spiegelman, B.M. Adipocytes as regulators of energy balance and glucose homeostasis. *Nature* **2006**, *444*, 847–853. [CrossRef] [PubMed]

61. Olefsky, J.M.; Glass, C.K. Macrophages, inflammation, and insulin resistance. *Annu. Rev. Physiol.* **2010**, *72*, 219–246. [CrossRef] [PubMed]

62. Troche, C.; Aydemir, T.B.; Cousins, R.J. Zinc transporter *Slc39a14* regulates inflammatory signaling associated with hypertrophic adiposity. *Am. J. Physiol. Endocrinol. Metab.* **2016**, *310*, E258–E268. [CrossRef] [PubMed]

63. Maxel, T.; Smidt, K.; Larsen, A.; Bennetzen, M.; Cullberg, K.; Fjeldborg, K.; Lund, S.; Pedersen, S.B.; Rungby, J. Gene expression of the zinc transporter ZIP14 (SLC39a14) is affected by weight loss and metabolic status and associates with PPARgamma in human adipose tissue and 3T3-L1 pre-adipocytes. *BMC Obes.* **2015**, *2*, 46. [CrossRef] [PubMed]

64. Kajimura, S.; Saito, M. A new era in brown adipose tissue biology: Molecular control of brown fat development and energy homeostasis. *Annu. Rev. Physiol.* **2014**, *76*, 225–249. [CrossRef] [PubMed]

65. Kajimura, S.; Spiegelman, B.M.; Seale, P. Brown and Beige Fat: Physiological Roles beyond Heat Generation. *Cell Metab.* **2015**, *22*, 546–559. [CrossRef] [PubMed]

66. Guo, L.; Li, X.; Tang, Q.Q. Transcriptional regulation of adipocyte differentiation: A central role for CCAAT/enhancer-binding protein (C/EBP) beta. *J. Biol. Chem.* **2015**, *290*, 755–761. [CrossRef] [PubMed]

67. Tang, Q.Q.; Lane, M.D. Adipogenesis: From stem cell to adipocyte. *Annu. Rev. Biochem.* **2012**, *81*, 715–736. [CrossRef] [PubMed]

68. Tominaga, K.; Kagata, T.; Johmura, Y.; Hishida, T.; Nishizuka, M.; Imagawa, M. SLC39A14, a LZT protein, is induced in adipogenesis and transports zinc. *FEBS J.* **2005**, *272*, 1590–1599. [CrossRef] [PubMed]

69. Schmidt, C.; Beyersmann, D. Transient peaks in zinc and metallothionein levels during differentiation of 3T3L1 cells. *Arch. Biochem. Biophys.* **1999**, *364*, 91–98. [CrossRef] [PubMed]

70. Tepaamorndech, S.; Kirschke, C.P.; Pedersen, T.L.; Keyes, W.R.; Newman, J.W.; Huang, L. Zinc transporter 7 deficiency affects lipid synthesis in adipocytes by inhibiting insulin-dependent Akt activation and glucose uptake. *FEBS J.* **2016**, *283*, 378–394. [CrossRef] [PubMed]

71. Seale, P.; Kajimura, S.; Yang, W.; Chin, S.; Rohas, L.M.; Uldry, M.; Tavernier, G.; Langin, D.; Spiegelman, B.M. Transcriptional control of brown fat determination by PRDM16. *Cell Metab.* **2007**, *6*, 38–54. [CrossRef] [PubMed]

72. Seale, P. Transcriptional Regulatory Circuits Controlling Brown Fat Development and Activation. *Diabetes* **2015**, *64*, 2369–2375. [CrossRef] [PubMed]

73. Fukada, T.; Civic, N.; Furuichi, T.; Shimoda, S.; Mishima, K.; Higashiyama, H.; Idaira, Y.; Asada, Y.; Kitamura, H.; Yamasaki, S.; et al. The zinc transporter SLC39A13/ZIP13 is required for connective tissue development; its involvement in BMP/TGF-beta signaling pathways. *PLoS ONE* **2008**, *3*, e3642. [CrossRef]

74. Fukunaka, A.; Fukada, T.; Bhin, J.; Suzuki, L.; Tsuzuki, T.; Takamine, Y.; Bin, B.H.; Yoshihara, T.; Ichinoseki-Sekine, N.; Naito, H.; et al. Zinc transporter ZIP13 suppresses beige adipocyte biogenesis and energy expenditure by regulating C/EBP-beta expression. *PLoS Genet.* **2017**, *13*, e1006950. [CrossRef] [PubMed]

75. Shinoda, K.; Luijten, I.H.; Hasegawa, Y.; Hong, H.; Sonne, S.B.; Kim, M.; Xue, R.; Chondronikola, M.; Cypess, A.M.; Tseng, Y.H.; et al. Genetic and functional characterization of clonally derived adult human brown adipocytes. *Nat. Med.* **2015**, *21*, 389–394. [CrossRef] [PubMed]

76. Jackson, M.J. Physiology of Zinc: General aspects. In *Zinc in Human Biology*; Mills, C.F., Ed.; Springer: London, UK, 1989; pp. 1–14.

77. Myers, S.A.; Nield, A.; Chew, G.S.; Myers, M.A. The zinc transporter, Slc39a7 (Zip7) is implicated in glycaemic control in skeletal muscle cells. *PLoS ONE* **2013**, *8*, e79316. [CrossRef] [PubMed]

78. Huang, L.; Kirschke, C.P.; Lay, Y.A.; Levy, L.B.; Lamirande, D.E.; Zhang, P.H. Znt7-null mice are more susceptible to diet-induced glucose intolerance and insulin resistance. *J. Biol. Chem.* **2012**, *287*, 33883–33896. [CrossRef] [PubMed]

79. Taniguchi, M.; Fukunaka, A.; Hagihara, M.; Watanabe, K.; Kamino, S.; Kambe, T.; Enomoto, S.; Hiromura, M. Essential role of the zinc transporter ZIP9/SLC39A9 in regulating the activations of Akt and Erk in B-cell receptor signaling pathway in DT40 cells. *PLoS ONE* **2013**, *8*, e58022. [CrossRef] [PubMed]

80. Taylor, K.M.; Hiscox, S.; Nicholson, R.I.; Hogstrand, C.; Kille, P. Protein kinase CK2 triggers cytosolic zinc signaling pathways by phosphorylation of zinc channel ZIP7. *Sci. Signal.* **2012**, *5*, ra11. [CrossRef] [PubMed]

81. Mitchell, W.K.; Williams, J.; Atherton, P.; Larvin, M.; Lund, J.; Narici, M. Sarcopenia, dynapenia, and the impact of advancing age on human skeletal muscle size and strength; a quantitative review. *Front. Physiol.* **2012**, *3*, 260. [CrossRef] [PubMed]

82. Cleasby, M.E.; Jamieson, P.M.; Atherton, P.J. Insulin resistance and sarcopenia: Mechanistic links between common co-morbidities. *J. Endocrinol.* **2016**, *229*, R67–R81. [CrossRef] [PubMed]

83. Summermatter, S.; Bouzan, A.; Pierrel, E.; Melly, S.; Stauffer, D.; Gutzwiller, S.; Nolin, E.; Dornelas, C.; Fryer, C.; Leighton-Davies, J.; et al. Blockade of Metallothioneins 1 and 2 Increases Skeletal Muscle Mass and Strength. *Mol. Cell. Biol.* **2017**, *37*, e00305–e00316. [CrossRef] [PubMed]

84. Aydemir, T.B.; Troche, C.; Kim, J.; Kim, M.H.; Teran, O.Y.; Leeuwenburgh, C.; Cousins, R.J. Aging amplifies multiple phenotypic defects in mice with zinc transporter Zip14 (Slc39a14) deletion. *Exp. Gerontol.* **2016**, *85*, 88–94. [CrossRef] [PubMed]

International Journal of
*Molecular Sciences*

MDPI

*Review*

# Zinc Signals and Immunity

**Martina Maywald, Inga Wessels and Lothar Rink ***

Institute of Immunology, RWTH Aachen University Hospital, Pauwelsstr. 30, 52074 Aachen, Germany;
martina.maywald@rwth-aachen.de (M.M.); iwessels@ukaachen.de (I.W.)
* Correspondence: lrink@ukaachen.de; Tel.: +49-241-80-80-208

Received: 27 September 2017; Accepted: 19 October 2017; Published: 24 October 2017

**Abstract:** Zinc homeostasis is crucial for an adequate function of the immune system. Zinc deficiency as well as zinc excess result in severe disturbances in immune cell numbers and activities, which can result in increased susceptibility to infections and development of especially inflammatory diseases. This review focuses on the role of zinc in regulating intracellular signaling pathways in innate as well as adaptive immune cells. Main underlying molecular mechanisms and targets affected by altered zinc homeostasis, including kinases, caspases, phosphatases, and phosphodiesterases, will be highlighted in this article. In addition, the interplay of zinc homeostasis and the redox metabolism in affecting intracellular signaling will be emphasized. Key signaling pathways will be described in detail for the different cell types of the immune system. In this, effects of fast zinc flux, taking place within a few seconds to minutes will be distinguish from slower types of zinc signals, also designated as "zinc waves", and late homeostatic zinc signals regarding prolonged changes in intracellular zinc.

**Keywords:** zinc flux; zinc wave; homeostatic zinc signal; signaling pathways; innate and adaptive immunity; zinc deficiency; immune function

## 1. Introduction

The metal zinc is nowadays well established to be essential for a well-operating immune system. However, knowledge about zinc homeostasis, zinc deficiency, and related diseases is comparatively new. In 1963, Dr. Prasad proved for the first time the existence of zinc deficiency in man [1]. Since then, knowledge about zinc evolved rapidly uncovering molecular mechanisms being indispensable for regulating zinc homeostasis in humans. Its significance as a structural component in proteins [2] and its participation in numerous cellular functions include, but are not limited to, cell proliferation and differentiation [3,4], RNA and DNA synthesis [5,6], stabilization of cell structures/membrane [7,8], as well as redox regulation [9,10], and apoptosis [11,12]. Zinc is involved in various metabolic and chronic diseases such as: type 1 diabetes, rheumatoid arthritis, cancer, neurodegenerative diseases, and depression [13–19]. Moreover, there is also strong evidence between zinc deficiency and several infectious diseases such as shigellosis, acute cutaneous leishmaniosis, malaria, human immunodeficiency virus (HIV), tuberculosis, measles, and pneumonia [20,21].

When zinc deficiency was first discovered, it was thought to be a rare disease. However, zinc deficiency is very common, with estimated two billion people worldwide being affected, and is identified as a major contributor to the burden of disease in developing countries. It is the 5th leading life-threatening factor, especially in developing countries [22]. In addition, industrial counties are affected by zinc deficiency, particularly the elderly population [23]. Despite zinc deficiency and related symptoms can easily be treated by proper zinc intake, suboptimal zinc status cannot simply diagnosed by reason of the lack of clinical signs and reliable biochemical indicators of zinc status. To date, no specific and reliable biomarker of zinc status is known, although serum/plasma zinc concentrations, hair zinc concentration, and urinary zinc excretion can be seen as potentially useful. Nevertheless, zinc status is highly impacted by the immune status itself (infection, inflammatory conditions), but also by diet, absorption and conserving mechanisms via gastrointestinal tract and kidneys [24]. Zinc uptake in the gastro intestinal (GI) tract is facilitated by an influx into the enterocyte, through the basolateral membrane and the transport into the portal circulation. Uptake mechanisms are not fully understood yet, however zinc transporters are mainly involved in zinc uptake or zinc efflux [25]. In this regard, Zrt-like, Irt-like protein (ZIP)4 is highly important since it is expressed along the entire GI tract acting as a major processor of zinc uptake into enterocytes from the apical membrane [26]. Moreover, zinc transporter (ZnT)3, is highly expressed in the human large and porcine small intestine and the esophagus [27,28]. Herein, its concrete function in the GI tract is largely unknown. However, studies in the esophagus uncovered its co-localization with sensory neuromediators and/or neuromodulators that are essential for the control of all functions of the GI tract either under physiological and pathological conditions as well as during diseases [27–29]. Hence, there is an ongoing need for the discovery of a reliable biological marker of zinc status.

Although the plasma pool is very small, it is highly important for cellular signaling since it is rapidly exchangeable and mobile. Consequently, intracellular zinc level can be altered resulting in altered cell function and differentiation [30,31]. The zinc-dependent regulation of the immune system is particularly interesting and will be discussed in more detail in this review. We will particularly focus on the importance of different types of zinc signals in innate as well as adaptive immunity, and highlight altered signaling pathways due to changed intracellular free zinc level.

## 2. Zinc Homeostasis and the Immune System: An Overview

With a total amount of 2–4 g, zinc is the second most abundant trace metal in the human body, iron being first. In contrast to the latter, zinc cannot be stored and has to be taken up via food daily to guarantee sufficient supply. A large number of especially inflammatory diseases, but also aging, pregnancy, lactation, and vegetarian or vegan lifestyles are associated with zinc deficiency. Thus, an important role of zinc in development and exacerbation of diseases can be assumed, as indicated earlier. This underlines the importance of a deep understanding of the various functions of zinc in the immune system and thereby for health and disease [21,32–34]. Already marginal-to-moderate zinc deficiency impairs immunity, delays wound healing, causes inflammation-independent low-grade production of inflammatory cytokines and increases oxidative stress [35]. Immune cells might even react more quickly to zinc deficiency than it is measurable in the plasma [36]. Imbalances including strong zinc deficiency but also zinc overload cause severe immune dysfunctions. Zinc intoxication is however rare and its symptoms mostly due to copper deficiency. Immunological hallmarks of zinc deficiency are thymic atrophy, lymphopenia, especially decreased $CD4^+$ T helper (Th) cell numbers, resulting in a decreased $CD4^+/CD8^+$ ratio [37]. In vitro data suggest that monopoiesis is increased during zinc deficiency [38], natural killer (NK) cell activity is decreased and monocyte cytotoxicity is increased [21]. Zinc dependent alterations in chemotaxis, phagocytosis, respiratory burst and formation of neutrophil extracellular traps by innate immune cells offer one explanation for the increased susceptibility to infections during zinc deficiency [39–41].

In humans, high zinc concentrations are found in retina (3.8 µg/g dry weight), choroid of the eye (274 µg/g) and in bone (100–250 µg/g), while only 1 µg/mL zinc is found in plasma, which equals around 0.1% of total body zinc. Within body fluids, zinc is predominantly bound to proteins including albumin, α2 macroglobulin (A2M), transferrin and others. Hypoalbuminemia can even result in zinc deficiency [42]. Zinc binding to those proteins can activate or inactivate their activity, or change characteristics important for substrate binding [43].

Zinc homeostasis is primarily controlled via the expression and action of 14 zinc transporters that increase cytoplasmic zinc (Zip1–14) and 10 zinc transporters that lower cytoplasmic zinc (ZnT1–10). Zrt-like, Irt-like proteins (Zip) are also named solute carrier family 39 (SLC39) A1–A14, while members of the zinc transporter (ZnT) family are also denoted SLC30A1–A10. Many of those transporters are found spanning the plasma membrane, but others are also located within mitochondrial, Golgi network, lysosomal, and vesicular or endoplasmic reticulum membranes. This implies that decreasing cytoplasmic zinc can describe export via ZnTs, but also the transport of zinc into one of those organelles [44]. The significance of zinc transportation via nicotinic acetylcholine receptors, voltage dependent calcium channels, transient receptor potential channels and glutamatergic receptors as well as of facilitated diffusion of zinc bound to amino acids compared to specific zinc transport through Zips and ZnTs remains to be defined in more detail [45]. Together, those zinc transporters regulate zinc homeostasis on tissue but also single cell and even intracellular level.

Physiological changes in body zinc homeostasis during acute phase response include the temporal transfer of serum zinc to the tissues, especially the liver, causing transient serum hypozincemia, which is rebalanced during resolution of the inflammatory response. The transient change in zinc homeostasis is proposed to act as a danger signal for immune cells [46]. In addition, pro-inflammatory acute phase proteins including interleukin-6 (IL-6) upregulate expression of zinc binding peptides such as metallothionein (MT) and A2M, augmenting zinc sequestration from extracellular microorganisms [43]. Intracellularly increased zinc can intoxicate engulfed pathogens and acts cytoprotective by promotion of neutralizing reactive oxygen and nitrogen species (ROS and NOS) as will be discussed later. Undernourishment and severe inflammatory diseases are paralleled with prolonged and severe forms of serum hypozincemia. An association of excessively elevated levels of inflammatory markers, reactive oxygen species and antimicrobial peptides such as calprotectin or matrix metalloproteases (MMP) that cause tissue injury especially in lung, liver and spleen with the augmented serum hypozincemia was suggested [46–49].

Measuring free intracellular zinc levels results in extremely low numbers in the pico- to nanomolar range, depending on the cell type. When zinc is transported into the cell, it is efficiently buffered by high affinity proteins. As affinity of zinc to metal binding sites of proteins is rather high and competitive towards other metal ions, free zinc concentrations fluctuate in a narrow range. Excess zinc ions are transported into subcellular stores such as the including interleukin (ER) or Golgi or into extracellular space [2]. Intracellular zinc-binding proteins include members of the MT family. If in addition to their strong binding activity to extracellular zinc, members of the S100 family can also chelate zinc intracellular is so contested [50]. As the decrease in intracellular zinc during monopoiesis is paralleled by an increase in S100A8 and S100A9 expression but not changes were found for MT-1 levels, binding of intracellular zinc by calprotectin in myeloid cells is likely [38].

MTs are proteins of 6–7 kDA, which can bind and quickly release up to 7 zinc ions, thereby sequestering up to 20% of intracellular zinc. Furthermore, they protect against various types of environmental stress, as MTs can chelate other heavy metals, decreasing their cytotoxicity and they can scavenge reactive oxygen species. Within the four MT classes, MT-1 and MT-2 are expressed ubiquitous throughout the body, whereas MT-3 and MT-4 are expressed cell type specific [44]. Within innate immune cells, MT is an important molecule for zinc regulation. MT deficiency impairs cytokine production and anti-microbial activities in macrophages after lipopolysaccharide (LPS) stimulation [51]. The S100 protein family includes 24 members, all acting calcium dependent and having calcium buffering capacities. Intracellular, they modulate apoptosis, transcription and enzyme activities; extracellular S100 proteins regulate chemotaxis and wound healing, but also proliferation and differentiation via surface receptors [52]. Zinc binding S100 proteins include S100B, S100A1, S100A2, S100A3, S100A5, S100A7, S100A8/9, S100A12 and S100A16. Calprotectin, a heterodimer of S100A8 and S100A9, is the most abundant protein of neutrophils. Increased plasma levels are found during severe inflammatory diseases paralleling serum hypozincemia, suggesting a link that has recently been explored [53,54]. Some low molecular binding partners of zinc have been found as well including adenosine triphosphate (ATP), glutathione, citrate, nicotinamine or bacillithiol [2].

## 3. Classification of Zinc Signals

Signaling cascades are highly complex and very sensitive to altered intracellular second messenger concentrations. Since zinc is established to act as second messenger comparable to calcium [55], it is obvious that cellular signals are altered due to changed intracellular zinc concentrations. Regarding this, free intracellular zinc level influence signaling pathways by binding reversibly to regulatory sites in signaling proteins, altering protein activity and stability [56]. Moreover, alterations can result from the transport of zinc through the plasma membrane, exchange with intracellular organelles or zinc binding proteins as MT and calprotectin, or a combination of these mechanisms.

Intracellular zinc signals can be differently classified. One way is by the time scale they occur, comprising: (1) a "zinc flux" occurring within seconds to minutes; (2) slightly slower zinc signals, described as "zinc wave" occurring within several minutes; and (3) homeostatic zinc signals occurring within several hours. Examples for each classification of zinc signals occurring in different cell types and signaling pathways are summarized in Table 1.

A zinc flux can arise by triggering receptors such as Toll like receptor (TLR)4 in, e.g., monocytes. In this connection, zinc acts as second messenger, comparable to calcium, by influencing signaling cascades in a direct manner. This kind of zinc signal is independent of synthesis of the zinc transporter proteins from the ZnT and Zip family responsible for zinc re-distribution and zinc uptake and is therefore classified as zinc flux. Activation of a zinc transporter to enable a sharp rise in cytoplasmic zinc is thus possible.

Second, zinc signals described as "zinc wave" were uncovered recently. Comparable to the zinc flux, zinc waves act also as second messenger, but are induced indirectly depending on calcium influx. This phenomenon was for instance observed in mast cells, when the Fc-epsilon receptor I (FcεRI) is cross-linked [55,57].

Late zinc signals occur on a timescale significantly longer than the others, accompany cellular differentiation, and last for several days. Those are usually involved in altered expression of proteins participating in zinc homeostasis. In this regard, maturation of for instance monocytes and dendritic cells (DC) as well as cytokine expression is highly dependent on zinc signals [38,58]. In general, altered zinc homeostasis results in changes in signaling cascades without acting as second messenger.

**Table 1.** Effect of zinc signals on Immune function: Altered immunological functions can be induced by different zinc signals, as zinc flux, zinc wave, and homeostatic zinc signal. Altered from [59].

| Zinc Signal | Duration | Effect |
|---|---|---|
| Zinc Flux | Seconds/ minutes | • Inhibition of PDE in monocytes/macrophages [60]<br>• Inhibition of MKP in monocytes/macrophages [61,62]<br>• Induction of PMA-triggered NET-formation in PMN [40]<br>• Induction of Lck recruitment to TCR [63,64]<br>• Zinc release from lysosomes in T cells [65]<br>• Triggering T cell activation by APC [66]<br>• Induction of Lck homodimerization/activation in T cells [67]<br>• Redistribution of zinc from nucleus/mitochondria to cytosol/microsomes [30] |
| Zinc wave | Minutes | • Zinc release from perinuclear area in mast cells [57] |
| homeostatic Zinc Signal | Hours | • Stabilization of MyD88 expression [5,68]<br>• Induction of A20 expression in T cells, monocytes [69]<br>• Negative regulation of IRAK signaling [70,71]<br>• MAPK activation in immune cells [72–74]<br>• Negative regulation of TRIF pathway in macrophages [5]<br>• Changed Zip and ZnT expression in DCs and T cells [58,72]<br>• Inhibition of AC transcription in T cells [75]<br>• Influence of cytokine production, e.g., IL-2 in T cells [72]<br>• Induction of Akt/ERK/p38 phosphorylation in T cells [73]<br>• Triggering of PTEN degradation in T cells [76]<br>• Inhibition of IL-6/IL-1 induced STAT3 phosphorylation [77] |

**Table 1.** *Cont.*

| Zinc Signal | Duration | Effect |
|---|---|---|
| | | • Alteration of M1/M2 differentiation by inhibition of STAT6 phosphorylation [78] |
| | | • Induction/stabilization of regulatory T cells [79,80] |
| | | • Inhibition of Th17 and Th9 cell differentiation [81,82] |
| | | • Reduced cytokine production, e.g., IFN-γ in T cells [83] |
| | | • Induction of NK cell killing and granzyme expression [84,85] |
| | | • Inhibition of caspase activity |
| | | • Inhibition of cAMP/cGMP hydrolysis [86,87] |
| | | • Inhibition of DNA fragmentation [6] |
| | | • Epigenetic modifications due to inhibition of SIRT1 [79] |
| | | • Inhibition of IL-1/TNF CRP expression in myeloid cells [53,69,88] |

AC: adenylate cyclase; cAMP: cyclic adenosine monophosphate; cGMP: cyclic guanine monophosphate; CRP: c-reactive protein; DC: dendritic cell; IL: interleukin; IFN: interferon; IRAK: interleukin-1 receptor-associated kinase; Lck: lymphocyte-specific protein tyrosine kinas; MAPK: mitogen-activated protein kinase; MKP: MAP-kinase phosphatase; MyD88: myeloid differentiation primary response gene 88; NET: neutrophil extracellular traps; NK: natural killer; PDE: phosphodiesterase; PMA: Phorbol-12-myristat-13-acetat; PTEN: phosphatase and tensin homolog deleted on chromosome 10; STAT: signal transducers and activators of transcription; TCR: T cell receptor; APC: Antigen presenting cell; Th: T helper; TNF: tumor necrosis factor; TRIF: Toll-interleukin-1 receptor (TIR) domain-containing adaptor-inducing interferon.

## 4. Zinc, Signaling and Immunity

In addition to its function as second messenger, another suggested mechanism how zinc is involved in signaling is that zinc alters ligand binding to receptors either by changing the ligand's or the receptor's affinity. Here, LPS revealed altered fluidity depending on zinc's availability and thereby binding characteristics to receptors such as TLR4 and CD14 were affected as well as subsequent signal transduction [89].

Another possible scenario is the alteration of cellular membrane composition and fluidity [90]. Here, zinc could influence the generation and stability of membrane complexes causing assembly or disassembly of receptors and probably altering their endocytosis, as is well described for neuronal cells, but has also been observed in immune cells [84,91]. Furthermore, alterations of extracellular zinc conditions affect the concentration of free intracellular zinc so that persistent changes in extracellular zinc modify cell metabolism on the long run. Recent data indicate, that there is a major difference in the effects of zinc on signaling in innate compared to adaptive immune cells. Therefore, data will be discussed separately.

### 4.1. Innate Immune Cell Functions

Cells of the innate immune system, including principally neutrophil granulocytes and monocytes/macrophages, but also mast cells, DCs, and NK cells are the first to encounter invading pathogens at the side of infection. Recognition of pathogens and initiation of their clearance needs to be fast, which is only possible if specificity is compromised. Efficient immune response is completed by adaptive immune cells, which need longer to be recruited and activated, but are highly specific.

Innate immune cells recognize pathogens via detection of general pathogen-associated molecular patterns (PAMPs). Receptors are simply denoted pattern recognition receptors (PRRs) and examples include TLR, retinoic acid-inducible gene-I-like receptors (RLR) and nucleotide-binding oligomerization domain-like receptors (NLR). At least 10 different TLRs exist in humans, enabling cells to distinguish groups of pathogens and their intracellular or extracellular location. Lipoprotein (gram positive bacteria), zymosan (fungi) and LPS (gram negative bacteria) are for example detected by TLR1, -2 and -4, respectively [92]. Binding of a PAMP activates signaling pathways leading to antimicrobial processes including cytokine production, degranulation, phagocytosis of the pathogens and the presentation of the antigen to other cells, including those of the adaptive immune system.

Ligation of PRRs generally leads to activation of interleukin-1 receptor-associated kinase (IRAK) family proteins through Myeloid differentiation primary response gene (MyD)88 signaling pathways as depicted in Figure 1 and can either induce gene expression via nuclear factor kappa B (NFκB) or mitogen activated protein kinases (MAPK). Binding to NLRs directly leads to activation of TAK1, NFκB and MAP kinase pathways [93]. Of note, most cells are equipped with the identical sets of signaling molecules, but choice of signaling pathway(s) activated depends on the type of pathogens, so that appropriate reaction to attack the invader is induced. First data connecting zinc-induced changes in immune cell function to regulation of intracellular signaling by zinc arose in the late 1970s [94,95]. Direct effects of zinc on signaling molecules or indirect effects via phosphatases, kinases and redox metabolism have been described since then.

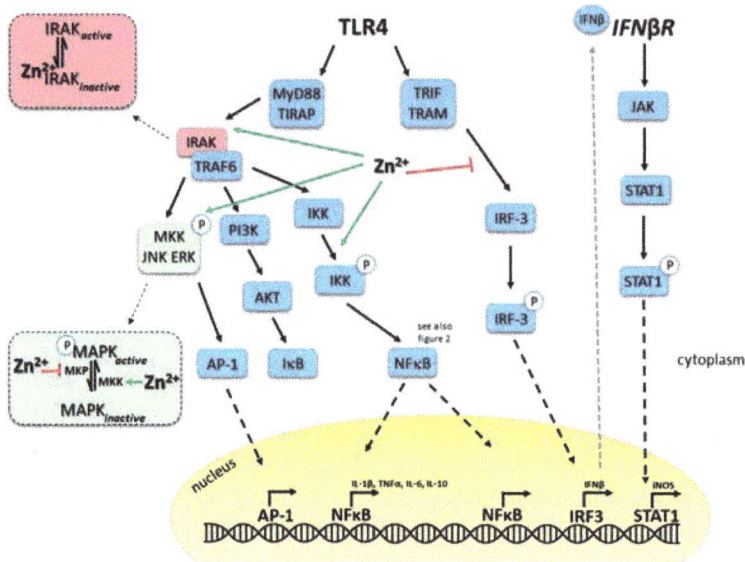

**Figure 1.** Zinc in TLR4 triggered signaling, illustrating explanations in the text. Black arrow: activation, green arrow: activating function of zinc, red T bar: inhibiting function of zinc, black dotted arrow, translocation of the molecule to nucleus, grey dotted arrow: secretion of the molecule. Abbreviations: ERK: extracellular Signal-regulated Kinase; IFN: interferon; IRAK: Interleukin-1 receptor-associated *kinase*; IκB: Inhibitor of NFκB; IKK: IκB kinase; IRF: interferon related factor; JAK: JNK janus kinase; JNK: c-Jun N-terminal Kinase; MAPK: mitogen activated protein kinases MEK: MAPK/Erk kinase; MKK: MAPK kinase; MKP: MAPK phosphatase; MyD88: Myeloid differentiation primary response gene 88; NFκB: nuclear factor (NF) κB. PI3K: phosphatidyl-inositol-3-phosphate; STAT: signal transducers and activators of transcription; TBK: Tank-binding kinase 1; TIRAP: toll-interleukin 1 receptor (TIR) domain containing adaptor protein; TLR: toll like receptor; TRAF: TNF receptor-associated factor; TRAM: TRIF-related adaptor molecule; TRIF: Toll-interleukin-1 receptor (TIR) domain-containing adaptor-inducing interferon.

## 4.2. Zinc and (De-)Phosphorylation

In the following, we will first discuss some general mechanisms of how zinc affects intracellular signaling molecules and their activity in innate immune cells. Subsequently, we will use this information to describe selected key receptor-induced signaling pathways for the different innate immune cell types. This is far from being inclusive for all the pathways investigated so far, but should give a basic idea of the state of research. One should carefully extrapolate known mechanisms within certain pathways from one cell type to another, over simplifying those mechanisms as other parameters such as environmental circumstances, activation status, maturity grade and age of a cell have been shown to matter as well.

In general, signal transduction is predominantly regulated by post-translational activation or inactivation of existing signaling molecules via changes in their phosphorylation status. Each signaling pathway involves phosphorylation of a whole cascade of molecules, where phosphorylation and thereby the signal can be passed on from one molecule to the other within a short period of time. Various residues within the molecules can be altered, but tyrosine and to a lesser extend serine and threonine phosphorylation are most common in immune cells. In leukocytes, especially Src family kinases are prevalent. As phosphorylation can be both at activating and inhibitory motifs, there is not a general consequence of elevated kinase activity. Phosphorylation by kinases can be removed by antagonizing phosphatases, balancing the extent of phosphorylation inside the cells. CD45, a transmembrane protein tyrosine phosphatase (PTPs) found generally on all leukocytes, is for example an important regulator of Src family kinase activity [65,73]. Zinc has a role in regulation of PTP and PT kinase (PTK) activity. Of note, kinase and phosphatase activity is not only affected by intracellular zinc homeostasis, but both are involved in changing expression of genes relevant for regulation of zinc homeostasis, as well.

Over 107 PTPs have been identified in humans. They are primarily regulated via the redox state of the cell, proteolysis, dimerization and phosphorylation. The $IC_{50}$ for zinc of important PTPs involved in immune cell signaling are summarized in Table 2. The table clearly shows that inhibition constants for PTPs are very low (nanomolar), so that cellular zinc levels are sufficient to alter enzymatic activity [96]. PTP1B, for example, has a putative affinity for inhibitory zinc as low as 3–17 nM [10,97]. Zinc competes with other co-factors of PTPs, which might explain their inhibition during high zinc conditions [98,99]. In addition, zinc directly acts on the highly conserved catalytic region of PTPs, suggesting that other enzymes from this family might be inhibited by zinc in a similar way [98,100]. SHP-2, phosphatase and tensin homolog deleted on chromosome 10 (PTEN) and PTP1B are discussed to be involved in LPS-induced signaling pathways in macrophages via different pathways and molecules [5,101]. This is in line with the observation that zinc prolonged the tyrosine phosphorylation of proteins in many immune pathways [55,99,102], involving all kinds of signaling molecules, underlining the complexity of zinc's effects.

In addition to PTPs, zinc is able to inhibit the activities of other molecules involved in signaling. However, most results were calculated from experiments using isolated enzymes or cell lysates. The resulting IC50 concentrations for example for PDEs in the µM range will hardly be reached within the cytosol of intact cells. However, a few studies indicate that high levels of zinc can be enriched in cellular compartments such as vesicles and lysosomes. Hennigar et al. found 20× higher zinc concentrations normalized to protein comparing cytosol and lysosomes [103]. Moreover, in catalytically active lysosomes the non-bound $Zn^{2+}$ concentration was measured to be close to the permissive concentration of 0.1 µM. In inactive vesicles at neutral pH the $Zn^{2+}$ concentration might be much higher as the driving force of the proton gradient is missing [104]. In addition, Vinkenborg et al. stated that free zinc(II) concentrations in vesicles are much higher than in the cytosol, due to different buffering capacities. Estimates are 1–100 µmol/L in insulin-storing granules for example [105]. More investigations on intracellular effects of zinc on activity of those signaling molecules are necessary.

Close relatives to PTPs, the MAPK phosphatases (MKP), can dephosphorylate threonine residues in addition to tyrosine in MAPK and might be zinc's target as well. The zinc flux has indeed been connected to inhibition of MKPs in neuronal cells, which should be tested for immune cells as well [61].

As zinc-induced secretion of cytokines by peripheral blood mononuclear cells (PBMCs) is abolished if cells are pre-incubated with the kinase inhibitor Herbimycin A, zinc might not only affect PTPs but also their antagonists, the tyrosine kinases (TK). Zinc-associated activation of tyrosine kinases has been noted by Vener et al., in cells from Alzheimer patients [106]. Results from Bennasroune et al. support this hypothesis, describing that zinc activates anaplastic lymphoma kinase (ALK), also a member of the receptor tyrosine kinase family [107]. Zinc was found within the molecular structure of other TKs including Bruton's (B)TK as well [108]. Here, zinc is probably involved in enzymatic activity or substrate binding. BTK is an important regulator of for example TLR4 signaling via Myeloid differentiation primary response gene (MyD)88 and also binds to TLR9, -8, and -6 [109]. Activation of mitochondrial Src tyrosine kinase by zinc indicates that effects of zinc are not limited to the cytoplasm [110].

Well-known substrates for Tyrosine (de-)phosphorylation are the MAPKs, such as MAPK/Erk kinase (MEK), Extracellular Signal-regulated Kinase (ERK), and p38, which are themselves involved in phosphorylation of other signaling molecules. A role of zinc in alteration of signal transducers and activators of transcription (STAT) phosphorylation has recently started to emerge [111,112]. Effects of zinc on MAPK activity have been described for all kinds of cells, and seem to be a very general molecular mechanism to translate the zinc signal into cellular function. PTP1 is known to alter the phosphorylation of JAK2 and dephosphorylates MAPKs in immune cells [97,113]. High zinc concentrations reduced phosphorylation of ERK in in rat glioma cells, while low zinc concentrations activated this MAPK [98]. Along with this p38 was activated in myeloid cells during zinc deficiency [48]. In contrast, zinc activated ERK in fibroblasts. Similar effects of severe zinc deficiency and high levels of zinc have been observed before, and might explain the variation of results. In addition, there might be a certain threshold of intracellular zinc concentrations decisive for its influence on signaling pathways.

**Table 2.** $IC_{50}$ values for signaling molecules in humans.

| Signaling Molecule | $IC_{50}$ | Reference |
|---|---|---|
| PTPRB | 98 pM | [114] |
| PTEN | 0.6 nM | [76] |
| PTP1B | 3–17 nM | [100] |
| SHP-1 | 92 nM | [115] |
| SHP-2 | 1–2 µM | [100] |
| TC PTP | 200 mM | [99] |
| Calcineurin | 250 nM–7 µM | [116] |
| Ca-dependent endonuclease | 1 µM (in the presence of 10 µM Ca) | [117] |
| LAR | 20 µM | [100] |
| PDE4A | >5 µM | [118] |
| PDE 5 | >1 µM | [119] |
| PDE 6 | >1 µM | [120] |
| IKK | 8.7 µM | [121] |
| Caspase 3 | <10 nM, 100 nM | [86,99] |

PTPRB: Protein phosphotyrosylphosphatase receptor type B; PTEN: Phosphatase and tensin homologue deleted on chromosome 10; PTP1B: Protein phosphotyrosylphosphatase 1B; SHP: SH2-containing phosphatase; TC PTP: T cell PTP; LAR: Leukocyte common antigen-related protein; PDE: phosphodiesterase; IKK: IκB kinase.

### 4.3. Zinc and Redox Metabolism: Two Strongly Intertwined Second Messengers

Balancing the redox state of a cell is important for regulation of its metabolism. Amongst others, ROS function as second messengers and interfere thereby with cellular signaling. Phorbol 12-myristate 13-acetate (PMA) activates the nicotinamide adenine dinucleotide phosphate (NADPH) oxidase via protein kinase C (PKC) resulting in ROS production. PMA is also a well-known inducer of a zinc flux [122]. As NADPH oxidase inhibition blocked the PMA-induced zinc signal, the release of zinc into the cytoplasm seems to be ROS-induced. $H_2O_2$ was shown to increase intracellular zinc in granulocytic cells as well, underlining the association of zinc homeostasis with redox metabolism. Both the zinc signal and the increased ROS levels are essential to subsequently activate key immune functions [40]. ROS can induce release of zinc from proteins such as MTs or PKC itself, but the source could also be a cellular compartment such as lysosomes as shown for T cells or the endoplasmatic reticulum [65,102,123–125]. The source of zinc probably depends on cell type, magnitude of ROS alteration and the kind of oxygen species ($O_2^-$, $H_2O_2$, HOCl or similar).

Interestingly, not only does the redox metabolism influence intracellular zinc homeostasis, but the zinc status of a cell also alters its redox state [126]. Pro-antioxidant functions have long been attributed to zinc for decades, and underlying mechanisms for this phenotypic observation are becoming increasingly clear. Although low concentrations of a zinc chelator TPEN ($N,N,N',N'$-Tetrakis(2-pyridylmethyl)ethylenediamine), 5 μM) were not able to affect PKC activity, using 100 μM of TPEN inhibited PKC's function and thereby ROS synthesis. The abrogation of PKC activation when zinc is chelated is suggested to be due to the requirements of zinc finger structures for PKC's activity, but also because translocation of the enzyme between membrane and cytoskeleton involves zinc [122,127,128]. MAPK phosphatase (MKP)-1 is possibly also involved in LPS-induced cytokine induction via PKC. As MKP-1 is directly inhibited by zinc, the role of the trace metal in regulating LPS-induced effects seems to be complex and involve several levels, PKC being rather upstream of other effects. In contrast, long lasting zinc deficiency increased basal ROS levels and augmented stimulation induced ROS synthesis in myeloid cells [48] underlining again the difference between long- and short-term zinc alterations. Regarding signaling in innate cells, analyses of the role of zinc in oxidative burst added nice data to the puzzle, but more data are still necessary as some discrepancies exist [40,41]. An increase of ROS production by granulocytes during zinc deficiency seems to be clearly documented, whereas the effect of high zinc conditions has not been clearly defined, yet [40,41,48,129–131].

### 4.4. Zinc, Cyclic Nucleotides and Proteinases A and G

The intracellular second messengers cyclic adenosine monophosphate (cAMP) and cyclic guanosine monophosphate (cGMP) are synthesized by adenylate cyclase (AC) and guanylate cyclase (GC). Both are degraded by cyclic nucleotide phosphodiesterases (PDE). Their main targets are protein kinase A (PKA) and PKG, respectively, mediating inflammatory gene expression. On the one hand, PDEs need zinc for their activity, as zinc is tightly bound to the catalytic center [118,120]. This was first observed in PDEs in Baker's yeast, indicating that this is a conserved mechanism [132]. On the other hand, PDEs have an inhibitory site, which binds zinc with lower affinity. If zinc is available in high concentrations, hydrolysis of cAMP and cGMP is inhibited, causing their intracellular increase. In human monocytes, PDE 1, 3 and 4 have been shown to be inhibited during high zinc conditions. Furthermore, zinc seems to be involved in PDE expression as well, blocking LPS-induced increase in PDE4B transcription in monocytes, adding one more way of how zinc regulates intracellular second messenger levels and thereby signaling [60]. Finally, elevated cGMP does not only target PKG but also cross activates PKA. Subsequently, PKA phosphorylates Raf-1 at serine 259, rendering it inactive and preventing its activation by serine 338 phosphorylation. Thereby the whole cascade of LPS-induced NFκB signaling pathway is suppressed [133]. Cyclic nucleotides are not only affected by zinc flux, but the homeostatic zinc signal has been shown to inhibit AC, causing decreased cAMP levels, while GC activity and cGMP levels are not affected [134,135]. Increased cAMP and cGMP levels have been associated with monopoiesis, which will be discussed later.

### 4.5. Zinc-Mediated Regulation of NFκB Signaling and A20 Expression

In addition to MAPKs, signaling in innate immune cells centers on NFκB. This molecule is involved in regulating a large set of genes, thereby affecting apoptosis, proliferation, cell adhesion, pro-inflammatory responses, tissue remodeling and stress-responses [21]. Important molecules within these signaling pathways include RelA (p65), RelB, c-Rel, p50/p105 (NFκB1) and p52/p100 (NFκB2), which can form various homo- and heterodimers. Inhibitors of NFκB (IκBs) can bind to single molecules and complexes, causing their sequestration in the cytoplasm, hindering the translocation to the nucleus as well as subsequent DNA binding and gene activation. Inhibitors are IκBα, IκBβ, IκBε, IκBγ, Bcl-3, p100 and p105 [36]. The complexity of this signaling process makes its investigation regarding an association to zinc homeostasis a difficult task. Existing studies show conflicting results and a general unilateral statement on the connection is difficult to formulate.

Diverse effects of zinc on NFκB signaling have been observed (see Figure 2) [136].

Chelating zinc using TPEN abrogated LPS-induced, NFκB-mediated signaling [102]. In contrast, zinc has been found to be inhibitory for NFκB signal transduction as well by others [137]. Clear mechanistic insights are not available, but suggestions include the effect of zinc on A20, a principally anti-inflammatory zinc-finger protein, important for tumor necrosis factor receptor (TNFR)- and TLR-induced signaling. A20 can de-ubiquitinate receptor interacting protein (RIP)-1, preventing its binding to I kappa B kinase (IKK)γ, which keeps sequestering NFκB in the cytoplasm. In addition, A20 can remove polyubiquitin chains from TNFR-associated factor (TRAF)6, thereby inhibiting TLR-mediated signaling.

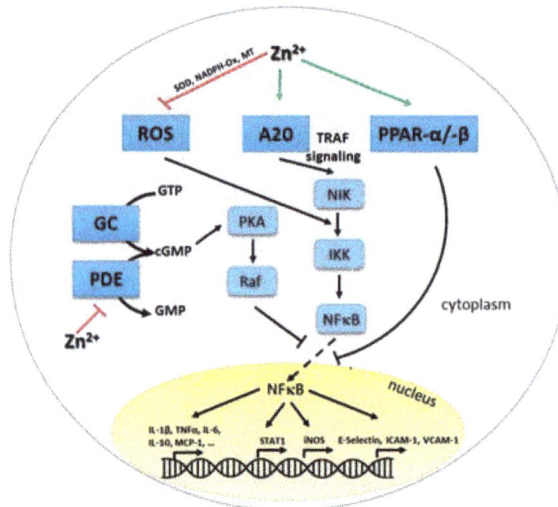

**Figure 2.** Complex impact of zinc on NFκB-centered signaling as illustration of signaling pathways described in the text. Red T bar: inhibitory function of zinc, green arrow: activating function of zinc, black T bar: inhibition, black arrow activation. Abbreviations: cGMP: cyclic guanine monophosphate; GC: guanine cyclase; GMP: guanine monophosphate; GTP: guanine triphosphate; ICAM: intercellular adhesion molecule; iNOS: inducible nitric oxide synthase; IKK: IκB kinase; IL: interleukin; MCP: monocyte chemoattractant protein; NIK: NFκB-inducing kinase; NFκB: nuclear factor (NF)κB; PDE: phosphodiesterase; PKA: protein kinase A; PPAR: Peroxisome proliferator-activated receptor; ROS: reactive oxygen species; STAT: Signal transducer and activator of transcription; TNF: tumor necrosis factor; VCAM: vascular cell adhesion molecule.

In endothelial and pre-monocytic cells, zinc-dependent induction of A20 mRNA expression and protein production was demonstrated, while information on the ubiquitination activity of the enzyme regarding zinc availability is not clear yet. Zinc signals upregulated mRNA expression of A20 in other immune cells as well [30,69,138]. Inhibition of gene expression by zinc via increased A20 and decreased NFκB activity was found for various molecules, including decreased expression of interleukin (IL)-1β and TNFα in myeloid cells and decreased C reactive protein (CRP) levels, lipid peroxidation and inflammatory cytokines in elderly subjects [53,88,138–140]. In addition to A20, peroxisome proliferator-activated receptor (PPAR) alpha is of interest as it alters the binding of NFκB to the DNA. Its zinc-induced expression results in decreased NFκB binding and down regulation of pro-inflammatory cytokine and adhesion molecule expression. [53]. NFκB is also inhibited by PKA via zinc-induced alteration of cGMP levels, as described earlier adding one more explanation for the decrease in NFκB induced gene expression during high zinc conditions and its increase during zinc deficiency [133]. The complex postulated effects of zinc on NFκB centered signaling pathways are summarized in Figure 2.

### 4.6. Zinc and Signaling Proteome

Most of the so far described effects of zinc in signaling depict consequences of immediate changes in zinc homeostasis. Long-term alterations in intracellular zinc are often related to altered expression of zinc transporters and zinc binding proteins. In a recent approach, Aude-Garcia et al. [68] cultured macrophages with zinc oxide nanoparticles or an equivalent dose of zinc acetate for at least 24 h and analyzed the proteome of the cells via 2D gel based analyses. In addition to changes regarding energy, mitochondrial, cytoskeletal and DNA-control metabolism, authors found altered levels of certain signaling molecules. Protein phosphatase 1a and 2a and S/T phosphatase were affected by zinc acetate. Zinc oxide nanoparticles changed the levels of Prostaglandin reductase 2, Phosphatidylinositol transfer protein (PITP)α, Acyl-protein thioesterase 2, Regulator of G-protein signaling 10, MOB kinase activator 1B and Inositol-3-phosphate synthase 1. Protein levels of MyD88 were affected by both the zinc ion and the nanoparticle in a similar manner [68]. If altered proteome is due to changes in gene expression, posttranscriptional mechanisms or if protein stability is reduced remains open. Furthermore, information on whether proteins are functional is missing. However, results suggest that certain signaling pathways are prioritized during high zinc conditions and are important to the cell. MyD88 might be stabilized to keep signaling pathways, important for the pro-inflammatory activity of the cell active during high zinc conditions. The study provides molecules to the list of zinc-regulated targets in signaling, but as protein species-based proteomics have some limitations, results have to be verified carefully [68]. Effects of zinc deficiency should be investigated using a similar approach as well.

### 4.7. Zinc and Transcription Factors

As a note, many transcription factors governing innate immune cell functions are zinc finger proteins, suggesting an additional direct or indirect role of zinc in intracellular signaling. Promyelocytic leukemia zinc-finger (PLZF) is a critical transcription factor for induced natural killer T cell (iNKT cell) development as is PU.1 for myelopoiesis [141,142]. GATA-4, -5, and -6 and krüppel like factor (KLF)-4 and KLF-5 are additional examples of zinc finger transcriptionfactors important during the development of innate immune cells [143,144]. The same is true for adaptive immune cells: zinc finger transcription factors (ZEB1, ZNF292, and ZNF644) were confirmed to play a role in CD4+ T cells in Rheumatoid Arthritis. A lot of genes involved in regulating zinc homeostasis reveal metal response elements in their promoters and can be induced by binding of metal-response element binding transcription factor (MTF)-1, which acts as an intracellular zinc sensor, suggesting feedback mechanisms [145]. After some initial analysis in this area of research, this topic has not been followed much recently, but would be a great area for future experiments.

## 5. From Altered Zinc Homeostasis via Signaling to Innate Immune Cell Functions

As described above, the effects of zinc homeostasis on signaling are complex. Knowing the effect of zinc on single signaling molecules does not enable prediction of the functional outcome of zinc deficiency or supplementation. In the upcoming paragraphs, examples will be described for some selected zinc-dependent key functions of innate immune cells. Predicted underlying molecular mechanism for the particular setting will be mentioned as well.

### 5.1. TLR4 Signaling in Monocytes

One of the best-studied examples for the role of zinc in innate immune signaling is TLR4 triggering by LPS. Pathways involved are illustrated in Figure 1 and molecules where zinc is potentially intervening in the signaling pathways are indicated. Long-term zinc deficiency augments pro-inflammatory cytokine production, which is reversible if zinc is reconstituted [48]. In contrast, within two minutes after LPS binding to the receptor complex a significant increase in intracellular zinc can be detected, which is involved in activation of MAPK signaling pathways. Amongst others, zinc inhibits dephosphorylation of MAPK and supports IKKα/β phosphorylation, NFκB translocation to the nucleus and gene activation [102,146]. Wan et al. recently separated zinc-dependent and -independent processes shortly after LPS triggering of TLR4 of murine macrophages. Depletion of zinc inhibited the LPS-induced activation of several kinases including ERK1/2, IKKβ, MKK3/6 and IκB in murine primary macrophages and cell lines. Interestingly, phosphorylation and ubiquitination of IRAK1 was not affected by zinc deficiency, at least not early after stimulation, suggesting a selective effect of zinc on certain signaling pathways, shortly after LPS-stimulation. However, degradation of ubiquitinated IRAK1 required zinc but did not involve proteasome [70]. Zinc chelation caused the accumulation of phosphorylated and ubiquitinated IRAK1. One can speculate, that inactivation of zinc-containing metalloproteases during zinc deficiency, prevents the destruction of IRAK1, which remains to be investigated. A membrane associated complex, which is phosphorylated after LPS stimulation, contains IRAK1 in addition to IKKα and IKKβ [147]. Degradation of IRAK1 is essential to release phosphorylated IKKs into the cytoplasm, which then degrade IκB and p105 activating NFκB and ERK signaling. However, zinc deficiency might affect phosphorylation of ERK and IκB via its effect on PTPs as well, which is not clear yet.

TLR4-induced pathways can be grouped into MyD88-dependent pathways, including MAPKs and NFκB, which induce expression of inflammatory cytokines and MyD88-independent/Toll-interleukin-1 receptor (TIR) domain-containing adaptor-inducing interferon (TRIF)-dependent pathways involving TRIF-related adaptor molecule (TRAM), TRIF and IRF-3 and triggering interferon (IFN)β expression (see Figure 1). The latter has an autocrine function and can bind to surface interferon receptor (IFNR) resulting in activation of the JAK-STAT pathway finally causing expression of CD40, CD80 and CD86, necessary for interaction of myeloid cells with T cells [5,148]. Zinc has been suggested to be involved in IFN signaling in hepatocytes recently, which should be tested for immune cells as well [149]. In addition, NFκB as well as the JAK-STAT pathway induce NO production via inducible nitric oxide synthase (iNOS), which is a potent anti-microbial mediator. While our knowledge on the role of zinc in MyD88-dependent pathways, especially early after receptor triggering, is broad and described in large detail [102], an association of zinc to TRIF pathways and NO synthesis is less explicit. In addition, most data originate from investigations of time points early after receptor triggering, representing investigation of the zinc flux. Pre-existing zinc deficiency augmented LPS-induced NO synthesis in murine macrophages and also increased LPS induced IFNβ expression as well as CD80 and CD86 (cell-line and bone marrow based). Effects were measured a few hours after the original stimulus and where therefore assigned to be regulated by homeostatic zinc signals. Reconstitution of zinc abolished the effect of the chelator. Interestingly, chelation of zinc caused decreased LPS-induced transcription of IL-1β, IL-6, and IL-10 in macrophages compared to zinc adequate controls. This indicates zinc chelation seems to inhibit the MyD88-depending pathways while triggering TRIF-dependent signaling. In line with this, a time-course experiment revealed, that adding the chelator at the time of TRIF activation caused the maximal increase in NO synthesis. This also indicates that TRIF-signaling is independent from the initial zinc signal, induced quickly after LPS binding to TLR4. MyD88-dependent pathways induce fast translocation of NFκB to the nucleus which was reduced, when cells were pre-incubated with TPEN. However, TRIF-dependent delayed NFκB activation was not affected by TPEN, underlining this hypothesis [5].

Regarding effects of persisting changes in intracellular zinc levels, the de-ubiquitination of TRAF6 by A20 is probably involved in turning off TLR4-induced signaling and pro-inflammatory activity of macrophages. Ubiquinated TRAF6 activates TAK1 a kinase phosphorylating IKKs and MKKs, resulting in gene expression which is thereby blocked [138,150,151]. Thus, balanced zinc homeostasis is important to prevent unwanted signaling disturbances. In macrophages, A20 is induced by pro-inflammatory cytokines, possibly as a feedback mechanism to terminate their expression [152]. Increased cGMPs levels due to high intracellular zinc could explain elevated phosphorylation of Raf-1, mediating its inactivation as well as inactivation of NFκB as can be seen in Figure 2. Investigation of

TLR7 and TLR3-induced signaling showed dependents of TRIF/IRF/IFNβ signaling as well, while NO production was not affected by zinc chelation after TLR3 activation.

Altogether, TLR4 signaling in macrophages is a great example of the combination of a short-lived zinc peak and modulation of basal intracellular zinc levels for longer periods to balance MyD88- and TRIF-dependent pathways, acting as a fine-tuning signal after TLR4 triggering. Using ligands for other TLRs including Pam3CSK4 (TLR1/2), Listeria monocytogenes (TLR2), flagellin (TLR5), FSL-1 (TLR6/2), ssRNA40 (TLR7) and ODN1826 (TLR9) all increased intracellular zinc in murine macrophages and primary human monocytes [5,102]. If this has similar consequences on signaling pathways remains to be investigated. Simplification of zinc's effect such as stating that zinc generally supports all TLR-induced signaling pathways, is difficult, as the whole environmental set up, including the current zinc supply, activation and maturity state of the cell, chronic inflammation, time after stimulation and more, can affect consequences of altered zinc levels.

## 5.2. Zinc Homeostasis, Hematopoiesis and STAT3

Long-term changes in zinc homeostasis usually involve changes in zinc transporter expression. In dendritic cells, stimulation with LPS changes the expression of Zip6 and Zip10 as well as of ZnT-1, -4 and -6 via pathways involving TRIF. An association of zinc with regulation of TRIF signaling has just been described above, explaining this observation. This causes a long-term reduction in intracellular zinc which is essential for cellular maturation [58]. This is just one of the examples explaining the often-observed effect of zinc on hematopoiesis. Monopoiesis has been shown to be paralleled by a decrease in intracellular zinc and similar data have just been generated for granulopoiesis [38,153].

IL-6 is one of the most potent inflammatory cytokines, activating immune cells via binding to its gp130 receptor found within the plasma membrane. Amongst others, IL-6 induces the expression of acute phase proteins (APP), which subsequently interact with TLRs and induce the NLRP3 inflammasome. Receptor ligation activates JAK-STAT3 and MAPK pathways. Recent years revealed that especially the phosphorylation of STAT3 is broadly affected by zinc homeostasis. Zinc deficiency increased IL-6-induced activation of JAK-STAT3 signaling pathways; reconstitution with zinc normalized the effect while zinc supplementation was shown to inhibit STAT3 phosphorylation induced by IL-6 and IL-1 [111]. Phosphorylation of STAT3 is known to be regulated by SHP1/2, whose activity is increased by zinc [154]. If this is the only mechanism how zinc is involved in JAK-STAT signal transduction remains to be further investigated. STAT3 seems to be an important player in regular myelopoiesis as well as in demand adapted emergency myelopoiesis, during inflammatory diseases such as sepsis [155]. The connection between disease-related serum hypozincemia, STAT3 activity and grade of differentiation needs to be explored.

In murine hematopoietic stem cells, zinc chloride promotes transient STAT3 phosphorylation (Tyr705), resulting in increased expression of pluripotency genes and inhibition of differentiation genes [112]. Expression of STAT3 was not altered. Thus, zinc plays a critical role in maintaining pluripotency and self-renewal of stem cells and can even replace leukemia inhibitory factor (LIF), which is normally essential for this process. Zinc disrupted differentiation induced by retinoic acid, at least for a certain period of time. Treatment of cells with zinc chloride and LIF did not augment pluripotency, suggesting that both factors regulate the same pathway, which is centered on STAT3. The observation that inhibition of STAT3 abrogates the effect of zinc chloride on the cells underlines the importance of STAT3 as a central molecule in hematopoiesis. Along with this, B cell proliferation, was increased during zinc deficiency [77]. Again, STAT3 phosphorylation was shown to be responsible here.

Neuronal development is also strongly affected by zinc via STAT3-centered signaling. Results from this very recent study add alteration in the redox-state as underlying mechanism for the effect of zinc on STAT3 phosphorylation. In this study, moderate zinc deficiency (1.5 µM) caused oxidative stress within 24 h, resulting in decreased tyrosine phosphorylation and retention of STAT3 in the cytosol. In addition to tyrosine (Y705), serine (S727) phosphorylation was analyzed. While only tyrosine phosphorylation was decreased during zinc deficiency when whole cell lysates were tested, in nuclear extracts phosphorylation of both residues was decreased [156]. If this is a cell-type specific effect or if it can be extrapolated to innate immune cells remains to be tested, but as redox and zinc metabolism are highly intertwined in innate immune cells, as described earlier, a similar mechanism is highly likely.

Thus far, an association of zinc homeostasis with the activity of STAT molecules other than STAT3 is barely investigated. The effect of zinc on phosphorylation of STAT1 and STAT6 was suggested to be relevant for lineage determination of macrophages. Zinc supplementation decreased STAT6 phosphorylation reducing the amount of M2 macrophages, while zinc deficiency increased phosphorylation of STAT2, thereby supporting M1 development [78]. While studies in T cells did not show an effect of zinc on STAT5 activity, recent results from myeloid cells suggest that zinc supplementation inhibits STAT5 phosphorylation [157]. This indicates a general mechanism of zinc affected STAT activity in cell differentiation, which should be investigates in more detail.

### 5.3. Zinc Alters Killing Activity of Natural Killer Cells

Although natural killer cells belong to the lymphoid lineage, they are counted as innate immune cells and kill infected or transformed cells. Regular human cells carry major histocompatibility complex (MHC)-I on their surface, which rescues them from being killed by binding to p58 killer cell inhibitory receptor (KIR) on NK cells. KIR can bind zinc which alters recognition of MHC-I and thereby the induction of subsequent signaling pathways. A zinc-dependent multimerization of KIR was suggested to be essential for formation of the clusters of KIR and human leukocyte antigen (HLA) molecules, also called "natural killer (NK) cell immune synapse" [84,158]. Effects of zinc deficiency on KIR multimerization have not been investigated, yet, but might explain why the lytic activity of NK cells is decreased when zinc supply is limited [159]. In general, existing studies for NK cells are limited to experiments, where zinc has been extracellularly added, whereas effects of altered intracellular zinc homeostasis can only be assumed, as measurements of intracellular zinc levels after stimulation are not available. Only some very recent data are available that showed a zinc flux shortly after IL-1 stimulation [85]. However, NK cells carry a variety of cell surface receptors, essential for fine tuning of their activities, which contain tyrosine phosphorylation sides such as immunoreceptor tyrosine-based activation motif (ITAM) or immunoreceptor tyrosine-based inhibitory motif (ITIM). As described earlier, the ability of zinc to inhibit PTPs is well accepted, so a role of intracellular zinc homeostasis in regulation of signaling pathways in NK cells is highly likely, but remains to be analyzed.

### 5.4. Mast Cell Degranulation Depends on Zinc Levels

Knowledge of a role of zinc in signaling in mast cells (MCs) is scarce. Cytoplasm and granules seem to represent two distinct zinc pools in this cell type. High amounts of zinc are stored in the granules, lost during immunglobulin (Ig)E-induced degranulation, and are quickly replenished thereafter. This zinc is probably not involved in signaling [160]. Zinc was however important for the process of degranulation, mediating granule translocation to the plasma membrane Zinc chelation inhibited degranulation. A transient increase in cytoplasmic zinc has been reported after cell activation [161]. This zinc signal affected translocation of NFκB to the nucleus and phosphorylation of ERK1/2, JNK1/2 and the translocation of PKC to the plasma membrane, as reported for other cells and stimuli [55,161]. Abrogation of the zinc wave via incubation with the zinc chelator TPEN abrogated activation of signaling pathways and blocked expression of IL-1β and TNFα. Zinc supplementation was able to prolong activation of the signaling pathways. Different probes were used to measure zinc here, enabling staining of the different zinc pools. It is well known, that low molecular weight probes highly differ in their exact intracellular location, which might explain the discrepancies [162].

## 6. The Adaptive Immunity

The adaptive immune system is built up of the humoral immune defense, represented by B lymphocytes (B cells), and the cellular immune defense, represented by T lymphocytes (T cells). Both cell types evolve off precursors and are educated to recognize their specific antigen in the thymus (T cells) or bone marrow (B cells), respectively. According to antigen contact, naïve lymphocytes differentiate into effector cells or memory cells, facilitating a stronger reaction to a known antigen as a secondary response due to immunological memory. The immune system is a highly active and proliferating organ influenced by a multitude of external factors such as nutrition. In this regard, the immune response is highly dependent on zinc, since zinc deficiency is accompanied with malfunctions and disorders, such as allergies, autoimmune diseases and cancer. In the past decade, studies on animal models have provided considerable knowledge about the underlying molecular mechanisms of the zinc-related modulation of the immune system, including new insights into how nutritional deficiency can alter immune cell homeostasis and function of the adaptive immunity. Zinc-related immunological changes in the adaptive immune response and altered signaling cascades are discussed in the following sections.

### 6.1. Zinc Homeostasis in Development of T and B Cells

For a long time, zinc is known to be essential especially for proper T cell and B cell development [163,164]. However, T cells are in particular much more sensitive to altered zinc signals resulting in altered differentiation or function, since zinc directly influence the biological activity of the thymic hormone thymulin by acting as essential co-factor. Thymulin, is produced by the thymus and released by thymic epithelial cells [165]. By binding to high-affinity receptors on T cells, it induces markers of differentiation in immature T cells, and promotes T cell function, including allogenic cytotoxicity, suppressor functions, and IL-2 production [166]. It is therefore not surprising that during zinc deficiency the recruitment of naïve Th cells and the percentage of cytotoxic T lymphocytes (CTL) precursors is diminished respectively [167]. The loss of developing T cells (pre-T cells) is ascribed to accelerated apoptosis consistent with the finding that pre-T cells express the lowest amount of the anti-apoptotic proteins B cell lymphoma (Bcl)-2 and Bcl-xL thereby increasing the vulnerability towards cell death. In general, zinc deficiency is described to affect for instance the hypothalamus-adrenal-pituitary axis, resulting in increased circulating glucocorticoids, being powerful apoptogens for immature lymphoid cells [168]. Additionally, caspase activity is known to be regulated in a zinc-dependent manner and zinc deficiency directly induce caspase 3-dependent cleavage of the cell cycle regulator p21 leading to an immediate induction of cyclin-dependent kinase (CDK)2 activity which may result in premature entry of the cells into S-phase and apoptotic cell death [169,170]. Since the thymus comprises of about 80% pre-T cells, its strong susceptibility towards zinc deficiency account for a main cause of disturbed T cell-mediated immune responses. Interestingly, in vitro and in vivo supplementation of serum zinc restored thymulin activity observed in moderately zinc-deficient mice and men, emphasizing a direct effect of serum zinc on thymulin activity [165,171].

Besides intrathymic functions on thymocytes and immature T cells, mature T cells in the periphery are also affected by thymulin. Herein, thymulin modulates cytokine release by PBMC, and proliferation of cytotoxic CD8-expressing CTL in combination with IL-2 responsible for T cell activation and proliferation [172]. Hence, it is obvious that zinc status has impact on immature as well as mature T cells through the direct activation of thymulin. As a consequence of zinc deficiency, T cell proliferation is decreased after mitogen stimulation, whereas zinc supplementation is able to reverse zinc deficiency-induced changes in the thymus and on peripheral cells [173].

In general, both mature T cell types: CD4-expressing Th cells as well as CD8-expressing CTL, are affected by zinc deficiency. On the one hand, the amount of CD8/CD73 co-expressing T cells is decreased during zinc deficiency. Those are predominantly precursors of cytotoxic T cells, needing CD73 expression for antigen recognition and proliferation as well as cytolytic process generation [167]. On the other hand, polarization of mature Th cells is disturbed. Polarization into Th1 cells is impaired and therefore changes in the Th1/Th2 ratio towards Th2 cells is observed leading to unbalanced cell-mediated immune responses [174]. Herein, Th1 cell products as IFNγ and IL-2 are decreased, whereas Th2 cell products like IL-4, IL-6, and IL-10 remain unaffected. Therefore, the incidence of infections and Th2-driven allergies is increased by zinc deficiency [23,175]. Furthermore, the functional impairment of T cell-mediated responses during zinc-deficiency is described to favor the development of autoimmune diseases [20,176]. In this regard, patients suffering from multiple sclerosis (MS) show lower plasma zinc level and lower zinc concentrations in chronic MS lesions [14,177,178]. Moreover, adequate zinc status is essential for transplant rejections since zinc supplementation was shown to beneficially influence allogeneic cardiac transplantation and intraportal islet transplantation [87,179]. Here, zinc as a pro-antioxidant is involved in the inhibition of apoptotic enzymes such as caspases and acts tolerogenic [79,87,180]. Another mechanism to lower transplant rejections by zinc administration in physiological doses is the capacity of zinc to induce and stabilize a specific subpopulation of Th cells, namely regulatory T cells (Treg) in vitro and in vivo [79–81]. This is of great importance, since the discovery of Treg offers a new paradigm for transplantation medicine since intra-graft Treg frequency seems to correlate with clinical graft acceptance, survival, and function [181,182]. Moreover, pro-inflammatory graft-related immune reactions are tried to impede that might be accomplished by Treg induction. To date, Treg gain more and more importance in combating diseases in clinical approaches and a multitude of studies elicited transforming growth factor (TGF)-β1 to be essential in conversion of Th cells into Treg in vitro [183–185]. In this connection, TGF-β1 seem to be important in establishing immunological tolerance by induction of Treg cells in mice [183,186,187] and humans [185,188] by triggering the TGF-β1-dependent Smad signaling pathway. Herein, studies suggest an essential role for Smad 2/3 in Forkhead-Box-Protein (FoxP)3 induction and cytokine suppression [189,190]. Interestingly, TGF-β-induced Smad signaling was shown to be intensified by zinc administration contributing to higher Treg cell induction (see Figure 3). Because Smad-binding elements were found in the conserved non-coding DNA sequence (CNS) 1 region of the FoxP3 promoter [189], and zinc promoted FoxP3 stability by preventing proteasomal degradation caused by the histone deacetylase Sirt1 [79], the synergistic effect of the combined treatment could result from both triggered mechanism.

Another mechanism controlling Treg function probably is CK2-mediated activation of zinc transporter Zip7 and thereby altering cellular zinc homeostasis. Zip7 is considered as gatekeeper of cytosolic zinc release from the ER [191], highlighting the involvement of zinc in rapid signaling. Protein kinase CK2 is a ubiquitously expressed tetrameric threonine/serine kinase composed of two catalytic α-subunits and β-subunits respectively [192]. CK2 particularly is involved in regulation of cell survival and proliferation, apoptosis and mitosis [193], by shuttling between the cytosol and nuclei of cells in order to support apoptosis or mitosis, respectively. Zip7 phosphorylation by CK2 results in Zip7-mediated zinc release from the ER and the subsequent activation of multiple downstream pathways that enhance cell proliferation and migration. In the context of Treg function, genetic ablation of the β-subunit of CK2 is addressed to insufficient potential to suppress allergic immune responses in the lungs mediated by Th2 cells in vivo [194]. Therefore, inappropriate CK2 function impairs cell-specific immunological tolerance that might be due, amongst others, to changed zinc homeostasis.

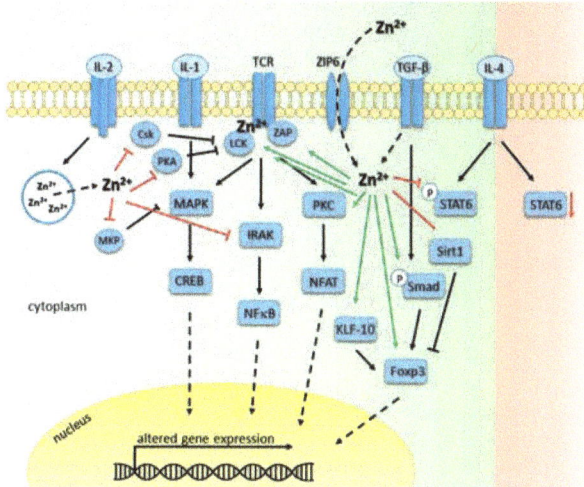

**Figure 3.** Influence of zinc signals on T cell signaling pathways. This figure presents an overview of T cell receptor (TCR)-, Interleukin-1 receptor (IL-1R)-, IL-2R-, IL-4R-, Transforming growth factor β1 receptor (TGF-β1R)-signaling in T cells, as well as zinc flux via zinc transporter ZIP6. Signaling pathways are described in detail in the text. Black arrow: activation, green arrow: activating function of zinc, red T bar: inhibiting function of zinc, black dotted arrow, translocation of the molecule to nucleus. Abbreviations: CREB: cyclic adenosine monophosphate response element-binding protein; Csk: c-src tyrosine kinase; Foxp3: forkhead-box-protein P3; IL: Interleukin; IRAK: interleukin-1 receptor-associated kinase; KLF-10: krüppel-like factor-10; Lck: lymphocyte-specific protein tyrosine kinase; MAPK: mitogen-activated protein kinase; MKP: MAP-kinase phosphatase; NFAT: nuclear factor of activated T cells; NFκB: nuclear factor kappa B; PKA: protein kinase A; PKC: protein kinase C; SIRT1: Sirtuin1; STAT: signal transducer and activator of transcription; TCR: T cell receptor; TGF-β1: transforming growth factor β1. ZAP: zeta-chain (TCR)-associated protein kinase.

In contrast to that, the ZIP8-zinc axis might have a negative impact on immune regulation. ZIP8 is known to be highly expressed in human T cells on the lysosomal membrane and is even more prominent when T cells are activated by TCR triggering [125]. Hence, after T cell activation free intralysosomal zinc is released into the cytoplasm increasing the zinc available, which leads to altered T cell function by modulation of distinct signaling pathways. In this connection, overexpression of ZIP8 was shown to trigger the pro-inflammatory immune responses by enhancing the IFN-γ production due to zinc-mediated reduction of CN phosphatase activity and prolonged phosphorylation of the transcription factor cyclic adenosine monophosphate response element-binding protein (CREB) [125]. Hence, the TCR-induced cytokine production is strengthened. In line with that, the NFκB activity is triggered by downregulation of the IκB kinase activity in pro-inflammatory responses observed in ZIP8 hypomorphic mice [137]. Hence, an overreacting immune system can lead to hyperinflammation, autoimmune diseases or sepsis.

Nevertheless, one has to keep in mind that T cell function is sensitively regulated by zinc concentration leading to cellular activation or deactivation respectively [31]. Additionally, results observed in different species and different cellular models needs to be compared with wariness.

In general, Treg development and survival is dependent on a high number of key factors, such as the transcription factor Foxp3 and additional signals, including IL-2, TGF-β, and co-stimulatory molecules like CD28. Moreover, transcription factors as IRF-1 and KLF-10 are described to influence Treg stability and function. One study elicited IRF-1 deficiency to result in a selective and marked increase in highly differentiated and activated FoxP3 expressing Treg cells in vivo [195]. IRF-1 plays a direct role in the generation and expansion of Treg cells by specifically repressing FoxP3 activity that can be dampened by zinc supplementation supporting the pro-tolerogenic immune reaction [196]. In contrast to IRF-1, KLF-10 seem to be indispensable for appropriate Treg function, because animals carrying a disruption in KLF-10 no longer show FoxP3 activation [197]. KLF-10-deficient Treg cells display impaired cell differentiation, altered cytokine profiles with enhanced Th1, Th2, and Th17 cytokine expression. Furthermore, a reduced capacity for suppression by wild-type co-cultured T effector cells, as well as accelerated atherosclerosis in immunodeficient atherosclerotic mice was exhibited [197].

During the last decade, the pro-inflammatory Th9 cell subpopulation is reported to be involved allergic asthma and autoimmune diseases like MS respectively [198,199]. Interestingly, comparable to the zinc-mediated reduction of pro-inflammatory Th17 cells in allogeneic immunoreactions [81,200], zinc supplementation in physiological doses facilitates a similar reduction of pro-inflammatory Th9 cells [82], subsequently leading to dampened allogeneic immunoreaction. Hence, an overreacting immune response can be beneficially influenced by zinc administration seeming to be promising to improve the life of patients suffering autoimmune or allergic diseases.

To date, zinc-related targets for the above-mentioned effects are still not fully known. However, some zinc-related targets for the above mentioned effects were found, comprising receptor proteins, kinases, phosphatases, caspases, and transcription factors. Those can be activated or inactivated by zinc administration or zinc deficiency depending on type of zinc signal and zinc level [4,30]. The impact of zinc signals on the adaptive immune system will be discussed in detail in the following paragraphs.

*6.2. Zinc Signals in T Cell Receptor-Triggered Signaling Cascades*

When T cell activation via T cell receptor (TCR) stimulation is examined into detail, already the TCR activating complex is influenced by zinc signals (see Figure 3). One has to keep in mind that TCR exhibit no intrinsic kinase activity and depend on Src-family tyrosine kinases for signal transduction, like lymphocyte protein tyrosine kinase (Lck). Lck is one of the first kinases activated, essential for T cell activation mediated either by proximal or distal Lck promoter activity, and is essential for phosphorylation of the 10 ITAM motifs of the T cell antigen receptor-signaling complex augmenting phosphorylation of the kinase ZAP70 [201,202]. Activation of ZAP70 phosphorylates downstream targets eventually inducing MAPK signaling pathway contributing to T cell activation. Zinc-mediated T cell activation is facilitated by either a zinc flux inducing direct or indirect Lck activation or SHP-1 deactivation, respectively, or homeostatic zinc signals altering gene expression, as for instance of distinct zinc transporter.

Lck activation occurs by zinc flux-dependent Lck homodimerization by stabilization the dimer interface of the SH3 domains [67]. Herein, the activation is highly complex by involving distinct tyrosine residues either in the so-called activation loop or at the C-terminal negative regulatory site [154,201]. On the other hand, Lck activation is facilitated by stabilization the interface site of Lck and the membrane proteins CD4 and CD8 respectively [63,64]. By forming a so-called "zinc clasp structure", zinc bridges two cysteine residues of each protein facilitating a stable interaction of Lck and CD4/CD8 [203]. Since the surface molecules CD8 and CD4 bind to MHC class I or II respectively, all components are recruited in close proximity to the TCR signaling complex, leading to T cell activation. On the other hand, Lck activity is influenced in an indirect manner due to reduced recruitment of the enzyme SHP-1. Since Lck homodimerization and phosphorylation, as well as TCR complex arrangement and signaling, is highly dependent on numerous PTP [204,205], all can be considered as potential targets for zinc-mediated regulation. Consequently, it cannot be precisely predicted whether a PTP regulation result in preferential dephosphorylation of an activating or inactivating tyrosine by zinc in vivo eventually facilitation signaling activation or inactivation.

T cell activation can furthermore be influenced by altered intracellular zinc signals and zinc concentrations mediated by specific zinc transporter, as for instance Zip6. Initiation of TCR signaling is initiated by specific interaction with an antigen-loaded MHC molecule on the surface of a neighboring antigen-presenting cell (APC) by formation of a functional immunologic synapse. This results in an immediately influx of zinc from the extracellular environment through the transporter Zip6 after T cell stimulation [66]. Hence, intracellular free zinc levels are altered influencing signaling pathways and subsequently T cell activation, maturation and differentiation.

In response to TCR stimulation, specific transcription factors and signaling pathways are triggered that are often zinc-regulated (see Figure 3). Nuclear factor of activated T cells (NFAT) mediates the expression of plenty of genes, such as IL-2. In resting T cells, NFAT proteins are constitutively phosphorylated remaining inactive and are located in the cytoplasm. In response to TCR/CD28-mediated calcium signaling, NFAT is dephosphorylated by calcineurin (CN), a calcium/calmodulin-dependent serine/threonine phosphatase, and eventually translocates into the nucleus [206]. Zinc and iron are essential cofactors for the catalytic domain of CN, containing a $Zn^{2+}$-$Fe^{2+}$ binuclear center. However, CN inhibition is only reported for zinc. Regarding this, physiologic zinc concentrations, ranging from 10 to 10 µM are described to exhibit a CN inhibition-capacity in vitro [116,207]. Zinc-dependent regulation of T cell activation, proliferation and differentiation is much more complex than modifying transcription factor activity as NFAT, since CN activity itself is regulated in a complex zinc-dependent manner. Herein, CN remains in its inactive phosphorylated form, due to phosphatidyl-inositol-3-kinase (PI3K) [206]. PI3K is a lipid kinase and generates phosphatidylinositol-3,4,5-trisphosphate (PI(3,4,5)P3).

PI(3,4,5)P3 acts as second messenger essential for the translocation of Akt to the plasma membrane where it is phosphorylated and activated by phosphoinositide-dependent kinase (PDK)1 and PDK2. Activation of Akt plays a pivotal role in fundamental cellular functions such as cell proliferation and survival by phosphorylating a variety of substrates [208]. In PI3K pathway, zinc signals are well reported to influence the signaling cascade in various cell types [209–212]. One possible explanation for this mechanism is an increased enzyme degradation of phosphatase and tensin homolog deleted on chromosome 10 (PTEN) [213,214]. PTEN in general functions as a dephosphorylase of PI(3,4,5)P3, a product of PI3K mediating the activation of PDK-1/Akt. More recently, a study examining the IL-2-induced PI3K/Akt signaling pathway uncovered an inhibition of PTEN. Here, homeostatic zinc signals seem to be necessary for the regulation the PI3K/Akt pathway, since the IL-2-induced Akt phosphorylation is diminished during zinc deficiency. Homeostatic zinc signals are known to upregulate phosphorylation of Akt at Ser473 and inhibit PTEN at subnanomolar concentrations ($IC_{50}$~0.59 nM). This inhibition seems to be mediated by zinc binding to cysteine thiol at position 124 (Cys124) essential for the catalytic activity of PTEN [76]. Thus, a modulation of this pathway occurs upstream of Akt, but down-stream of Jak1, because STAT-5 signaling is not influenced by zinc signals [65]. These results suggest a comparable activation of PI3K via zinc in T cells. Zinc-mediated inhibition of CN results in NFAT inactivation and reduction of TCR-mediated transcription, whereas activation of PI3K signaling act agonistic.

Zinc signals as mediator of T cell signal transduction is known since the first reports of a potential role of this ion in signaling were described. Its interaction with PKC is identified as biochemical basis of these observations [215]. The PKC family members are serine/threonine kinases comprising of several isoforms are known to play crucial roles in intracellular signal transduction elicited by various extracellular stimuli, like growth factors, hormones, and neurotransmitters. PKC members can be subdivided into: (1) classical PKCs, activated by cofactors like $Ca^{2+}$ and diacylglycerol (DAC); (2) novel PKC that bind DAC but no $Ca^{2+}$; and (3) atypical PKCs that interact with neither $Ca^{2+}$ nor DAC. Several isoforms are well described to be important in T cell function. Following T cell activation via TCR/CD28 triggering, PKCθ is involved in the activation of several transcription factors. Additionally, PKCα is involved in T cell proliferation and IL-2 production. Furthermore, several PKC isoforms are involved in survival of B cells, pre-B cell development, and induction of tolerance toward self-antigens [216]. PKC is described as a zinc metallo-enzyme. Atomic absorption measurements on the intact enzyme indicated that four zinc atoms ($4.2 \pm 0.5$) are bound per PKCα molecule. Similar stoichiometric ratios were determined for PKCβII and PKCγ [217]. PKC comprise of four commonly conserved domains (C1-C4). Herein, DAC binds to C1 domain in the N-terminal regulatory part of PKCα1, which contains two homologous regions containing six cysteine (Cys) and two histidine (His) residues, forming a total of four Cys3His zinc binding motifs [218]. No information about a differential effect of zinc on the different isoforms is available, and varying forms of C1 domains are present in conventional, novel, and atypical PKCs, pointing to probable zinc binding to all known PKC isoforms [216].

Late zinc signals are mentioned to affect multiple steps during PKC activation, as for instance augmented PKC kinase activity, increased affinity to phorbol esters, and enhanced binding to the cytoskeleton and plasma membrane [219]. Inhibition of the above mentioned events can be seen during zinc deficiency in vitro, facilitated by membrane-permeable zinc chelators, as TPEN. Interestingly, PKC itself can be a source for zinc release, thus interaction between PKC and zinc is not limited to an effect of zinc on the PKC activation. PKC activation by lipid second messengers or thiol oxidation lead to measurable zinc release from the regulatory domain [220,221]. In addition, PKC regulates the intracellular free zinc concentration and distribution. In T cells, phorbol ester treatment lead to a zinc flux resulting in redistribution of zinc from the nucleus and mitochondria to the cytosol and microsomes [222].

Besides activating signaling cascades, also inhibition of TCR signaling and subsequent inhibition of overall T cell activation is facilitated by phosphorylation status of different amino acids. One of those is phosphorylation of the inactivating tyrosine 505 of Lck via c-src tyrosine kinase COOH-terminal Srk kinase (Csk) [223]. The zinc flux interferes with this event in different ways: (1) zinc signals lead to an inhibition of Csk [75]; and (2) activation of Csk in T cells can be observed via phosphorylation by PKA [224]. Although zinc has no direct impact on PKA activity itself, it inhibit AC transcription by a homeostatic zinc signal thereby inhibiting the formation of the PKA activator cAMP. Furthermore, cAMP is influenced by zinc via inhibition of PDE that can block degradation of cyclic nucleotides, leading to activation of PKA [60]. An inhibition of Csk and AC promote TCR signaling, while PDE inhibition antagonizes it. However, the outcome on TCR signaling resulting from the modulation of this pathway by zinc in vivo remains to be elucidated yet.

### 6.3. Zinc Signals in Interleukin Receptor Signaling Pathways in T Cells

Besides PI3K and PTEN modulation downstream of TCR activation, zinc signals also affects signals originating from the IL-1 receptor (IL-1R). High zinc concentrations of about 100 µM inhibit IL-1β-stimulated IFNγ production in primary human T cells and IL-1 dependent proliferation of murine T cells. Zinc incubation leads to a reduced activity of IRAK that is a central kinase in the signaling pathways downstream of the IL-1 receptor. IRAK comprise of four different genes existing in the human genome (IRAK1, IRAK2, IRAK3, and IRAK4), and studies with transgenic mice, have revealed distinct, non-redundant biological roles [225]. Interestingly, IRAK4 is reported to be essential for the TLR signaling in innate immunity as well as for TCR-mediated signaling in adaptive immunity leading to the activation of NFκB [226], however this is still matter of debate [227]. Since zinc signals show modulating capacity on TCR as well as on IL-1-dependent signaling via IRAK inhibition, two central signaling pathways are uncovered in negative regulation of T cell activation. Nevertheless, one has to keep in mind that IRAK function and zinc-mediated regulation is not limited to T cells. High similarity between TLR4 and IL-1R signal transduction suggests IRAK inhibition for another mechanism by which zinc negatively influences TLR4 signaling in innate immunity as discussed before.

Moreover, triggering of interleukin receptor (IL-2R) also results in altered intracellular zinc signals. Following IL-2R stimulation, zinc signal is releases of from lysosomal compartment called zincosomes within several minutes after T cell stimulation [65]. Thus, an intracellular zinc flux is triggered influencing PTP, as for instance dual-specificity phosphatases (DUSP) or Protein phosphatase (PP)2A activity in a direct manner. Subsequently, ultimately signal transduction and protein expression, such as diminished dephosphorylation of MAPK MEK and ERK, or elevated transcription of zinc transporter such as Zip6, can be modulated by zinc.

Not only is kinase activity regulated by zinc signals, but cytokine production is also influenced. During zinc deficiency zinc administration leads to a fast rise in intracellular zinc levels, i.e., zinc flux, due to an increased expression of cell membrane-located zinc transporters Zip10 and Zip12 because of the former zinc deficiency. Due to altered intracellular zinc homeostasis, activity of signaling molecules as MAPK p38, and NFκB p65 are affected consequently leading to highly increased IL-2 mRNA expression and IL-2 cytokine production [72].

In general, a forecast of zinc-related effects on T cell function is impossible due to the high number of signaling pathways modulated by zinc signals. Several studies concerning zinc status and T cell function in vivo, stated an increase of the delayed type hypersensitivity reaction upon correction of zinc deficiency [228]. By contrast, zinc administration diminishes the allogeneic reaction in the mixed lymphocyte culture (MLC) [229] and stabilizes regulatory T cell function in vitro and in vivo [79,230] indicating that zinc may have multiple, opposing functions, depending on its concentration and interaction with multiple other environmental factors.

## 6.4. Zinc Signals in B Cell Maturation, Survival and Function

B cells represent the specific humoral immunity and differentiate to antibody-producing plasma cells after stimulation. Nearly all of the zinc regulated signaling pathways discussed above in T cells are also important in B cells maturation and differentiation, such as tyrosine phosphorylation, PKC, MAPK signaling, and activation of the transcription factors NFAT and NFκB. In contrast to other cell types discussed before, mature B cell proliferation and function is not as dependent on the organisms' zinc status. Hence, the impact of zinc deficiency on B cell maturation and function and its subsequent impact on the immune system is not as profound as it is in the context of T cells [231]. Zinc signals itself seem to have no impact on B cell activity in a direct manner. However, zinc deficiency affect B cell lymphopoiesis in different ways. Following acute zinc deficiency, the B cell compartment is largely reduced (about 50% to 70%) and the composition of surviving B cells within the B cell compartment markedly changed [232]. In contrast, chronic zinc deficiency is reported to merely influence T cell development and B cell development remains sustained [164].

During cellular development, lymphocytes are sensitive for apoptosis signals due to positive and negative selection mechanisms in the primary lymphoid organs. Mature cells are apoptosis resistant and getting inactivated in the case of autoreactivity known as anergy. A strict selection guarantees adequate cellular function, thus avoiding cellular autoreactivity by eliminating the majority of newly formed cells by apoptosis. Zinc deficiency increases apoptosis within the B cell population and leads to cell depletion [11,233]. Thus, there are fewer naïve B cells during zinc deficiency that can react to neoantigens. Contrarily to zinc deficiency, low zinc levels have no influence on the cell cycle status of precursor B cells and only modest influence on cycling pro-B cells. Taking into account that T cell numbers are reduced during zinc deficiency as well, and that recognition of most antigens is T cell dependent, it is probable that the body is unable to respond with antibody production in response to neoantigens. This assumption is consistent with findings showing a disturbed antibody production by B cells during zinc depletion [234]. This disturbed feedback mechanism might also be an explanation for zinc deficient patients, like elderly and hemodialysis patients, showing a reduced response to vaccination [235–237]. Another possible mechanism impairing adequate immune responses are epigenetic changes of DNA. The epigenome is highly influenced by environmental changes, but also by nutrition [238]. Since histone modifying enzymes as histonedeacetylases are known to be regulated in a zinc-dependent manner [79], the organism's zinc status is important for adequate epigenetic modification of DNA.

Studies uncovered B cell activity and antibody production to be impaired during zinc deficiency respectively leading to higher risk of parasitic infection [239]. On a molecular mechanism this might be due to diminished STAT6 phosphorylation during zinc deficiency. Hence, zinc signals are important for proper IL-4 induced STAT6 phosphorylation [77]. Since IL-4 promotes activation of early B cells and immunoglobulin class switch towards IgE, and thereby the further antibody specification, this mechanism explains the higher risk of parasitic infection during zinc deficiency. Furthermore, zinc signals are important for IL-6 induced STAT3 phosphorylation shown to be increased due to zinc deficiency. IL-6 mediates the activation and final differentiation of B cells into plasma cells and IL-6 overproduction is associated with autoantibody production. Diseases such as rheumatoid arthritis and plasma cell neoplasia go along with reduced serum zinc levels, indicating potential co-effects of IL-6 overproduction and enhanced susceptibility of B cells due to zinc deficiency. Those observations indicate that a strict regulation and proper zinc homeostasis is necessary to keep the immune system balanced [77].

Studies reveal that antibody production as consequence to T cell-dependent antigen recognition is more sensitive to zinc deficiency than antibody production in response to T cell-independent antigens [240]. Zinc deficient mice show reduced antibody recall responses to antigens they have been immunized with before. This effect was observed in T cell-independent as well as in T cell-dependent systems. Thus, immunologic memory also seems to be influenced by zinc [241]. However, mature B cells still are more resistant to zinc deficiency-induced apoptosis due to high Bcl2 level. Thus, B cell memory is less affected than the primary response [233]. Homeostatic zinc signals influence several regulatory proteins, such as those from the Bcl/Bax family [11] and furthermore several aspects of apoptotic signal transduction. In this connection, the calcium-dependent endonuclease, which mediates DNA fragmentation, is inhibited by zinc. However, this target is beyond the point of no return for programmed cell death, and an inhibition could explain a suppression of DNA fragmentation during apoptosis, but not the effect on cellular survival. Another important group of enzymes in apoptosis is cysteine-aspartic acid proteases, also known as caspases. Inactive pro-caspases first are activated by proteolytic cleavage and second form a signaling cascade to transduce initial apoptotic signals to effector enzymes mediating the organized cellular destruction characteristic for programmed cell death. Low micromolar zinc concentrations are reported to inhibit caspases-3, -6, and -8 [242]. A half maximal inhibitory concentration for caspase-3 was found below 10 nM [99]. This value is within the physiological range of free intracellular zinc level, leading to the suggestion that endogenous zinc can inhibit caspase-3 activity subsequently altering apoptosis signals and cellular survival. More recently, similar results were observed in an in vivo rat heterotrophic heart transplant model that might contribute to an increased allograft survival [87].

Even though both cell types of the adaptive immune system utilize the same signaling pathways, B cell function seems to be affected by zinc to a lesser magnitude. Even the reduced antibody production during zinc deficiency is based on reduced B cell numbers whereas it remains unaffected on a per-cell basis [243]. This indicates an effect on cellular development rather than on function. One reason might be a difference in zinc homeostasis, making mature B cells less susceptible to conditions of limited zinc availability. Although B cells are highly susceptible to apoptosis during development, and zinc is one factor that influences these signals, mature B cells can tolerate comparable conditions due to changes in zinc-regulating proteins, but also by changing the expression patterns of several other factors that regulate the cellular responsiveness to apoptotic signals. One highly important zinc-regulating protein in B cells is the zinc importer Zip10.

Zip10 is essential for cell survival during early B-cell development as well as for adequate B-cell receptor (BCR) signaling [244,245]. A loss of Zip10 during an early B cell stage was reported to specifically abrogated cell survival. Eventually, this leads to an absence of mature B cells in vivo, provoking spleno-atrophy and reduced Ig level. Moreover, the absence of Zip10 causes an impaired T cell-dependent and -independent immune response respectively. Additionally, due to the dysregulated BCR signaling, mature B cells proliferated poorly in response to BCR crosslinking. On the molecular level, this is probable due to a disturbed JAK-STAT-Zip10 signaling axis, since the JAK-STAT pathway is known to modulate expression of Zip10 [244]. Furthermore, a disturbed regulation of the BCR signal strength is likely. This might be due to the positive regulator function of Zip10 regarding CD45R phosphatase activity. In the case of Zip10 malfunction, CD45R phosphatase activity is reduced resulting in hyperactivation of tyrosine-protein kinase Lyn that consequently provoke an altered threshold for BCR signaling [245].

## 7. Conclusions

Summing up, all immune cells are directly affected by zinc signals. Most prominent pathologic changes are found during zinc deficiency indicating that zinc is a main regulator of cellular function and signal transduction. However, precise underlying molecular mechanisms still need to be investigated in further detail, to guarantee beneficial effects of clinical zinc application to patients suffering of distinct diseases.

**Author Contributions:** Inga Wessels wrote Sections 2, 4, and 5; and designed Figures 1 and 2, and Table 2. Martina Maywald wroteSections 1, 3, 6, and 7; and designed Figure 3, and Table 1. Lothar Rink made substantial contributions to the conception and design of the manuscript and substantively revised it.

**Conflicts of Interest:** The authors declare that they have no conflict of interest.

## References

1. Prasad, A.S.; Miale, A., Jr.; Farid, Z.; Sandstead, H.H.; Schulert, A.R. Zinc metabolism in patients with the syndrome of iron deficiency anemia, hepatosplenomegaly, dwarfism, and hypognadism. *J. Lab. Clin. Med.* **1963**, *61*, 537–549. [PubMed]
2. Krezel, A.; Maret, W. The biological inorganic chemistry of zinc ions. *Arch. Biochem. Biophys.* **2016**, *611*, 3–19. [CrossRef] [PubMed]
3. Pfaender, S.; Fohr, K.; Lutz, A.K.; Putz, S.; Achberger, K.; Linta, L.; Liebau, S.; Boeckers, T.M.; Grabrucker, A.M. Cellular zinc homeostasis contributes to neuronal differentiation in human induced pluripotent stem cells. *Neural Plast.* **2016**, *2016*, 3760702. [CrossRef] [PubMed]
4. Haase, H.; Rink, L. Multiple impacts of zinc on immune function. *Metallomics Integr. Biomet. Sci.* **2014**, *6*, 1175–1180. [CrossRef] [PubMed]
5. Brieger, A.; Rink, L.; Haase, H. Differential regulation of tlr-dependent myd88 and trif signaling pathways by free zinc ions. *J. Immunol.* **2013**, *191*, 1808–1817. [CrossRef] [PubMed]
6. Duke, R.C.; Chervenak, R.; Cohen, J.J. Endogenous endonuclease-induced DNA fragmentation: An early event in cell-mediated cytolysis. *Proc. Natl. Acad. Sci. USA* **1983**, *80*, 6361–6365. [CrossRef] [PubMed]
7. Finamore, A.; Massimi, M.; Conti Devirgiliis, L.; Mengheri, E. Zinc deficiency induces membrane barrier damage and increases neutrophil transmigration in Caco-2 cells. *J. Nutr.* **2008**, *138*, 1664–1670. [PubMed]
8. Miyoshi, Y.; Tanabe, S.; Suzuki, T. Cellular zinc is required for intestinal epithelial barrier maintenance via the regulation of claudin-3 and occludin expression. *Am. J. Physiol. Gastrointest. Liver Physiol.* **2016**, *311*, G105–G116. [CrossRef] [PubMed]
9. Maret, W. Zinc coordination environments in proteins as redox sensors and signal transducers. *Antioxid. Redox Signal.* **2006**, *8*, 1419–1441. [CrossRef] [PubMed]
10. Bellomo, E.; Hogstrand, C.; Maret, W. Redox and zinc signalling pathways converging on protein tyrosine phosphatases. *Free Radic. Biol. Med.* **2014**, *75* (Suppl. S1), S9. [CrossRef] [PubMed]
11. Truong-Tran, A.Q.; Carter, J.; Ruffin, R.E.; Zalewski, P.D. The role of zinc in caspase activation and apoptotic cell death. *Biometals* **2001**, *14*, 315–330. [CrossRef] [PubMed]
12. Sunderman, F.W., Jr. The influence of zinc on apoptosis. *Ann. Clin. Lab. Sci.* **1995**, *25*, 134–142. [PubMed]
13. Maret, W. Zinc in pancreatic islet biology, insulin sensitivity, and diabetes. *Prev. Nutr. Food Sci.* **2017**, *22*, 1–8. [CrossRef] [PubMed]
14. Bredholt, M.; Frederiksen, J.L. Zinc in multiple sclerosis: A systematic review and meta-analysis. *ASN Neuro* **2016**, *8*. [CrossRef] [PubMed]
15. Xin, L.; Yang, X.; Cai, G.; Fan, D.; Xia, Q.; Liu, L.; Hu, Y.; Ding, N.; Xu, S.; Wang, L.; et al. Serum levels of copper and zinc in patients with rheumatoid arthritis: A meta-analysis. *Biol. Trace Elem. Res.* **2015**, *168*, 1–10. [CrossRef] [PubMed]
16. Stelmashook, E.V.; Isaev, N.K.; Genrikhs, E.E.; Amelkina, G.A.; Khaspekov, L.G.; Skrebitsky, V.G.; Illarioshkin, S.N. Role of zinc and copper ions in the pathogenetic mechanisms of alzheimer's and parkinson's diseases. *Biochem. Mosc.* **2014**, *79*, 391–396. [CrossRef] [PubMed]

17. Szewczyk, B. Zinc homeostasis and neurodegenerative disorders. *Front. Aging Neurosci.* **2013**, *5*, 33. [CrossRef] [PubMed]

18. Alder, H.; Taccioli, C.; Chen, H.; Jiang, Y.; Smalley, K.J.; Fadda, P.; Ozer, H.G.; Huebner, K.; Farber, J.L.; Croce, C.M.; et al. Dysregulation of miR-31 and miR-21 induced by zinc deficiency promotes esophageal cancer. *Carcinogenesis* **2012**, *33*, 1736–1744. [CrossRef] [PubMed]

19. Ressnerova, A.; Raudenska, M.; Holubova, M.; Svobodova, M.; Polanska, H.; Babula, P.; Masarik, M.; Gumulec, J. Zinc and copper homeostasis in head and neck cancer: Review and meta-analysis. *Curr. Med. Chem.* **2016**, *23*, 1304–1330. [CrossRef] [PubMed]

20. Overbeck, S.; Rink, L.; Haase, H. Modulating the immune response by oral zinc supplementation: A single approach for multiple diseases. *Arch. Immunol. Ther. Exp. Warsz.* **2008**, *56*, 15–30. [CrossRef] [PubMed]

21. Gammoh, N.Z.; Rink, L. Zinc in infection and inflammation. *Nutrients* **2017**, *9*. [CrossRef]

22. World Health Organization (WHO). *World Health Organization—The World Health Report*; WHO: Geneva, Switzerland, 2002; Volume 83.

23. Rink, L. *Zinc in Human Health*; IOS Press: Amsterdam, The Netherlands, 2011; p. 596.

24. Lowe, N.M.; Dykes, F.C.; Skinner, A.L.; Patel, S.; Warthon-Medina, M.; Decsi, T.; Fekete, K.; Souverein, O.W.; Dullemeijer, C.; Cavelaars, A.E.; et al. Eurreca-estimating zinc requirements for deriving dietary reference values. *Crit. Rev. Food Sci. Nutr.* **2013**, *53*, 1110–1123. [CrossRef] [PubMed]

25. Krebs, N.F. Overview of zinc absorption and excretion in the human gastrointestinal tract. *J. Nutr.* **2000**, *130* (Suppl. 5S), 1374s–1377s. [PubMed]

26. Cousins, R.J. Gastrointestinal factors influencing zinc absorption and homeostasis. *Int. J. Vitam. Nutr. Res.* **2010**, *80*, 243–248. [CrossRef] [PubMed]

27. Wojtkiewicz, J.; Makowska, K.; Bejer-Olenska, E.; Gonkowski, S. Zinc transporter 3 (znt3) as an active substance in the enteric nervous system of the porcine esophagus. *J. Mol. Neurosci.* **2017**, *61*, 315–324. [CrossRef] [PubMed]

28. Wojtkiewicz, J.; Rytel, L.; Makowska, K.; Gonkowski, S. Co-localization of zinc transporter 3 (Znt3) with sensory neuromediators and/or neuromodulators in the enteric nervous system of the porcine esophagus. *Biometals* **2017**, *30*, 393–403. [CrossRef] [PubMed]

29. Skrovanek, S.; DiGuilio, K.; Bailey, R.; Huntington, W.; Urbas, R.; Mayilvaganan, B.; Mercogliano, G.; Mullin, J.M. Zinc and gastrointestinal disease. *World J. Gastrointest. Pathophysiol.* **2014**, *5*, 496–513. [CrossRef] [PubMed]

30. Haase, H.; Rink, L. Functional significance of zinc-related signaling pathways in immune cells. *Annu. Rev. Nutr.* **2009**, *29*, 133–152. [CrossRef] [PubMed]

31. Hojyo, S.; Fukada, T. Roles of zinc signaling in the immune system. *J. Immunol. Res.* **2016**, *2016*, 6762343. [CrossRef] [PubMed]

32. Brieger, A.; Rink, L. Zink und immunfunktionen. *Ernährung Med.* **2010**, *25*, 156–160. [CrossRef]

33. Mocchegiani, E.; Costarelli, L.; Giacconi, R.; Cipriano, C.; Muti, E.; Tesei, S.; Malavolta, M. Nutrient–gene interaction in ageing and successful ageing: A single nutrient (Zinc) and some target genes related to inflammatory/immune response. *Mech. Ageing Dev.* **2006**, *127*, 517–525. [CrossRef] [PubMed]

34. Freeland-Graves, J.H.; Bodzy, P.W.; Eppright, M.A. Zinc status of vegetarians. *J. Am. Diet. Assoc.* **1980**, *77*, 655–661. [PubMed]

35. Sandström, B.; Cederblad, Å.; Lindblad, B.S.; Lönnerdal, B. Acrodermatitis enteropathica, zinc metabolism, copper status, and immune function. *Arch. Pediatr. Adolesc. Med.* **1994**, *148*, 980–985. [CrossRef] [PubMed]

36. Jarosz, M.; Olbert, M.; Wyszogrodzka, G.; Młyniec, K.; Librowski, T. Antioxidant and anti-inflammatory effects of zinc. Zinc-dependent NF-κB signaling. *Inflammopharmacology* **2017**, *25*, 11–24. [CrossRef] [PubMed]

37. Rink, L.; Haase, H. Zinc homeostasis and immunity. *Trends Immunol.* **2007**, *28*, 1–4. [CrossRef] [PubMed]

38. Dubben, S.; Honscheid, A.; Winkler, K.; Rink, L.; Haase, H. Cellular zinc homeostasis is a regulator in monocyte differentiation of HL-60 cells by 1 α,25-dihydroxyvitamin D3. *J. Leukoc. Biol.* **2010**, *87*, 833–844. [CrossRef] [PubMed]

39. Weston, W.L.; Huff, J.C.; Humbert, J.R.; Hambidge, K.M.; Neldner, K.H.; Walravens, P.A. Zinc correction of defective chemotaxis in acrodermatitis enteropathica. *Arch. Dermatol.* **1977**, *113*, 422–425. [CrossRef] [PubMed]

40. Hasan, R.; Rink, L.; Haase, H. Zinc signals in neutrophil granulocytes are required for the formation of neutrophil extracellular traps. *Innate Immun.* **2013**, *19*, 253–264. [CrossRef] [PubMed]

41. Hasan, R.; Rink, L.; Haase, H. Chelation of free $Zn^{2+}$ impairs chemotaxis, phagocytosis, oxidative burst, degranulation, and cytokine production by neutrophil granulocytes. *Biol. Trace Elem. Res.* **2016**, *171*, 79–88. [CrossRef] [PubMed]

42. Livingstone, C. Zinc: Physiology, deficiency, and parenteral nutrition. *Nutr. Clin. Pract.* **2015**, *30*, 371–382. [CrossRef] [PubMed]

43. Mocchegiani, E.; Malavolta, M. Zinc dyshomeostasis, ageing and neurodegeneration: Implications of A2M and inflammatory gene polymorphisms. *J. Alzheimer's Dis.* **2007**, *12*, 101–109. [CrossRef]

44. Kimura, T.; Kambe, T. The Functions of Metallothionein and ZIP and ZnT Transporters: An Overview and Perspective. *Int. J. Mol. Sci.* **2016**, *17*, 336. [CrossRef] [PubMed]

45. Inoue, K.; O'Bryant, Z.; Xiong, Z.-G. Zinc-Permeable Ion Channels: Effects on Intracellular Zinc Dynamics and Potential Physiological/Pathophysiological Significance. *Curr. Med. Chem.* **2015**, *22*, 1248–1257. [CrossRef] [PubMed]

46. Wessels, I.; Cousins, R.J. Zinc dyshomeostasis during polymicrobial sepsis in mice involves zinc transporter Zip14 and can be overcome by zinc supplementation. *Am. J. Physiol. Gastrointest. Liver Physiol.* **2015**, *309*, G768–G778. [CrossRef] [PubMed]

47. Knoell, D.L.; Julian, M.W.; Bao, S.; Besecker, B.; Macre, J.E.; Leikauf, G.D.; DiSilvestro, R.A.; Crouser, E.D. Zinc deficiency increases organ damage and mortality in a murine model of polymicrobial sepsis. *Crit. Care Med.* **2009**, *37*, 1380. [CrossRef] [PubMed]

48. Wessels, I.; Haase, H.; Engelhardt, G.; Rink, L.; Uciechowski, P. Zinc deficiency induces production of the proinflammatory cytokines IL-1β and TNFα in promyeloid cells via epigenetic and redox-dependent mechanisms. *J. Nutr. Biochem.* **2013**, *24*, 289–297. [CrossRef] [PubMed]

49. Prasad, A.S. Impact of the discovery of human zinc deficiency on health. *J. Trace Elem. Med. Biol.* **2014**, *28*, 357–363. [CrossRef] [PubMed]

50. Brophy, M.B.; Hayden, J.A.; Nolan, E.M. Calcium ion gradients modulate the zinc affinity and antibacterial activity of human calprotectin. *J. Am. Chem. Soc.* **2012**, *134*, 18089–18100. [CrossRef] [PubMed]

51. Sugiura, T.; Kuroda, E.; Yamashita, U. Dysfunction of macrophages in metallothionein-knock out mice. *J. UOEH* **2004**, *26*, 193–205. [CrossRef] [PubMed]

52. Donato, R.; R Cannon, B.; Sorci, G.; Riuzzi, F.; Hsu, K.; J Weber, D.; L Geczy, C. Functions of S100 proteins. *Curr. Mol. Med.* **2013**, *13*, 24–57. [CrossRef] [PubMed]

53. Bao, B.; Prasad, A.S.; Beck, F.W.; Fitzgerald, J.T.; Snell, D.; Bao, G.W.; Singh, T.; Cardozo, L.J. Zinc decreases c-reactive protein, lipid peroxidation, and inflammatory cytokines in elderly subjects: A potential implication of zinc as an atheroprotective agent. *Am. J. Clin. Nutr.* **2010**, *91*, 1634–1641. [CrossRef] [PubMed]

54. Lienau, S.; Engelhardt, G.; Rink, L.; Weßels, I. The role of zinc in calprotectin expression in human monocytic cells. Unpublished work. 2017.

55. Yamasaki, S.; Sakata-Sogawa, K.; Hasegawa, A.; Suzuki, T.; Kabu, K.; Sato, E.; Kurosaki, T.; Yamashita, S.; Tokunaga, M.; Nishida, K.; et al. Zinc is a novel intracellular second messenger. *J. Cell Biol.* **2007**, *177*, 637–645. [CrossRef] [PubMed]

56. Maret, W. Metals on the move: Zinc ions in cellular regulation and in the coordination dynamics of zinc proteins. *Biometals* **2011**, *24*, 411–418. [CrossRef] [PubMed]

57. Yamasaki, S.; Hasegawa, A.; Hojyo, S.; Ohashi, W.; Fukada, T.; Nishida, K.; Hirano, T. A novel role of the L-type calcium channel α1D subunit as a gatekeeper for intracellular zinc signaling: Zinc wave. *PLoS ONE* **2012**, *7*, e39654. [CrossRef] [PubMed]

58. Kitamura, H.; Morikawa, H.; Kamon, H.; Iguchi, M.; Hojyo, S.; Fukada, T.; Yamashita, S.; Kaisho, T.; Akira, S.; Murakami, M.; et al. Toll-like receptor-mediated regulation of zinc homeostasis influences dendritic cell function. *Nat. Immunol.* **2006**, *7*, 971–977. [CrossRef] [PubMed]

59. Fukada, T.; Kambe, T. *Zinc Signals in Cellular Functions and Disorders*; Springer: Tokyo, Japan, 2014.

60. Von Bülow, V.; Rink, L.; Haase, H. Zinc-mediated inhibition of cyclic nucleotide phosphodiesterase activity and expression suppresses TNF-α and IL-1β production in monocytes by elevation of guanosine 3′,5′-cyclic monophosphate. *J. Immunol.* **2005**, *175*, 4697–4705. [CrossRef] [PubMed]

61. Ho, Y.; Samarasinghe, R.; Knoch, M.E.; Lewis, M.; Aizenman, E.; DeFranco, D.B. Selective inhibition of mitogen-activated protein kinase phosphatases by zinc accounts for extracellular signal-regulated kinase 1/2-dependent oxidative neuronal cell death. *Mol. Pharmacol.* **2008**, *74*, 1141–1151. [CrossRef] [PubMed]

62. Barford, D.; Das, A.K.; Egloff, M.P. The structure and mechanism of protein phosphatases: Insights into catalysis and regulation. *Annu. Rev. Biophys. Biomol. Struct.* **1998**, *27*, 133–164. [CrossRef] [PubMed]
63. Huse, M.; Eck, M.J.; Harrison, S.C. A $Zn^{2+}$ ion links the cytoplasmic tail of CD4 and the N-terminal region of Lck. *J. Biol. Chem.* **1998**, *273*, 18729–18733. [CrossRef] [PubMed]
64. Lin, R.S.; Rodriguez, C.; Veillette, A.; Lodish, H.F. Zinc is essential for binding of p56(lck) to CD4 and CD8α. *J. Biol. Chem.* **1998**, *273*, 32878–32882. [CrossRef] [PubMed]
65. Kaltenberg, J.; Plum, L.M.; Ober-Blobaum, J.L.; Honscheid, A.; Rink, L.; Haase, H. Zinc signals promote IL-2-dependent proliferation of T cells. *Eur. J. Immunol.* **2010**, *40*, 1496–1503. [CrossRef] [PubMed]
66. Yu, M.; Lee, W.W.; Tomar, D.; Pryshchep, S.; Czesnikiewicz-Guzik, M.; Lamar, D.L.; Li, G.; Singh, K.; Tian, L.; Weyand, C.M.; et al. Regulation of T cell receptor signaling by activation-induced zinc influx. *J. Exp. Med.* **2011**, *208*, 775–785. [CrossRef] [PubMed]
67. Romir, J.; Lilie, H.; Egerer-Sieber, C.; Bauer, F.; Sticht, H.; Muller, Y.A. Crystal structure analysis and solution studies of human Lck-SH3; zinc-induced homodimerization competes with the binding of proline-rich motifs. *J. Mol. Biol.* **2007**, *365*, 1417–1428. [CrossRef] [PubMed]
68. Aude-Garcia, C.; Dalzon, B.; Ravanat, J.-L.; Collin-Faure, V.; Diemer, H.; Strub, J.M.; Cianferani, S.; Van Dorsselaer, A.; Carrière, M.; Rabilloud, T. A combined proteomic and targeted analysis unravels new toxic mechanisms for zinc oxide nanoparticles in macrophages. *J. Proteom.* **2016**, *134*, 174–185. [CrossRef] [PubMed]
69. Prasad, A.S.; Bao, B.; Beck, F.W.; Sarkar, F.H. Zinc-suppressed inflammatory cytokines by induction of A20-mediated inhibition of nuclear factor-κB. *Nutrition* **2011**, *27*, 816–823. [CrossRef] [PubMed]
70. Wan, Y.; Petris, M.J.; Peck, S.C. Separation of zinc-dependent and zinc-independent events during early LPS-stimulated TLR4 signaling in macrophage cells. *FEBS Lett.* **2014**, *588*, 2928–2935. [CrossRef] [PubMed]
71. Wellinghausen, N.; Martin, M.; Rink, L. Zinc inhibits interleukin-1-dependent T cell stimulation. *Eur. J. Immunol.* **1997**, *27*, 2529–2535. [CrossRef] [PubMed]
72. Daaboul, D.; Rosenkranz, E.; Uciechowski, P.; Rink, L. Repletion of zinc in zinc-deficient cells strongly up-regulates IL-1β-induced IL-2 production in T-cells. *Metallomics* **2012**, *4*, 1088–1097. [CrossRef] [PubMed]
73. Honscheid, A.; Dubben, S.; Rink, L.; Haase, H. Zinc differentially regulates mitogen-activated protein kinases in human T cells. *J. Nutr. Biochem.* **2012**, *23*, 18–26. [CrossRef] [PubMed]
74. Azriel-Tamir, H.; Sharir, H.; Schwartz, B.; Hershfinkel, M. Extracellular zinc triggers ERK-dependent activation of $Na^+/H^+$ exchange in colonocytes mediated by the zinc-sensing receptor. *J. Biol. Chem.* **2004**, *279*, 51804–51816. [CrossRef] [PubMed]
75. Klein, C.; Sunahara, R.K.; Hudson, T.Y.; Heyduk, T.; Howlett, A.C. Zinc inhibition of cAMP signaling. *J. Biol. Chem.* **2002**, *277*, 11859–11865. [CrossRef] [PubMed]
76. Plum, L.M.; Brieger, A.; Engelhardt, G.; Hebel, S.; Nessel, A.; Arlt, M.; Kaltenberg, J.; Schwaneberg, U.; Huber, M.; Rink, L.; et al. PTEN-inhibition by zinc ions augments interleukin-2-mediated Akt phosphorylation. *Metallomics* **2014**, *6*, 1277–1287. [CrossRef] [PubMed]
77. Gruber, K.; Maywald, M.; Rosenkranz, E.; Haase, H.; Plumakers, B.; Rink, L. Zinc deficiency adversely influences interleukin-4 and interleukin-6 signaling. *J. Biol. Regul. Homeost. Agents* **2013**, *27*, 661–671. [PubMed]
78. Dierichs, L.; Kloubert, V.; Rink, L. Cellular zinc homeostasis modulates polarization of THP-1-derived macrophages. *Eur. J. Nutr.* **2017**, 1–9. [CrossRef] [PubMed]
79. Rosenkranz, E.; Metz, C.H.; Maywald, M.; Hilgers, R.D.; Wessels, I.; Senff, T.; Haase, H.; Jager, M.; Ott, M.; Aspinall, R.; et al. Zinc supplementation induces regulatory T cells by inhibition of Sirt-1 deacetylase in mixed lymphocyte cultures. *Mol. Nutr. Food Res.* **2016**, *60*, 661–671. [CrossRef] [PubMed]
80. Maywald, M.; Meurer, S.K.; Weiskirchen, R.; Rink, L. Zinc supplementation augments TGF-β1-dependent regulatory T cell induction. *Mol. Nutr. Food Res.* **2017**, *61*. [CrossRef] [PubMed]
81. Rosenkranz, E.; Maywald, M.; Hilgers, R.D.; Brieger, A.; Clarner, T.; Kipp, M.; Plumakers, B.; Meyer, S.; Schwerdtle, T.; Rink, L. Induction of regulatory T cells in Th1-/Th17-driven experimental autoimmune encephalomyelitis by zinc administration. *J. Nutr. Biochem.* **2016**, *29*, 116–123. [CrossRef] [PubMed]
82. Maywald, M.; Rink, L. Zinc supplementation dampens T helper 9 differentiation in allogeneic immune reactions in vitro. Unpublished work. 2017.
83. Campo, C.A.; Wellinghausen, N.; Faber, C.; Fischer, A.; Rink, L. Zinc inhibits the mixed lymphocyte culture. *Biol. Trace Elem. Res.* **2001**, *79*, 15–22. [PubMed]

84. Kumar, S.; Rajagopalan, S.; Sarkar, P.; Dorward, D.W.; Peterson, M.E.; Liao, H.-S.; Guillermier, C.; Steinhauser, M.L.; Vogel, S.S.; Long, E.O. Zinc-induced polymerization of killer-cell Ig-like receptor into filaments promotes its inhibitory function at cytotoxic immunological synapses. *Mol. Cell* **2016**, *62*, 21–33. [CrossRef] [PubMed]

85. Rolles, B.; Maywald, M.; Rink, L. Influence of zinc deficiency and supplementation on nk cell cytotoxicity. Unpublished work. 2017.

86. Perry, D.K.; Smyth, M.J.; Stennicke, H.R.; Salvesen, G.S.; Duriez, P.; Poirier, G.G.; Hannun, Y.A. Zinc is a potent inhibitor of the apoptotic protease, caspase-3. A novel target for zinc in the inhibition of apoptosis. *J. Biol. Chem.* **1997**, *272*, 18530–18533. [CrossRef] [PubMed]

87. Kown, M.H.; van der Steenhoven, T.J.; Jahncke, C.L.; Mari, C.; Lijkwan, M.A.; Koransky, M.L.; Blankenberg, F.G.; Strauss, H.W.; Robbins, R.C. Zinc chloride-mediated reduction of apoptosis as an adjunct immunosuppressive modality in cardiac transplantation. *J. Heart Lung Transpl.* **2002**, *21*, 360–365. [CrossRef]

88. Morgan, C.I.; Ledford, J.R.; Zhou, P.; Page, K. Zinc supplementation alters airway inflammation and airway hyperresponsiveness to a common allergen. *J. Inflamm.* **2011**, *8*, 36. [CrossRef] [PubMed]

89. Wellinghausen, N.; Schromm, A.B.; Seydel, U.; Brandenburg, K.; Luhm, J.; Kirchner, H.; Rink, L. Zinc enhances lipopolysaccharide-induced monokine secretion by alteration of fluidity state of lipopolysaccharide. *J. Immunol.* **1996**, *157*, 3139–3145. [PubMed]

90. Chvapil, M. Effect of zinc on cells and biomembranes. *Med. Clin. N. Am.* **1976**, *60*, 799–812. [CrossRef]

91. Hansen, K.B.; Furukawa, H.; Traynelis, S.F. Control of assembly and function of glutamate receptors by the amino-terminal domain. *Mol. Pharmacol.* **2010**, *78*, 535–549. [CrossRef] [PubMed]

92. Tartey, S.; Takeuchi, O. Pathogen recognition and Toll-like receptor targeted therapeutics in innate immune cells. *Int. Rev. Immunol.* **2017**, *36*, 57–73. [CrossRef] [PubMed]

93. Futosi, K.; Fodor, S.; Mócsai, A. Reprint of Neutrophil cell surface receptors and their intracellular signal transduction pathways. *Int. Immunopharmacol.* **2013**, *17*, 1185–1197. [CrossRef] [PubMed]

94. Brewer, G.J.; Aster, J.C.; Knutsen, C.A.; Kruckeberg, W.C. Zinc inhibition of calmodulin: A proposed molecular mechanism of zinc action on cellular functions. *Am. J. Hematol.* **1979**, *7*, 53–60. [CrossRef] [PubMed]

95. Wellinghausen, N.; Rink, L. The significance of zinc for leukocyte biology. *J. Leukoc. Biol.* **1998**, *64*, 571–577. [PubMed]

96. Mustelin, T.; Vang, T.; Bottini, N. Protein tyrosine phosphatases and the immune response. *Nat. Rev. Immunol.* **2005**, *5*, 43–57. [CrossRef] [PubMed]

97. Medgyesi, D.; Hobeika, E.; Biesen, R.; Kollert, F.; Taddeo, A.; Voll, R.E.; Hiepe, F.; Reth, M. The protein tyrosine phosphatase PTP1B is a negative regulator of CD40 and BAFF-R signaling and controls B cell autoimmunity. *J. Exp. Med.* **2014**, *211*, 427–440. [CrossRef] [PubMed]

98. Haase, H.; Maret, W. Protein tyrosine phosphatases as targets of the combined insulinomimetic effects of zinc and oxidants. *Biometals* **2005**, *18*, 333–338. [CrossRef] [PubMed]

99. Maret, W.; Jacob, C.; Vallee, B.L.; Fischer, E.H. Inhibitory sites in enzymes: Zinc removal and reactivation by thionein. *Proc. Natl. Acad. Sci. USA* **1999**, *96*, 1936–1940. [CrossRef] [PubMed]

100. Haase, H.; Maret, W. Fluctuations of cellular, available zinc modulate insulin signaling via inhibition of protein tyrosine phosphatases. *J. Trace Elem. Med. Biol.* **2005**, *19*, 37–42. [CrossRef] [PubMed]

101. Sly, L.M.; Rauh, M.J.; Kalesnikoff, J.; Büchse, T.; Krystal, G. SHIP, SHIP2, and PTEN activities are regulated in vivo by modulation of their protein levels: SHIP is up-regulated in macrophages and mast cells by lipopolysaccharide. *Exp. Hematol.* **2003**, *31*, 1170–1181. [CrossRef] [PubMed]

102. Haase, H.; Ober-Blobaum, J.L.; Engelhardt, G.; Hebel, S.; Heit, A.; Heine, H.; Rink, L. Zinc signals are essential for lipopolysaccharide-induced signal transduction in monocytes. *J. Immunol.* **2008**, *181*, 6491–6502. [CrossRef] [PubMed]

103. Hennigar, S.R.; Seo, Y.A.; Sharma, S.; Soybel, D.I.; Kelleher, S.L. ZnT2 is a critical mediator of lysosomal-mediated cell death during early mammary gland involution. *Sci. Rep.* **2015**, *5*, 8033. [CrossRef] [PubMed]

104. Lockwood, T.D. Lysosomal metal, redox and proton cycles influencing the CysHis cathepsin reaction. *Metallomics* **2013**, *5*, 110–124. [CrossRef] [PubMed]

105. Vinkenborg, J.L.; Nicolson, T.J.; Bellomo, E.A.; Koay, M.S.; Rutter, G.A.; Merkx, M. Genetically encoded FRET sensors to monitor intracellular $Zn^{2+}$ homeostasis. *Nat. Methods* **2009**, *6*, 737–740. [CrossRef] [PubMed]

106. Vener, A.V.; Aksenova, M.V.; Burbaeva, G.S. Drastic reduction of the zinc-and magnesium-stimulated protein tyrosine kinase activities in Alzheimer's disease hippocampus. *FEBS Lett.* **1993**, *328*, 6–8. [CrossRef]

107. Bennasroune, A.; Mazot, P.; Boutterin, M.-C.; Vigny, M. Activation of the orphan receptor tyrosine kinase ALK by zinc. *Biochem. Biophys. Res. Commun.* **2010**, *398*, 702–706. [CrossRef] [PubMed]

108. Baraldi, E.; Carugo, K.D.; Hyvönen, M.; Surdo, P.L.; Riley, A.M.; Potter, B.V.; O'Brien, R.; Ladbury, J.E.; Saraste, M. Structure of the PH domain from Bruton's tyrosine kinase in complex with inositol 1,3,4,5-tetrakisphosphate. *Structure* **1999**, *7*, 449–460. [CrossRef]

109. Arbibe, L.; Jean-Paul, M.; Teusch, N.; Kline, L.; Guha, M.; Mackman, N.; Godowski, P.J.; Ulevitch, R.J.; Knaus, U.G. Toll-like receptor 2-mediated NF-κB activation requires a Rac1-dependent pathway. *Nat. Immunol.* **2000**, *1*, 533. [CrossRef] [PubMed]

110. Zhang, Y.; Xing, F.; Zheng, H.; Xi, J.; Cui, X.; Xu, Z. Roles of mitochondrial Src tyrosine kinase and zinc in nitric oxide-induced cardioprotection against ischemia/reperfusion injury. *Free Radic. Res.* **2013**, *47*, 517–525. [CrossRef] [PubMed]

111. Liu, M.-J.; Bao, S.; Napolitano, J.R.; Burris, D.L.; Yu, L.; Tridandapani, S.; Knoell, D.L. Zinc regulates the acute phase response and serum amyloid A production in response to sepsis through JAK-STAT3 signaling. *PLoS ONE* **2014**, *9*, e94934. [CrossRef] [PubMed]

112. Hu, J.; Yang, Z.; Wang, J.; Yu, J.; Guo, J.; Liu, S.; Qian, C.; Song, L.; Wu, Y.; Cheng, J. Zinc Chloride Transiently Maintains Mouse Embryonic Stem Cell Pluripotency by Activating Stat3 Signaling. *PLoS ONE* **2016**, *11*, e0148994. [CrossRef] [PubMed]

113. Bellomo, E.; Massarotti, A.; Hogstrand, C.; Maret, W. Zinc ions modulate protein tyrosine phosphatase 1B activity. *Metallomics* **2014**, *6*, 1229–1239. [CrossRef] [PubMed]

114. Wilson, M.; Hogstrand, C.; Maret, W. Picomolar concentrations of free zinc (II) ions regulate receptor protein-tyrosine phosphatase β activity. *J. Biol. Chem.* **2012**, *287*, 9322–9326. [CrossRef] [PubMed]

115. Haase, H.; Maret, W. Intracellular zinc fluctuations modulate protein tyrosine phosphatase activity in insulin/insulin-like growth factor-1 signaling. *Exp. Cell Res.* **2003**, *291*, 289–298. [CrossRef]

116. Takahashi, K.; Akaishi, E.; Abe, Y.; Ishikawa, R.; Tanaka, S.; Hosaka, K.; Kubohara, Y. Zinc inhibits calcineurin activity in vitro by competing with nickel. *Biochem. Biophys. Res. Commun.* **2003**, *307*, 64–68. [CrossRef]

117. Lohmann, R.D.; Beyersmann, D. Cadmium and zinc mediated changes of the $Ca^{2+}$-dependent endonuclease in apoptosis. *Biochem. Biophys. Res. Commun.* **1993**, *190*, 1097–1103. [CrossRef] [PubMed]

118. Percival, M.D.; Yeh, B.; Falgueyret, J.-P. Zinc dependent activation of cAMP-specific phosphodiesterase (PDE4A). *Biochem. Biophys. Res. Commun.* **1997**, *241*, 175–180. [CrossRef] [PubMed]

119. Francis, S.H.; Colbran, J.L.; McAllister-Lucas, L.M.; Corbin, J.D. Zinc interactions and conserved motifs of the cGMP-binding cGMP-specific phosphodiesterase suggest that it is a zinc hydrolase. *J. Biol. Chem.* **1994**, *269*, 22477–22480. [PubMed]

120. He, F.; Seryshev, A.B.; Cowan, C.W.; Wensel, T.G. Multiple zinc binding sites in retinal rod cGMP phosphodiesterase, PDE6αβ. *J. Biol. Chem.* **2000**, *275*, 20572–20577. [CrossRef] [PubMed]

121. Jeon, K.I.; Jeong, J.Y.; Jue, D.M. Thiol-reactive metal compounds inhibit NF-κB activation by blocking I κB kinase. *J. Immunol.* **2000**, *164*, 5981–5989. [CrossRef] [PubMed]

122. Haase, H.; Hebel, S.; Engelhardt, G.; Rink, L. Flow cytometric measurement of labile zinc in peripheral blood mononuclear cells. *Anal. Biochem.* **2006**, *352*, 222–230. [CrossRef] [PubMed]

123. Krężel, A.; Hao, Q.; Maret, W. The zinc/thiolate redox biochemistry of metallothionein and the control of zinc ion fluctuations in cell signaling. *Arch. Biochem. Biophys.* **2007**, *463*, 188–200. [CrossRef] [PubMed]

124. Korichneva, I. Redox regulation of cardiac protein kinase C. *Exp. Clin. Cardiol.* **2005**, *10*, 256–261. [PubMed]

125. Aydemir, T.B.; Liuzzi, J.P.; McClellan, S.; Cousins, R.J. Zinc transporter ZIP8 (SLC39A8) and zinc influence IFN-γ expression in activated human T cells. *J. Leukoc. Biol.* **2009**, *86*, 337–348. [CrossRef] [PubMed]

126. Slepchenko, K.G.; Lu, Q.; Li, Y.V. Zinc wave during the treatment of hypoxia is required for initial reactive oxygen species activation in mitochondria. *Int. J. Physiol. Pathophysiol. Pharmacol.* **2016**, *8*, 44. [PubMed]

127. Zalewski, P.; Forbes, I.; Giannakis, C.; Cowled, P.; Betts, W. Synergy between zinc and phorbol ester in translocation of protein kinase C to cytoskeleton. *FEBS Lett.* **1990**, *273*, 131–134. [CrossRef]

128. Beyersmann, D.; Haase, H. Functions of zinc in signaling, proliferation and differentiation of mammalian cells. *Biometals* **2001**, *14*, 331–341. [CrossRef] [PubMed]

129. Lindahl, M.; Leanderson, P.; Tagesson, C. Novel aspect on metal fume fever: Zinc stimulates oxygen radical formation in human neutrophils. *Hum. Exp. Toxicol.* **1998**, *17*, 105–110. [CrossRef] [PubMed]

130. Powell, S.R. The antioxidant properties of zinc. *J. Nutr.* **2000**, *130* (Suppl. 5S), 1447S–1454S. [PubMed]
131. Freitas, M.; Porto, G.; Lima, J.L.; Fernandes, E. Zinc activates neutrophils' oxidative burst. *Biometals* **2010**, *23*, 31. [CrossRef] [PubMed]
132. Londesborough, J.; Suoranta, K. Zinc-containing cyclic nucleotide phosphodiesterases from bakers' yeast. *Methods Enzymol.* **1988**, *159*, 777–785. [PubMed]
133. Von Bülow, V.; Dubben, S.; Engelhardt, G.; Hebel, S.; Plümakers, B.; Heine, H.; Rink, L.; Haase, H. Zinc-dependent suppression of TNF-α production is mediated by protein kinase A-induced inhibition of Raf-1, IκB Kinase β, and NF-κB. *J. Immunol.* **2007**, *179*, 4180–4186. [CrossRef] [PubMed]
134. Klein, C.; Heyduk, T.; Sunahara, R.K. Zinc inhibition of adenylyl cyclase correlates with conformational changes in the enzyme. *Cell Signal.* **2004**, *16*, 1177–1185. [CrossRef] [PubMed]
135. Gao, X.; Du, Z.; Patel, T.B. Copper and Zinc Inhibit Gαs Function A Nucleotide-Free State of Gαs Induced by $Cu^{2+}$ and $Zn^{2+}$. *J. Biol. Chem.* **2005**, *280*, 2579–2586. [CrossRef] [PubMed]
136. Uzzo, R.G.; Crispen, P.L.; Golovine, K.; Makhov, P.; Horwitz, E.M.; Kolenko, V.M. Diverse effects of zinc on NF-κB and AP-1 transcription factors: Implications for prostate cancer progression. *Carcinogenesis* **2006**, *27*, 1980–1990. [CrossRef] [PubMed]
137. Liu, M.-J.; Bao, S.; Gálvez-Peralta, M.; Pyle, C.J.; Rudawsky, A.C.; Pavlovicz, R.E.; Killilea, D.W.; Li, C.; Nebert, D.W.; Wewers, M.D. ZIP8 regulates host defense through zinc-mediated inhibition of NF-κB. *Cell Rep.* **2013**, *3*, 386–400. [CrossRef] [PubMed]
138. Prasad, A.S.; Bao, B.; Beck, F.W.; Kucuk, O.; Sarkar, F.H. Antioxidant effect of zinc in humans. *Free Radic. Biol. Med.* **2004**, *37*, 1182–1190. [CrossRef] [PubMed]
139. Yan, Y.-W.; Fan, J.; Bai, S.-L.; Hou, W.-J.; Li, X.; Tong, H. Zinc prevents abdominal aortic aneurysm formation by induction of A20-mediated suppression of NF-κB pathway. *PLoS ONE* **2016**, *11*, e0148536. [CrossRef] [PubMed]
140. Li, C.; Guo, S.; Gao, J.; Guo, Y.; Du, E.; Lv, Z.; Zhang, B. Maternal high-zinc diet attenuates intestinal inflammation by reducing DNA methylation and elevating h3k9 acetylation in the a20 promoter of offspring chicks. *J. Nutri. Biochem.* **2015**, *26*, 173–183. [CrossRef] [PubMed]
141. Staitieh, B.S.; Fan, X.; Neveu, W.; Guidot, D.M. Nrf2 regulates PU. 1 expression and activity in the alveolar macrophage. *Am. J. Physiol. Lung Cell. Mol. Physiol.* **2015**, *308*, L1086–L1093. [CrossRef] [PubMed]
142. Zhang, S.; Laouar, A.; Denzin, L.K.; Sant'Angelo, D.B. Zbtb16 (PLZF) is stably suppressed and not inducible in non-innate T cells via T cell receptor-mediated signaling. *Sci. Rep.* **2015**, *5*, 12113. [CrossRef] [PubMed]
143. Molkentin, J.D. The zinc finger-containing transcription factors GATA-4,-5, and-6 ubiquitously expressed regulators of tissue-specific gene expression. *J. Biol. Chem.* **2000**, *275*, 38949–38952. [CrossRef] [PubMed]
144. Ghaleb, A.M.; Nandan, M.O.; Chanchevalap, S.; Dalton, W.B.; Hisamuddin, I.M. Krüppel-like factors 4 and 5: The yin and yang regulators of cellular proliferation. *Cell Res.* **2005**, *15*, 92. [CrossRef] [PubMed]
145. Laity, J.H.; Andrews, G.K. Understanding the mechanisms of zinc-sensing by metal-response element binding transcription factor-1 (MTF-1). *Arch. Biochem. Biophys.* **2007**, *463*, 201–210. [CrossRef] [PubMed]
146. Cross, J.L.; Johnson, P. Tyrosine phosphorylation in immune cells: Direct and indirect effects on toll-like receptor-induced proinflammatory cytokine production. *Crit. Rev. Immunol.* **2009**, *29*, 347–367.
147. Cho, J.; Tsichlis, P.N. Phosphorylation at Thr-290 regulates Tpl2 binding to NF-κB1/p105 and Tpl2 activation and degradation by lipopolysaccharide. *Proc. Natl. Acad. Sci. USA* **2005**, *102*, 2350–2355. [CrossRef] [PubMed]
148. Guo, J.; Friedman, S.L. Toll-like receptor 4 signaling in liver injury and hepatic fibrogenesis. *Fibrogenesis Tissue Repair* **2010**, *3*, 21. [CrossRef] [PubMed]
149. Read, S.A.; O'Connor, K.S.; Suppiah, V.; Ahlenstiel, C.L.; Obeid, S.; Cook, K.M.; Cunningham, A.; Douglas, M.W.; Hogg, P.J.; Booth, D. Zinc is a potent and specific inhibitor of IFN-γ3 signalling. *Nat. Commun.* **2017**, *8*. [CrossRef] [PubMed]
150. Denk, A.; Wirth, T.; Baumann, B. NF-κB transcription factors: Critical regulators of hematopoiesis and neuronal survival. *Cytokine Growth Factor Rev.* **2000**, *11*, 303–320. [CrossRef]
151. Wang, C.; Deng, L.; Hong, M.; Akkaraju, G.R.; Inoue, J.-I.; Chen, Z.J. TAK1 is a ubiquitin-dependent kinase of MKK and IKK. *Nature* **2001**, *412*, 346. [CrossRef] [PubMed]
152. Boone, D.L.; Turer, E.E.; Lee, E.G.; Regina-Celeste, A.; Wheeler, M.T.; Tsui, C.; Hurley, P.; Chien, M.; Chai, S.; Hitotsumatsu, O. The ubiquitin-modifying enzyme A20 is required for termination of toll-like receptor responses. *Nat. Immunol.* **2004**, *5*, 1052. [CrossRef] [PubMed]

153. Tillmann, N.; Engelhardt, G.; Rink, L.; Weßels, I. Zinc in granulopoiesis. Unpublished work. 2017.

154. Bellomo, E.; Birla Singh, K.; Massarotti, A.; Hogstrand, C.; Maret, W. The metal face of protein tyrosine phosphatase 1B. *Coord. Chem. Rev.* **2016**, *327–328*, 70–83. [CrossRef] [PubMed]

155. Manz, M.G.; Boettcher, S. Emergency granulopoiesis. *Nat. Rev. Immunol.* **2014**, *14*, 302–314. [CrossRef] [PubMed]

156. Supasai, S.; Aimo, L.; Adamo, A.; Mackenzie, G.; Oteiza, P. Zinc deficiency affects the STAT1/3 signaling pathways in part through redox-mediated mechanisms. *Redox Biol.* **2017**, *11*, 469–481. [CrossRef] [PubMed]

157. Aster, I.; Engelhardt, G.; Rink, L.; Weßels, I. The influence of zinc on granulocyte-macrophage colony stimulating factor-induced signaling in U937 cells. Unpublished work. 2017.

158. Rajagopalan, S.; Long, E.O. Zinc bound to the killer cell-inhibitory receptor modulates the negative signal in human NK cells. *J. Immunol.* **1998**, *161*, 1299–1305. [PubMed]

159. Rajagopalan, S.; Winter, C.C.; Wagtmann, N.; Long, E.O. The Ig-related killer cell inhibitory receptor binds zinc and requires zinc for recognition of HLA-C on target cells. *J. Immunol.* **1995**, *155*, 4143–4146. [PubMed]

160. Ho, L.H.; Ruffin, R.E.; Murgia, C.; Li, L.; Krilis, S.A.; Zalewski, P.D. Labile zinc and zinc transporter ZnT4 in mast cell granules: Role in regulation of caspase activation and NF-κB translocation. *J. Immunol.* **2004**, *172*, 7750–7760. [CrossRef] [PubMed]

161. Kabu, K.; Yamasaki, S.; Kamimura, D.; Ito, Y.; Hasegawa, A.; Sato, E.; Kitamura, H.; Nishida, K.; Hirano, T. Zinc is required for Fc epsilon RI-mediated mast cell activation. *J. Immunol.* **2006**, *177*, 1296–1305. [CrossRef] [PubMed]

162. Ollig, J.; Kloubert, V.; Weßels, I.; Haase, H.; Rink, L. Parameters Influencing Zinc in Experimental Systems in Vivo and in Vitro. *Metals* **2016**, *6*, 71. [CrossRef]

163. Osati-Ashtiani, F.; King, L.E.; Fraker, P.J. Variance in the resistance of murine early bone marrow B cells to a deficiency in zinc. *Immunology* **1998**, *94*, 94–100. [CrossRef] [PubMed]

164. King, L.E.; Frentzel, J.W.; Mann, J.J.; Fraker, P.J. Chronic zinc deficiency in mice disrupted T cell lymphopoiesis and erythropoiesis while B cell lymphopoiesis and myelopoiesis were maintained. *J. Am. Coll. Nutr.* **2005**, *24*, 494–502. [CrossRef] [PubMed]

165. Dardenne, M.; Savino, W.; Wade, S.; Kaiserlian, D.; Lemonnier, D.; Bach, J.F. In vivo and in vitro studies of thymulin in marginally zinc-deficient mice. *Eur. J. Immunol.* **1984**, *14*, 454–458. [CrossRef] [PubMed]

166. Saha, A.R.; Hadden, E.M.; Hadden, J.W. Zinc induces thymulin secretion from human thymic epithelial cells in vitro and augments splenocyte and thymocyte responses in vivo. *Int. J. Immunopharmacol.* **1995**, *17*, 729–733. [CrossRef]

167. Beck, F.W.; Kaplan, J.; Fine, N.; Handschu, W.; Prasad, A.S. Decreased expression of CD73 (ecto-5′-nucleotidase) in the CD8+ subset is associated with zinc deficiency in human patients. *J. Lab. Clin. Med.* **1997**, *130*, 147–156. [CrossRef]

168. Fraker, P.J. Roles for cell death in zinc deficiency. *J. Nutr.* **2005**, *135*, 359–362. [PubMed]

169. Chai, F.; Truong-Tran, A.Q.; Evdokiou, A.; Young, G.P.; Zalewski, P.D. Intracellular zinc depletion induces caspase activation and p21 Waf1/Cip1 cleavage in human epithelial cell lines. *J. Infect. Dis.* **2000**, *182* (Suppl. S1), S85–S92. [CrossRef] [PubMed]

170. King, K.L.; Cidlowski, J.A. Cell cycle regulation and apoptosis. *Annu. Rev. Physiol.* **1998**, *60*, 601–617. [CrossRef] [PubMed]

171. Prasad, A.S.; Meftah, S.; Abdallah, J.; Kaplan, J.; Brewer, G.J.; Bach, J.F.; Dardenne, M. Serum thymulin in human zinc deficiency. *J. Clin. Investig.* **1988**, *82*, 1202–1210. [CrossRef] [PubMed]

172. Coto, J.A.; Hadden, E.M.; Sauro, M.; Zorn, N.; Hadden, J.W. Interleukin 1 regulates secretion of zinc-thymulin by human thymic epithelial cells and its action on T-lymphocyte proliferation and nuclear protein kinase C. *Proc. Natl. Acad. Sci. USA* **1992**, *89*, 7752–7756. [CrossRef] [PubMed]

173. Dowd, P.S.; Kelleher, J.; Guillou, P.J. T-lymphocyte subsets and interleukin-2 production in zinc-deficient rats. *Br. J. Nutr.* **1986**, *55*, 59–69. [CrossRef] [PubMed]

174. Prasad, A.S. Effects of zinc deficiency on Th1 and Th2 cytokine shifts. *J. Infect. Dis.* **2000**, *182* (Suppl. S1), S62–S68. [CrossRef] [PubMed]

175. Richter, M.; Bonneau, R.; Girard, M.A.; Beaulieu, C.; Larivee, P. Zinc status modulates bronchopulmonary eosinophil infiltration in a murine model of allergic inflammation. *Chest* **2003**, *123* (Suppl. S3), 446S. [CrossRef]

176. Honscheid, A.; Rink, L.; Haase, H. T-lymphocytes: A target for stimulatory and inhibitory effects of zinc ions. *Endocr. Metab. Immune Disord. Drug Targets* **2009**, *9*, 132–144. [CrossRef] [PubMed]

177. Socha, K.; Karpinska, E.; Kochanowicz, J.; Soroczynska, J.; Jakoniuk, M.; Wilkiel, M.; Mariak, Z.D.; Borawska, M.H. Dietary habits; concentration of copper, zinc, and Cu-to-Zn ratio in serum and ability status of patients with relapsing-remitting multiple sclerosis. *Nutrition* **2017**, *39–40*, 76–81. [CrossRef] [PubMed]

178. Popescu, B.F.; Frischer, J.M.; Webb, S.M.; Tham, M.; Adiele, R.C.; Robinson, C.A.; Fitz-Gibbon, P.D.; Weigand, S.D.; Metz, I.; Nehzati, S.; et al. Pathogenic implications of distinct patterns of iron and zinc in chronic MS lesions. *Acta Neuropathol.* **2017**, *134*, 45–64. [CrossRef] [PubMed]

179. Okamoto, T.; Kuroki, T.; Adachi, T.; Ono, S.; Hayashi, T.; Tajima, Y.; Eguchi, S.; Kanematsu, T. Effect of zinc on early graft failure following intraportal islet transplantation in rat recipients. *Ann. Transpl.* **2011**, *16*, 114–120. [CrossRef]

180. Chimienti, F.; Seve, M.; Richard, S.; Mathieu, J.; Favier, A. Role of cellular zinc in programmed cell death: Temporal relationship between zinc depletion, activation of caspases, and cleavage of Sp family transcription factors. *Biochem. Pharmacol.* **2001**, *62*, 51–62. [CrossRef]

181. Hanidziar, D.; Koulmanda, M. Inflammation and the balance of Treg and Th17 cells in transplant rejection and tolerance. *Curr. Opin. Organ Transpl.* **2010**, *15*, 411–415. [CrossRef] [PubMed]

182. Graca, L.; Thompson, S.; Lin, C.Y.; Adams, E.; Cobbold, S.P.; Waldmann, H. Both CD4$^+$CD25$^+$ and CD4$^+$CD25$^-$ regulatory cells mediate dominant transplantation tolerance. *J. Immunol.* **2002**, *168*, 5558–5565. [CrossRef] [PubMed]

183. Chen, W.; Jin, W.; Hardegen, N.; Lei, K.J.; Li, L.; Marinos, N.; McGrady, G.; Wahl, S.M. Conversion of peripheral CD4$^+$CD25$^-$ naive T cells to CD4$^+$CD25$^+$ regulatory T cells by TGF-β induction of transcription factor Foxp3. *J. Exp. Med.* **2003**, *198*, 1875–1886. [CrossRef] [PubMed]

184. Huter, E.N.; Stummvoll, G.H.; DiPaolo, R.J.; Glass, D.D.; Shevach, E.M. Cutting edge: Antigen-specific TGF β-induced regulatory T cells suppress Th17-mediated autoimmune disease. *J. Immunol.* **2008**, *181*, 8209–8213. [CrossRef] [PubMed]

185. Davidson, T.S.; DiPaolo, R.J.; Andersson, J.; Shevach, E.M. Cutting Edge: IL-2 is essential for TGF-β-mediated induction of Foxp3+ T regulatory cells. *J. Immunol.* **2007**, *178*, 4022–4026. [CrossRef] [PubMed]

186. Kretschmer, K.; Apostolou, I.; Hawiger, D.; Khazaie, K.; Nussenzweig, M.C.; von Boehmer, H. Inducing and expanding regulatory T cell populations by foreign antigen. *Nat. Immunol.* **2005**, *6*, 1219–1227. [CrossRef] [PubMed]

187. Kasagi, S.; Zhang, P.; Che, L.; Abbatiello, B.; Maruyama, T.; Nakatsukasa, H.; Zanvit, P.; Jin, W.; Konkel, J.E.; Chen, W. In vivo-generated antigen-specific regulatory T cells treat autoimmunity without compromising antibacterial immune response. *Sci. Transl. Med.* **2014**, *6*, 241ra78. [CrossRef] [PubMed]

188. Horwitz, D.A.; Zheng, S.G.; Wang, J.; Gray, J.D. Critical role of IL-2 and TGF-β in generation, function and stabilization of Foxp3$^+$CD4$^+$ Treg. *Eur. J. Immunol.* **2008**, *38*, 912–915. [CrossRef] [PubMed]

189. Takaki, H.; Ichiyama, K.; Koga, K.; Chinen, T.; Takaesu, G.; Sugiyama, Y.; Kato, S.; Yoshimura, A.; Kobayashi, T. STAT6 Inhibits TGFβ1-mediated Foxp3 induction through direct binding to the Foxp3 promoter, which is reverted by retinoic acid receptor. *J. Biol. Chem.* **2008**, *283*, 14955–14962. [CrossRef] [PubMed]

190. Tone, Y.; Furuuchi, K.; Kojima, Y.; Tykocinski, M.L.; Greene, M.I.; Tone, M. Smad3 and NFAT cooperate to induce Foxp3 expression through its enhancer. *Nat. Immunol.* **2008**, *9*, 194–202. [CrossRef] [PubMed]

191. Hogstrand, C.; Kille, P.; Nicholson, R.I.; Taylor, K.M. Zinc transporters and cancer: A potential role for ZIP7 as a hub for tyrosine kinase activation. *Trends Mol. Med.* **2009**, *15*, 101–111. [CrossRef] [PubMed]

192. Niefind, K.; Raaf, J.; Issinger, O.G. Protein kinase CK2 in health and disease: Protein kinase CK2: From structures to insights. *Cell. Mol. Life Sci.* **2009**, *66*, 1800–1816. [CrossRef] [PubMed]

193. St-Denis, N.A.; Litchfield, D.W. Protein kinase CK2 in health and disease: From birth to death: The role of protein kinase CK2 in the regulation of cell proliferation and survival. *Cell. Mol. Life Sci.* **2009**, *66*, 1817–1829. [CrossRef] [PubMed]

194. Ulges, A.; Klein, M.; Reuter, S.; Gerlitzki, B.; Hoffmann, M.; Grebe, N.; Staudt, V.; Stergiou, N.; Bohn, T.; Bruhl, T.J.; et al. Protein kinase CK2 enables regulatory T cells to suppress excessive Th2 responses in vivo. *Nat. Immunol.* **2015**, *16*, 267–275. [CrossRef] [PubMed]

195. Fragale, A.; Gabriele, L.; Stellacci, E.; Borghi, P.; Perrotti, E.; Ilari, R.; Lanciotti, A.; Remoli, A.L.; Venditti, M.; Belardelli, F.; et al. IFN regulatory factor-1 negatively regulates CD4$^+$ CD25$^+$ regulatory T cell differentiation by repressing Foxp3 expression. *J. Immunol.* **2008**, *181*, 1673–1682. [CrossRef] [PubMed]

196. Maywald, M.; Rink, L. Zinc supplementation induces CD4+CD25+Foxp3+ antigen-specific regulatory T cells and suppresses IFN-γ production by upregulation of Foxp3 and KLF-10 and downregulation of IRF-1. *Eur. J. Nutr.* **2017**, *56*, 1859–1869. [CrossRef] [PubMed]

197. Cao, Z.; Wara, A.K.; Icli, B.; Sun, X.; Packard, R.R.; Esen, F.; Stapleton, C.J.; Subramaniam, M.; Kretschmer, K.; Apostolou, I.; et al. Kruppel-like factor KLF10 targets transforming growth factor-β1 to regulate CD4$^+$CD25$^-$ T cells and T regulatory cells. *J. Biol. Chem.* **2009**, *284*, 24914–24924. [CrossRef] [PubMed]

198. Schutze, N.; Trojandt, S.; Kuhn, S.; Tomm, J.M.; von Bergen, M.; Simon, J.C.; Polte, T. Allergen-Induced IL-6 Regulates IL-9/IL-17A Balance in CD4+ T Cells in Allergic Airway Inflammation. *J. Immunol.* **2016**, *197*, 2653–2664. [CrossRef] [PubMed]

199. Elyaman, W.; Khoury, S.J. Th9 cells in the pathogenesis of EAE and multiple sclerosis. *Semin. Immunopathol.* **2017**, *39*, 79–87. [CrossRef] [PubMed]

200. Kitabayashi, C.; Fukada, T.; Kanamoto, M.; Ohashi, W.; Hojyo, S.; Atsumi, T.; Ueda, N.; Azuma, I.; Hirota, H.; Murakami, M.; et al. Zinc suppresses Th17 development via inhibition of STAT3 activation. *Int. Immunol.* **2010**, *22*, 375–386. [CrossRef] [PubMed]

201. Palacios, E.H.; Weiss, A. Function of the Src-family kinases, Lck and Fyn, in T-cell development and activation. *Oncogene* **2004**, *23*, 7990–8000. [CrossRef] [PubMed]

202. Chiang, Y.J.; Hodes, R.J. T-cell development is regulated by the coordinated function of proximal and distal Lck promoters active at different developmental stages. *Eur. J. Immunol.* **2016**, *46*, 2401–2408. [CrossRef] [PubMed]

203. Kim, P.W.; Sun, Z.Y.; Blacklow, S.C.; Wagner, G.; Eck, M.J. A zinc clasp structure tethers Lck to T cell coreceptors CD4 and CD8. *Science* **2003**, *301*, 1725–1728. [CrossRef] [PubMed]

204. Mustelin, T.; Tasken, K. Positive and negative regulation of T-cell activation through kinases and phosphatases. *Biochem. J.* **2003**, *371*, 15–27. [CrossRef] [PubMed]

205. Furlan, G.; Minowa, T.; Hanagata, N.; Kataoka-Hamai, C.; Kaizuka, Y. Phosphatase CD45 both positively and negatively regulates T cell receptor phosphorylation in reconstituted membrane protein clusters. *J. Biol. Chem.* **2014**, *289*, 28514–28525. [CrossRef] [PubMed]

206. Macian, F. NFAT proteins: Key regulators of T-cell development and function. *Nat. Rev. Immunol.* **2005**, *5*, 472–484. [CrossRef] [PubMed]

207. Huang, J.; Zhang, D.; Xing, W.; Ma, X.; Yin, Y.; Wei, Q.; Li, G. An approach to assay calcineurin activity and the inhibitory effect of zinc ion. *Anal. Biochem.* **2008**, *375*, 385–387. [CrossRef] [PubMed]

208. Osaki, M.; Oshimura, M.; Ito, H. PI3K-Akt pathway: Its functions and alterations in human cancer. *Apoptosis* **2004**, *9*, 667–676. [CrossRef] [PubMed]

209. Liang, D.; Yang, M.; Guo, B.; Cao, J.; Yang, L.; Guo, X.; Li, Y.; Gao, Z. Zinc inhibits H$_2$O$_2$-induced MC3T3-E1 cells apoptosis via MAPK and PI3K/AKT pathways. *Biol. Trace Elem. Res.* **2012**, *148*, 420–429. [CrossRef] [PubMed]

210. Baek, S.H.; Kim, M.Y.; Mo, J.S.; Ann, E.J.; Lee, K.S.; Park, J.H.; Kim, J.Y.; Seo, M.S.; Choi, E.J.; Park, H.S. Zinc-induced downregulation of Notch signaling is associated with cytoplasmic retention of Notch1-IC and RBP-Jk via PI3k-Akt signaling pathway. *Cancer Lett.* **2007**, *255*, 117–126. [CrossRef] [PubMed]

211. Eom, S.J.; Kim, E.Y.; Lee, J.E.; Kang, H.J.; Shim, J.; Kim, S.U.; Gwag, B.J.; Choi, E.J. Zn$^{2+}$ induces stimulation of the c-Jun N-terminal kinase signaling pathway through phosphoinositide 3-Kinase. *Mol. Pharmacol.* **2001**, *59*, 981–986. [PubMed]

212. Tang, X.; Shay, N.F. Zinc has an insulin-like effect on glucose transport mediated by phosphoinositol-3-kinase and Akt in 3T3-L1 fibroblasts and adipocytes. *J. Nutr.* **2001**, *131*, 1414–1420. [PubMed]

213. Wu, W.; Wang, X.; Zhang, W.; Reed, W.; Samet, J.M.; Whang, Y.E.; Ghio, A.J. Zinc-induced PTEN protein degradation through the proteasome pathway in human airway epithelial cells. *J. Biol. Chem.* **2003**, *278*, 28258–28263. [CrossRef] [PubMed]

214. Kwak, Y.D.; Wang, B.; Pan, W.; Xu, H.; Jiang, X.; Liao, F.F. Functional interaction of phosphatase and tensin homologue (PTEN) with the E3 ligase NEDD4-1 during neuronal response to zinc. *J. Biol. Chem.* **2010**, *285*, 9847–9857. [CrossRef] [PubMed]

215. Csermely, P.; Szamel, M.; Resch, K.; Somogyi, J. Zinc can increase the activity of protein kinase C and contributes to its binding to plasma membranes in T lymphocytes. *J. Biol. Chem.* **1988**, *263*, 6487–6490. [PubMed]

216. Tan, S.L.; Parker, P.J. Emerging and diverse roles of protein kinase C in immune cell signalling. *Biochem. J.* **2003**, *376*, 545–552. [CrossRef] [PubMed]

217. Quest, A.F.; Bloomenthal, J.; Bardes, E.S.; Bell, R.M. The regulatory domain of protein kinase C coordinates four atoms of zinc. *J. Biol. Chem.* **1992**, *267*, 10193–10197. [PubMed]

218. Hubbard, S.R.; Bishop, W.R.; Kirschmeier, P.; George, S.J.; Cramer, S.P.; Hendrickson, W.A. Identification and characterization of zinc binding sites in protein kinase C. *Science* **1991**, *254*, 1776–1779. [CrossRef] [PubMed]

219. Forbes, I.J.; Zalewski, P.D.; Giannakis, C.; Petkoff, H.S.; Cowled, P.A. Interaction between protein kinase C and regulatory ligand is enhanced by a chelatable pool of cellular zinc. *Biochim. Biophys. Acta* **1990**, *1053*, 113–117. [CrossRef]

220. Knapp, L.T.; Klann, E. Superoxide-induced stimulation of protein kinase C via thiol modification and modulation of zinc content. *J. Biol. Chem.* **2000**, *275*, 24136–24145. [CrossRef] [PubMed]

221. Korichneva, I.; Hoyos, B.; Chua, R.; Levi, E.; Hammerling, U. Zinc release from protein kinase C as the common event during activation by lipid second messenger or reactive oxygen. *J. Biol. Chem.* **2002**, *277*, 44327–44331. [CrossRef] [PubMed]

222. Csermely, P.; Gueth, S.; Somogyi, J. The tumor promoter tetradecanoyl-phorbol-acetate (TPA) elicits the redistribution of zinc in subcellular fractions of rabbit thymocytes measured by X-ray fluorescence. *Biochem. Biophys. Res. Commun.* **1987**, *144*, 863–868. [CrossRef]

223. Chow, L.M.; Fournel, M.; Davidson, D.; Veillette, A. Negative regulation of T-cell receptor signalling by tyrosine protein kinase p50csk. *Nature* **1993**, *365*, 156–160. [CrossRef] [PubMed]

224. Vang, T.; Torgersen, K.M.; Sundvold, V.; Saxena, M.; Levy, F.O.; Skalhegg, B.S.; Hansson, V.; Mustelin, T.; Tasken, K. Activation of the COOH-terminal Src kinase (Csk) by cAMP-dependent protein kinase inhibits signaling through the T cell receptor. *J. Exp. Med.* **2001**, *193*, 497–507. [CrossRef] [PubMed]

225. Rhyasen, G.W.; Starczynowski, D.T. IRAK signalling in cancer. *Br. J. Cancer* **2015**, *112*, 232–237. [CrossRef] [PubMed]

226. Suzuki, N.; Suzuki, S.; Millar, D.G.; Unno, M.; Hara, H.; Calzascia, T.; Yamasaki, S.; Yokosuka, T.; Chen, N.J.; Elford, A.R.; et al. A critical role for the innate immune signaling molecule IRAK-4 in T cell activation. *Science* **2006**, *311*, 1927–1932. [CrossRef] [PubMed]

227. Kawagoe, T.; Sato, S.; Jung, A.; Yamamoto, M.; Matsui, K.; Kato, H.; Uematsu, S.; Takeuchi, O.; Akira, S. Essential role of IRAK-4 protein and its kinase activity in Toll-like receptor-mediated immune responses but not in TCR signaling. *J. Exp. Med.* **2007**, *204*, 1013–1024. [CrossRef] [PubMed]

228. Haase, H.; Mocchegiani, E.; Rink, L. Correlation between zinc status and immune function in the elderly. *Biogerontology* **2006**, *7*, 421–428. [CrossRef] [PubMed]

229. Faber, C.; Gabriel, P.; Ibs, K.H.; Rink, L. Zinc in pharmacological doses suppresses allogeneic reaction without affecting the antigenic response. *Bone Marrow Transpl.* **2004**, *33*, 1241–1246. [CrossRef] [PubMed]

230. Rosenkranz, E.; Hilgers, R.D.; Uciechowski, P.; Petersen, A.; Plumakers, B.; Rink, L. Zinc enhances the number of regulatory T cells in allergen-stimulated cells from atopic subjects. *Eur. J. Nutr.* **2017**, *56*, 557–567. [CrossRef] [PubMed]

231. Fraker, P.J.; Telford, W.G. A reappraisal of the role of zinc in life and death decisions of cells. *Proc. Soc. Exp. Biol. Med.* **1997**, *215*, 229–236. [CrossRef] [PubMed]

232. Fraker, P.J.; King, L.E. Reprogramming of the immune system during zinc deficiency. *Annu. Rev. Nutr.* **2004**, *24*, 277–298. [CrossRef] [PubMed]

233. Fraker, P.J.; King, L.E.; Laakko, T.; Vollmer, T.L. The dynamic link between the integrity of the immune system and zinc status. *J. Nutr.* **2000**, *130* (Suppl. 5S), 1399S–1406S. [PubMed]

234. DePasquale-Jardieu, P.; Fraker, P.J. Interference in the development of a secondary immune response in mice by zinc deprivation: Persistence of effects. *J. Nutr.* **1984**, *114*, 1762–1769. [PubMed]

235. Cakman, I.; Rohwer, J.; Schutz, R.M.; Kirchner, H.; Rink, L. Dysregulation between Th1 and Th2 T cell subpopulations in the elderly. *Mech. Ageing Dev.* **1996**, *87*, 197–209. [CrossRef]

236. Maywald, M.; Rink, L. Zinc homeostasis and immunosenescence. *J. Trace Elem. Med. Biol.* **2015**, *29*, 24–30. [CrossRef] [PubMed]

237. Bonomini, M.; Di Paolo, B.; De Risio, F.; Niri, L.; Klinkmann, H.; Ivanovich, P.; Albertazzi, A. Effects of zinc supplementation in chronic haemodialysis patients. *Nephrol. Dial. Transpl.* **1993**, *8*, 1166–1168.

238. Wessels, I. Epigenetics and Metal Deficiencies. *Curr. Nutr. Rep.* **2014**, *3*, 196–203. [CrossRef]

239. Kopf, M.; Le Gros, G.; Bachmann, M.; Lamers, M.C.; Bluethmann, H.; Kohler, G. Disruption of the murine IL-4 gene blocks Th2 cytokine responses. *Nature* **1993**, *362*, 245–248. [CrossRef] [PubMed]

240. Moulder, K.; Steward, M.W. Experimental zinc deficiency: Effects on cellular responses and the affinity of humoral antibody. *Clin. Exp. Immunol.* **1989**, *77*, 269–274. [PubMed]

241. Fraker, P.J.; Gershwin, M.E.; Good, R.A.; Prasad, A. Interrelationships between zinc and immune function. *Fed. Proc.* **1986**, *45*, 1474–1479. [PubMed]

242. Stennicke, H.R.; Salvesen, G.S. Biochemical characteristics of caspases-3, -6, -7, and -8. *J. Biol. Chem.* **1997**, *272*, 25719–25723. [CrossRef] [PubMed]

243. Cook-Mills, J.M.; Fraker, P.J. Functional capacity of the residual lymphocytes from zinc-deficient adult mice. *Br. J. Nutr.* **1993**, *69*, 835–848. [CrossRef] [PubMed]

244. Miyai, T.; Hojyo, S.; Ikawa, T.; Kawamura, M.; Irie, T.; Ogura, H.; Hijikata, A.; Bin, B.H.; Yasuda, T.; Kitamura, H.; et al. Zinc transporter SLC39A10/ZIP10 facilitates antiapoptotic signaling during early B-cell development. *Proc. Natl. Acad. Sci. USA* **2014**, *111*, 11780–11785. [CrossRef] [PubMed]

245. Hojyo, S.; Miyai, T.; Fujishiro, H.; Kawamura, M.; Yasuda, T.; Hijikata, A.; Bin, B.H.; Irie, T.; Tanaka, J.; Atsumi, T.; et al. Zinc transporter SLC39A10/ZIP10 controls humoral immunity by modulating B-cell receptor signal strength. *Proc. Natl. Acad. Sci. USA* **2014**, *111*, 11786–11791. [CrossRef] [PubMed]

International Journal of
*Molecular Sciences*

MDPI

*Review*

# Elemental Ingredients in the Macrophage Cocktail: Role of ZIP8 in Host Response to *Mycobacterium tuberculosis*

**Charlie J. Pyle [1], Abul K. Azad [2], Audrey C. Papp [3], Wolfgang Sadee [3], Daren L. Knoell [4],\* and Larry S. Schlesinger [2],\***

[1]   Department of Molecular Genetics and Microbiology, Duke University, Durham, NC 27710, USA;
      charlie.pyle@gmail.com
[2]   Texas Biomedical Research Institute, San Antonio, TX 78227, USA; AAzad@txbiomed.org
[3]   Center for Pharmacogenomics, Department of Cancer Biology and Genetics, College of Medicine,
      The Ohio State University Wexner Medical Center, Columbus, OH 43085, USA; papp.2@osu.edu (A.C.P.);
      Wolfgang.Sadee@osumc.edu (W.S.)
[4]   College of Pharmacy, The University of Nebraska Medical Center, Omaha, NE 68198-6120, USA
\*    Correspondence: daren.knoell@unmc.edu (D.L.K.); LSchlesinger@txbiomed.org (L.S.S.);
      Tel.: +1-402-559-9016 (D.L.K.); +1-210-258-9419 (L.S.S.)

Received: 6 October 2017; Accepted: 6 November 2017; Published: 9 November 2017

**Abstract:** Tuberculosis (TB) is a global epidemic caused by the infection of human macrophages with the world's most deadly single bacterial pathogen, *Mycobacterium tuberculosis* (*M.tb*). *M.tb* resides in a phagosomal niche within macrophages, where trace element concentrations impact the immune response, bacterial metal metabolism, and bacterial survival. The manipulation of micronutrients is a critical mechanism of host defense against infection. In particular, the human zinc transporter Zrt-/Irt-like protein 8 (ZIP8), one of 14 ZIP family members, is important in the flux of divalent cations, including zinc, into the cytoplasm of macrophages. It also has been observed to exist on the membrane of cellular organelles, where it can serve as an efflux pump that transports zinc into the cytosol. ZIP8 is highly inducible in response to *M.tb* infection of macrophages, and we have observed its localization to the *M.tb* phagosome. The expression, localization, and function of ZIP8 and other divalent cation transporters within macrophages have important implications for TB prevention and dissemination and warrant further study. In particular, given the importance of zinc as an essential nutrient required for humans and *M.tb*, it is not yet clear whether ZIP-guided zinc transport serves as a host protective factor or, rather, is targeted by *M.tb* to enable its phagosomal survival.

**Keywords:** zinc; zinc transporter; tuberculosis; lung; macrophage; innate immunity

---

## 1. Introduction

Tuberculosis (TB) is a major cause of global morbidity and mortality. One in three people are infected with the pathogen responsible for TB, *Mycobacterium tuberculosis* (*M.tb*) [1]. Primary infection is established in the lungs, following inhalation of aerosolized respiratory droplets expelled from a contagious person [2]. Infection results in clinical latency in most healthy human hosts but may reemerge as a potentially fatal pneumonia if immune competence is disrupted [3]. *M.tb* is a facultative intracellular bacterium of macrophages that gains cellular entry through phagocytosis and resides within distinctive phagosomes. The successful intraphagosomal survival of *M.tb* is predicated on its circumvention of the mechanisms evolved to destroy phagocytosed pathogens [4]. Macrophage trace element redistribution is a critical host defense strategy against *M.tb* [5].

*M.tb* residence and growth in mononuclear phagocytes depends on its ability to acquire host-derived nutrients within a suitable range. Macrophages and *M.tb* compete for control of elemental cationic micronutrients, which are essential for mycobacterial growth but also toxic at elevated concentrations [5–7]. The manipulation of trace element flux is the function of many microbial virulence factors and host immune responses. Due to their charge, cationic micronutrients require specialized transport mechanisms to penetrate the phospholipid bilayers of both the plasma membrane and the phagosome. Infection by *M.tb* alters the battery of membrane spanning ion channels present in macrophages [8–11]. Clearly a "tug of war" for the control of trace elements between host and microbe exists, and transmembrane spanning metal ion transporters serve as the primary conduit of micronutrient biodistribution during infection. Over the past decade, several examples have emerged and will first be reviewed before a detailed discussion of zinc and zinc transporters.

## 2. Overview of Metal Metabolism at the Host-Pathogen Interface

The term nutritional immunity was coined to describe the anti-microbial benefits associated with redistribution of iron from the vascular space to intracellular compartments [12]. However, it has come to encompass both systemic and cellular nutrient deprivation of multiple trace elements, including iron, manganese, and zinc, from extracellular or intracellular pathogens. A cadre of innate immune effector cells mount that response with the production of trace element binding proteins, cellular importers, and their associated regulatory factors, following pathogen recognition [13]. Macrophages accumulate iron, copper, and zinc during mycobacterial infection [14].

TB is associated with anemia, which results from macrophage iron retention [15]. Although protective against extracellular pathogen growth, iron loading of macrophages may be beneficial to *M.tb* by providing access to essential nutrition. Impaired access to intracellular labile iron reduces the growth of *M.tb* in macrophages from patients with hereditary hemochromatosis [16], a disease that disrupts iron accumulation due to elevated ferroportin-1 (IREG1) export across the plasma membrane [17]. Macrophages have evolved a complex system of intracellular iron redistribution to counter microbial exploitation of cellular iron internalization. The primary mechanism of that defense is the modulation of intraphagosomal iron content. IREG1 is also localized to the mycobacterial phagosome and may serve to sequester iron away from bacteria [9]. Iron is essential for *M.tb* growth [18] and enters the *M.tb* phagosome from intracellular and extracellular stores [19,20]. It can be captured from transferrin or lactoferrin by mycobacterial siderophores, including carboxymycobactins and exochelin, or by heme import [21].

Previously described in detail, macrophage transporters hyper-concentrate trace elements within the phagosome in order to limit mycobacterial growth [5]. Natural resistance-associated macrophage protein 1 (NRAMP1) is a proton/divalent cation antiporter [22] with broad substrate specificity, including iron, zinc, copper, and manganese [23]. Polymorphisms in NRAMP1 are associated with increased susceptibility to pulmonary tuberculosis [24]. In murine models, NRAMP1 is rapidly localized to the phagosome [25,26] and is associated with resistance to intracellular pathogens [27]. NRAMP1 actively acidifies the bacterial phagosome in mice [28]. It increases the translocation of the proton ATPase to the phagosome, following interferon gamma (IFN-γ) activation, leading to the generation of Fenton-mediated free radical production [29,30]. NRAMP1 is capable of shuttling metals bi-directionally against a proton gradient, whereby the direction of transport is determined by proton and divalent cation concentrations [22]. When phagosomal pH and iron levels are lower than those of the cytoplasm, iron is imported into the phagosome through NRAMP1, resulting in the generation of reactive oxygen species (ROS) through the Fenton and Haber-Weiss reactions [7,8]. Alternatively, in instances in which intraphagosomal concentrations are higher than those of the cytosol, NRAMP1 can export iron and manganese and import protons into the phagosome, increasing acidity and depriving pathogens of those essential nutrients [30,31].

Macrophages also use a strategy involving the phagosomal hyper-concentration of copper [14] through a separate set of copper transporters [32]. IFN-$\gamma$ activation of murine macrophages induces the expression of the plasma membrane copper transporter CTR1, leading to copper uptake. Subsequent translocation of the copper importer ATP7A to mycobacterial phagosomes leads to increased intraphagosomal concentrations of copper and thereby generation of bactericidal Fenton free radicals [10,33]. As a countermeasure, *M.tb* actively up-regulates the mycobacterial copper transport protein B (MctB), which rescues it from copper toxicity [34].

## 3. Regulation of Zinc Balance between Host and Pathogen

The transient hyper-accumulation of zinc in the phagosomes of human macrophages reduces the survival of phagocytosed extracellular pathogens. Zinc accumulation causes the up-regulation of bacterial cation efflux pumps, which are critical for the adaptation of intracellular pathogens, including *Salmonella typhimurium* and *M.tb* [35,36]. The mycobacterial manganese efflux pump Metal cation-transporting p-type ATPase C (CtpC) is required for *M.tb* survival at high zinc conditions [37]. Although no definitive mechanism for zinc-associated toxicity has yet been identified in these models, there are several potential avenues through which elevated zinc concentrations may be toxic to *M.tb*. Those mechanisms include the displacement of iron from sulfhydryl moieties of bacterial enzymes [38] or the disruption of manganese uptake, which reduces bacterial free radical tolerance [39]. The mycobacterial transcriptional repressors Zur and IdeR sense elevations in zinc and iron, respectively, leading to reduced expression of the gene cluster for the type VII secretion system 6 kDa early secretory antigenic target protein family secretion system-3 (ESX-3) [40,41]. Metal-dependent suppression of the critical mycobacterial virulence factors EsxG and EsxH in the ESX3 locus thereby reduces *M.tb* survival [42].

The mechanism through which zinc traverses macrophage membranes in response to *M.tb* infection remains an area of active investigation. Twenty-four dedicated zinc transport proteins are primarily responsible for zinc biodistribution. Each transporter has distinct induction patterns, expression profiles, subcellular localization, and tissue distribution, providing each transporter with a unique role in zinc metabolism. Ten solute carrier 30A (SLC30A) family members, the zinc transport proteins (ZnTs), remove zinc from the cytoplasm across the plasma membrane or into cytosolic organelles. Conversely, fourteen solute carrier 39A (SLC39A) family members, the Zrt-Irt-like-Proteins (ZIPs) move zinc into the cytoplasm from the extracellular environment or out of intracellular vesicles [43,44]. Individually, some ZIP and ZnT proteins have been shown to traffic other divalent cations [45–47]. Further, other nondedicated divalent cation transporters have the capacity to transport zinc, including NRAMP1, IREG1, and divalent metal transporter 1 (DMT1) [23,48,49].

There is constitutive mRNA expression above a threshold of one relative copy number (RCN), for ZIPs 1, 6, 8, and 10, as well as ZnTs 1, 5, 6, 7, and 9, in resting human monocyte-derived macrophages (MDMs). The infection of MDMs with virulent *M.tb* H$_{37}$R$_v$ for 8 h alters the expression pattern of several ZIPs and ZnTs, indicating a global disruption of zinc homeostasis (Figure 1A). In RPMI media supplemented with autologous human serum, infection does not significantly alter the expression of ZIP1 or 6 but decreases ZnTs 5, 6, and 9 and slightly increases ZnT7. ZnT7 expression has previously been shown to increase in response to infection by intracellular fungal pathogens [50]. The resting expression of ZnT1, ZIP8, and ZIP10 is the highest among the 24 zinc transporters. *M.tb* infection significantly increases ZnT1 and ZIP8 but decreases ZIP10. Although infection alters the expression of multiple ZIPs and ZnTs, ZIP8 is unique as it is the sole ZIP zinc importer induced by *M.tb* in MDMs. Human alveolar macrophages (hAMs) reside within a unique microenvironmental niche in the alveolus (gas exchange apparatus) and are the phagocytic cells initially targeted by *M.tb* during airborne infection [51]. As might be expected, the resting expression pattern of zinc transporters in hAMs varies substantially from that of MDMs. There is constitutive mRNA expression above a threshold of 10 reads per million (RPM), determined as described [52], for ZIPs 1, 4, 7, 8, and 9, as well as ZnTs 1, 7, and 9 (Figure 1B). The infection of hAMs with *M.tb* H$_{37}$R$_v$ also alters ZIP and ZnT mRNA expression. As compared to ZIP8 in MDMs, ZIP1 is the most highly expressed ZIP in resting hAMs, and its expression is reduced following 72 h of *M.tb* infection. Additionally, infection results in the reduced expression of ZIPs 4 and 9 within 24 h, as well as the reduced expression of ZIP7 and ZnT9 after 72 h. ZIP8, ZnT7, and ZnT1 are increased during infection. Although there is a comparative delay in the induction of ZIP8 expression in hAMs, it again emerges as the most responsive zinc transporter to *M.tb* infection. The alterations we observed in ZIP and ZnT expression in each of our in vitro models likely impact zinc metabolism within particular macrophage subsets in distinct ways, with each transporter contributing to specific aspects of cumulative cellular zinc flux. The expression profile of ZIP8 is unique as it has high constitutive expression and is the most responsive zinc transporter to *M.tb* infection in MDMs and hAMs. ZIP8 is the only zinc importer increased by *M.tb* in MDMs and emerges as the dominant ZIP expressed in hAMs during infection, although ZIPs 12, 13, and 14 do increase from low resting levels in that model. Overall, the stimulation of ZIP8 expression by *M.tb* is a prominent feature of the macrophage response to infection, which should be viewed in the context of a generalized shift in cellular zinc metabolism.

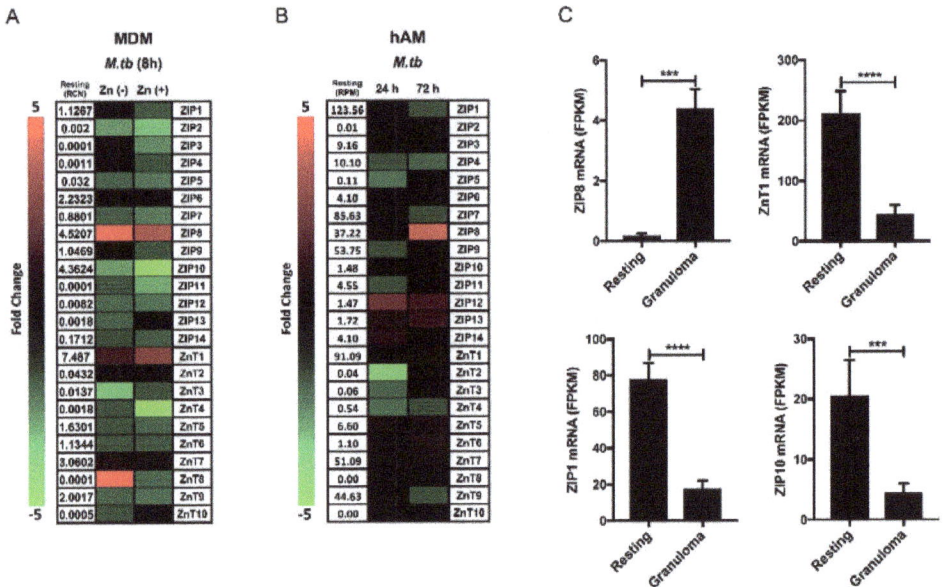

**Figure 1.** Macrophage zinc transporter mRNA expression during *M.tb* infection. Zrt-Irt-like-Protein (ZIP) and zinc transport protein (ZnT) mRNA expression is altered by infection with *Mycobacterium tuberculosis* (*M.tb*) H$_{37}$R$_{v}$ using a multiplicity of infection (MOI) of 5:1 for (**A**) 8 h in monocyte-derived macrophages (MDMs) in the absence or presence of ZnSO$_4$ 18 µM, as determined by qRT-PCR relative to GAPDH (*n* = 3) or (**B**) for 24 or 72 h in human alveolar macrophages (hAMs) infected by *M.tb* H$_{37}$R$_{v}$, as determined by AmpliSeq Transcriptome analysis (*n* = 6). (**C**) The mRNA expression of ZIP8 is increased and ZnT1, ZIP1, and ZIP10 are decreased in *M.marinum*-infected zebrafish granulomas compared to resting macrophages, as determined by RNA-Seq. (A and B are unpublished data; C was generated using supplementary data published in Cronan et al. [53]) (mean ± SEM; *** *p* < 0.001; **** *p* < 0.0001; Prism-7: one-tailed Students *t*-test). MDM [54] and hAM [55] isolation, culture, and infection with *M.tb*, as well as the assay of human zinc transporters by qRT-PCR in MDMs [56] and AmpliSeq Transcriptome analysis in hAMs [52], were performed as previously described.

In an important study, Botella et al. [36] revealed that *M.tb* infection of human macrophages induces mRNA expression of metallothionein (MT) intracellular zinc binding proteins and ZnT1 by activating the metal responsive transcription factor MTF-1. Based upon these observations, a model of ZnT1 phagosomal localization as a mechanism for the hyper-accumulation of vesicular zinc was proposed [36]. The contribution of phagosomal NRAMP1 relative to increases in zinc should also be considered [23,25]. The activation of MTF-1 and subsequent MT transcription indicates that zinc levels also increase within the cytosolic compartment during *M.tb* infection. In cell culture models with limited extracellular zinc, MTF-1 transcriptional activation by zinc is likely exclusively due to intracellular redistribution [36]. Knowing that cellular zinc trafficking within the human physiologic range, both high and low, is an essential aspect of macrophage metal metabolism, future studies of zinc flux across the plasma membrane during *M.tb* infection should include zinc levels that simulate the human condition. Physiologically relevant zinc supplementation [56] of *M.tb*-infected MDMs in vitro alters their mRNA expression profile. The addition of zinc during infection results in the reduced expression of most zinc transporters (Figure 1A). It reduces the extent of ZIP8 induction but increases ZnT1 expression and leads to further repression of ZIP1 and ZIP10 in response to infection. Cumulatively these changes indicate a shift toward cytosolic zinc efflux. The increased transcriptional activation of ZnT1 [57,58], coupled with the suppression of ZIP10 expression [59] during supplementation, indicates that the extent of MTF-1 activation in infected macrophages is dependent on extracellular zinc import.

In a recent study using the zebrafish model of TB pathogenesis, Cronan et al. evaluated the impact of mycobacterial reprogramming of granuloma macrophages and provided a granuloma-specific macrophage transcriptomic signature [53]. Parsing of that data revealed that ZIP8 is highly induced and that ZIP1 and ZIP10 expression is reduced in mycobacterial granulomas (Figure 1C). Surprisingly, ZnT1 expression is significantly reduced in granuloma macrophages, which may reflect altered zinc metabolism during mesenchymal-epithelial transition or species-specific differences. However, ZIP8 induction during mycobacterial infection appears to be a critical, evolutionarily conserved response that is maintained during granuloma formation and among the lineages of human macrophages that are central to TB pathogenesis (Figure 1A,B).

## 4. ZIP8 in Macrophage Infection by *M.tb*

The manipulation of intracellular zinc through altered macrophage zinc transporter expression impacts the growth of intracellular yeast [60,61], fungal [50,62], and bacterial [35] pathogens. Zinc redistribution into the bacterial phagosome is Tol-like receptor (TLR)-dependent [35]. TLR4 activation of human macrophages induces the production of ZIP8 through Nuclear factor kappa-light-chain-enhancer of activated B cells (NF-κB) [63], and *M.tb* infection transiently induces NF-κB [55,64]. ZIP8 was discovered due to its production in human monocytes in response to *Mycobacterium bovis* Bacillus Calmette Guérin (BCG) cell wall cytoskeletal extract and BCG infection [65]. It is expressed in human macrophages in response to infection with nonpathogenic and virulent mycobacterial species, including *M.tb*, as well as to gram-negative and gram-positive bacteria [11]. In particular, MDM infection with virulent *M.tb* $H_{37}R_v$ results in the robust induction of ZIP8 mRNA for at least 24 h post infection (Figure 2A). Consistent with our previously published results involving other cell types [66], ZIP8 induction results in the production of a membrane bound, glycosylated 140 kDa protein. ZIP8 protein in MDMs is elevated within 24 h, following infection with *M.tb* $H_{37}R_v$ or BCG, and remains elevated for at least 72 h (Figure 2B).

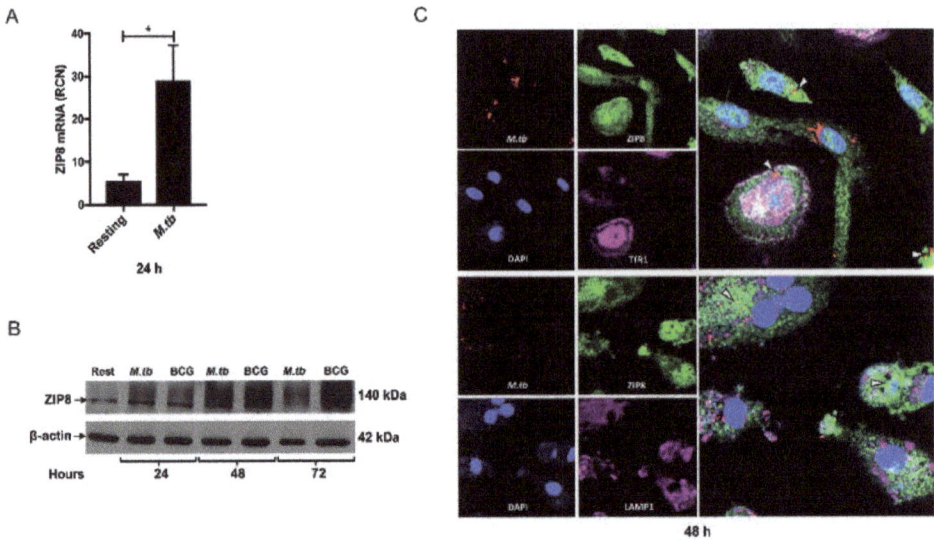

**Figure 2.** ZIP8 protein is induced and localizes with *M.tb* in human macrophages. MDM production of ZIP8: (**A**) mRNA is significantly induced for 24 h, following infection with *M.tb* $H_{37}R_v$ (MOI 5:1), as determined by qRT-PCR relative to GAPDH ($n = 3$) and (**B**) ZIP8 protein is robustly increased by infection with *M.tb* $H_{37}R_v$ or *M.bovis* BCG (MOI 5:1) between 24 and 72 h, as determined by Western blot relative to β-actin ($n = 3$). (**C**) The infection of MDMs with mCherry expressing *M.tb* $H_{37}R_v$ (MOI 5:1) for 48 h leads to the extensive co-localization of ZIP8 with *M.tb* (yellow; indicated by arrow heads) and TfR1 (abundant white; in merged upper panel) but very limited co-localization with LAMP-1 (negligible white; in merged lower panel) (A, B, and C are unpublished data) (mean ± SEM; * $p < 0.05$; Prism-7: one-tailed Students *t*-test). MDM isolation, culture, and infection [54]; qRT-PCR and Western blot of ZIP8 in MDMs [56]; and confocal fluorescence microscopy using an Olympus FV1000-Spectral System at 60× magnification in infected MDMs [67] were performed as previously described. Rabbit polyclonal antiserum anti-peptide to amino acid residues 225 to 243 of human ZIP8 was purchased from Covance (Princeton, NJ, USA). Mouse anti-human monoclonal β-actin (#69101) antibody was purchased from MP Biomedicals (Santa Ana, CA, USA). Mouse anti-human monoclonal CD71 (#334102) antibody was purchased from Biolegend (San Diego, CA, USA). Mouse anti-human monoclonal LAMP-1 (#ab25630) antibody was purchased from Abcam (Cambridge, UK).

Cytosolic zinc import in activated macrophages is ZIP8-dependent [56]. ZIP8 is present on the plasma membrane and on intracellular vesicles in primary human macrophages, epithelial cells, T-cells, and cell lines [63,65,66,68,69], indicating a role for ZIP8 in cytosolic cation increase through cellular influx and vesicular efflux. Viral induced over-expression of ZIP8 in murine chondrocytes increases MTF-1 nuclear localization and transcriptional activity [70]. Slc39a8 hypomorphic mouse fetal fibroblasts have reduced MT expression in response to tumor necrosis factor alpha (TNFα) [63], indicating that ZIP8-dependent zinc increases the transcription of MTF-1 target genes. Furthermore, ZIP8 regulates ZnT1 expression in primary human macrophages [56]. Based upon these observations, it is plausible that macrophage zinc loading during *M.tb* infection [14] is a function of zinc import through ZIP8, which then leads to observed elevations in macrophage MT and ZnT1 mRNA expression through the activation of MTF-1.

ZIP8 has multiple glycosylation sites and potential protein-binding partners that may influence membrane orientation and localization [66,71,72]. Further, zinc deprivation of the bacterial phagosome in macrophages and dendritic cells through the up-regulation and trafficking of ZIP8 to the phagosome-lysosome pathway has been proposed [73]. Knowing this, we determined the cellular localization of ZIP8 [63] in relation to early phagosome marker transferrin receptor-1 (TfR1), late endosome/lysosome marker LAMP-1, and *M.tb* in macrophages by fluorescence confocal microscopy, as previously described [67]. We observed that ZIP8 becomes abundant within the phagosome and co-localizes with *M.tb* (Figure 2C). Further, the association of ZIP8 and the pathogen is durable and persists within the phagosome over an extended time frame. Co-localization studies with TfR1 and LAMP-1 indicate that ZIP8 resides primarily within the phagosome, akin to TfR1, and not the phagolysosome in macrophages (Figure 2C). In consideration of the zinc-poisoning paradigm that was previously highlighted, this result indicates a potential role for ZIP8 in both macrophage zinc loading and eventual phagosomal detoxification, following the initial super-concentration of vacuolar zinc that has been observed within 24 h of infection [14,36]. The coordination of zinc efflux through the paired induction of *M.tb* CtpC and macrophage ZIP8 expression may generate a complimentary safeguard in favor of *M.tb* against zinc poisoning. In this context, the induction of ZIP8 expression during *M.tb* infection may serve as a host susceptibility factor.

Alternatively, it is important to consider that all microbes require zinc for survival [74] and that zinc at appropriate concentrations enhances mycobacterial growth [75]. Zinc has the capacity to interact with many proteins in both eukaryotic and prokaryotic cells. The antioxidant properties of zinc afford the protection of vulnerable sulfhydryl groups from damage by ROS [76], which are akin to those generated by high phagosomal concentrations of iron [8] or copper [34]. Thus, high phagosomal zinc concentrations may actually benefit *M.tb* in some regards by enhancing access and limiting the damage incurred from free radical production by other trace elements. Therefore the ZIP8-dependent sequestration of zinc away from *M.tb*, as with intracellular fungal pathogens [50,62], may actually have some host protective effects.

Cation transport by ZIP8 is pH dependent and electroneutral, indicating that it facilitates the co-transport of other ionic species as well [47,69,71]. ZIP8 participates in the cytosolic influx of manganese, cadmium, iron, zinc, and selenite [71,77]. Iron and zinc inhibit the ZIP8-mediated uptake of each other [69]. ZIP proteins contain binuclear metal centers, where metal binding at one site affects the transporter metal selectivity at the second site [78]. Zinc uptake by ZIP8 is competitively inhibited by both iron and cadmium and non-competitively by cobalt, nickel, and copper but is not inhibited by magnesium or manganese [79]. Given that ZIP8 has a relatively high substrate promiscuity, along with a directional transport and localization profile, it is reasonable to expect that, in combination with other metal transporters, ZIP8 contributes to the flux of multiple divalent cations toward and or away from *M.tb* across a number of macrophage membranes, including the mycobacterial phagosome (Figure 3).

ZIP8 and NRAMP1 share multiple substrates, raising the possibility that there may also exist dynamic interplay between the two transporters on the *M.tb* phagosome for the regulation of iron and zinc. Further, iron and zinc within the phagosome may antagonize the transport of one another in a similar way to what occurs in the intestine [80]. ZIP8-dependent iron transport across the phagosomal membrane has the potential to contribute to the previously proposed models of phagosomal iron deprivation [15], involving other transporters such as NRAMP1 [30] or IREG1 [9]. That efflux could counteract the host protective iron-dependent generation of intraphagosomal ROS [8]. ZIP8 activity is pH dependent and potentially drives bicarbonate flux [46]; therefore, it may also impact intraphagosomal pH, which is critical to the maintenance of the intracellular mycobacterial niche [81]. Ultimately the impact of ZIP8 on mycobacterial growth and survival within the *M.tb* phagosome depends on a complex array of variables, including but not limited to host nutritional status and genetic variation [82–84], and the co-expression and localization of other trace element transporters, as well as mycobacterial responses to metal flux.

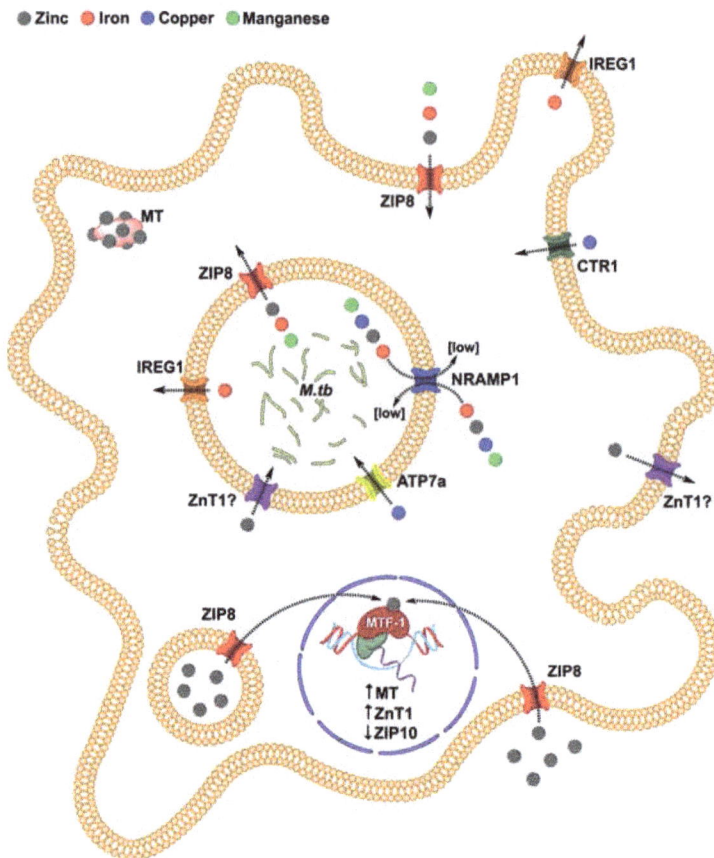

**Figure 3.** The hypothetical impact of ZIP8 on MTF-1 expression, as well as the localization and direction of trace element transport in macrophages during early infection with *M.tb*. Arrows indicate the direction of trace element transport.

## 5. Implications of ZIP8 Induction on TB

ZIP8 expression is highly induced in human and murine tissues and circulating cells during systemic and local inflammation [70,85–87]. The increased intracellular sequestration of zinc, similar to iron, is a mechanism of nutritional immunity in which vascular trace element deprivation limits the extracellular growth of invading pathogens [12]. Zinc, as a vital commodity in times of need, is also therefore redirected to biosynthetic processes that bolster host immune defense. Zinc mobilization from the vasculature into the vital organs predominantly occurs due to ZIP14 induction in the liver [88]. Our group and others have shown that several ZIPs are involved in zinc redistribution into vital organs other than the liver, which is in part due to sequestration of zinc folowing ZIP8 [63], as well as ZIP14 [89] induction in tissue macrophages. ZIP8 is elevated in circulating peripheral blood monocytes during the acute phase response, which is associated with reduced serum zinc concentrations [86]. Similarly, circulating levels of zinc in plasma or serum are reduced in patients with active TB [90–92]. In time, zinc levels recover with antibiotic therapy in the absence of zinc supplementation [90,93], indicating that, similar to iron, the intracellular redistribution of vascular zinc occurs in TB. Based on this and our findings, we propose that the ZIP8-dependent iron and zinc loading of macrophages is likely a driving force in systemic nutritional immunity during active TB.

A balance between pro-inflammatory and immune suppressive cytokines is critical for the control of *M.tb* [94,95]. ZIP8 activity modulates the inflammatory response as an intracellular second messenger, thereby altering the production of pro-inflammatory cytokines and IL-10 [56,63,68,70]. IL-10 is an important immune modulator during *M.tb* infection. It is associated with TB progression [96] and reactivation [97] in murine models and reduced macrophage host defense capabilities [98–100] or, alternatively, may enhance the control of the infection [101]. ZIP8-dependent zinc import skews cytokine signaling towards a pro-inflammatory profile in activated macrophages, particularly through the inhibition of IL-10 production [56]. Conversely, in activated monocytes, ZIP8 induction leads to negative feedback inhibition of NF-κB and the reduction of pro-inflammatory cytokines [63] that, at balanced levels, are critical for the control of *M.tb* [4]. Further investigations are needed to clarify the impact of ZIP8 on cytokine responses in TB.

Observations that *M.tb* infection increases macrophage intracellular zinc accumulation, leading to MTF-1 nuclear localization and transcriptional activity 24 h after infection [36], are likely due, in part, to ZIP8-dependent zinc influx. In a murine *Slc39a8* floxed; *Col2a1-Cre* chondrocyte-specific ZIP8 knockout model, ZIP8-dependent zinc increased intracellular zinc concentrations and the MTF-1 dependent transcription of matrix metalloproteinases (MMPs) during inflammation [70]. Subsequent MMP activity resulted in the degradation of the surrounding tissue. *M.tb* infection induces MMP production by macrophages and surrounding cells, which results in pulmonary tissue destruction [102,103]. In animal models, during the early phase of mycobacterial granuloma formation, MMP release enhances macrophage recruitment to the site of infection [104], which is associated with increased macrophage infection and dissemination [105]. Furthermore MMP catalytic activity and function requires zinc [106]. The antibiotic doxycycline is the only known Food and Drug Administration-approved MMP inhibitor and has been suggested as an adjunct antibiotic therapy because it reduces *M.tb* growth in vitro and in vivo [107]. In light of these findings, it is intriguing to speculate that ZIP8 induction and subsequent zinc influx increases susceptibility to *M.tb* by driving the MTF-1 transcription of MMPs, resulting in tissue destruction, increased macrophage recruitment, and bacterial growth.

ZIP8 is an interesting target for host-directed pharmacotherapy for the treatment and prevention of TB. To our knowledge, there are no specifically targeted pharmacological antagonists or agonists of ZIP transporters currently available. If the induction of ZIP8 is confirmed to be a pathogenic process during mycobacterial infection, the development of small molecule inhibitors or biopharmaceuticals could prove valuable. However, within the framework of global zinc homeostasis, the selective targeting of ZIP8 may have myriad effects. Alternatively, a strategy of zinc supplementation in an effort to nullify the potential antagonism of zinc poisoning by phagosomal ZIP8 could have some utility as a preventative therapy for high risk populations, although, to date, zinc supplementation has not been shown to have clinical benefit during active TB [108].

## 6. Conclusions

Tuberculosis is the world's most deadly infectious disease caused by a single pathogen. Globally it is estimated that over 10 million people are infected with *M.tb* every year. Unfortunately, TB treatment is limited by lack of universal availability of effective medications, drug toxicity, and lack of compliance, the last resulting in the increasing incidence of multiple drug resistant strains. This is further complicated because individuals with chronically depressed immunity have a much higher risk of developing TB [1]. Zinc deficiency is a major cause of immune dysfunction and infection [109]. *M.tb* infection influences human zinc metabolism. Understanding the complexities of macrophage responses to *M.tb* is particularly important, given its substantial impact on global human health. An alteration in the trafficking of divalent cations is an established host defense strategy against extracellular and intracellular pathogens, including *M.tb*. Multiple metal binding proteins and transporters contribute to those responses. ZIP8 is induced in response to *M.tb* infection and localizes to the plasma membrane and intracellular vesicles, including the *M.tb* phagosome. In concert with other established cation transporters, ZIP8 may be positioned in the phagosome to render a fundamentally important impact on macrophage host defense and TB pathogenesis. Further, ZIP8 has been shown to have potent immunomodulatory functions that are influenced by systemic and cellular zinc status. The impact of zinc metabolism on macrophage host defense functions during *M.tb* infection remains an underexplored area of research with promising potential for the generation of translationally applicable future findings.

**Acknowledgments:** Larry S. Schlesinger and Daren L. Knoell would like to thank Claire Dodd and Sheng Ying Bao for their technical assistance. Charlie J. Pyle would like to thank the National Heart Lung and Blood Institute (T32 Fellowship HL007946) and the American Foundation for Pharmaceutical Education (Pre-Doctoral Fellowship in Pharmaceutical Sciences) for their support. Partial support was provided by the Bill and Melinda Gates Foundation and the National Institutes of Health.

**Author Contributions:** Charlie J. Pyle, Abul K. Azad, Audrey C. Papp, Wolfgang Sadee, Daren L. Knoell and Larry S. Schlesinger conceived and designed experiments. Charlie J. Pyle, Abul K. Azad and Audrey C. Papp performed experiments. Charlie J. Pyle performed the statistical analysis. Charlie J. Pyle, Daren L. Knoell, and Larry S. Schlesinger wrote the manuscript.

**Conflicts of Interest:** The authors declare no conflict of interest.

## Abbreviations

| | |
|---|---|
| TB | Tuberculosis |
| *M.tb* | *Mycobacterium tuberculosis* |
| ZIP | Zrt-/Irt-like protein |
| IREG1 | Ferroportin-1 |
| NRAMP1 | Natural resistance-associated macrophage protein 1 |
| IFN-$\gamma$ | Interferon gamma |
| ROS | Reactive oxygen species |
| CTR1 | High affinity copper uptake protein-1 |
| ATP7A | ATPase copper transporting alpha |
| MctB | Mycobacterial copper transport protein B |
| CtpC | Metal cation-transporting p-type ATPase C |
| Zur | Zinc uptake repressor |
| IdeR | Iron-dependent repressor |
| ESX-3 | 6 kDa early secretory antigenic target protein family secretion system-3 |
| EsxG | ESAT-6-like protein G |
| EsxH | ESAT-6-like protein H |
| ZnT | Zinc transport protein |
| DMT1 | Divalent metal transporter 1 |
| MDM | Monocyte-derived macrophage |
| hAM | Human alveolar macrophage |
| MT | Metallothionein |
| MTF-1 | Metal responsive transcription factor-1 |
| TLR | Tol-like receptor |
| NF-$\kappa$B | Nuclear factor kappa-light-chain-enhancer of activated B cells |
| BCG | *Mycobacterium bovis* Bacillus Calmette Guérin |
| kDa | kiloDalton |
| qRT-PCR | Quantitative Real-Time polymerase chain reaction |
| MOI | Multiplicity of infection |
| TfR1 | Transferrin receptor-1 |
| LAMP-1 | Lysosome-associated membrane protein-1 |
| TNF$\alpha$ | Tumor Necrosis Factor alpha |
| IL-10 | Interleukin 10 |
| MMP | Metalloproteinase |

## References

1.  WHO. *Global Tuberculosis Report*; World Health Organization: Geneva, Switzerland, 2016.
2.  Torrelles, J.B.; Schlesinger, L.S. Integrating Lung Physiology, Immunology, and Tuberculosis. *Trends Microbiol.* **2017**, *25*, 688–697. [CrossRef] [PubMed]
3.  Dodd, C.E.; Schlesinger, L.S. New concepts in understanding latent tuberculosis. *Curr. Opin. Infect. Dis.* **2017**, *30*, 316–321. [CrossRef] [PubMed]
4.  Rajaram, M.V.; Ni, B.; Dodd, C.E.; Schlesinger, L.S. Macrophage immunoregulatory pathways in tuberculosis. *Semin. Immunol.* **2014**, *26*, 471–485. [CrossRef] [PubMed]
5.  Neyrolles, O.; Wolschendorf, F.; Mitra, A.; Niederweis, M. Mycobacteria, metals, and the macrophage. *Immunol. Rev.* **2015**, *264*, 249–263. [CrossRef] [PubMed]
6.  Mertz, W. The essential trace elements. *Science* **1981**, *213*, 1332–1338. [CrossRef] [PubMed]
7.  Kuhn, D.E.; Baker, B.D.; Lafuse, W.P.; Zwilling, B.S. Differential iron transport into phagosomes isolated from the RAW264.7 macrophage cell lines transfected with Nramp1Gly169 or Nramp1Asp169. *J. Leukoc. Biol.* **1999**, *66*, 113–119. [PubMed]
8.  Zwilling, B.S.; Kuhn, D.E.; Wikoff, L.; Brown, D.; Lafuse, W. Role of iron in Nramp1-mediated inhibition of mycobacterial growth. *Infect. Immun.* **1999**, *67*, 1386–1392. [PubMed]
9.  Van Zandt, K.E.; Sow, F.B.; Florence, W.C.; Zwilling, B.S.; Satoskar, A.R.; Schlesinger, L.S.; Lafuse, W.P. The iron export protein ferroportin 1 is differentially expressed in mouse macrophage populations and is present in the mycobacterial-containing phagosome. *J. Leukoc. Biol.* **2008**, *84*, 689–700. [CrossRef] [PubMed]
10. Rowland, J.L.; Niederweis, M. Resistance mechanisms of *Mycobacterium tuberculosis* against phagosomal copper overload. *Tuberculosis (Edinb.)* **2012**, *92*, 202–210. [CrossRef] [PubMed]
11. Blischak, J.D.; Tailleux, L.; Mitrano, A.; Barreiro, L.B.; Gilad, Y. Mycobacterial infection induces a specific human innate immune response. *Sci. Rep.* **2015**, *5*, 16882. [CrossRef] [PubMed]
12. Weinberg, E.D. Nutritional immunity. Host's attempt to withold iron from microbial invaders. *JAMA* **1975**, *231*, 39–41. [CrossRef] [PubMed]
13. Hood, M.I.; Skaar, E.P. Nutritional immunity: Transition metals at the pathogen-host interface. *Nat. Rev. Microbiol.* **2012**, *10*, 525–537. [CrossRef] [PubMed]
14. Wagner, D.; Maser, J.; Lai, B.; Cai, Z.; Barry, C.E., 3rd; Honer Zu Bentrup, K.; Russell, D.G.; Bermudez, L.E. Elemental analysis of *Mycobacterium avium-*, *Mycobacterium tuberculosis-*, and *Mycobacterium smegmatis-*containing phagosomes indicates pathogen-induced microenvironments within the host cell's endosomal system. *J. Immunol.* **2005**, *174*, 1491–1500. [PubMed]
15. Nairz, M.; Schroll, A.; Sonnweber, T.; Weiss, G. The struggle for iron—A metal at the host-pathogen interface. *Cell. Microbiol.* **2010**, *12*, 1691–1702. [CrossRef] [PubMed]
16. Olakanmi, O.; Schlesinger, L.S.; Britigan, B.E. Hereditary hemochromatosis results in decreased iron acquisition and growth by *Mycobacterium tuberculosis* within human macrophages. *J. Leukoc. Biol.* **2007**, *81*, 195–204. [CrossRef] [PubMed]
17. Pietrangelo, A. Hereditary hemochromatosis. *Biochim. Biophys. Acta* **2006**, *1763*, 700–710. [CrossRef] [PubMed]
18. Kurthkoti, K.; Amin, H.; Marakalala, M.J.; Ghanny, S.; Subbian, S.; Sakatos, A.; Livny, J.; Fortune, S.M.; Berney, M.; Rodriguez, G.M. The Capacity of *Mycobacterium tuberculosis* To Survive Iron Starvation Might Enable It To Persist in Iron-Deprived Microenvironments of Human Granulomas. *mBio* **2017**, *8*. [CrossRef] [PubMed]
19. Olakanmi, O.; Schlesinger, L.S.; Ahmed, A.; Britigan, B.E. Intraphagosomal *Mycobacterium tuberculosis* acquires iron from both extracellular transferrin and intracellular iron pools. Impact of interferon-gamma and hemochromatosis. *J. Biol. Chem.* **2002**, *277*, 49727–49734. [CrossRef] [PubMed]
20. Olakanmi, O.; Schlesinger, L.S.; Ahmed, A.; Britigan, B.E. The nature of extracellular iron influences iron acquisition by *Mycobacterium tuberculosis* residing within human macrophages. *Infect. Immun.* **2004**, *72*, 2022–2028. [CrossRef] [PubMed]
21. Fang, Z.; Sampson, S.L.; Warren, R.M.; Gey van Pittius, N.C.; Newton-Foot, M. Iron acquisition strategies in mycobacteria. *Tuberculosis (Edinb.)* **2015**, *95*, 123–130. [CrossRef] [PubMed]

22. Blackwell, J.M.; Goswami, T.; Evans, C.A.; Sibthorpe, D.; Papo, N.; White, J.K.; Searle, S.; Miller, E.N.; Peacock, C.S.; Mohammed, H.; et al. SLC11A1 (formerly NRAMP1) and disease resistance. *Cell. Microbiol.* **2001**, *3*, 773–784. [CrossRef] [PubMed]

23. Gunshin, H.; Mackenzie, B.; Berger, U.V.; Gunshin, Y.; Romero, M.F.; Boron, W.F.; Nussberger, S.; Gollan, J.L.; Hediger, M.A. Cloning and characterization of a mammalian proton-coupled metal-ion transporter. *Nature* **1997**, *388*, 482–488. [CrossRef] [PubMed]

24. Yuan, L.; Ke, Z.; Guo, Y.; Xi, X.; Luo, Z. NRAMP1 D543N and INT4 polymorphisms in susceptibility to pulmonary tuberculosis: A meta-analysis. *Infect. Genet. Evol.* **2017**, *54*, 91–97. [CrossRef] [PubMed]

25. Gruenheid, S.; Pinner, E.; Desjardins, M.; Gros, P. Natural resistance to infection with intracellular pathogens: The Nramp1 protein is recruited to the membrane of the phagosome. *J. Exp. Med.* **1997**, *185*, 717–730. [CrossRef] [PubMed]

26. Searle, S.; Bright, N.A.; Roach, T.I.; Atkinson, P.G.; Barton, C.H.; Meloen, R.H.; Blackwell, J.M. Localisation of Nramp1 in macrophages: Modulation with activation and infection. *J. Cell Sci.* **1998**, *111*, 2855–2866. [PubMed]

27. Skamene, E.; Schurr, E.; Gros, P. Infection genomics: Nramp1 as a major determinant of natural resistance to intracellular infections. *Annu. Rev. Med.* **1998**, *49*, 275–287. [CrossRef] [PubMed]

28. Hackam, D.J.; Rotstein, O.D.; Zhang, W.; Gruenheid, S.; Gros, P.; Grinstein, S. Host resistance to intracellular infection: Mutation of natural resistance-associated macrophage protein 1 (Nramp1) impairs phagosomal acidification. *J. Exp. Med.* **1998**, *188*, 351–364. [CrossRef] [PubMed]

29. Fritsche, G.; Nairz, M.; Werner, E.R.; Barton, H.C.; Weiss, G. Nramp1-functionality increases iNOS expression via repression of IL-10 formation. *Eur. J. Immunol.* **2008**, *38*, 3060–3067. [CrossRef] [PubMed]

30. Nairz, M.; Fritsche, G.; Crouch, M.L.; Barton, H.C.; Fang, F.C.; Weiss, G. Slc11a1 limits intracellular growth of Salmonella enterica sv. Typhimurium by promoting macrophage immune effector functions and impairing bacterial iron acquisition. *Cell. Microbiol.* **2009**, *11*, 1365–1381. [CrossRef] [PubMed]

31. Jabado, N.; Jankowski, A.; Dougaparsad, S.; Picard, V.; Grinstein, S.; Gros, P. Natural resistance to intracellular infections: Natural resistance-associated macrophage protein 1 (Nramp1) functions as a pH-dependent manganese transporter at the phagosomal membrane. *J. Exp. Med.* **2000**, *192*, 1237–1248. [CrossRef] [PubMed]

32. Samanovic, M.I.; Ding, C.; Thiele, D.J.; Darwin, K.H. Copper in microbial pathogenesis: Meddling with the metal. *Cell Host Microbe* **2012**, *11*, 106–115. [CrossRef] [PubMed]

33. White, C.; Lee, J.; Kambe, T.; Fritsche, K.; Petris, M.J. A role for the ATP7A copper-transporting ATPase in macrophage bactericidal activity. *J. Biol. Chem.* **2009**, *284*, 33949–33956. [CrossRef] [PubMed]

34. Wolschendorf, F.; Ackart, D.; Shrestha, T.B.; Hascall-Dove, L.; Nolan, S.; Lamichhane, G.; Wang, Y.; Bossmann, S.H.; Basaraba, R.J.; Niederweis, M. Copper resistance is essential for virulence of *Mycobacterium tuberculosis*. *Proc. Natl. Acad. Sci. USA* **2011**, *108*, 1621–1626. [CrossRef] [PubMed]

35. Kapetanovic, R.; Bokil, N.J.; Achard, M.E.; Ong, C.L.; Peters, K.M.; Stocks, C.J.; Phan, M.D.; Monteleone, M.; Schroder, K.; Irvine, K.M.; et al. Salmonella employs multiple mechanisms to subvert the TLR-inducible zinc-mediated antimicrobial response of human macrophages. *FASEB J.* **2016**, *30*, 1901–1912. [CrossRef] [PubMed]

36. Botella, H.; Peyron, P.; Levillain, F.; Poincloux, R.; Poquet, Y.; Brandli, I.; Wang, C.; Tailleux, L.; Tilleul, S.; Charriere, G.M.; et al. Mycobacterial p(1)-type ATPases mediate resistance to zinc poisoning in human macrophages. *Cell Host Microbe* **2011**, *10*, 248–259. [CrossRef] [PubMed]

37. Padilla-Benavides, T.; Long, J.E.; Raimunda, D.; Sassetti, C.M.; Arguello, J.M. A novel P(1B)-type $Mn^{2+}$-transporting ATPase is required for secreted protein metallation in mycobacteria. *J. Biol. Chem.* **2013**, *288*, 11334–11347. [CrossRef] [PubMed]

38. Xu, F.F.; Imlay, J.A. Silver(I), mercury(II), cadmium(II), and zinc(II) target exposed enzymic iron-sulfur clusters when they toxify Escherichia coli. *Appl. Environ. Microbiol.* **2012**, *78*, 3614–3621. [CrossRef] [PubMed]

39. Eijkelkamp, B.A.; Morey, J.R.; Ween, M.P.; Ong, C.L.; McEwan, A.G.; Paton, J.C.; McDevitt, C.A. Extracellular zinc competitively inhibits manganese uptake and compromises oxidative stress management in Streptococcus pneumoniae. *PLoS ONE* **2014**, *9*, e89427. [CrossRef] [PubMed]

40. Maciag, A.; Dainese, E.; Rodriguez, G.M.; Milano, A.; Provvedi, R.; Pasca, M.R.; Smith, I.; Palu, G.; Riccardi, G.; Manganelli, R. Global analysis of the *Mycobacterium tuberculosis* Zur (FurB) regulon. *J. Bacteriol.* **2007**, *189*, 730–740. [CrossRef] [PubMed]

41. Rodriguez, G.M.; Voskuil, M.I.; Gold, B.; Schoolnik, G.K.; Smith, I. ideR, An essential gene in *mycobacterium tuberculosis*: Role of IdeR in iron-dependent gene expression, iron metabolism, and oxidative stress response. *Infect. Immun.* **2002**, *70*, 3371–3381. [CrossRef] [PubMed]

42. Tinaztepe, E.; Wei, J.R.; Raynowska, J.; Portal-Celhay, C.; Thompson, V.; Philips, J.A. Role of Metal-Dependent Regulation of ESX-3 Secretion in Intracellular Survival of *Mycobacterium tuberculosis*. *Infect. Immun.* **2016**, *84*, 2255–2263. [CrossRef] [PubMed]

43. Lichten, L.A.; Cousins, R.J. Mammalian zinc transporters: Nutritional and physiologic regulation. *Annu. Rev. Nutr.* **2009**, *29*, 153–176. [CrossRef] [PubMed]

44. Kimura, T.; Kambe, T. The Functions of Metallothionein and ZIP and ZnT Transporters: An Overview and Perspective. *Int. J. Mol. Sci.* **2016**, *17*, 336. [CrossRef] [PubMed]

45. Leyva-Illades, D.; Chen, P.; Zogzas, C.E.; Hutchens, S.; Mercado, J.M.; Swaim, C.D.; Morrisett, R.A.; Bowman, A.B.; Aschner, M.; Mukhopadhyay, S. SLC30A10 is a cell surface-localized manganese efflux transporter, and parkinsonism-causing mutations block its intracellular trafficking and efflux activity. *J. Neurosci.* **2014**, *34*, 14079–14095. [CrossRef] [PubMed]

46. Liu, Z.; Li, H.; Soleimani, M.; Girijashanker, K.; Reed, J.M.; He, L.; Dalton, T.P.; Nebert, D.W. Cd2+ versus Zn2+ uptake by the ZIP8 HCO3–dependent symporter: Kinetics, electrogenicity and trafficking. *Biochem. Biophys. Res. Commun.* **2008**, *365*, 814–820. [CrossRef] [PubMed]

47. Girijashanker, K.; He, L.; Soleimani, M.; Reed, J.M.; Li, H.; Liu, Z.; Wang, B.; Dalton, T.P.; Nebert, D.W. Slc39a14 gene encodes ZIP14, a metal/bicarbonate symporter: Similarities to the ZIP8 transporter. *Mol. Pharmacol.* **2008**, *73*, 1413–1423. [CrossRef] [PubMed]

48. Espinoza, A.; Le Blanc, S.; Olivares, M.; Pizarro, F.; Ruz, M.; Arredondo, M. Iron, copper, and zinc transport: Inhibition of divalent metal transporter 1 (DMT1) and human copper transporter 1 (hCTR1) by shRNA. *Biol. Trace Elem. Res.* **2012**, *146*, 281–286. [CrossRef] [PubMed]

49. Mitchell, C.J.; Shawki, A.; Ganz, T.; Nemeth, E.; Mackenzie, B. Functional properties of human ferroportin, a cellular iron exporter reactive also with cobalt and zinc. *Am. J. Physiol. Cell Physiol.* **2014**, *306*, C450–C459. [CrossRef] [PubMed]

50. Subramanian Vignesh, K.; Landero Figueroa, J.A.; Porollo, A.; Caruso, J.A.; Deepe, G.S., Jr. Granulocyte macrophage-colony stimulating factor induced Zn sequestration enhances macrophage superoxide and limits intracellular pathogen survival. *Immunity* **2013**, *39*, 697–710. [CrossRef] [PubMed]

51. Dodd, C.E.; Pyle, C.J.; Glowinski, R.; Rajaram, M.V.; Schlesinger, L.S. CD36-Mediated Uptake of Surfactant Lipids by Human Macrophages Promotes Intracellular Growth of *Mycobacterium tuberculosis*. *J. Immunol.* **2016**, *197*, 4727–4735. [CrossRef] [PubMed]

52. Lavalett, L.; Rodriguez, H.; Ortega, H.; Sadee, W.; Schlesinger, L.S.; Barrera, L.F. Alveolar macrophages from tuberculosis patients display an altered inflammatory gene expression profile. *Tuberculosis (Edinb.)* **2017**, *107*, 156–167. [CrossRef] [PubMed]

53. Cronan, M.R.; Beerman, R.W.; Rosenberg, A.F.; Saelens, J.W.; Johnson, M.G.; Oehlers, S.H.; Sisk, D.M.; Jurcic Smith, K.L.; Medvitz, N.A.; Miller, S.E.; et al. Macrophage Epithelial Reprogramming Underlies Mycobacterial Granuloma Formation and Promotes Infection. *Immunity* **2016**, *45*, 861–876. [CrossRef] [PubMed]

54. Schlesinger, L.S. Macrophage phagocytosis of virulent but not attenuated strains of *Mycobacterium tuberculosis* is mediated by mannose receptors in addition to complement receptors. *J. Immunol.* **1993**, *150*, 2920–2930. [PubMed]

55. Brooks, M.N.; Rajaram, M.V.; Azad, A.K.; Amer, A.O.; Valdivia-Arenas, M.A.; Park, J.H.; Nunez, G.; Schlesinger, L.S. NOD2 controls the nature of the inflammatory response and subsequent fate of *Mycobacterium tuberculosis* and M. bovis BCG in human macrophages. *Cell. Microbiol.* **2011**, *13*, 402–418. [CrossRef] [PubMed]

56. Pyle, C.J.; Akhter, S.; Bao, S.; Dodd, C.E.; Schlesinger, L.S.; Knoell, D.L. Zinc Modulates Endotoxin-Induced Human Macrophage Inflammation through ZIP8 Induction and C/EBPβ Inhibition. *PLoS ONE* **2017**, *12*, e0169531. [CrossRef] [PubMed]

57. Langmade, S.J.; Ravindra, R.; Daniels, P.J.; Andrews, G.K. The transcription factor MTF-1 mediates metal regulation of the mouse ZnT1 gene. *J. Biol. Chem.* **2000**, *275*, 34803–34809. [CrossRef] [PubMed]

58. Hardyman, J.E.; Tyson, J.; Jackson, K.A.; Aldridge, C.; Cockell, S.J.; Wakeling, L.A.; Valentine, R.A.; Ford, D. Zinc sensing by metal-responsive transcription factor 1 (MTF1) controls metallothionein and ZnT1 expression to buffer the sensitivity of the transcriptome response to zinc. *Metallomics* **2016**, *8*, 337–343. [CrossRef] [PubMed]

59. Lichten, L.A.; Ryu, M.S.; Guo, L.; Embury, J.; Cousins, R.J. MTF-1-mediated repression of the zinc transporter Zip10 is alleviated by zinc restriction. *PLoS ONE* **2011**, *6*, e21526. [CrossRef] [PubMed]

60. Dos Santos, F.M.; Piffer, A.C.; Schneider, R.O.; Ribeiro, N.S.; Garcia, A.W.A.; Schrank, A.; Kmetzsch, L.; Vainstein, M.H.; Staats, C.C. Alterations of zinc homeostasis in response to *Cryptococcus neoformans* in a murine macrophage cell line. *Future Microbiol.* **2017**, *12*, 491–504. [CrossRef] [PubMed]

61. Ribeiro, N.S.; Dos Santos, F.M.; Garcia, A.W.A.; Ferrareze, P.A.G.; Fabres, L.F.; Schrank, A.; Kmetzsch, L.; Rott, M.B.; Vainstein, M.H.; Staats, C.C. Modulation of Zinc Homeostasis in Acanthamoeba castellanii as a Possible Antifungal Strategy against *Cryptococcus gattii. Front. Microbiol.* **2017**, *8*, 1626. [CrossRef] [PubMed]

62. Winters, M.S.; Chan, Q.; Caruso, J.A.; Deepe, G.S., Jr. Metallomic analysis of macrophages infected with Histoplasma capsulatum reveals a fundamental role for zinc in host defenses. *J. Infect. Dis.* **2010**, *202*, 1136–1145. [CrossRef] [PubMed]

63. Liu, M.J.; Bao, S.; Galvez-Peralta, M.; Pyle, C.J.; Rudawsky, A.C.; Pavlovicz, R.E.; Killilea, D.W.; Li, C.; Nebert, D.W.; Wewers, M.D.; et al. ZIP8 regulates host defense through zinc-mediated inhibition of NF-κB. *Cell Rep.* **2013**, *3*, 386–400. [CrossRef] [PubMed]

64. Giacomini, E.; Remoli, M.E.; Scandurra, M.; Gafa, V.; Pardini, M.; Fattorini, L.; Coccia, E.M. Expression of proinflammatory and regulatory cytokines via NF-κB and MAPK-dependent and IFN regulatory factor-3-independent mechanisms in human primary monocytes infected by *Mycobacterium tuberculosis. Clin. Dev. Immunol.* **2011**, *2011*, 841346. [CrossRef] [PubMed]

65. Begum, N.A.; Kobayashi, M.; Moriwaki, Y.; Matsumoto, M.; Toyoshima, K.; Seya, T. *Mycobacterium bovis* BCG cell wall and lipopolysaccharide induce a novel gene, BIGM103, encoding a 7-TM protein: Identification of a new protein family having Zn-transporter and Zn-metalloprotease signatures. *Genomics* **2002**, *80*, 630–645. [CrossRef] [PubMed]

66. Besecker, B.; Bao, S.; Bohacova, B.; Papp, A.; Sadee, W.; Knoell, D.L. The human zinc transporter SLC39A8 (Zip8) is critical in zinc-mediated cytoprotection in lung epithelia. *Am. J. Physiol. Lung Cell. Mol. Physiol.* **2008**, *294*, L1127–L1136. [CrossRef] [PubMed]

67. Sweet, L.; Singh, P.P.; Azad, A.K.; Rajaram, M.V.; Schlesinger, L.S.; Schorey, J.S. Mannose receptor-dependent delay in phagosome maturation by *Mycobacterium avium* glycopeptidolipids. *Infect. Immun.* **2010**, *78*, 518–526. [CrossRef] [PubMed]

68. Aydemir, T.B.; Liuzzi, J.P.; McClellan, S.; Cousins, R.J. Zinc transporter ZIP8 (SLC39A8) and zinc influence IFN-γ expression in activated human T cells. *J. Leukoc. Biol.* **2009**, *86*, 337–348. [CrossRef] [PubMed]

69. Wang, C.Y.; Jenkitkasemwong, S.; Duarte, S.; Sparkman, B.K.; Shawki, A.; Mackenzie, B.; Knutson, M.D. ZIP8 is an iron and zinc transporter whose cell-surface expression is up-regulated by cellular iron loading. *J. Biol. Chem.* **2012**, *287*, 34032–34043. [CrossRef] [PubMed]

70. Kim, J.H.; Jeon, J.; Shin, M.; Won, Y.; Lee, M.; Kwak, J.S.; Lee, G.; Rhee, J.; Ryu, J.H.; Chun, C.H.; et al. Regulation of the catabolic cascade in osteoarthritis by the zinc-ZIP8-MTF1 axis. *Cell* **2014**, *156*, 730–743. [CrossRef] [PubMed]

71. He, L.; Girijashanker, K.; Dalton, T.P.; Reed, J.; Li, H.; Soleimani, M.; Nebert, D.W. ZIP8, member of the solute-carrier-39 (SLC39) metal-transporter family: Characterization of transporter properties. *Mol. Pharmacol.* **2006**, *70*, 171–180. [CrossRef] [PubMed]

72. Jenkitkasemwong, S.; Wang, C.Y.; Mackenzie, B.; Knutson, M.D. Physiologic implications of metal-ion transport by ZIP14 and ZIP8. *Biometals* **2012**, *25*, 643–655. [CrossRef] [PubMed]

73. Kehl-Fie, T.E.; Skaar, E.P. Nutritional immunity beyond iron: A role for manganese and zinc. *Curr. Opin. Chem. Biol.* **2010**, *14*, 218–224. [CrossRef] [PubMed]

74. Sugarman, B. Zinc and infection. *Rev. Infect. Dis.* **1983**, *5*, 137–147. [CrossRef] [PubMed]

75. Patterson, D.S. Influence of cobalt and zinc ions on the growth and porphyrin production of *Mycobacterium Tuberculosis avium. Nature* **1960**, *185*, 57. [CrossRef] [PubMed]

76. Prasad, A.S.; Bao, B.; Beck, F.W.; Kucuk, O.; Sarkar, F.H. Antioxidant effect of zinc in humans. *Free Radic. Biol. Med.* **2004**, *37*, 1182–1190. [CrossRef] [PubMed]

77. McDermott, J.R.; Geng, X.; Jiang, L.; Galvez-Peralta, M.; Chen, F.; Nebert, D.W.; Liu, Z. Zinc- and bicarbonate-dependent ZIP8 transporter mediates selenite uptake. *Oncotarget* **2016**, *7*, 35327–35340. [CrossRef] [PubMed]

78. Zhang, T.; Liu, J.; Fellner, M.; Zhang, C.; Sui, D.; Hu, J. Crystal structures of a ZIP zinc transporter reveal a binuclear metal center in the transport pathway. *Sci. Adv.* **2017**, *3*, e1700344. [CrossRef] [PubMed]

79. Koike, A.; Sou, J.; Ohishi, A.; Nishida, K.; Nagasawa, K. Inhibitory effect of divalent metal cations on zinc uptake via mouse Zrt-/Irt-like protein 8 (ZIP8). *Life Sci.* **2017**, *173*, 80–85. [CrossRef] [PubMed]

80. Rossander-Hulten, L.; Brune, M.; Sandstrom, B.; Lonnerdal, B.; Hallberg, L. Competitive inhibition of iron absorption by manganese and zinc in humans. *Am. J. Clin. Nutr.* **1991**, *54*, 152–156. [PubMed]

81. Sturgill-Koszycki, S.; Schlesinger, P.H.; Chakraborty, P.; Haddix, P.L.; Collins, H.L.; Fok, A.K.; Allen, R.D.; Gluck, S.L.; Heuser, J.; Russell, D.G. Lack of acidification in *Mycobacterium* phagosomes produced by exclusion of the vesicular proton-ATPase. *Science* **1994**, *263*, 678–681. [CrossRef] [PubMed]

82. Boycott, K.M.; Beaulieu, C.L.; Kernohan, K.D.; Gebril, O.H.; Mhanni, A.; Chudley, A.E.; Redl, D.; Qin, W.; Hampson, S.; Kury, S.; et al. Autosomal-Recessive Intellectual Disability with Cerebellar Atrophy Syndrome Caused by Mutation of the Manganese and Zinc Transporter Gene SLC39A8. *Am. J. Hum. Genet.* **2015**, *97*, 886–893. [CrossRef] [PubMed]

83. Park, J.H.; Hogrebe, M.; Gruneberg, M.; DuChesne, I.; von der Heiden, A.L.; Reunert, J.; Schlingmann, K.P.; Boycott, K.M.; Beaulieu, C.L.; Mhanni, A.A.; et al. SLC39A8 Deficiency: A Disorder of Manganese Transport and Glycosylation. *Am. J. Hum. Genet.* **2015**, *97*, 894–903. [CrossRef] [PubMed]

84. Park, J.H.; Hogrebe, M.; Fobker, M.; Brackmann, R.; Fiedler, B.; Reunert, J.; Rust, S.; Tsiakas, K.; Santer, R.; Gruneberg, M.; et al. SLC39A8 deficiency: Biochemical correction and major clinical improvement by manganese therapy. *Genet. Med.* **2017**. [CrossRef] [PubMed]

85. Galvez-Peralta, M.; Wang, Z.; Bao, S.; Knoell, D.L.; Nebert, D.W. Tissue-Specific Induction of Mouse ZIP8 and ZIP14 Divalent Cation/Bicarbonate Symporters by, and Cytokine Response to, Inflammatory Signals. *Int. J. Toxicol.* **2014**, *33*, 246–258. [CrossRef] [PubMed]

86. Besecker, B.Y.; Exline, M.C.; Hollyfield, J.; Phillips, G.; Disilvestro, R.A.; Wewers, M.D.; Knoell, D.L. A comparison of zinc metabolism, inflammation, and disease severity in critically ill infected and noninfected adults early after intensive care unit admission. *Am. J. Clin. Nutr.* **2011**, *93*, 1356–1364. [CrossRef] [PubMed]

87. Raymond, A.D.; Gekonge, B.; Giri, M.S.; Hancock, A.; Papasavvas, E.; Chehimi, J.; Kossenkov, A.V.; Nicols, C.; Yousef, M.; Mounzer, K.; et al. Increased metallothionein gene expression, zinc, and zinc-dependent resistance to apoptosis in circulating monocytes during HIV viremia. *J. Leukoc. Biol.* **2010**, *88*, 589–596. [CrossRef] [PubMed]

88. Liuzzi, J.P.; Lichten, L.A.; Rivera, S.; Blanchard, R.K.; Aydemir, T.B.; Knutson, M.D.; Ganz, T.; Cousins, R.J. Interleukin-6 regulates the zinc transporter Zip14 in liver and contributes to the hypozincemia of the acute-phase response. *Proc. Natl. Acad. Sci. USA* **2005**, *102*, 6843–6848. [CrossRef] [PubMed]

89. Sayadi, A.; Nguyen, A.T.; Bard, F.A.; Bard-Chapeau, E.A. Zip14 expression induced by lipopolysaccharides in macrophages attenuates inflammatory response. *Inflamm. Res.* **2013**, *62*, 133–143. [CrossRef] [PubMed]

90. Halsted, J.A.; Smith, J.C., Jr. Plasma-zinc in health and disease. *Lancet* **1970**, *1*, 322–324. [CrossRef]

91. Wang, G.Q.; Lin, M.Y. Serum trace element levels in tuberculous pleurisy. *Biol. Trace Elem. Res.* **2011**, *141*, 86–90. [CrossRef] [PubMed]

92. Ghulam, H.; Kadri, S.M.; Manzoor, A.; Waseem, Q.; Aatif, M.S.; Khan, G.Q.; Manish, K. Status of zinc in pulmonary tuberculosis. *J. Infect. Dev. Ctries* **2009**, *3*, 365–368. [PubMed]

93. Cernat, R.I.; Mihaescu, T.; Vornicu, M.; Vione, D.; Olariu, R.I.; Arsene, C. Serum trace metal and ceruloplasmin variability in individuals treated for pulmonary tuberculosis. *Int. J. Tuberc. Lung Dis.* **2011**, *15*, 1239–1245. [CrossRef] [PubMed]

94. Tobin, D.M.; Vary, J.C., Jr.; Ray, J.P.; Walsh, G.S.; Dunstan, S.J.; Bang, N.D.; Hagge, D.A.; Khadge, S.; King, M.C.; Hawn, T.R.; et al. The lta4h locus modulates susceptibility to mycobacterial infection in zebrafish and humans. *Cell* **2010**, *140*, 717–730. [CrossRef] [PubMed]

95. Tobin, D.M.; Roca, F.J.; Oh, S.F.; McFarland, R.; Vickery, T.W.; Ray, J.P.; Ko, D.C.; Zou, Y.; Bang, N.D.; Chau, T.T.; et al. Host genotype-specific therapies can optimize the inflammatory response to mycobacterial infections. *Cell* **2012**, *148*, 434–446. [CrossRef] [PubMed]

96. Beamer, G.L.; Flaherty, D.K.; Assogba, B.D.; Stromberg, P.; Gonzalez-Juarrero, M.; de Waal Malefyt, R.; Vesosky, B.; Turner, J. Interleukin-10 promotes *Mycobacterium tuberculosis* disease progression in CBA/J mice. *J. Immunol.* **2008**, *181*, 5545–5550. [CrossRef] [PubMed]

97. Turner, J.; Gonzalez-Juarrero, M.; Ellis, D.L.; Basaraba, R.J.; Kipnis, A.; Orme, I.M.; Cooper, A.M. In vivo IL-10 production reactivates chronic pulmonary tuberculosis in C57BL/6 mice. *J. Immunol.* **2002**, *169*, 6343–6351. [CrossRef] [PubMed]

98. Cyktor, J.C.; Turner, J. Interleukin-10 and immunity against prokaryotic and eukaryotic intracellular pathogens. *Infect. Immun.* **2011**, *79*, 2964–2973. [CrossRef] [PubMed]

99. O'Leary, S.; O'Sullivan, M.P.; Keane, J. IL-10 blocks phagosome maturation in *Mycobacterium tuberculosis*-infected human macrophages. *Am. J. Respir. Cell Mol. Biol.* **2011**, *45*, 172–180. [CrossRef] [PubMed]

100. Queval, C.J.; Song, O.R.; Deboosere, N.; Delorme, V.; Debrie, A.S.; Iantomasi, R.; Veyron-Churlet, R.; Jouny, S.; Redhage, K.; Deloison, G.; et al. STAT3 Represses Nitric Oxide Synthesis in Human Macrophages upon *Mycobacterium tuberculosis* Infection. *Sci. Rep.* **2016**, *6*, 29297. [CrossRef] [PubMed]

101. Arcos, J.; Sasindran, S.J.; Moliva, J.I.; Scordo, J.M.; Sidiki, S.; Guo, H.; Venigalla, P.; Kelley, H.V.; Lin, G.; Diangelo, L.; et al. *Mycobacterium tuberculosis* cell wall released fragments by the action of the human lung mucosa modulate macrophages to control infection in an IL-10-dependent manner. *Mucosal Immunol.* **2017**, *10*, 1248–1258. [CrossRef] [PubMed]

102. Ong, C.W.; Elkington, P.T.; Friedland, J.S. Tuberculosis, pulmonary cavitation, and matrix metalloproteinases. *Am. J. Respir. Crit. Care Med.* **2014**, *190*, 9–18. [CrossRef] [PubMed]

103. Belton, M.; Brilha, S.; Manavaki, R.; Mauri, F.; Nijran, K.; Hong, Y.T.; Patel, N.H.; Dembek, M.; Tezera, L.; Green, J.; et al. Hypoxia and tissue destruction in pulmonary TB. *Thorax* **2016**, *71*, 1145–1153. [CrossRef] [PubMed]

104. Volkman, H.E.; Pozos, T.C.; Zheng, J.; Davis, J.M.; Rawls, J.F.; Ramakrishnan, L. Tuberculous granuloma induction via interaction of a bacterial secreted protein with host epithelium. *Science* **2010**, *327*, 466–469. [CrossRef] [PubMed]

105. Davis, J.M.; Ramakrishnan, L. The role of the granuloma in expansion and dissemination of early tuberculous infection. *Cell* **2009**, *136*, 37–49. [CrossRef] [PubMed]

106. Yang, H.; Makaroff, K.; Paz, N.; Aitha, M.; Crowder, M.W.; Tierney, D.L. Metal Ion Dependence of the Matrix Metalloproteinase-1 Mechanism. *Biochemistry* **2015**, *54*, 3631–3639. [CrossRef] [PubMed]

107. Walker, N.F.; Clark, S.O.; Oni, T.; Andreu, N.; Tezera, L.; Singh, S.; Saraiva, L.; Pedersen, B.; Kelly, D.L.; Tree, J.A.; et al. Doxycycline and HIV infection suppress tuberculosis-induced matrix metalloproteinases. *Am. J. Respir. Crit. Care Med.* **2012**, *185*, 989–997. [CrossRef] [PubMed]

108. Grobler, L.; Nagpal, S.; Sudarsanam, T.D.; Sinclair, D. Nutritional supplements for people being treated for active tuberculosis. *Cochrane Database Syst. Rev.* **2016**, *6*, CD006086. [CrossRef]

109. Bailey, R.L.; West, K.P., Jr.; Black, R.E. The epidemiology of global micronutrient deficiencies. *Ann. Nutr. Metab.* **2015**, *66*, 22–33. [CrossRef] [PubMed]

International Journal of
*Molecular Sciences*

MDPI

*Review*

# Impact of Labile Zinc on Heart Function: From Physiology to Pathophysiology

Belma Turan * and Erkan Tuncay

Department of Biophysics, Ankara University, Faculty of Medicine, 06100 Ankara, Turkey;
erkan.tuncay@medicine.ankara.edu.tr
* Correspondence: belma.turan@medicine.ankara.edu.tr; Tel.: +90312-595-8137

Received: 14 October 2017; Accepted: 8 November 2017; Published: 12 November 2017

**Abstract:** Zinc plays an important role in biological systems as bound and histochemically reactive labile $Zn^{2+}$. Although $Zn^{2+}$ concentration is in the nM range in cardiomyocytes at rest and increases dramatically under stimulation, very little is known about precise mechanisms controlling the intracellular distribution of $Zn^{2+}$ and its variations during cardiac function. Recent studies are focused on molecular and cellular aspects of labile $Zn^{2+}$ and its homeostasis in mammalian cells and growing evidence clarified the molecular mechanisms underlying $Zn^{2+}$-diverse functions in the heart, leading to the discovery of novel physiological functions of labile $Zn^{2+}$ in parallel to the discovery of subcellular localization of $Zn^{2+}$-transporters in cardiomyocytes. Additionally, important experimental data suggest a central role of intracellular labile $Zn^{2+}$ in excitation-contraction coupling in cardiomyocytes by shaping $Ca^{2+}$ dynamics. Cellular labile $Zn^{2+}$ is tightly regulated against its adverse effects through either $Zn^{2+}$-transporters, $Zn^{2+}$-binding molecules or $Zn^{2+}$-sensors, and, therefore plays a critical role in cellular signaling pathways. The present review summarizes the current understanding of the physiological role of cellular labile $Zn^{2+}$ distribution in cardiomyocytes and how a remodeling of cellular $Zn^{2+}$-homeostasis can be important in proper cell function with $Zn^{2+}$-transporters under hyperglycemia. We also emphasize the recent investigations on $Zn^{2+}$-transporter functions from the standpoint of human heart health to diseases together with their clinical interest as target proteins in the heart under pathological condition, such as diabetes.

**Keywords:** zinc transporters; intracellular labile zinc; heart failure; endoplasmic reticulum stress; left ventricle

---

## 1. Introduction

Zinc is a redox inactive element and presents in almost all biological tissues. Zinc is accepted to be a component of antioxidant defense system and contributes to maintain the cell redox balance through different mechanisms [1]. There is a close relation between cellular labile zinc ($Zn^{2+}$), endogenous processes, and biological macromolecules under physiological condition, mainly due to its unique property being their structural component or major regulator of macromolecules [2]. Labile $Zn^{2+}$ plays an important role as a signaling molecule in large number of cells and tissues being a constituent of many proteins and enzymes in human body [3–5]. For this reason, even mild zinc-deficiency may impact numerous aspects of human health, including heart function [6]. Although $Zn^{2+}$ has traditionally been regarded as relatively nontoxic, recent studies have shown how high intracellular labile $Zn^{2+}$ level ($[Zn^{2+}]_i$) can be a potent killer of numerous cell types [7] including cardiomyocytes [8–10]. In cardiomyocytes, $[Zn^{2+}]_i$ is measured to be less than one nanomolar under physiological conditions being much less than the intracellular free $Ca^{2+}$ ($[Ca^{2+}]_i$) [8,11]. Moreover, oxidants caused about 30-fold increase in $[Zn^{2+}]_i$ but only 2-fold in $[Ca^{2+}]_i$ in cardiomyocytes [8]. Therefore, any increase in $[Zn^{2+}]_i$ may be much more toxic biologically than is generally realized. Furthermore, any evolution of biomolecules to scavenge $[Zn^{2+}]_i$ is crucially important in ameliorating the cellular toxicity. Supporting these above statements, early studies showed that any disruption in $[Zn^{2+}]_i$ homeostasis could be associated with severe disorders, including injuries to cardiac tissues [6,12]. Another interesting action of $[Zn^{2+}]_i$ has been shown its insulino-mimetic activity in diabetic patients [13].

It has been shown that $[Zn^{2+}]_i$ in mammalian cardiomyocytes plays an important role in excitation-contraction coupling [8,9] and in excitation-transcription coupling [14]. In mammalian cells, $[Zn^{2+}]_i$ homeostasis is regulated with different ways. Mainly, cellular $[Zn^{2+}]_i$ is controlled via its pools. It can be released from metalloproteins (a structural component) or metalloenzymes (a cofactor) under pathological conditions. As up to 15% of the total cellular zinc can be bound to metallothioneins (MT), they can represent a significant pool of $Zn^{2+}$ [15,16]. Other type storage for $Zn^{2+}$ is intracellular compartments such as organelles and vesicles. $Zn^{2+}$ transport is mediated by $Zn^{2+}$-dependent proteins, named $Zn^{2+}$ transporters, which are localized to both sarcolemma and intracellular membranes [17,18]. MTs and two $Zn^{2+}$ transporter families, solute carrier 39A [SLC39A] or Zrt-, Irt-like proteins 7ZIPs and solute carrier 30A (SLC30A) or ZnTs play crucial roles to maintain the cellular $Zn^{2+}$ homeostasis [19–23]. Although there are many recent very good review articles focused on recent progress to describe the physiological and biological functions of ZIP and ZnT transporters and provided a better understanding of $Zn^{2+}$ biology, in here, we aimed to review the recent data on $Zn^{2+}$ signaling by $Zn^{2+}$ transporters in the heart under physiological and pathological conditions, and therefore, to update our current understanding on the role of cellular $[Zn^{2+}]_i$ regulation in heart function.

## 2. Role of Zinc in Human Health

Zinc, as one of important micronutrients in human nutrition, has very prominent role in the maintenance of tissue functions. As mentioned in a number of studies and review articles, its recognition for biological systems mostly depends on its being an essential component of many enzymes and its role in many physiological or metabolic processes in the mammalian body due to it being the most abundant intracellular metal ion found in cytosol, vesicles, organelles, and in the nucleus [2–4,7,24–30]. Therefore, trace element zinc has greatly attracted the attention of biological scientists for its importance in clinical medicine. Furthermore, its role in several nutritional disorders has been clearly established [2,31–34]. Although the demonstration of zinc essentiality for the growth of *Aspergillus niger* is very early in the literature [35], a recognition of its importance for the growth of plants and animals was not appreciated until the 1950–60s, what was considered improbable until 1960. It was believed that this element could be toxic if it was over its physiological level with high zinc intakes [36–38], although zinc was essential for human health and that its deficiency in humans would never occur [39].

Although $Zn^{2+}$ is the most abundant intracellular metal ion found in cytosol, vesicles, organelles, and in the nucleus of mammalian cells including cardiomyocytes [11,40], even a small deficiency is a disaster to human health due to the number of biological dysfunctions [39]. In the literature, there are about 7000 articles searched using the key words "zinc deficiency and humans." However, there is an early study on this topic published in *Nature* related with role of zinc deficiency in fruit trees in Britain by Bold C et al. [41]. Following this work, most studies were performed in animals, until the first observation in humans by Prasad et al. [39]. Important roles of $Zn^{2+}$ associated with health implications and pharmacological targets have evoked further interest regarding its status in human health and nutrition.

Several review articles besides original works emphasized the central role of zinc content in human health starting from maternal period using documents from both underdeveloped and developing countries [27]. Zinc deficiency in humans seems to be widespread throughout the world. Generally, it is assumed that zinc deficiency in humans is due to inadequate dietary intake, gut malabsorption, or defective metabolism whereas at the cellular level, decrease zinc level can be caused a large series of morbid processes [5,8,29]. Recent clinical studies focusing on population-based groups demonstrated that there are conflicting results associated with zinc deficiency and related symptoms in different populations [42–44]. Moreover, some studies showed a positive correlation between serum zinc and type 2 diabetes risks in either middle-aged and older Finnish men or the Norwegian population as well as in development of liver fibrosis in the Miami adults [45–48]. Additionally, Kunutsor and Laukkanen [49] performed a population-based cohort study in the same country and, due to their data, they suggested that a higher serum zinc concentration is positively and independently associated with incident hypertension in men. However, clinical diagnosis of marginal zinc deficiency in humans remains problematic [50].

It is now well accepted that a well-controlled $Zn^{2+}$ homeostasis via $Zn^{2+}$ regulatory mechanisms can control influx/efflux of zinc to prevent toxic zinc accumulation into cells, particularly endogenous labile $Zn^{2+}$ plays a significant role in cytotoxic events at single cell level, including cardiomyocytes [8–10,40,51]. As mentioned in many articles, $Zn^{2+}$ is essential for growth and development for both plants and mammalians. At the cellular level, it is critically involved in proliferation, differentiation, and apoptosis [52–56]. Indeed, $Zn^{2+}$ was known as relatively harmless comparison to several other metal ions with similar chemical properties, while there is not much interest to $Zn^{2+}$ toxicity in biological systems although it has been known to be hazard in industrial aspects [36,57,58]. It is believed that overt symptoms of zinc toxicity require its large amount ingestion by the body. On the other hand, it has been known how zinc compounds as supplementary agents in daily diets are important for humans, particularly under some special conditions.

However, there are still widespread controversies and ambiguities with respect to the toxic effects and mechanisms of excess zinc in humans. Authors using different zinc-compounds, particularly metallic nanoparticles including zinc, demonstrated that the solubility and the size of zinc compounds have a major role in the induced toxic responses, whereas uptake of their large ones inside the cells was likely to play a key role in the detected cell cycle arrest [59]. In a recent interesting study demonstrated another important role of heavy metals including zinc in pathological biomineralization of cancer tissue due to their important roles in morphogenesis of tumors considering their ability to enter into covalent bonds with calcium salt molecules. Romanjuk et al. [60] examined the role of microelements in breast cancer calcifications. Their data demonstrated how excess heavy metals (such as iron, copper, chromium, and nickel) in the body could lead to pathologies in the tissues/organs, at most, via progressively increasing rate of degenerative/necrotic alterations. However, Hoang et al. [61] discussed in a widespread manner how zinc is an important metal as a possible preventive and therapeutic agent in pancreatic, prostate, and breast cancer in humans.

As summary of this part, the body zinc level in humans as well as zinc intake with nutrition, are receiving increasing attention, also due to its putative role in the development of different pathological conditions, there are serious conflicts between publications.

## 3. Labile $Zn^{2+}$ in Cardiac Physiology and Pathology

As discussed in many review articles, zinc is a vital nutrient for human health via its incredibly important roles in physiology and pathology of many organ functions, at most, due to its important roles in proteins and enzymes [62–64]. As a consequence, zinc is required for the function of organs, including the heart. Furthermore, impairment of $Zn^{2+}$ homeostasis is associated with a variety of health problems such as cardiovascular disorders [65–68]. Previously, several authors have documented the importance of zinc in cardiovascular function and diseases in many good review articles [1,66,69–73]. In these articles, it has been focused on the critical role of intracellular labile $Zn^{2+}$ in the redox signaling pathway, where certain triggers lead to release of $Zn^{2+}$ from proteins/intracellular pools and cause myocardial damage [8,14,40,51,74]. Thus, the area of $Zn^{2+}$-homeostasis seems to be emerging in cardiovascular disease. However, it has been demonstrated that labile $Zn^{2+}$ outside a narrow concentration range are toxic to a variety of cells [69], including cardiomyocytes [8–10,14,40,51,75].

It has been known, at cellular levels, that certain pathological conditions including hyperglycemia and/or changes in specific signaling pathways can stimulate the production of molecules related with the redox-state of the cells. Among these changes, the alterations in the voltage-dependent ionic channels, ion transporters and some ion exchanges [76]. Similar to $[Ca^{2+}]_i$ homeostasis, any alteration in $[Zn^{2+}]_i$-homeostasis, including redox-status of the cells under hyperglycemia, can be involved in development of cardiac dysfunction [8,9,77–81]. In that regard, it is clear that a well-controlled redox-status in cardiomyocytes can be very beneficial, in part, via mediation of either $[Ca^{2+}]_i$ homeostasis, $[Zn^{2+}]_i$ homeostasis or both for a cardioprotective approach in diabetic cardiomyopathy as well as other pathologies in patients with heart disease [51,82]. Therefore, it can be suggested that a well-controlled regulation of $[Zn^{2+}]_i$ homeostasis at cellular level, similar to $[Ca^{2+}]_i$ homeostasis, can have important strategy to protect heart against redox-unbalanced pathologies [12]. In this regard, in an early study by Kamalov et al. [83], it has been mentioned that $[Zn^{2+}]_i/[Ca^{2+}]_i$ ratio in cardiomyocytes and mitochondria have optimal ranges, having important roles to control the redox-status as well as oxidative stress status of cells. As shown in many articles, any enhancement of antioxidant defenses in cells are providing benefits against these pathological conditions, including the control of $[Zn^{2+}]_i$ associated control of $[Ca^{2+}]_i$ homeostasis, particularly, via RyR2 [40,51,82,84,85]. As summary, ours and others' studies have documented that, at tissue level, $[Zn^{2+}]_i$-homeostasis in heart is impaired by different signaling mechanisms, including oxidative stress in the heart, and thereby, plays important role in the pathogenesis of myocardial injury. Our current understanding of the roles of $[Zn^{2+}]_i$ homeostasis and $[Zn^{2+}]_i$ signaling in human myocardial injury is yet limited under any pathological stimulus.

The importance and complexity of $Zn^{2+}$ action has been presumed to parallel the degree of $Ca^{2+}$'s participation in cellular processes [9,51,72,86]. At cellular level, $Zn^{2+}$ homeostasis is regulated through $Zn^{2+}$ transporters, $Zn^{2+}$-binding molecules, and $Zn^{2+}$ sensors. Interestingly, most of studies related with the role of intracellular labile $Zn^{2+}$ in cell function are associated with its toxicity, most probably, due to its service as an important secondary messenger in various intracellular signal transduction pathways [79,87].

As mentioned, in previous paragraphs, the cellular toxicity of exogenous and a redox-inert labile $Zn^{2+}$ is characterized generally by cellular responses such as mitochondrial dysfunction, elevated production of reactive oxygen species/reactive nitrogen species ROS/RNS, loss of signaling quiescence leading to apoptosis in cells, cell death and increased expression of adaptive and inflammatory genes [2,4–6,8,9,14,23,88]. Central to the molecular effects of $Zn^{2+}$ are its interactions with cysteinyl-thiols of proteins, which alter their functionality by modulating their reactivity and participation in redox reactions, as well as its a cis-acting element that is the binding site for metal-responsive transcription factor-1 (MTF-1). When cytosolic labile $Zn^{2+}$ is increased, it can lead to an increase in MTF-1 activity, which in turn leads to an increase in activation of *MTF-1* target genes [8,9,51,78,89].

Ongoing studies together with early studies demonstrated that both $Zn^{2+}$ deficiency and excess are detrimental to cells, causing growth retardation, important metabolic disorders and, particularly, inducing an impaired excitation-contraction cycling in cardiomyocytes [8,9,14,90]. Although total cellular $Zn^{2+}$ is about 200 μM, the cytosolic labile $[Zn^{2+}]_i$ in cardiomyocytes is less than 1 nM under physiological conditions [8,11], whereas it can increase either ~30-fold with acute oxidant exposure or ~2-fold under chronic hyperglycemia [8,9,51]. Therefore, in general, not only can $Zn^{2+}$ deficiency be detrimental, causing depressed growth and serious metabolic disorders, but excess $Zn^{2+}$ can also be toxic to many cells [8,70,91,92].

## 4. Role of Cellular Labile $Zn^{2+}$ in Electrical Properties of Cardiomyocytes

It is well accepted that a controlled $Ca^{2+}$ signaling is key essential mechanism for regular function of cardiomyocytes. As mentioned in Section 3, the intracellular $Ca^{2+}$ accumulation induced by increased production of ROS can cause injury and dysfunction of cardiomyocytes [93–95].

The sarcoplasmic reticulum (SR) is a main intracellular $Ca^{2+}$ store in cardiomyocytes and ryanodine receptors (RyR2s) on the membrane of the SR play a central role in modulating $Ca^{2+}$

signaling. These SR Ca$^{2+}$ release channels contain many cysteine residues in their regulatory domain and putative Ca$^{2+}$ pore region. These residues are susceptible to modification by oxidants [94,96,97].

Supporting the previous data, Woodier and co-workers [98] demonstrated nicely how cytosolic Zn$^{2+}$ can act as a high affinity activator of RyR2 while their experimental approach enabled the study of RyR2 function under tight control of the chemical environment. Importantly, it has been widely discussed later how Zn$^{2+}$ at 1 nM concentration has an ability to directly activate RyR2, which have a much higher affinity for Zn$^{2+}$ than Ca$^{2+}$ (about three-fold) [99]. Moreover, their data provided important information on the role of Zn$^{2+}$ on RyR2 modulatory function in the absence of Ca$^{2+}$ and presented a paradigm shift in our general understanding RyR2 activation during excitation-contraction coupling. Therefore, the already known data together with these new data provided an important mechanistic explanation associated with [Zn$^{2+}$]$_i$ dishomeostasis and certain cardiomyopathies, including diabetic cardiomyopathy [51,82,98,99].

In our early study, we, for the time, have shown that the intracellular labile Zn$^{2+}$ level, [Zn$^{2+}$]$_i$ in rabbit cardiomyocytes, is less than 1 nM in physiological conditions [8], about 100-fold less than [Ca$^{2+}$]$_i$. At various concentrations, labile Zn$^{2+}$ leads to the release of toxic ROS [79]. More importantly, our data demonstrated that [Zn$^{2+}$]$_i$ was increased markedly (about 30-fold but only doubled [Ca$^{2+}$]$_i$) with thiol-reactive oxidants and contributed to oxidant-induced alterations of excitation-contraction coupling ECC under in vitro conditions. Therefore, that information emphasized the importance of [Zn$^{2+}$]$_i$ measurement such as, how it could lead to significant overestimation of [Ca$^{2+}$]$_i$, if it was overlooked [8]. In later studies, under in vivo conditions, [Zn$^{2+}$]$_i$ was increased by 70% in diabetes [51,78] and over 200% in aldosteronism [100]. Interestingly, Xie and Zhu [101] aimed to understand better the modulation of RyR2s during oxidative stress and showed how RyR2 in rat ventricular myocytes was modulated biphasically by sulfhydryl oxidation, contribution of increased [Zn$^{2+}$]$_i$ besides increased [Ca$^{2+}$]$_i$.

Although it has been shown the resting level of [Zn$^{2+}$]$_i$ in cardiomyocytes, there was very little information about precise mechanisms controlling intracellular distribution of Zn$^{2+}$ and its variations during cardiac function. Therefore, we aimed to investigate the rapid changes in Zn$^{2+}$ homeostasis in detailed and using the Zn$^{2+}$-specific fluorescent dye, FluoZin-3, in comparison to Ca$^{2+}$-dependent Fluo-3 fluorescence, we, for the first time, vizualised the existence of Zn$^{2+}$ sparks and Zn$^{2+}$ transients, in quiescent and electrically-stimulated cardiomyocytes, similarly to known rapid Ca$^{2+}$ changes, while both Zn$^{2+}$ sparks and Zn$^{2+}$ transients required Ca$^{2+}$ entry [9]. Inhibiting the SR-Ca$^{2+}$ release, or increasing the Ca$^{2+}$ load in a low-Na$^+$ solution, suppressed or increased Zn$^{2+}$ movements, respectively. Moreover, oxidation by H$_2$O$_2$ facilitated, and acidic pH inhibited the Ca$^{2+}$-dependent Zn$^{2+}$ release in freshy isolated rat ventricular cardiomyocytes.

In historical background, studies show that most experimental studies on the role of [Zn$^{2+}$]$_i$ were performed in nervous system, at most, due to the localization of Zn$^{2+}$ in nerve terminals and synaptic vesicles of excitatory neurons in the central nervous system [102,103]. The [Zn$^{2+}$]$_i$ homeostasis in mammalian cells results from a coordinated regulation by different proteins involved in the uptake, excretion, and intracellular storage/trafficking of Zn$^{2+}$ [104]. It has been shown the Zn$^{2+}$ influx via the L-type Ca$^{2+}$ channels in heart cells [8,14,75] while it was via L-type and N-type Ca$^{2+}$ channels in neurons [105]. As can be seen in Figure 1, here, we reinvestigated the effects of extracellular and intracellular Zn$^{2+}$ on the L-type Ca$^{2+}$ current. In the presence of external Ca$^{2+}$, the L-type Ca$^{2+}$ current is inhibited by external Zn$^{2+}$ (ZnCl$_2$) and intracellular Zn$^{2+}$ loading (with Zn$^{2+}$-pyrithione, ZnPT) also reduces the L-type Ca$^{2+}$ current as a concentration- and time-dependent manner (Figure 1A). Although both effects are washable, the intracellular Zn$^{2+}$ loading induced a marked leftward shift in inactivation of the channels (Figure 1C) without any effect under external Zn$^{2+}$ exposure (Figure 1B). Similar to ours, Alvarez-Collazo et al. [106] demonstrated the modulation of transmembrane Ca$^{2+}$ movements and their regulation by β-adrenergic stimulation with both basal intracellular and extracellular Zn$^{2+}$, emphasizing the importance of well-controlled cellular [Zn$^{2+}$]$_i$-homeostasis for prevention/treatment of cardiac dysfunction, whereas Traynelis et al. [107] showed that both T-type Ca$^{2+}$ current and

L-type $Ca^{2+}$ current are inhibited by excess external $Zn^{2+}$ via inhibition of N-methyl-D-aspartate (NMDA) receptors. Although the exact underlying mechanisms of $Zn^{2+}$ effects are not clear yet, surface charge effects could be invoked to explain some of the $Zn^{2+}$ actions. However, as for other divalent metal cations, most of the effects of $Zn^{2+}$ could be well explained by changes in the gating of ion channels [108–110] that we have shown in Figure 1 similar to our previous data [75]. Furthermore, squid $K^+$ channels are far more sensitive to $Zn^{2+}$ than $Na^+$ channels but the interactions of $Zn^{2+}$ with gating charges appear similar in both cases [109]. In that regard, Aras [111] reported that, during sublethal ischemia, the early rise in neuronal $Zn^{2+}$, preceding the rise in intracellular $Ca^{2+}$, was responsible for the hyperpolarizing shift in the voltage dependency of the delayed rectifier Kv2.1 channels.

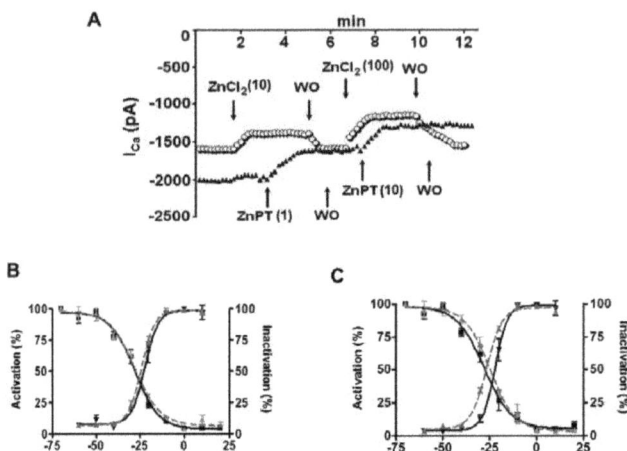

**Figure 1.** A patch-clamp study on the time course of L-type $Ca^{2+}$-current ($I_{Ca}$), depressed by $Zn^{2+}$ exposed either extracellulary (ZnCl$_2$) or intracellularly (i.e., loaded with $Zn^{2+}$-ionophore pyrithione, ZnPT). (**A**) The $I_{Ca}$ was evoked at 0 mV from a holding potential $-80$ mV and $I_{Ca}$ was recorded in the presence of either ZnCl$_2$ (10 and 100 μM) or ZnPT (1 and 10 μM) in whole-cell patch-clamped ventricular cardiomyocytes in the presence of 1.8 mM external $Ca^{2+}$. WO represents the washout of applications. Corresponding availability curves of the $I_{Ca}$ by either ZnCl$_2$ (**B**) or ZnPT (**C**). Note about 10 mV left-shift by ZnPT application in availability curve of $I_{Ca}$.

Labile $Zn^{2+}$, with even picomolar concentrations, modulates many cellular processes via either inhibiting or activating many proteins, enzymes, kinases and phosphatases, particularly, at Ser/Thr or Tyr sites [2,29,112–114]. Therefore, these findings indicate clearly that cellular labile $Zn^{2+}$ level has critical importance for cellular physiological function besides its physiopathological or toxicological role, and strongly suggest its modulatory activity of signal transduction processes [106,115]. Furthermore, it is known that fluctuations of $[Zn^{2+}]_i$ participate in important cellular functions of not only breast cancer cells [116] but also mammalian cardiomyocyte [51,82,89]. In diabetic rat heart, an important actor in contractile machinery family, RyR2 and its accessory kinases protein kinase A (PKA) and calmodulin-dependent protein kinase II (CaMKII) are phosphorylated, significantly, at most, due to increased oxidative stress via increased $[Zn^{2+}]_i$. Interestingly, these changes could be prevented with antioxidant treatment under either in vivo or in vitro conditions. Indeed, either excess $Zn^{2+}$ exposure to cardiomyocytes or labile $Zn^{2+}$ loading of cardiomyocytes with zinc-ionophore could induce marked phoshorylation in RyR2, PKA and CaMKII as well as transcription factors such as nuclear factor κB (NFκB) and glycogen synthase kinase-3 (GSK) and other endogenous actors such as protein kinase B also known as Akt [89]. Parallel to our data, it has been shown that labile $Zn^{2+}$ inhibits the activity of adenylyl cyclases as well as the hormone and forskolin stimulation of cAMP synthesis in N18TG2 cells [117], and, by preventing guanosine-5'-triphosphate (GTP) binding to the GTPase [118]. Also, in the presence of $Ca^{2+}$ and calmodulin, increasing concentrations (in micromolars) of $Zn^{2+}$ caused a progressive inhibition of substrate phosphorylation by CaMKII such as to produce a concentration-dependent inhibition of phospholamban phosphorylation [119]. However, Yi and coworkers [120] examined the role of extracellular $Zn^{2+}$ exposure on cardiomyocyte contraction-relaxation function by using molecular and cellular techniques and showed that RyR2 and phospholamban were markedly dephosphorylated after permeating the hearts with 50 μM $Zn^{2+}$. The different results associated with RyR2 phosphorylation level with $Zn^{2+}$-exposures, most probably depending on the differences between $Zn^{2+}$-exposure periods. One group experiments were performed in hours, while others in minutes.

Recently, since M-type (Kv7, *KCNQ*) potassium channels are important for the control of the excitability of neurons and muscle cells, Gao et al. [121] studied the effect of intracellular labile $Zn^{2+}$ on M-type (Kv7, KCNQ) $K^+$-channels. Their results reported that $[Zn^{2+}]_i$ directly and reversibly augments the activity of recombinant and native M channels, being mechanistically distinct from the known redox-dependent KCNQ channel potentiation.

Taken into consideration the facts of $[Zn^{2+}]_i$ in many cell types such as operation of many physiological and pathological mechanisms of cell excitation via the suppression of activity or expression of ion-channels, transporters, pumps, and receptors or pharmacological augmentation of their activities as a recognized strategy for the treatment of hyper-excitability disorders, it would be very helpful to understand well the action of $[Zn^{2+}]_i$ in cardiomyocytes. However, physiological mechanisms resulting in ionic channel potentiation are rare. As short due to already known data, a large amount of $Zn^{2+}$-proteins that are modulated by or contain $Zn^{2+}$ can directly or indirectly affect the many cellular processes, at most, due to labile $Zn^{2+}$ action on cell redox-balance [9,76]. At the cellular level, the effects of labile $Zn^{2+}$ in cardiomyocytes via its action in membrane receptors, transporters and ionic channels as well as endogenous accessory proteins of contractile machinery and some transcription factors are summarized in Tables 1 and 2, respectively.

**Table 1.** Effect of excess $Zn^{2+}$ on cardiomyocyte electrical and mechanical activity.

| Parameters | Excess $Zn^{2+}$ | Parameters | Excess $Zn^{2+}$ |
|---|---|---|---|
| *Electrical activity* | | *Mechanical activity* | |
| Resting membrane potential | ↔ | Muscle Contraction | ↓ |
| Action potential duration (APD) | ↔ | Contraction rate | ↓ |
| Time to peak AP amplitude (TP) | ↑ | Relaxation rate | ↓ |
| $Ca^{2+}$ transients | ↓ | Time to peak contraction | ↔ |
| L-type $Ca^{2+}$ currents | ↓ | Time at 50% of relaxation | ↔ |
| Mitochondrial membrane potential | ↑ | | |

Arrows represent the increased (↑) decreased (↓) or unchanged (↔) electrical and mechanical activity as time or amplitude of the parameters.

**Table 2.** Effects of excess $Zn^{2+}$ on biochemical and ultrastructural parameters in cardiomycoytes.

| Parameters | Excess $Zn^{2+}$ | Parameters | Excess $Zn^{2+}$ |
|---|---|---|---|
| *Biochemical parameters* | | | |
| pRyR2/RyR2 | ↑ | Promyelocytic leukemia(PML) | ↑ |
| pPKA/PKA | ↑ | Bcl-2/BAX | ↓ |
| FK506-binding protein(FKBP12.6) | ↔ | pAkt/Akt | ↑ |
| pCaMKII/CaMKII | ↑ | pNFκB/NFκB | ↑ |
| Calregulin | ↑ | pGSK/GSK | ↑ |
| Glucose regulated protein (GRP78) | ↑ | | |
| *Ultrastructure parameters* | | | |
| Morphological changes in mitochondria | ↑ | Electron density of Z-lines | ↓ |
| Number of lysosomes | ↑ | | |

Arrows represent the increased (↑), decreased (↓), or unchanged (↔) protein expression levels in biochemical parameters or number of lysosomes, Z-lines, and morphological changes in ultrastructure parameters.

## 5. Labile $Zn^{2+}$ Pools in Cardiomyocytes

As described in previous sections, labile $Zn^{2+}$ regulates the expression and activation of biological molecules such as ion channels, transcription factors, enzymes, adapters, and growth factors, along with their receptors in many cell types, including cardiomyocytes. Excess $Zn^{2+}$ can be detrimental to cells, particularly that of cardiomyocytes [8–10,14]. As an intracellular signal transducer in multiple cellular functions, it has been shown that intracellular labile $Zn^{2+}$ has an important role in ECC in cardiomyocytes by shaping $Ca^{2+}$ dynamics [9,51,98], while it acts as a neuromodulator in synaptic transmissions [122,123]. In the regulation of cellular labile $Zn^{2+}$, subcellular pools as well as metalloproteins are important actors, which also include many$Zn^{2+}$-transporters [29,87,124].

In our previous study performed with freshly isolated ventricular cardiomyocytes, we demonstrated that rapid changes in labile $Zn^{2+}$ mostly resulted from $Zn^{2+}$ displacement by $Ca^{2+}$ ions from intracellular binding sites that were highly sensitive to the redox status of the cardiomyocytes by using $Zn^{2+}$-specific fluorescence dye, FluoZin-3 [9]. In order to examine the physiological importance of the protein-bound $Zn^{2+}$ pools, similar to other studies [125] by applying acidic pH to the cardiomyocytes or by causing an oxidative stress with $H_2O_2$, we induced chemical modifications of the thiol groups in the proteins and demonstrated that $Zn^{2+}$ binding to metallothioneins decreased at acid pH and significantly reduced contraction without altering $[Ca^{2+}]_i$.

Our testing S(E)R as an possible intracellular $Zn^{2+}$ pool by using ryanodine application, we observed significantly and simultaneously decreases in the intensities of both $Zn^{2+}$ transients and $Ca^{2+}$ transients in field-stimulated cardiomyocytes without affecting their basal levels [9]. Furthermore, we performed additional experiments with caffeine. As can be seen in Figure 2, we observed two different responses in Fluo-3 or FluoZin-3 loaded cells. In Fluo-3 loaded cells, there was fast transitory and short-lived large increase as response to caffeine application, whereas there was a slow initial large increase followed with a stable long-lived plateau in FluoZin-3 loaded cells. These data support the hypothesis of sarco(endo)plasmic reticulum, S(E)R could be a $Zn^{2+}$ pool similar to $Ca^{2+}$ in cardiomyocytes. Furthermore, recently by using Förster resonance energy transfer (FRET)-based recombinant-targeted $Zn^{2+}$-probes [11], we have shown that $[Zn^{2+}]_i$ in cardiomyocytes is calculated less than 1 nM, while ~5-fold higher in S(E)R, and less than cytosolic-level in mitochondria. Elevated cytosolic $Zn^{2+}$ appears to contribute to deleterious changes in many cellular signaling-pathways including hyperglycemia-challenged cardiomyocytes [8–10,19,51]. Moreover, we also demonstrated that elevated cytosolic $Zn^{2+}$ appears to be associated with loss of S(E)R $Zn^{2+}$ via hyperphosphorylation of $Zn^{2+}$-transporter ZIP7, which further induces endoplasmic reticulum (ER) stress in cardiomyocytes in the diabetic rat heart [40]. Of note, in eukaryotes, Ellis et al. [126] demonstrated the $Zn^{2+}$-deficiency associated disruption in ER such as alteration in its function and induction of ER stress.

Additional studies also pointed out mitochondria to be another intracellular $Zn^{2+}$-pool in cardiomyocytes [9]. Expose of FluoZin-3 loaded cardiomyocytes either to a mitochondrial complex I inhibitor or to carbonyl cyanide 4-(trifluoromethoxy) phenylhydrazone (FCCP) a mitochondrial protonophore induced rapid and significant inhibitory effects on the $Zn^{2+}$-changes with only mild initial effects on the Fluo-3 loaded cells. In this regard, in cortical neurons, it was proposed that the source of $Ca^{2+}$-dependent $Zn^{2+}$ release appears largely to be mitochondria [127]. In cardiomyocytes, since mitochondria constitute the major source of intracellular ROS production [128], mitochondria-related excessive ROS production has been implicated in the pathogenesis of many cardiovascular diseases. Indeed, the $Ca^{2+}$-dependency of the glutamate mobilization of intracellular $Zn^{2+}$ in neurons is attributed to the generation of ROS arising from both cytosolic and mitochondrial sources [129]. In cardiomyocytes, $Ca^{2+}$ influx might trigger transitory change in ROS production leading to $Zn^{2+}$ transients even on this time scale. Such a hypothesis could account for our recent observations [40].

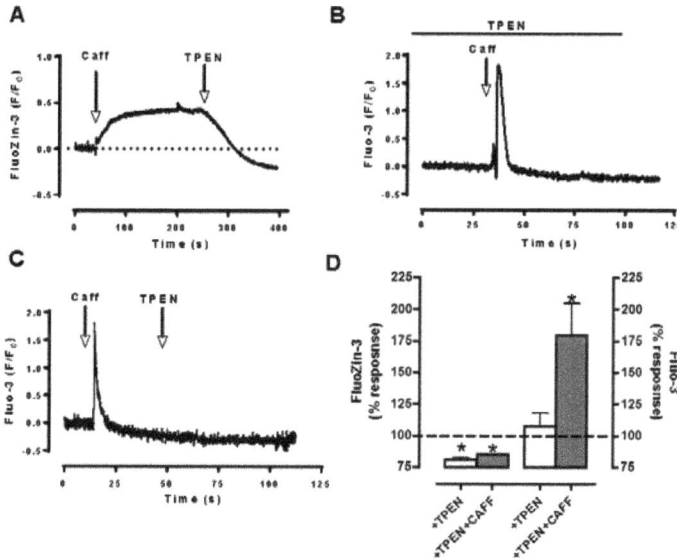

**Figure 2.** FluoZin-3 and Fluo-3 responses to caffeine application demonstrate sarco(endo)plasmic reticulum as an intracellular labile $Zn^{2+}$ pool in cardiomyocytes performed with confocal imaging. We used either a $Zn^{2+}$-specific fluorescence dye, FluoZin-3 (**A**) or a $Ca^{2+}$-specific fluorescence dye, Fluo-3 (**B**) loaded ventricular cardiomyocytes isolated from rat heart and stimulated them with 10 mM caffeine (Caff) in the either absence (**A** and **B**, respectively) or presence of a membrane-permeant $Zn^{2+}$ chelator TPEN (*N,N,N′,N′*-tetrakis(2-pyridylmethyl) ethylenediamine, 30-μM) (**C**), respectively. Normalized caffeine responses are given as F/F0, where F is the fluorescence signal and $F_0$ is the diastolic fluorescence. The mean ($\pm$ standard error of mean (SEM)) values are given for the protocols in (**D**), $n = 5$–6 for hearts/protocol. * $p < 0.05$ vs. before application.

Besides voltage-dependent ionic channels, transient receptor potential (TRP) channels are a large family associated with multi-signal transducers and play important roles in different organ function, including heart function. The functional and structural control of TRP channels by trace metal ions, including $Zn^{2+}$, has been demonstrated in different cell types [130]. For example, the high expression levels of both transient receptor potential cation channel3(TRPC3) and TRPC6 have been shown in the heart and could participate in the pathogenesis of cardiac hypertrophy and heart failure as a pathological response to chronic mechanical stress [131]. Additionally, the activation of transient receptor potential cation channel subfamily M member 4 (TRPM4), a $Ca^{2+}$-activated, but $Ca^{2+}$-impermeable non-selective cation channel, has been also demonstrated to have role in conduction block and other arrhythmic propensities associated with cardiac remodeling and injury [132]. Furthermore, Uchida and coworkers [133] have documented that extracellular $Zn^{2+}$ regulates TRPM5 channel activation [133]. On the other hand, Lambert and coworkers demonstrated that extracellular $Zn^{2+}$ exposure to HEK293 cells did inhibit TRPM5 and TRPM1 activity [134]. Moreover, Abiria and coworkers [135] recently showed that the majority of TRPM7 is localized in abundant intracellular vesicles in HEK293 cells and ROS-mediated TRPM7 activation releases $Zn^{2+}$ from these vesicles following $Zn^{2+}$ overload. They emphasized the important role of TRPM7-mediated $Zn^{2+}$ release and the regulation of ROS signaling processes during postnatal stress/injury. Therefore, one can emphasized how will be very important to understand the roles of TRP channels as detailed in the regard of their contribution to the key procedures for the development of cardiovascular disorders.

Furthermore, it seems that they may provide basic scientific knowledge for the development of new preventive and therapeutic approaches to manage patients with cardiovascular diseases [136].

On this basis, the effects of increased intracellular labile $Zn^{2+}$ on the structure of cardiomyocytes particularly focused on ultrastructure of mitochondria were examined by electron microscopy by using short-term ZnPT incubation (0.1 or 1 μM for 15–20 min) in freshly isolated ventricular cells. As can be seen in Figure 3A–C, marked irregular mitochondrial cristae and significantly clustered and degenerated mitochondria between the myofibrils together with electron-dense matrix were observed in labile $Zn^{2+}$ loaded cells. Additionally, there were an important amount of fragmented mitochondria and rounding and swelling in mitochondrion. Although the sarcomere showed normal structural appearance with regular myofibrils and mitochondrial structure in the control group cells (Figure 3A), there were dramatic signs of injury in the form of condensation, increased matrix density, and deposits of electron-dense material in the loaded cardiomyocytes. These $Zn^{2+}$ loading effects in mitochondria can support its $Zn^{2+}$ sensing pool in cardiomyocytes. However, the early data by Jang et al. [137] showed that NO mobilizes intracellular $Zn^{2+}$ via Cyclic guanosine monophosphate/Protein Kinase G (cGMP/PKG) signaling pathway and prevents mitochondrial oxidant damage in cardiomyocytes.

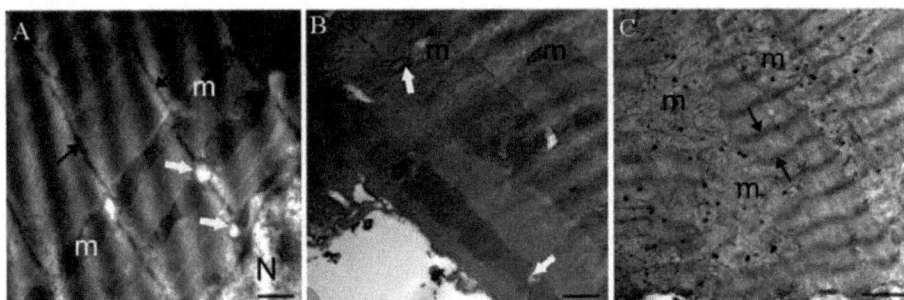

**Figure 3.** An electron microscopy examination of the effect of increased intracellular labile $Zn^{2+}$ on ultrastructure of cardiomyocytes. The microscopic examination of freshly isolated cardiomyocytes under physiological condition (**A**) showing regular myofibrils and mitochondrial (m) structure and under 0.1 μM ZnPT incubation for 15–20 min (**B**) showing irregular mitochondrial cristae (white arrows). Magnification is ×21, 560 and bars represent 500 nm. On the right, cardiomyocytes showing clustered and degenerated mitochondria (m) under 1 μM ZnPT incubation for 15–20 min(C). Black arrows are showing irregular Z-lines. Magnification is ×12,930 and bar represents 1000 nm.

## 6. $Zn^{2+}$ Transporters in Cardiomyocytes

Recent review articles well summarized the already published data, performed in different mammalian cells except cardiomyocytes, which showed the regulation of $Zn^{2+}$ homeostasis via a number of $Zn^{2+}$-transporters and how they are crucial for proper cellular functions [22,23,29,138–142]. In the review by Kambe [143], it has been pointed out how the impaired $Zn^{2+}$ transporter functions into and out of cells strongly linked to clinical human diseases [144]. Since the membrane transporters having the great potential for drug targets [145,146], hence, cytosolic labile $Zn^{2+}$ and $Zn^{2+}$ transporters should be considered as novel therapeutic targets for diseases, including heart diseases. In this section, we aimed to describe the physiological and molecular functions of $Zn^{2+}$-transporters, which regulate $[Zn^{2+}]_i$ homeostasis and are involved in signal transduction and heart diseases, particularly such as diabetes.

Even in early reviews, tightly control of $[Zn^{2+}]_i$ homeostasis is defined due to existence of specific $Zn^{2+}$-transporters, including the coordinated regulation of $[Zn^{2+}]_i$ homeostasis in terms of its uptake, efflux, distribution, and storage, which are documented in several review articles, nicely [20,147]. In general, $Zn^{2+}$ transporters belong to a family of transmembrane proteins that control the flux of $Zn^{2+}$

across cellular membranes into cytosol (ZIPs) and out of cytosol (ZnTs) in many types of cells, therefore, contribute to the distribution, storage, and compartmentalization of $Zn^{2+}$. Additionally, these all predict proteins with multiple membrane spanning regions, and most have a histidine-rich intracellular loop. The first described $Zn^{2+}$ transporter in mammalian cells is ZnT-1 in kidney cells [148] while the second one is ZnT3 in regions of the brain [149]. The ZnT proteins (solute-linked carrier 30, SLC30) and the ZIP (zinc-regulated trans- porte /iron-regulated transporter Zrt/Irt)-like, solute-linked carrier, 39, *SLC39*) have been identified in mammalian tissues [22,23,150]. In mammals, there are 10 members of the $Zn^{2+}$ efflux transporters (ZnT1–10) and 14 members of the zinc influx transporters (ZIP1–14).

ZIP proteins are thought to form homodimers to transport $Zn^{2+}$ across the cellular membrane [151], while the conserved hydrophilic residue seems to sense metal specificity [152]. Supporting this statement, it has been demonstrated that ZIP8 and ZIP14, possessing glutamic acid instead of the conserved histidine, can efficiently transport $Fe^{2+}$, $Mn^{2+}$, and $Cd^{2+}$ in addition to $Zn^{2+}$. Human genetic disorders caused by mutations and single-nucleotide polymorphisms in $Zn^{2+}$ transporter genes were summarized by Kambe et al. [144,153]. They documented that a number of genetic disorders are caused by mutations in the genes encoding ZIPs and ZnTs, such as ZIP4 in acrodermatitis enteropathica, ZIP13 in the spondylocheiro dysplastic form of Ehlers-Danlos syndrome, ZnT2 in transient neonatal zinc-deficiency, ZnT8 in type 1 and 2 diabetes mellitus, and ZnT10 in Parkinsonism and dystonia [22,23,141,142].

Furthermore, in the recent review article by Hara et al. [29], mechanisms of $Zn^{2+}$-transporter expression and modification have been documented, very widely in different mammalian cells/tissues, focusing on their physiological roles from their molecular basis to genetic importance. In that review, they presented the role of ZnTs such as ZnT2, ZnT3, ZnT4, and ZnT8, which localize to acidic compartments and to vesicles such as endosomes/lysosomes, synaptic vesicles, and insulin granules as well as the ZIPs such as ZIP4, ZIP5, ZIP6, ZIP7, ZIP8, ZIP10, ZIP12, ZIP13, and ZIP14 in different cells. A number of cellular proteins, enzymes, kinases and phosphatases interact with $Zn^{2+}$ for their biological functions. Studies have revealed that $Zn^{2+}$ acts not only as an accessory molecule for proteins but also as a signaling molecule, much like $Ca^{2+}$ [15,29].

Among others, ZnT7 has been demonstrated to play an important role in both growth and the accumulation of body fat in mice [154] as well as the association between its deficiency and metabolic disorders such as insulin and glucose intolerance and hyperglycemia [155]. Furthermore, studies on ZnT8 showed that the ZnT8 is strongly related to type 1 and 2 diabetes [156,157]. In this regard, later studies emphasized that ZnT8 expressed in pancreatic β-cells is involved in secreting insulin, forming crystals [158–160], and eliminating insulin by the liver [161]. In this field, the recent review articles by Chabosseau and Rutter [162] and by Rutter et al. [163] reviewed the regulation and roles of $Zn^{2+}$ in islet cells and the mechanisms by which ZnT8 variants might affect glucose homeostasis and diabetes risk. Correspondingly, they presented that genetic variants in the ZnT8 gene, which encodes the diabetes-associated granule-resident ZnT8, are associated with an altered risk of type 2 diabetes. Additionally, they discussed the effects on insulin secretion and action of deleting or over-expressing ZnT8 highly selectively in the pancreatic β-cell, and the role of $Zn^{2+}$ in insulin signaling. Due to their own data together with the others' data, it has been concluded that maintenance of glucose homeostasis, and therefore lower diabetes risk, due to a proper intake level of dietary zinc at systemic level and a well-controlled $[Zn^{2+}]_i$ homeostasis at cellular level is provided with both insulin release and insulin action at physiological levels.

In the content of ZIPs' roles, there are a limited number of studies in mammalian cells in the literature without any in cardiomyocytes except our study [40]. However, Ellis et al. [126] demonstrated that the zinc deficiency in ER leads to an unfolded protein response (UPR) in human cells. In a later study, Huang et al. [155] proposed that ZIP7 is localized to Golgi apparatus in Chinese-Hamster Ovary-cells, allowing $Zn^{2+}$-release from Golgi lumen into cytosol. It has been also suggested that ZIP7 facilitates release of $Zn^{2+}$ from ER [164] and behaves as a critical component in sub-cellular re-distribution of $Zn^{2+}$ in other systems [165]. Additionally, it has been hypothesized that protein kinase-2 (CK2) triggers cytosolic $Zn^{2+}$-signaling-pathways by phosphorylating ZIP7 [116], while some studies have also highlighted its important contribution to $Zn^{2+}$-homeostasis under pathological conditions [166–168]. In addition, in a recent study, it was demonstrated that ZIP7, which predominantly localizes to the ER membrane, promotes rapid cell proliferation in intestinal crypts by maintaining ER function. They also found that mice with an intestinalepithelium-specific ZIP7 deletion exhibited extensive apoptosis in the stem-cell-derived transit-amplifying cells due to increased ER stress. They further showed that UPR signaling upregulates ZIP7, which maintains $[Zn^{2+}]_i$ homeostasis under ER stress and facilitates epithelial proliferation. Therefore, ZIP7 is considered as a novel regulator of $[Zn^{2+}]_i$ homeostasis of the intestinal epithelium [169].

Although studies have shown the presence of weakly expressed ZIP7 and ZnT7 in mammalian heart [170,171], their subcellular localizations and functional roles in cardiomyocytes were not yet known well. In that regard, we hypothesized that disruption of $Zn^{2+}$-transporters and $Zn^{2+}$-axis such as ZIP7 and ZnT7 might contribute to deleterious changes in diabetic cardiomyocytes. Therefore, we first clarified their subcellular localizations into S(E)R and then explored their functional roles in $Zn^{2+}$ homeostasis, particularly under hyperglycemia. Additionally, we tested their roles in cytosolic $Zn^{2+}$ re-distribution and development of ER-stress in hyperglycemic conditions, at most due to activation of casein kinase 2 alpha (CK2$\alpha$) [40]. We observed markedly increased mRNA and protein levels of ZIP7 in ventricular cardiomyocytes from diabetic rats or high glucose-treated H9c2 cells whilst ZnT7 expression was low comparison to those of controls. Additionally, we observed increased ZIP7 phosphorylation in response to high glucose in vivo and in vitro in ventricular cardiomyocytes. Using recombinant targeted FRET-based sensors, we showed that hyperglycemia induced a marked redistribution of cellular labile $Zn^{2+}$, increasing cytosolic labile $Zn^{2+}$ and lowering labile $Zn^{2+}$ in the S(E)R. These changes involve alterations in ZIP7-phosphorylation and were suppressed by siRNA-mediated silencing of CK2$\alpha$. Due to our whole data, we, for the first time, demonstrated that opposing changes in the expression of ZIP7 and ZnT7 observed in hyperglycemia is very important for development of ER stress in the heart. In addition, we also pointed out an importance of sub-cellular labile $Zn^{2+}$ re-distribution in the hyperglycemic heart, which is resulting from altered ZIP7 and ZnT7 activity and contributing to cardiac dysfunction in diabetes [40]. Furthermore, Myers [140] previously discussed very widely the roles of $Zn^{2+}$ transporters and $Zn^{2+}$ signaling by using recent new roles of $Zn^{2+}$ and its transporters in the synthesis, secretion, and action of insulin are dependent on zinc and the transporters in type 2 diabetes. Author, particularly, emphasized the role of cellular $Zn^{2+}$'s dynamic as a "intracellular second messenger" to control insulin signaling and glucose homeostasis. Therefore, it was raised extensively a new research field into the pathophysiology of insulin resistance and possibility of new this-field related drug targets in diabetes [13,172–175].

In conclusion, the already known data associated with the role of ZIPs and ZnTs provide novel insights into regulation of cellular-$Zn^{2+}$ and its role in the heart under pathological conditions, including hyperglycemia/diabetes-associated cardiac dysfunction. Additionally, all findings can provide new targets such as cellular $[Zn^{2+}]_i$-regulation via mediation of $Zn^{2+}$-transporters and suggest that modulation of some endogenous kinases such as CK2$\alpha$ may provide a novel means to correct cardiac dysfunction under any pathological condition.

## 7. Concluding Remarks

Both early and recent studies strongly emphasized how $[Zn^{2+}]_i$ homeostasis is tightly controlled by the coordinated regulation of its uptake, efflux, distribution, and storage in mammalian cells. A number of proteins involved in different signaling pathways, mitochondrial metabolism, and ion channels, which are also common targets of labile $Zn^{2+}$, play pivotal roles in controlling cardiac contractility. The already known documents associated with the role of zinc in cardiac function are summarized in Figure 4. However, these regulatory actions of $Zn^{2+}$ are not limited to the function of the heart, but also extend to numerous other organ systems in mammalians. In this review, the regulation of cellular labile $Zn^{2+}$ levels, $Zn^{2+}$-mediated signal transduction, impacts of $Zn^{2+}$ on ionic channels and

S(E)R, and finally, the roles of $Zn^{2+}$ transporters in healthy and diseased heart, including diabetic heart, were outlined to help widen the current understanding of the versatile and complex roles of $Zn^{2+}$. Although much has been learned from recent studies, revealed important relationships between $Zn^{2+}$ transporters and heart diseases and indicating the potential of $Zn^{2+}$ transporters as therapeutic targets, their precise physiological functions are not clear. Given the multiple roles of $Zn^{2+}$ in various cell types and the detailed research development of $Zn^{2+}$-containing new markers/sensors will improve the ways to handle heart failure in humans.

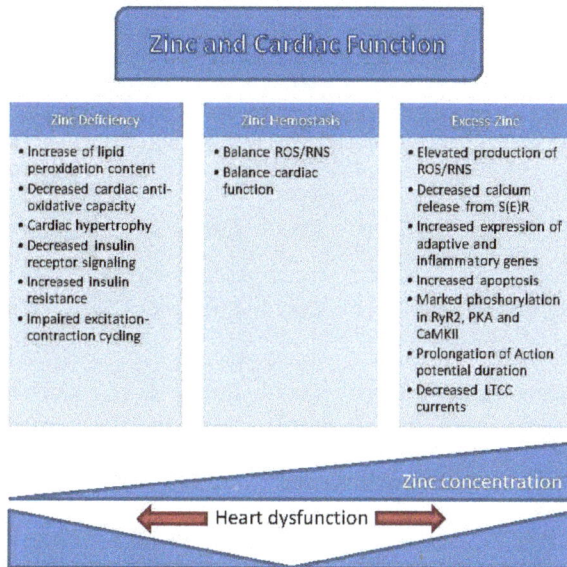

**Figure 4.** A summary of zinc and its role in cardiac function. Basic mechanisms affected with either zinc-deficiency or zinc-excess and, the, in turn, underline the heart dysfunction, mainly as a zinc-concentration dependent manner in the heart.

**Acknowledgments:** We thank our many colleagues for their excellent works. This work is supported through grant TUBITAK SBAG-113S466.

**Conflicts of Interest:** The authors declare no conflict of interest.

## References

1.  Oteiza, P.I. Zinc and the modulation of redox homeostasis. *Free Radic. Biol. Med.* **2012**, *53*, 1748–1759. [CrossRef] [PubMed]
2.  Vallee, B.L.; Falchuk, K.H. The biochemical basis of Zinc physiology. *Physiol. Rev.* **1993**, *73*, 79–118. [PubMed]
3.  Drinker, H.S. Concerning Modern Corporate Mortgages. *Univ. Pa. Law Rev. Am. Law* **1926**, *74*, 360–366. [CrossRef]
4.  Murakami, M.; Hirano, T. Intracellular Zinc homeostasis and Zinc signaling. *Cancer Sci.* **2008**, *99*, 1515–1522. [CrossRef] [PubMed]
5.  Maret, W. Zinc and human disease. *Met. Ions Life Sci.* **2013**, *13*, 389–414. [PubMed]
6.  Coudray, C.; Charlon, V.; de Leiris, J.; Favier, A. Effect of Zinc deficiency on lipid peroxidation status and infarct size in rat hearts. *Int. J. Cardiol.* **1993**, *41*, 109–113. [CrossRef]
7.  Maret, W. Zinc biochemistry: From a single Zinc enzyme to a key element of life. *Adv. Nutr. (Bethesda, Md.)* **2013**, *4*, 82–91. [CrossRef] [PubMed]

8.  Turan, B.; Fliss, H.; Desilets, M. Oxidants increase intracellular free $Zn^{2+}$ concentration in rabbit ventricular myocytes. *Am. J. Physiol.* **1997**, *272*, H2095–H2106. [PubMed]
9.  Tuncay, E.; Bilginoglu, A.; Sozmen, N.N.; Zeydanli, E.N.; Ugur, M.; Vassort, G.; Turan, B. Intracellular free Zinc during cardiac excitation-contraction cycle: Calcium and redox dependencies. *Cardiovasc. Res.* **2011**, *89*, 634–642. [CrossRef] [PubMed]
10. Tuncay, E.; Turan, B. Intracellular $Zn^{2+}$ increase in cardiomyocytes induces both electrical and mechanical dysfunction in heart via endogenous generation of reactive nitrogen species. *Biol. Trace Element Res.* **2016**, *169*, 294–302. [CrossRef] [PubMed]
11. Chabosseau, P.; Tuncay, E.; Meur, G.; Bellomo, E.A.; Hessels, A.; Hughes, S.; Johnson, P.R.; Bugliani, M.; Marchetti, P.; Turan, B.; et al. Mitochondrial and ER-targeted eCALWY probes reveal high levels of free $Zn^{2+}$. *ACS Chem. Biol.* **2014**, *9*, 2111–2120. [CrossRef] [PubMed]
12. Crawford, A.J.; Bhattacharya, S.K. Excessive intracellular Zinc accumulation in cardiac and skeletal muscles of dystrophic hamsters. *Exp. Neurol.* **1987**, *95*, 265–276. [CrossRef]
13. Jansen, J.; Rosenkranz, E.; Overbeck, S.; Warmuth, S.; Mocchegiani, E.; Giacconi, R.; Weiskirchen, R.; Karges, W.; Rink, L. Disturbed Zinc homeostasis in diabetic patients by in vitro and in vivo analysis of insulinomimetic activity of Zinc. *J. Nutr. Biochem.* **2012**, *23*, 1458–1466. [CrossRef] [PubMed]
14. Atar, D.; Backx, P.H.; Appel, M.M.; Gao, W.D.; Marban, E. Excitation-transcription coupling mediated by Zinc influx through voltage-dependent Calcium channels. *J. Biol. Chem.* **1995**, *270*, 2473–2477. [CrossRef] [PubMed]
15. Maret, W. Metals on the move: Zinc ions in cellular regulation and in the coordination dynamics of Zinc proteins. *Biometals* **2011**, *24*, 411–418. [CrossRef] [PubMed]
16. Coyle, P.; Philcox, J.C.; Carey, L.C.; Rofe, A.M. Metallothionein: The multipurpose protein. *Cell. Mol. Life Sci.* **2002**, *59*, 627–647. [CrossRef] [PubMed]
17. Hennigar, S.R.; Kelleher, S.L. Zinc networks: The cell-specific compartmentalization of Zinc for specialized functions. *Biol. Chem.* **2012**, *393*, 565–578. [CrossRef] [PubMed]
18. Kambe, T. An overview of a wide range of functions of ZnT and ZIP Zinc transporters in the secretory pathway. *Biosci. Biotechnol. Biochem.* **2011**, *75*, 1036–1043. [CrossRef] [PubMed]
19. Kambe, T.; Tsuji, T.; Hashimoto, A.; Itsumura, N. The physiological, biochemical, and molecular roles of Zinc transporters in Zinc homeostasis and metabolism. *Physiol. Rev.* **2015**, *95*, 749–784. [CrossRef] [PubMed]
20. Kambe, T.; Yamaguchi-Iwai, Y.; Sasaki, R.; Nagao, M. Overview of mammalian Zinc transporters. *Cell. Mol. Life Sci.* **2004**, *61*, 49–68. [CrossRef] [PubMed]
21. Eide, D.J. The SLC39 family of metal ion transporters. *Pflugers Arch. Eur. J. Physiol.* **2004**, *447*, 796–800. [CrossRef] [PubMed]
22. Lichten, L.A.; Cousins, R.J. Mammalian Zinc transporters: Nutritional and physiologic regulation. *Ann. Rev. Nutr.* **2009**, *29*, 153–176. [CrossRef] [PubMed]
23. Fukada, T.; Kambe, T. Molecular and genetic features of zinc transporters in physiology and pathogenesis. *Met. Integr. Biomet. Sci.* **2011**, *3*, 662–674. [CrossRef] [PubMed]
24. Todd, W.R.; Elvehjem, C.A.; Hart, E.B. Zinc in the nutrition of the rat. *Am. J. Physiol.* **1934**, *107*, 146–156. [CrossRef]
25. Prasad, A.S.; Miale, A., Jr.; Farid, Z.; Sandstead, H.H.; Schulert, A.R. Zinc metabolism in patients with the syndrome of iron deficiency anemia, hepatosplenomegaly, dwarfism, and hypognadism. *J. Lab. Clin. Med.* **1963**, *61*, 537–549. [PubMed]
26. Buamah, P.K.; Russell, M.; Bates, G.; Ward, A.M.; Skillen, A.W. Maternal zinc status: A determination of central nervous system malformation. *Br. J. Obstet. Gynaecol.* **1984**, *91*, 788–790. [CrossRef] [PubMed]
27. Prasad, A.S. Impact of the discovery of human Zinc deficiency on health. *J. Trace Elements Med. Biol. Organ Soc. Miner. Trace Elements (GMS)* **2014**, *28*, 357–363. [CrossRef] [PubMed]
28. Chen, X.P.; Zhang, Y.Q.; Tong, Y.P.; Xue, Y.F.; Liu, D.Y.; Zhang, W.; Deng, Y.; Meng, Q.F.; Yue, S.C.; Yan, P.; et al. Harvesting more grain Zinc of wheat for human health. *Sci. Rep.* **2017**, *7*, 7016. [CrossRef] [PubMed]
29. Hara, T.; Takeda, T.A.; Takagishi, T.; Fukue, K.; Kambe, T.; Fukada, T. Physiological roles of Zinc transporters: Molecular and genetic importance in Zinc homeostasis. *J. Physiol. Sci.* **2017**, *67*, 283–301. [CrossRef] [PubMed]
30. Arquilla, E.R.; Packer, S.; Tarmas, W.; Miyamoto, S. The effect of Zinc on insulin metabolism. *Endocrinology* **1978**, *103*, 1440–1449. [CrossRef] [PubMed]

31. Prasad, A.S. Zinc deficiency and effects of Zinc supplementation on sickle cell anemia subjects. *Progress Clin. Biol. Res.* **1981**, *55*, 99–122.

32. Savin, J.A. Skin disease: The link with Zinc. *Br. Med. J. (Clin. Res. Ed.)* **1984**, *289*, 1476–1477. [CrossRef]

33. Simmer, K.; Thompson, R.P. Maternal Zinc and intrauterine growth retardation. *Clin. Sci.* **1985**, *68*, 395–399. [CrossRef] [PubMed]

34. Allen, L.H. Zinc and micronutrient supplements for children. *Am. J. Clin. Nutr.* **1998**, *68*, 495s–498s. [PubMed]

35. Raulin, J. Chemical studies on vegetation. *J. Ann. Sci. Nat.* **1869**, *11*, 93–99. (In French)

36. Fosmire, G.J. Zinc toxicity. *Am. J. Clin. Nutr.* **1990**, *51*, 225–227. [PubMed]

37. Jones, M.M.; Schoenheit, J.E.; Weaver, A.D. Pretreatment and heavy metal LD50 values. *Toxicol. Appl. Pharmacol.* **1979**, *49*, 41–44. [CrossRef]

38. Bentley, P.J.; Grubb, B.R. Effects of a Zinc-deficient diet on tissue Zinc concentrations in rabbits. *J. Anim. Sci.* **1991**, *69*, 4876–4882. [CrossRef] [PubMed]

39. Prasad, A.S.; Walker, D.G.; Dehgani, A.; Halsted, J.A. Cirrhosis of the liver in Iran. *Arch. Intern. Med.* **1961**, *108*, 100–108. [CrossRef] [PubMed]

40. Tuncay, E.; Bitirim, V.C.; Durak, A.; Carrat, G.R.J.; Taylor, K.M.; Rutter, G.A.; Turan, B. Hyperglycemia-induced changes in ZIP7 and ZnT7 expression cause $Zn^{2+}$ Release From the sarco(endo)plasmic reticulum and mediate ER stress in the heart. *Diabetes* **2017**, *66*, 1346–1358. [CrossRef] [PubMed]

41. Bould, C.; Nicholas, D.J.; Tolhurst, J.A.H.; Wallace, T.; Potter, J.M.S. Zinc deficiency of fruit trees in Britain. *Nature* **1949**, *164*, 801. [CrossRef] [PubMed]

42. Stelmach, I.; Grzelewski, T.; Bobrowska-Korzeniowska, M.; Kopka, M.; Majak, P.; Jerzynska, J.; Stelmach, W.; Polanska, K.; Sobala, W.; Gromadzinska, J.; et al. The role of Zinc, Copper, plasma glutathione peroxidase enzyme, and vitamins in the development of allergic diseases in early childhood: The Polish mother and child cohort study. *Allergy Asthma Proc.* **2014**, *35*, 227–232. [CrossRef] [PubMed]

43. Markiewicz-Zukowska, R.; Gutowska, A.; Borawska, M.H. Serum Zinc concentrations correlate with mental and physical status of nursing home residents. *PLoS ONE* **2015**, *10*, e0117257. [CrossRef] [PubMed]

44. Yang, H.K.; Lee, S.H.; Han, K.; Kang, B.; Lee, S.Y.; Yoon, K.H.; Kwon, H.S.; Park, Y.M. Lower serum Zinc levels are associated with unhealthy metabolic status in normal-weight adults: The 2010 korea national health and nutrition examination survey. *Diabetes Metab.* **2015**, *41*, 282–290. [CrossRef] [PubMed]

45. Yary, T.; Virtanen, J.K.; Ruusunen, A.; Tuomainen, T.P.; Voutilainen, S. Serum Zinc and risk of type 2 diabetes incidence in men: The kuopio ischaemic heart disease risk factor study. *J. Trace Elements Med. Biol. Organ Soc. Miner. Trace Elements (GMS)* **2016**, *33*, 120–124. [CrossRef] [PubMed]

46. Hansen, A.F.; Simic, A.; Asvold, B.O.; Romundstad, P.R.; Midthjell, K.; Syversen, T.; Flaten, T.P. Trace elements in early phase type 2 diabetes mellitus-A population-based study. The HUNT study in Norway. *J. Trace Elements Med. Biol. Organ Soc. Miner. Trace Elements (GMS)* **2017**, *40*, 46–53. [CrossRef] [PubMed]

47. Simic, A.; Hansen, A.F.; Asvold, B.O.; Romundstad, P.R.; Midthjell, K.; Syversen, T.; Flaten, T.P. Trace element status in patients with type 2 diabetes in Norway: The HUNT3 Survey. *J. Trace Elements Med. Biol. Organ Soc. Miner. Trace Elements (GMS)* **2017**, *41*, 91–98. [CrossRef] [PubMed]

48. Martinez, S.S.; Campa, A.; Li, Y.; Fleetwood, C.; Stewart, T.; Ramamoorthy, V.; Baum, M.K. Low plasma Zinc is associated with higher mitochondrial oxidative stress and faster liver fibrosis development in the miami adult studies in HIV cohort. *J. Nutr.* **2017**, *147*, 556–562. [CrossRef] [PubMed]

49. Kunutsor, S.K.; Laukkanen, J.A. Serum Zinc concentrations and incident hypertension: New findings from a population-based cohort study. *J. Hypertens.* **2016**, *34*, 1055–1061. [CrossRef] [PubMed]

50. Lee, Y.M.; Wolf, P.; Hauner, H.; Skurk, T. Effect of a fermented dietary supplement containing chromium and Zinc on metabolic control in patients with type 2 diabetes: A randomized, placebo-controlled, double-blind cross-over study. *Food Nutr. Res.* **2016**, *60*, 30298. [CrossRef] [PubMed]

51. Tuncay, E.; Okatan, E.N.; Vassort, G.; Turan, B. β-blocker timolol prevents arrhythmogenic $Ca^{2+}$ release and normalizes $Ca^{2+}$ and $Zn^{2+}$ dyshomeostasis in hyperglycemic rat heart. *PLoS ONE* **2013**, *8*, e71014. [CrossRef] [PubMed]

52. Fraker, P.J.; Telford, W.G. A reappraisal of the role of zinc in life and death decisions of cells. *Proc. Soc. Exp. Biol. Med. Soc. Exp. Biol. Med. (New York, NY)* **1997**, *215*, 229–236. [CrossRef]

53. Truong-Tran, A.Q.; Carter, J.; Ruffin, R.E.; Zalewski, P.D. The role of Zinc in caspase activation and apoptotic cell death. *Biometals* **2001**, *14*, 315–330. [CrossRef] [PubMed]

54. McLaughlin, B.; Pal, S.; Tran, M.P.; Parsons, A.A.; Barone, F.C.; Erhardt, J.A.; Aizenman, E. p38 activation is required upstream of potassium current enhancement and caspase cleavage in thiol oxidant-induced neuronal apoptosis. *J. Neurosci.* **2001**, *21*, 3303–3311. [PubMed]

55. Wiseman, D.A.; Wells, S.M.; Wilham, J.; Hubbard, M.; Welker, J.E.; Black, S.M. Endothelial response to stress from exogenous $Zn^{2+}$ resembles that of NO-mediated nitrosative stress, and is protected by MT-1 overexpression. *Am. J. Physiol. Cell Physiol.* **2006**, *291*, C555–C568. [CrossRef] [PubMed]

56. Cummings, J.E.; Kovacic, J.P. The ubiquitous role of Zinc in health and disease. *J. Vet. Emerg. Crit. Care (San Antonio, Tex. 2001)* **2009**, *19*, 215–240. [CrossRef] [PubMed]

57. Hambidge, K.M.; Olivarasbach, J.; Jacobs, M.; Purcell, S.; Statland, C.; Poirier, J. Randomized study of Zinc supplementation during pregnancy. *Fed. Proc.* **1986**, *45*, 974.

58. Brown, M.A.; Thom, J.V.; Orth, G.L.; Cova, P.; Juarez, J. Food poisoning involving zinc contamination. *Arch. Environ. Health* **1964**, *8*, 657–660. [CrossRef] [PubMed]

59. Uski, O.; Torvela, T.; Sippula, O.; Karhunen, T.; Koponen, H.; Peraniemi, S.; Jalava, P.; Happo, M.; Jokiniemi, J.; Hirvonen, M.R.; et al. In vitro toxicological effects of Zinc containing nanoparticles with different physico-chemical properties. *Toxicol. In Vitro* **2017**, *42*, 105–113. [CrossRef] [PubMed]

60. Romanjuk, A.; Lyndin, M.; Moskalenko, R.; Gortinskaya, O.; Lyndina, Y. The role of heavy metal salts in pathological biomineralization of breast cancer tissue. *Adv. Clin. Exp. Med.* **2016**, *25*, 907–910. [CrossRef] [PubMed]

61. Hoang, B.X.; Han, B.; Shaw, D.G.; Nimni, M. Zinc as a possible preventive and therapeutic agent in pancreatic, prostate, and breast cancer. *Eur. J. Cancer Prev.* **2016**, *25*, 457–461. [CrossRef] [PubMed]

62. Livingstone, C. Zinc: Physiology, deficiency, and parenteral nutrition. *Nutr. Clin. Pract.* **2015**, *30*, 371–382. [CrossRef] [PubMed]

63. Cabrera, A.J. Zinc, aging, and immunosenescence: An overview. *Pathobiol. Aging Age Relat. Dis.* **2015**, *5*, 25592. [CrossRef] [PubMed]

64. Tsuji, T.; Kurokawa, Y.; Chiche, J.; Pouyssegur, J.; Sato, H.; Fukuzawa, H.; Nagao, M.; Kambe, T. Dissecting the process of activation of cancer-promoting Zinc-requiring ectoenzymes by Zinc metalation mediated by ZNT transporters. *J. Biol. Chem.* **2017**, *292*, 2159–2173. [CrossRef] [PubMed]

65. Aamodt, R.L.; Rumble, W.F.; Johnston, G.S.; Foster, D.; Henkin, R.I. Zinc metabolism in humans after oral and intravenous administration of Zn-69m. *Am. J. Clin. Nutr.* **1979**, *32*, 559–569. [PubMed]

66. Little, P.J.; Bhattacharya, R.; Moreyra, A.E.; Korichneva, I.L. Zinc and cardiovascular disease. *Nutrition* **2010**, *26*, 1050–1057. [CrossRef] [PubMed]

67. Gimelli, A.; Menichetti, F.; Soldati, E.; Liga, R.; Vannozzi, A.; Marzullo, P.; Bongiorni, M.G. Relationships between cardiac innervation/perfusion imbalance and ventricular arrhythmias: Impact on invasive electrophysiological parameters and ablation procedures. *Eur. J. Nucl. Med. Mol. Imaging* **2016**, *43*, 2383–2391. [CrossRef] [PubMed]

68. Wang, S.; Wang, B.; Wang, Y.; Tong, Q.; Liu, Q.; Sun, J.; Zheng, Y.; Cai, L. Zinc prevents the development of diabetic cardiomyopathy in db/db mice. *Int. J. Mol. Sci.* **2017**, *18*, 580. [CrossRef] [PubMed]

69. Bozym, R.A.; Chimienti, F.; Giblin, L.J.; Gross, G.W.; Korichneva, I.; Li, Y.; Libert, S.; Maret, W.; Parviz, M.; Frederickson, C.J.; et al. Free Zinc ions outside a narrow concentration range are toxic to a variety of cells in vitro. *Exp. Biol. Med. (Maywood, NJ)* **2010**, *235*, 741–750. [CrossRef] [PubMed]

70. Efeovbokhan, N.; Bhattacharya, S.K.; Ahokas, R.A.; Sun, Y.; Guntaka, R.V.; Gerling, I.C.; Weber, K.T. Zinc and the prooxidant heart failure phenotype. *J. Cardiovasc. Pharmacol.* **2014**, *64*, 393–400. [CrossRef] [PubMed]

71. Hashemian, M.; Poustchi, H.; Mohammadi-Nasrabadi, F.; Hekmatdoost, A. Systematic review of Zinc biochemical indicators and risk of coronary heart disease. *ARYA Atheroscler.* **2015**, *11*, 357–365. [PubMed]

72. Lee, S.R.; Noh, S.J.; Pronto, J.R.; Jeong, Y.J.; Kim, H.K.; Song, I.S.; Xu, Z.; Kwon, H.Y.; Kang, S.C.; Sohn, E.H.; et al. The critical roles of Zinc: Beyond impact on myocardial signaling. *Korean J. Physiol. Pharmacol.* **2015**, *19*, 389–399. [CrossRef] [PubMed]

73. Chu, A.; Foster, M.; Samman, S. Zinc Status and risk of cardiovascular diseases and type 2 diabetes mellitus-a systematic review of prospective cohort studies. *Nutrients* **2016**, *8*, 707. [CrossRef] [PubMed]

74. Tatsumi, T.; Fliss, H. Hypochlorous acid and chloramines increase endothelial permeability: Possible involvement of cellular zinc. *Am. J. Physiol.* **1994**, *267 Pt 2*, H1597–H1607.

75. Turan, B. Zinc-induced changes in ionic currents of cardiomyocytes. *Biol. Trace Element Res.* **2003**, *94*, 49–60. [CrossRef]

76. Zima, A.V.; Blatter, L.A. Redox regulation of cardiac calcium channels and transporters. *Cardiovasc. Res.* **2006**, *71*, 310–321. [CrossRef] [PubMed]

77. Dineley, K.E.; Richards, L.L.; Votyakova, T.V.; Reynolds, I.J. Zinc causes loss of membrane potential and elevates reactive oxygen species in rat brain mitochondria. *Mitochondrion* **2005**, *5*, 55–65. [CrossRef] [PubMed]

78. Ayaz, M.; Turan, B. Selenium prevents diabetes-induced alterations in [Zn$^{2+}$]$_i$ and metallothionein level of rat heart via restoration of cell redox cycle. *Am. J. Physiol. Heart Circ. Physiol.* **2006**, *290*, H1071–H1080. [CrossRef] [PubMed]

79. Maret, W. Molecular aspects of human cellular zinc homeostasis: Redox control of zinc potentials and zinc signals. *Biometals* **2009**, *22*, 149–157. [CrossRef] [PubMed]

80. Kuster, G.M.; Lancel, S.; Zhang, J.; Communal, C.; Trucillo, M.P.; Lim, C.C.; Pfister, O.; Weinberg, E.O.; Cohen, R.A.; Liao, R.; et al. Redox-mediated reciprocal regulation of SERCA and Na$^+$-Ca$^{2+}$ exchanger contributes to sarcoplasmic reticulum Ca$^{2+}$ depletion in cardiac myocytes. *Free Radic. Biol. Med.* **2010**, *48*, 1182–1187. [CrossRef] [PubMed]

81. Ranasinghe, P.; Pigera, S.; Galappatthy, P.; Katulanda, P.; Constantine, G.R. Zinc and diabetes mellitus: Understanding molecular mechanisms and clinical implications. *Daru* **2015**, *23*, 44. [CrossRef] [PubMed]

82. Tuncay, E.; Okatan, E.N.; Toy, A.; Turan, B. Enhancement of cellular antioxidant-defence preserves diastolic dysfunction via regulation of both diastolic Zn$^{2+}$ and Ca$^{2+}$ and prevention of RyR2-leak in hyperglycemic cardiomyocytes. *Oxid. Med. Cell. Longev.* **2014**, *2014*, 290381. [CrossRef] [PubMed]

83. Kamalov, G.; Ahokas, R.A.; Zhao, W.; Zhao, T.; Shahbaz, A.U.; Johnson, P.L.; Bhattacharya, S.K.; Sun, Y.; Gerling, I.C.; Weber, K.T. Uncoupling the coupled calcium and zinc dyshomeostasis in cardiac myocytes and mitochondria seen in aldosteronism. *J. Cardiovasc. Pharmacol.* **2010**, *55*, 248–254. [CrossRef] [PubMed]

84. Reilly-O'Donnell, B.; Robertson, G.B.; Karumbi, A.; McIntyre, C.; Bal, W.; Nishi, M.; Takeshima, H.; Stewart, A.J.; Pitt, S.J. Dysregulated Zn$^{2+}$ homeostasis impairs cardiac type-2 ryanodine receptor and mitsugumin 23 functions, leading to sarcoplasmic reticulum Ca$^{2+}$ leakage. *J. Biol. Chem.* **2017**, *292*, 13361–13373. [CrossRef] [PubMed]

85. Brugger, D.; Windisch, W.M. Short-term subclinical Zinc deficiency in weaned piglets affects cardiac redox metabolism and Zinc concentration. *J. Nutr.* **2017**, *147*, 521–527. [CrossRef] [PubMed]

86. Permyakov, E.A.; Kretsinger, R.H. Cell signaling, beyond cytosolic calcium in eukaryotes. *J. Inorg. Biochem.* **2009**, *103*, 77–86. [CrossRef] [PubMed]

87. Hirano, T.; Murakami, M.; Fukada, T.; Nishida, K.; Yamasaki, S.; Suzuki, T. Roles of Zinc and Zinc signaling in immunity: Zinc as an intracellular signaling molecule. *Adv. Immunol.* **2008**, *97*, 149–176. [PubMed]

88. Cicek, F.A.; Tokcaer-Keskin, Z.; Ozcinar, E.; Bozkus, Y.; Akcali, K.C.; Turan, B. Di-peptidyl peptidase-4 inhibitor sitagliptin protects vascular function in metabolic syndrome: Possible role of epigenetic regulation. *Mol. Biol. Rep.* **2014**, *41*, 4853–4863. [CrossRef] [PubMed]

89. Billur, D.; Tuncay, E.; Okatan, E.N.; Olgar, Y.; Durak, A.T.; Degirmenci, S.; Can, B.; Turan, B. Interplay between cytosolic free Zn$^{2+}$ and mitochondrion morphology changes in rat ventricular cardiomyocytes. *Biol. Trace Element Res.* **2016**, *174*, 177–188. [CrossRef] [PubMed]

90. Wang, S.; Luo, M.; Zhang, Z.; Gu, J.; Chen, J.; Payne, K.M.; Tan, Y.; Wang, Y.; Yin, X.; Zhang, X.; et al. Zinc deficiency exacerbates while Zinc supplement attenuates cardiac hypertrophy in high-fat diet-induced obese mice through modulating p38 MAPK-dependent signaling. *Toxicol. Lett.* **2016**, *258*, 134–146. [CrossRef] [PubMed]

91. Wu, W.; Bromberg, P.A.; Samet, J.M. Zinc ions as effectors of environmental oxidative lung injury. *Free Radic. Biol. Med.* **2013**, *65*, 57–69. [CrossRef] [PubMed]

92. Plum, L.M.; Rink, L.; Haase, H. The essential toxin: Impact of Zinc on human health. *Int. J. Environ. Res. Public Health* **2010**, *7*, 1342–1365. [CrossRef] [PubMed]

93. Yaras, N.; Ugur, M.; Ozdemir, S.; Gurdal, H.; Purali, N.; Lacampagne, A.; Vassort, G.; Turan, B. Effects of diabetes on ryanodine receptor Ca release channel (RyR2) and Ca$^{2+}$ homeostasis in rat heart. *Diabetes* **2005**, *54*, 3082–3088. [CrossRef] [PubMed]

94. Ermak, G.; Davies, K.J. Calcium and oxidative stress: From cell signaling to cell death. *Mol. Immunol.* **2002**, *38*, 713–721. [CrossRef]

95. Chakraborti, T.; Ghosh, S.K.; Michael, J.R.; Batabyal, S.K.; Chakraborti, S. Targets of oxidative stress in cardiovascular system. *Mol. Cell. Biochem.* **1998**, *187*, 1–10. [CrossRef] [PubMed]

96. Kawakami, M.; Okabe, E. Superoxide anion radical-triggered Ca$^{2+}$ release from cardiac sarcoplasmic reticulum through ryanodine receptor Ca$^{2+}$ channel. *Mol. Pharmacol.* **1998**, *53*, 497–503. [PubMed]
97. Boraso, A.; Williams, A.J. Modification of the gating of the cardiac sarcoplasmic reticulum Ca$^{2+}$-release channel by H$_2$O$_2$ and dithiothreitol. *Am. J. Physiol.* **1994**, *267 Pt 2*, H1010–H1016.
98. Woodier, J.; Rainbow, R.D.; Stewart, A.J.; Pitt, S.J. Intracellular Zinc modulates cardiac ryanodine receptor-mediated Calcium release. *J. Biol. Chem.* **2015**, *290*, 17599–17610. [CrossRef] [PubMed]
99. Stewart, A.J.; Pitt, S.J. Zinc controls RyR2 activity during excitation-contraction coupling. *Channels* **2015**, *9*, 227–229. [CrossRef] [PubMed]
100. Kamalov, G.; Ahokas, R.A.; Zhao, W.; Shahbaz, A.U.; Bhattacharya, S.K.; Sun, Y.; Gerling, I.C.; Weber, K.T. Temporal responses to intrinsically coupled Calcium and Zinc dyshomeostasis in cardiac myocytes and mitochondria during aldosteronism. *Am. J. Physiol. Heart Circ. Physiol.* **2010**, *298*, H385–H394. [CrossRef] [PubMed]
101. Xie, H.; Zhu, P.H. Biphasic modulation of ryanodine receptors by sulfhydryl oxidation in rat ventricular myocytes. *Biophys. J.* **2006**, *91*, 2882–2891. [CrossRef] [PubMed]
102. Perez-Clausell, J.; Danscher, G. Intravesicular localization of Zinc in rat telencephalic boutons. A histochemical study. *Brain Res.* **1985**, *337*, 91–98. [CrossRef]
103. Slomianka, L. Neurons of origin of Zinc-containing pathways and the distribution of Zinc-containing boutons in the hippocampal region of the rat. *Neuroscience* **1992**, *48*, 325–352. [CrossRef]
104. Sekler, I.; Sensi, S.L.; Hershfinkel, M.; Silverman, W.F. Mechanism and regulation of cellular Zinc transport. *Mol. Med.* **2007**, *13*, 337–343. [CrossRef] [PubMed]
105. Kerchner, G.A.; Canzoniero, L.M.; Yu, S.P.; Ling, C.; Choi, D.W. Zn$^{2+}$ current is mediated by voltage-gated Ca$^{2+}$ channels and enhanced by extracellular acidity in mouse cortical neurones. *J. Physiol.* **2000**, *528 Pt 1*, 39–52. [CrossRef]
106. Alvarez-Collazo, J.; Diaz-Garcia, C.M.; Lopez-Medina, A.I.; Vassort, G.; Alvarez, J.L. Zinc modulation of basal and β-adrenergically stimulated L-type Ca$^{2+}$ current in rat ventricular cardiomyocytes: Consequences in cardiac diseases. *Pflugers Arch.* **2012**, *464*, 459–470. [CrossRef] [PubMed]
107. Traynelis, S.F.; Burgess, M.F.; Zheng, F.; Lyuboslavsky, P.; Powers, J.L. Control of voltage-independent Zinc inhibition of NMDA receptors by the NR1 subunit. *J. Neurosci.* **1998**, *18*, 6163–6175. [PubMed]
108. Zhang, S.; Kehl, S.J.; Fedida, D. Modulation of Kv1.5 potassium channel gating by extracellular Zinc. *Biophys. J.* **2001**, *81*, 125–136. [CrossRef]
109. Gilly, W.F.; Armstrong, C.M. Slowing of sodium channel opening kinetics in squid axon by extracellular Zinc. *J. Gen. Physiol.* **1982**, *79*, 935–964. [CrossRef] [PubMed]
110. Gilly, W.F.; Armstrong, C.M. Divalent cations and the activation kinetics of potassium channels in squid giant axons. *J. Gen. Physiol.* **1982**, *79*, 965–996. [CrossRef] [PubMed]
111. Aras, M.A.; Saadi, R.A.; Aizenman, E. Zn$^{2+}$ regulates Kv2.1 voltage-dependent gating and localization following ischemia. *Eur. J. Neurosci.* **2009**, *30*, 2250–2257. [CrossRef] [PubMed]
112. Haase, H.; Maret, W. Intracellular Zinc fluctuations modulate protein tyrosine phosphatase activity in insulin/insulin-like growth factor-1 signaling. *Exp. Cell Res.* **2003**, *291*, 289–298. [CrossRef]
113. von Bulow, V.; Rink, L.; Haase, H. Zinc-mediated inhibition of cyclic nucleotide phosphodiesterase activity and expression suppresses TNF-α and IL-1 β production in monocytes by elevation of guanosine 3′,5′-cyclic monophosphate. *J. Immunol.* **2005**, *175*, 4697–4705. [CrossRef] [PubMed]
114. van der Heyden, M.A.; Wijnhoven, T.J.; Opthof, T. Molecular aspects of adrenergic modulation of cardiac L-type Ca$^{2+}$ channels. *Cardiovasc. Res.* **2005**, *65*, 28–39. [CrossRef] [PubMed]
115. Wilson, M.; Hogstrand, C.; Maret, W. Picomolar concentrations of free zinc(II) ions regulate receptor protein-tyrosine phosphatase beta activity. *J. Biol. Chem.* **2012**, *287*, 9322–9326. [CrossRef] [PubMed]
116. Taylor, K.M.; Hiscox, S.; Nicholson, R.I.; Hogstrand, C.; Kille, P. Protein kinase CK2 triggers cytosolic Zinc signaling pathways by phosphorylation of Zinc channel ZIP7. *Sci. Signal.* **2012**, *5*, ra11. [CrossRef] [PubMed]
117. Klein, C.; Sunahara, R.K.; Hudson, T.Y.; Heyduk, T.; Howlett, A.C. Zinc inhibition of cAMP signaling. *J. Biol. Chem.* **2002**, *277*, 11859–11865. [CrossRef] [PubMed]
118. Gao, X.; Du, Z.; Patel, T.B. Copper and Zinc inhibit Galphas function: A nucleotide-free state of Galphas induced by Cu$^{2+}$ and Zn$^{2+}$. *J. Biol. Chem.* **2005**, *280*, 2579–2586. [CrossRef] [PubMed]

119. Baltas, L.G.; Karczewski, P.; Bartel, S.; Krause, E.G. The endogenous cardiac sarcoplasmic reticulum Ca$^{2+}$/calmodulin-dependent kinase is activated in response to β-adrenergic stimulation and becomes Ca$^{2+}$-independent in intact beating hearts. *FEBS Lett.* **1997**, *409*, 131–136. [CrossRef]

120. Yi, T.; Vick, J.S.; Vecchio, M.J.; Begin, K.J.; Bell, S.P.; Delay, R.J.; Palmer, B.M. Identifying cellular mechanisms of Zinc-induced relaxation in isolated cardiomyocytes. *Am. J. Physiol. Heart Circ. Physiol.* **2013**, *305*, H706–H715. [CrossRef] [PubMed]

121. Gao, H.; Boillat, A.; Huang, D.; Liang, C.; Peers, C.; Gamper, N. Intracellular zinc activates KCNQ channels by reducing their dependence on phosphatidylinositol 4,5-bisphosphate. *Proc. Natl. Acad. Sci. USA* **2017**, *114*, E6410–E6419. [CrossRef] [PubMed]

122. Sensi, S.L.; Paoletti, P.; Bush, A.I.; Sekler, I. Zinc in the physiology and pathology of the CNS. *Nat. Rev. Neurosci.* **2009**, *10*, 780–791. [CrossRef] [PubMed]

123. Frederickson, C.J.; Koh, J.Y.; Bush, A.I. The neurobiology of Zinc in health and disease. *Nat. Rev. Neurosci.* **2005**, *6*, 449–462. [CrossRef] [PubMed]

124. Maret, W. Zinc coordination environments in proteins as redox sensors and signal transducers. *Antioxid. Redox Signal.* **2006**, *8*, 1419–1441. [CrossRef] [PubMed]

125. Dineley, K.E.; Devinney, M.J., 2nd; Zeak, J.A.; Rintoul, G.L.; Reynolds, I.J. Glutamate mobilizes [Zn$^{2+}$] through Ca$^{2+}$-dependent reactive oxygen species accumulation. *J. Neurochem.* **2008**, *106*, 2184–2193. [PubMed]

126. Ellis, C.D.; Wang, F.; MacDiarmid, C.W.; Clark, S.; Lyons, T.; Eide, D.J. Zinc and the MSC2 Zinc transporter protein are required for endoplasmic reticulum function. *J. Cell Biol.* **2004**, *166*, 325–335. [CrossRef] [PubMed]

127. Sensi, S.L.; Ton-That, D.; Weiss, J.H. Mitochondrial sequestration and Ca$^{2+}$-dependent release of cytosolic Zn$^{2+}$ loads in cortical neurons. *Neurobiol. Dis.* **2002**, *10*, 100–108. [CrossRef] [PubMed]

128. Chance, B.; Sies, H.; Boveris, A. Hydroperoxide metabolism in mammalian organs. *Physiol. Rev.* **1979**, *59*, 527–605. [PubMed]

129. Dedkova, E.N.; Blatter, L.A. Characteristics and function of cardiac mitochondrial nitric oxide synthase. *J. Physiol.* **2009**, *587 Pt 4*, 851–872. [CrossRef]

130. Bouron, A.; Kiselyov, K.; Oberwinkler, J. Permeation, regulation and control of expression of TRP channels by trace metal ions. *Pflugers Arch.* **2015**, *467*, 1143–1164. [CrossRef] [PubMed]

131. Yamaguchi, Y.; Iribe, G.; Nishida, M.; Naruse, K. Role of TRPC3 and TRPC6 channels in the myocardial response to stretch: Linking physiology and pathophysiology. *Progress Biophys. Mol. Biol.* **2017**. [CrossRef] [PubMed]

132. Hu, Y.; Duan, Y.; Takeuchi, A.; Hai-Kurahara, L.; Ichikawa, J.; Hiraishi, K.; Numata, T.; Ohara, H.; Iribe, G.; Nakaya, M.; et al. Uncovering the arrhythmogenic potential of TRPM4 activation in atrial-derived HL-1 cells using novel recording and numerical approaches. *Cardiovasc. Res.* **2017**, *113*, 1243–1255. [CrossRef] [PubMed]

133. Uchida, K.; Tominaga, M. Extracellular Zinc ion regulates transient receptor potential melastatin 5 (TRPM5) channel activation through its interaction with a pore loop domain. *J. Biol. Chem.* **2013**, *288*, 25950–25955. [CrossRef] [PubMed]

134. Lambert, S.; Drews, A.; Rizun, O.; Wagner, T.F.; Lis, A.; Mannebach, S.; Plant, S.; Portz, M.; Meissner, M.; Philipp, S.E.; et al. Transient receptor potential melastatin 1 (TRPM1) is an ion-conducting plasma membrane channel inhibited by Zinc ions. *J. Biol. Chem.* **2011**, *286*, 12221–12233. [CrossRef] [PubMed]

135. Abiria, S.A.; Krapivinsky, G.; Sah, R.; Santa-Cruz, A.G.; Chaudhuri, D.; Zhang, J.; Adstamongkonkul, P.; DeCaen, P.G.; Clapham, D.E. TRPM7 senses oxidative stress to release Zn$^{2+}$ from unique intracellular vesicles. *Proc. Natl. Acad. Sci. USA* **2017**, *114*, E6079–E6088. [CrossRef] [PubMed]

136. Ma, S.; Jiang, Y.; Huang, W.; Li, X.; Li, S. Role of transient receptor potential channels in heart transplantation: A potential novel therapeutic target for cardiac allograft vasculopathy. *Med. Sci. Monit.* **2017**, *23*, 2340–2347. [CrossRef] [PubMed]

137. Jang, Y.; Wang, H.; Xi, J.; Mueller, R.A.; Norfleet, E.A.; Xu, Z. NO mobilizes intracellular Zn$^{2+}$ via cGMP/PKG signaling pathway and prevents mitochondrial oxidant damage in cardiomyocytes. *Cardiovasc. Res.* **2007**, *75*, 426–433. [CrossRef] [PubMed]

138. Kambe, T.; Fukada, T.; Toyokuni, S. Editorial: The cutting edge of zinc biology. *Arch. Biochem. Biophys.* **2016**, *611*, 1–2. [CrossRef] [PubMed]

139. Kimura, T.; Kambe, T. The functions of metallothionein and ZIP and ZnT transporters: An overview and perspective. *Int. J. Mol. Sci.* **2016**, *17*, 336. [CrossRef] [PubMed]

140. Myers, S.A. Zinc transporters and Zinc signaling: New insights into their role in type 2 diabetes. *Int. J. Endocrinol.* **2015**, *2015*, 167503. [CrossRef] [PubMed]

141. Huang, L.; Tepaamorndech, S. The SLC30 family of zinc transporters-a review of current understanding of their biological and pathophysiological roles. *Mol. Aspects Med.* **2013**, *34*, 548–560. [CrossRef] [PubMed]

142. Jeong, J.; Eide, D.J. The SLC39 family of Zinc transporters. *Mol. Aspects Med.* **2013**, *34*, 612–619. [CrossRef] [PubMed]

143. Kambe, T. Methods to evaluate Zinc transport into and out of the secretory and endosomal-lysosomal compartments in DT40 cells. *Methods Enzymol.* **2014**, *534*, 77–92. [PubMed]

144. Kambe, T.; Hashimoto, A.; Fujimoto, S. Current understanding of ZIP and ZnT zinc transporters in human health and diseases. *Cell. Mol. Life Sci. CMLS* **2014**, *71*, 3281–3295. [CrossRef] [PubMed]

145. Shima, T.; Jesmin, S.; Matsui, T.; Soya, M.; Soya, H. Differential effects of type 2 diabetes on brain glycometabolism in rats: Focus on glycogen and monocarboxylate transporter 2. *J. Physiol. Sci.* **2016**. [CrossRef] [PubMed]

146. Bu, H.M.; Yang, C.Y.; Wang, M.L.; Ma, H.J.; Sun, H.; Zhang, Y. K(ATP) channels and MPTP are involved in the cardioprotection bestowed by chronic intermittent hypobaric hypoxia in the developing rat. *J. Physiol. Sci.* **2015**, *65*, 367–376. [CrossRef] [PubMed]

147. King, J.C.; Shames, D.M.; Woodhouse, L.R. Zinc homeostasis in humans. *J. Nutr.* **2000**, *130* (Suppl. S5), 1360s–1366s. [PubMed]

148. Palmiter, R.D.; Findley, S.D. Cloning and functional characterization of a mammalian Zinc transporter that confers resistance to Zinc. *EMBO J.* **1995**, *14*, 639–649. [PubMed]

149. Wenzel, H.J.; Cole, T.B.; Born, D.E.; Schwartzkroin, P.A.; Palmiter, R.D. Ultrastructural localization of Zinc transporter-3 (ZnT-3) to synaptic vesicle membranes within mossy fiber boutons in the hippocampus of mouse and monkey. *Proc. Natl. Acad. Sci. USA* **1997**, *94*, 12676–12681. [CrossRef] [PubMed]

150. Kambe, T.; Weaver, B.P.; Andrews, G.K. The genetics of essential metal homeostasis during development. *Genesis* **2008**, *46*, 214–228. [CrossRef] [PubMed]

151. Bin, B.H.; Fukada, T.; Hosaka, T.; Yamasaki, S.; Ohashi, W.; Hojyo, S.; Miyai, T.; Nishida, K.; Yokoyama, S.; Hirano, T. Biochemical characterization of human ZIP13 protein: A homo-dimerized zinc transporter involved in the spondylocheiro dysplastic Ehlers-Danlos syndrome. *J. Biol. Chem.* **2011**, *286*, 40255–40265. [CrossRef] [PubMed]

152. Taylor, K.M.; Morgan, H.E.; Smart, K.; Zahari, N.M.; Pumford, S.; Ellis, I.O.; Robertson, J.F.; Nicholson, R.I. The emerging role of the LIV-1 subfamily of Zinc transporters in breast cancer. *Mol. Med.* **2007**, *13*, 396–406. [CrossRef] [PubMed]

153. Kambe, T.; Geiser, J.; Lahner, B.; Salt, D.E.; Andrews, G.K. *Slc39a1* to 3 (subfamily II) Zip genes in mice have unique cell-specific functions during adaptation to Zinc deficiency. *Am. J. Physiol. Regul. Integr. Comp. Physiol.* **2008**, *294*, R1474–R1481. [CrossRef] [PubMed]

154. Huang, L.; Yu, Y.Y.; Kirschke, C.P.; Gertz, E.R.; Lloyd, K.K. Znt7 (Slc30a7)-deficient mice display reduced body Zinc status and body fat accumulation. *J. Biol. Chem.* **2007**, *282*, 37053–37063. [CrossRef] [PubMed]

155. Huang, L.; Kirschke, C.P.; Zhang, Y.; Yu, Y.Y. The ZIP7 gene (*Slc39a7*) encodes a Zinc transporter involved in Zinc homeostasis of the Golgi apparatus. *J. Biol. Chem.* **2005**, *280*, 15456–15463. [CrossRef] [PubMed]

156. Sladek, R.; Rocheleau, G.; Rung, J.; Dina, C.; Shen, L.; Serre, D.; Boutin, P.; Vincent, D.; Belisle, A.; Hadjadj, S.; et al. A genome-wide association study identifies novel risk loci for type 2 diabetes. *Nature* **2007**, *445*, 881–885. [CrossRef] [PubMed]

157. Wenzlau, J.M.; Juhl, K.; Yu, L.; Moua, O.; Sarkar, S.A.; Gottlieb, P.; Rewers, M.; Eisenbarth, G.S.; Jensen, J.; Davidson, H.W.; et al. The cation efflux transporter ZnT8 (Slc30A8) is a major autoantigen in human type 1 diabetes. *Proc. Natl. Acad. Sci. USA* **2007**, *104*, 17040–17045. [CrossRef] [PubMed]

158. Lemaire, K.; Ravier, M.A.; Schraenen, A.; Creemers, J.W.; Van de Plas, R.; Granvik, M.; Van Lommel, L.; Waelkens, E.; Chimienti, F.; Rutter, G.A.; et al. Insulin crystallization depends on Zinc transporter ZnT8 expression, but is not required for normal glucose homeostasis in mice. *Proc. Natl. Acad. Sci. USA* **2009**, *106*, 14872–14877. [CrossRef] [PubMed]

159. Nicolson, T.J.; Bellomo, E.A.; Wijesekara, N.; Loder, M.K.; Baldwin, J.M.; Gyulkhandanyan, A.V.; Koshkin, V.; Tarasov, A.I.; Carzaniga, R.; Kronenberger, K.; et al. Insulin storage and glucose homeostasis in mice null for the granule zinc transporter ZnT8 and studies of the type 2 diabetes-associated variants. *Diabetes* **2009**, *58*, 2070–2083. [CrossRef] [PubMed]

160. Wijesekara, N.; Dai, F.F.; Hardy, A.B.; Giglou, P.R.; Bhattacharjee, A.; Koshkin, V.; Chimienti, F.; Gaisano, H.Y.; Rutter, G.A.; Wheeler, M.B. β cell-specific ZnT8 deletion in mice causes marked defects in insulin processing, crystallisation and secretion. *Diabetologia* **2010**, *53*, 1656–1668. [CrossRef] [PubMed]

161. Tamaki, M.; Fujitani, Y.; Hara, A.; Uchida, T.; Tamura, Y.; Takeno, K.; Kawaguchi, M.; Watanabe, T.; Ogihara, T.; Fukunaka, A.; et al. The diabetes-susceptible gene *Slc30A8/ZnT8* regulates hepatic insulin clearance. *J. Clin. Investig.* **2013**, *123*, 4513–4524. [CrossRef] [PubMed]

162. Chabosseau, P.; Rutter, G.A. Zinc and diabetes. *Arch. Biochem. Biophys.* **2016**, *611*, 79–85. [CrossRef] [PubMed]

163. Rutter, G.A.; Chabosseau, P.; Bellomo, E.A.; Maret, W.; Mitchell, R.K.; Hodson, D.J.; Solomou, A.; Hu, M. Intracellular Zinc in insulin secretion and action: A determinant of diabetes risk? *Proc. Nutr. Soc.* **2016**, *75*, 61–72. [CrossRef] [PubMed]

164. Hogstrand, C.; Kille, P.; Nicholson, R.I.; Taylor, K.M. Zinc transporters and cancer: A potential role for ZIP7 as a hub for tyrosine kinase activation. *Trends Mol. Med.* **2009**, *15*, 101–111. [CrossRef] [PubMed]

165. Taylor, K.M.; Vichova, P.; Jordan, N.; Hiscox, S.; Hendley, R.; Nicholson, R.I. ZIP7-mediated intracellular Zinc transport contributes to aberrant growth factor signaling in antihormone-resistant breast cancer Cells. *Endocrinology* **2008**, *149*, 4912–4920. [CrossRef] [PubMed]

166. Liu, Y.; Batchuluun, B.; Ho, L.; Zhu, D.; Prentice, K.J.; Bhattacharjee, A.; Zhang, M.; Pourasgari, F.; Hardy, A.B.; Taylor, K.M.; et al. Characterization of Zinc influx transporters (ZIPs) in pancreatic β cells: Roles in regulating cytosolic Zinc homeostasis and insulin secretion. *J. Biol. Chem.* **2015**, *290*, 18757–18769. [CrossRef] [PubMed]

167. Grubman, A.; Lidgerwood, G.E.; Duncan, C.; Bica, L.; Tan, J.L.; Parker, S.J.; Caragounis, A.; Meyerowitz, J.; Volitakis, I.; Moujalled, D.; et al. Deregulation of subcellular biometal homeostasis through loss of the metal transporter, ZIP7, in a childhood neurodegenerative disorder. *Acta Neuropathol. Commun.* **2014**, *2*, 25. [CrossRef] [PubMed]

168. Groth, C.; Sasamura, T.; Khanna, M.R.; Whitley, M.; Fortini, M.E. Protein trafficking abnormalities in Drosophila tissues with impaired activity of the ZIP7 Zinc transporter Catsup. *Development* **2013**, *140*, 3018–3027. [CrossRef] [PubMed]

169. Ohashi, W.; Kimura, S.; Iwanaga, T.; Furusawa, Y.; Irie, T.; Izumi, H.; Watanabe, T.; Hijikata, A.; Hara, T.; Ohara, O.; et al. Zinc Transporter Slc39A7/ZIP7 promotes intestinal epithelial self-renewal by resolving ER stress. *PLoS Genet.* **2016**, *12*, e1006349. [CrossRef] [PubMed]

170. Kirschke, C.P.; Huang, L. ZnT7, a novel mammalian Zinc transporter, accumulates Zinc in the Golgi apparatus. *J. Biol. Chem.* **2003**, *278*, 4096–4102. [CrossRef] [PubMed]

171. Yang, J.; Zhang, Y.; Cui, X.; Yao, W.; Yu, X.; Cen, P.; Hodges, S.E.; Fisher, W.E.; Brunicardi, F.C.; Chen, C.; et al. Gene profile identifies Zinc transporters differentially expressed in normal human organs and human pancreatic cancer. *Curr. Mol. Med.* **2013**, *13*, 401–409. [PubMed]

172. Haase, H.; Maret, W. Fluctuations of cellular, available zinc modulate insulin signaling via inhibition of protein tyrosine phosphatases. *J. Trace Elements Med. Biol. Organ Soc. Miner. Trace Elements (GMS)* **2005**, *19*, 37–42. [CrossRef] [PubMed]

173. Yamasaki, S.; Sakata-Sogawa, K.; Hasegawa, A.; Suzuki, T.; Kabu, K.; Sato, E.; Kurosaki, T.; Yamashita, S.; Tokunaga, M.; Nishida, K.; et al. Zinc is a novel intracellular second messenger. *J. Cell Biol.* **2007**, *177*, 637–645. [CrossRef] [PubMed]

174. Myers, S.A.; Nield, A.; Myers, M. Zinc transporters, mechanisms of action and therapeutic utility: Implications for type 2 diabetes mellitus. *J. Nutr. Metab.* **2012**, *2012*, 173712. [CrossRef] [PubMed]

175. Jansen, J.; Karges, W.; Rink, L. Zinc and diabetes—Clinical links and molecular mechanisms. *J. Nutr. Biochem.* **2009**, *20*, 399–417. [CrossRef] [PubMed]

International Journal of
*Molecular Sciences*

MDPI

*Review*

# Dietary Zinc Acts as a Sleep Modulator

**Yoan Cherasse \*** and **Yoshihiro Urade**

International Institute for Integrative Sleep Medicine (WPI-IIIS), University of Tsukuba, 305-8575 Tsukuba, Japan; urade.yoshihiro.ft@u.tsukuba.ac.jp
\* Correspondence: cherasse.yoan.fm@u.tsukuba.ac.jp; Tel.: +81-29-853-3773

Received: 3 October 2017; Accepted: 2 November 2017; Published: 5 November 2017

**Abstract:** While zinc is known to be important for many biological processes in animals at a molecular and physiological level, new evidence indicates that it may also be involved in the regulation of sleep. Recent research has concluded that zinc serum concentration varies with the amount of sleep, while orally administered zinc increases the amount and the quality of sleep in mice and humans. In this review, we provide an exhaustive study of the literature connecting zinc and sleep, and try to evaluate which molecular mechanism is likely to be involved in this phenomenon. A better understanding should provide critical information not only about the way zinc is related to sleep but also about how sleep itself works and what its real function is.

**Keywords:** sleep; zinc; nutrition; brain; randomized controlled trial

## 1. Introduction

Zinc is the second most abundant trace metal in the human body, and is essential for many biological processes. Nevertheless, many new functions remain to be discovered for this unique divalent cation. A very recent body of evidence suggests that zinc is involved in the regulation of sleep, one of the most essential physiological functions in the entire animal kingdom. The aim of this review is to provide an overview of the current data suggesting that zinc influences sleep in mice and humans. We first provide a brief background on what sleep is, and what we know of its regulation. We then explore the identified roles of zinc in the brain, as well as how food regulates sleep. Furthermore, we perform a comprehensive review of the literature connecting zinc and sleep. Finally, we discuss possible molecular mechanisms by which zinc can act in the brain and regulate sleep.

## 2. Sleep

Sleep is defined as a natural periodic state of rest, in which the eyes remain closed and consciousness is completely or partially abolished, so that there is a decrease in bodily movement and responsiveness to external stimuli. However, an important aspect of sleep, unlike coma or anesthesia, is that it must be easily and immediately reversible. While everybody experiences sleep, 2500 years of research since Alcmaeon, Hippocrates and Aristotle (450–350 B.C.) could not yet clearly elucidate the most simplest question as "why do we need to sleep?". To finally answer this question, today's sleep research efforts are mostly focused on understanding how sleep is regulated.

Sleep is a process common to the whole animal kingdom, from *Caenorhabditis elegans* to *Drosophila melanogaster*, zebrafish and of course mammals [1]. Sleeplessness has a huge impact on the human physiology and is commonly associated with metabolic disorders (obesity and diabetes), cardiovascular diseases (hypertension) and mental disorders (anxiety and depression). Furthermore, insomnia has also recently been associated with neurodegenerative diseases such as Alzheimer's disease [2], while large cohort studies demonstrated that short (5 h or less) and long sleepers (more than 9 h) live shorter than people with an appropriate amount of sleep (6 to 8 h per night) [3–5]. In electroencephalogram (EEG) recordings, there are two distinct stages characterizing sleep: non-rapid eye movement (NREM) sleep, sometimes also called slow wave sleep, and rapid eye movement (REM) sleep or paradoxical sleep. NREM sleep is characterized by slow but relatively high-amplitude oscillations, while REM sleep exhibits an EEG with higher frequency but lower amplitude, similar to (but distinct from) that of wakefulness. Sleep has an essential function to allow the human body to physically restore and heal itself. It is especially important to maintain an efficient immune system and avoid metabolic and cardiovascular disorders associated with insomnia. In the brain, sleep is also essential in some memory consolidation processes [6] and possibly for brain detoxification [7,8].

The regulation of sleep and wakefulness involves many regions and cellular subtypes in the brain. Indeed, the ascending arousal system promotes wakefulness through a network composed of the monaminergic neurons in the locus coeruleus (LC), histaminergic neurons in the tuberomammilary nucleus (TMN), glutamatergic neurons in the parabrachial nucleus (PB) and orexinergic neurons in the lateral hypothalamus, among others. On the other hand, only a handful of regions able to promote sleep have been identified so far. The ventrolateral pre-optic area (VLPO) was the first "sleep center" to be identified by Saper's team [9], and is considered as the master regulator for the so-called "wake/sleep flip-flop switch" [10]. More recently, Fuller's laboratory also discovered that sleep can be promoted by the activation of a gamma-aminobutyric acid-ergic (GABAergic) population of neurons located in the parafacial zone [11,12], while the role of the GABAergic $A_{2A}R$-expressing neurons of the nucleus accumbens [13] and the striatum has just been revealed [14,15]. In total, more than 50 neurotransmitters and their respective receptors are involved in the process of controlling the vigilance state of the brain.

## 3. Zinc and the Central Nervous System

The trace metal zinc is an essential cofactor for more than 300 enzymes and 1000 transcription factors [16]. A moderate deficiency of zinc is sometimes observed in humans, and is responsible for growth retardation, male hypogonadism, taste alteration, inefficient wound healing and immune system, as well as mental retardation. In the central nervous system, zinc is the second most abundant trace metal and is involved in many processes. In addition to its role in enzymatic activity, it also plays a major role in cell signaling and modulation of neuronal activity. Zinc finger proteins, a huge family of zinc-containing proteins, play key roles in the mechanisms of DNA replication and transcription regulation [17–20]. Zinc has also been implicated in neurodegenerative diseases. Some Alzheimer's disease patients exhibit a systemic deficiency in zinc [21], however, it has also been proven that amyloid plaques are highly enriched in zinc. It is possible that the amyloid plaques immobilize the pool of zinc in the brain and therefore reduce the bioavailability into the neurons.

Zinc is utilized by tissue as a function of zinc transporters. Zinc transporters are playing an important role in the homeostasis of zinc and are tightly controlling concentration of this ion in the different organs in order to allow proper biological functions, while impaired zinc transporter function correlates with clinical human diseases [22]. Interestingly, a research in Drosophila studied the molecular polymorphisms of the gene *Catecholamines up*, strongly associated with day sleep [23], and characterized it as the Drosophila ortholog of the mammalian *ZIP7* zinc transporter [24].

In addition to its role as a cofactor, zinc is also a modulator of neuronal activity in the brain. While the majority of zinc is protein-bound, some specific subpopulations of neurons contain vesicles filled with weakly bound or free zinc ions ($Zn^{2+}$). These zinc-containing neurons were first identified in the mossy fibers of the hippocampus [25]. The first reported and most abundant population of zinc-containing neurons is glutamatergic (sometimes called "gluzinergic" neurons), and zinc released from the vesicles of these neurons into the synaptic cleft could modulate N-methyl-D-aspartate receptor (NMDAR) activity in a dose-dependent and reversible manner [26–28]. However, zinc can also modulate the activity of other glutamate receptors, such as α-amino-3-hydroxy-5-methyl-4-isoxazolepropionic acid (AMPA) [29], metabotropic receptors [30] as well as the receptors for other neurotransmitters [31] such as adenosine [32], dopamine [33] and serotonin [34]. Furthermore, zinc can also decrease the uptake of glutamate [35] and dopamine [36] transporters, and it exhibits various effects on calcium [37], potassium [38], sodium [39] and chloride [40] channels, while recent evidence demonstrates that zinc can also be released into glycinergic synapses [41].

One of the best-characterized physiological functions of zinc after its release into the synaptic cleft involves the modification of hippocampus-dependent memory by the amygdala. The lateral nucleus of the amygdala, a component of the limbic system that is essential for emotion, receives massive projections from the entorhinal cortex. Kodirov et al. suggested that synaptically released $Zn^{2+}$ in that location was responsible for long-term potentiation (LTP) by depressing feed-forward GABAergic inhibition of the post-synaptic neurons and thus serves as an essential mechanism for the acquisition and storage of spatial memory in a learning task [42].

For a long time, measuring zinc concentration in the synaptic cleft remained approximate at best, and reported results could vary by several orders of magnitude [43]. Finally, a recent in vivo study determined that the resting concentration of synaptic zinc is extremely low (<10 nM), but after being released from pre-synaptic vesicles after ischemic stroke zinc concentration rises quickly (within a few milliseconds) and remains in the nanomolar range, which is sufficient to activate high-affinity receptors (such as NMDA Receptor 2A "GluN2A") but not low-affinity receptors [44].

After zinc has been released from gluzinergic neurons into the synaptic cleft, its concentration is rapidly decreased through several mechanisms. First, a very efficient mechanism of zinc reuptake is in charge to remove available zinc and reconstitute zinc vesicles [44]. Second, unlike conventional neurotransmitters, zinc can be translocated from the synaptic cleft (or even the pre-synaptic vesicle) into post-synaptic neurons through zinc-permeable gated channels such as NMDAR [45]. Furthermore, an unknown amount of zinc is expected to diffuse away from the synaptic cleft due to the concentration difference between the cleft and the cerebrospinal fluid (CSF). It is also hypothesized that glial cells play a critical role not only in the removal of released zinc but also in the integration of synaptic transmission modulated by zinc [46]; however, the precise mechanisms remain elusive.

## 4. "Sleep as You Eat" or How Food Can Regulate Sleep

Eating and sleeping are two intrinsic essential activities in animals. Numerous studies have reported how our sleep status can modulate the way we eat. For example, total sleep deprivation for one night in healthy adults resulted in an increase of desire for highly palatable food compared to control non-sleep deprived subjects [47]. Even partial but chronic sleep deprivation was sufficient to increase food consumption beyond the physiological balance and induce weight gain [48]. Interestingly, recent experiments on mice demonstrated that a partial inhibition of REM sleep increased the absorption of high-calorie food; however, blocking neuronal activity in the medial prefontal cortex could reverse the effect on sucrose but not fat consumption [49]. On the other hand, a growing number of studies has reported the opposite effect, where diet regulates sleep [50–52]. In a pioneering clinical study, Phillips et al. provided an isocaloric high carbohydrate/low fat (HC/LF) or low carbohydrate/high fat (LC/HF) diet to eight healthy young men. They observed a significant decrease of NREM sleep after consumption of the HC/LF diet, and an increase of REM sleep for both diets compared to a balanced control diet [53].

Not only the quantity of sleep, but also its quality can be modulated by our eating behavior. A recent study demonstrated the surprising beneficial effects of kiwifruit consumption on sleep quality in a four-week trial on 24 subjects. The score of the Chinese version of the Pittsburgh Sleep Quality Index (CPSQI) auto-evaluation test, the waking time after sleep onset, as well as the sleep onset latency were significantly decreased, while the total sleep time and the sleep efficiency were significantly increased [54]. While the mechanisms involved in such effects remain elusive, recent efforts permit the identification of active compounds and their molecular mechanisms in sleep-promoting foods and natural compounds including saffron [55,56], honokiol [57], magnolol [58], phlorotannin [59], sake yeast [60,61] or ashwagandha leaf extracts [62]. Finally, Grandner et al. analyzed data from the National Health and Nutrition Examination Survey on 5587 American citizens to determine the dietary nutrients

*Int. J. Mol. Sci.* **2017**, *18*, 2334

associated with short and long sleep duration [63]. They identified several vitamins and minerals whose dietary intake correlated with a modification of sleep amount, and notably characterized zinc as one of them. According to their results, very short sleepers (<5 h) ingested significantly less zinc than did normal or long sleepers. These results are in accordance with the very limited number of studies that have compared zinc amount in humans and sleep patterns. The difference of zinc consumption observed in this study might be the result of the consumption of food more or less rich in zinc between subjects such as oyster, other seafood and meat. It would also be interesting to measure to which extent blood-zinc concentration correlates with zinc consumption and to determine if this parameter also varies with the amount of sleep.

## 5. Sleep Regulation, an Unexpected Function of Zinc

### 5.1. Clinical Studies

In 2009, a population study on 890 healthy Jinan residents in China evaluated the relationship between zinc/copper serum concentrations and several physiological factors such as sex, age, drinking and smoking behavior, and sleep [64]. Regarding sleep, the mean concentration of serum copper remained constant regardless of the amount of sleep; however, the highest concentration of serum zinc was found in subjects sleeping a "normal" amount of 7 to 9 h per night (1.337–1.442 mg/L), compared to short (<7 h) and long (>9 h) sleepers (0.789–0.934 mg/L). A later cross-sectional study measured zinc and copper content in the serum and hair of 126 adult Korean women [65]. This time, the group of women with the highest serum and hair zinc/copper ratio had the highest percentage of optimal amount of sleep (7–7.9 h). Recently, another Chinese cohort study compared blood zinc concentration and sleep quality in 1295 children from the Jintan Child Cohort [66]. Blood sampling was performed on the same children twice: at preschool (3–5 years old) and several years later during their 6th grade (11–15 years old). No significant association between zinc status and sleep could be found in these children in their younger age; however, blood zinc concentration correlated with sleep duration and sleep quality (CPSQI test) in their pre-adolescent age. Furthermore, a longitudinal association between the first and second sampling periods demonstrated that zinc blood concentration at preschool age predicted the development of poor sleep quality and efficiency several years later.

As well as these reports, to the best of our knowledge only three studies have more or less directly evaluated the effect of zinc supplementation on sleep in humans. In the first study, the authors focused on the effect of a 12-month iron and zinc supplementation on sleep in 877 infants from Zanzibar and 567 infants from Nepal, both groups being vastly subjected to malnutrition [67]. Infants from Zanzibar not suffering from iron deficiency anemia (IDA) and who received supplemental zinc slept an extra 1.3 h at night and a total of 1.7 h extra per day (night sleep + naps) compared to infants receiving a placebo. Zinc supplementation also resulted in sleep time increase in Nepalese infants with IDA, albeit to a smaller extent. In another double-blind placebo-controlled clinical trial, the authors evaluated the effect of a triple supplementation of melatonin, magnesium and zinc on 43 residents of a long-term care facility in Italy who exhibited primary insomnia [68]. Patients ingested daily a combination of melatonin (5 mg), magnesium (225 mg) and zinc (11.25 mg) mixed in 100 g of pear pulp for 60 days, one hour before bedtime. Patients that received this mineral supplement exhibited a remarkable improvement of sleep quality with a Pittsburgh Sleep Quality Index reduced from 12.7 ± 2.6 to 5.5 ± 1.9. On the other hand, placebo-treated patients did not show any sleep quality improvement (12.3 ± 3.6 and 12.0 ± 4.4, respectively, before and after placebo treatment). Finally, the most recent study determined the effect of zinc supplementation from natural sources (zinc-rich oysters and zinc-containing yeast extracts) on 120 healthy subjects in a randomized controlled trial in Japan [69]. Compared with the placebo group, individuals treated for three months with daily zinc supplements demonstrated an improved sleep onset latency and sleep efficiency compared to control subjects (Figure 1A).

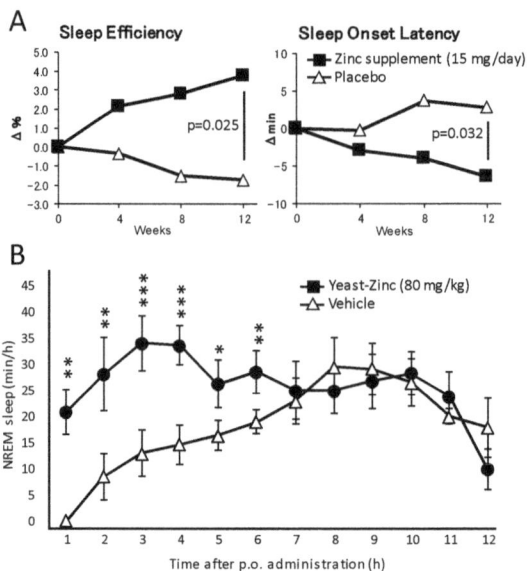

**Figure 1.** Dietary zinc improves sleep quality in humans and increases NREM sleep in mice. (**A**) Two groups of 30 volunteers absorbed daily 15 mg of zinc (in 40 g of Pacific oysters) or placebo (40 g of scallops). After 12 weeks of supplementation, sleep efficiency and sleep onset latency improved in the group treated with zinc compared to the control group. (**B**) Oral administration of zinc-containing yeast extract (80 mg/kg) in mice at the onset of dark time increased the amount of NREM sleep for 6 h compared to mice receiving vehicle. * $p < 0.05$, ** $p < 0.01$, *** $p < 0.001$ compared with vehicle treatment.

### 5.2. Experimental Evidence

All these human studies measure the amount or the quality of sleep in correlation with zinc supplementation, and conclude with an improvement of the sleep pattern of the tested subjects. However, because zinc is not provided alone, the effects might arguably arise from different compounds administered at the same time as zinc. For instance, zinc supplementation was complemented with iron in the Zanzibar and Nepal study, while clinical examination demonstrated iron deficiency in a substantial amount of the children before the beginning of the study. Providing iron to IDA infants might account for the improvement of sleep observed by the authors. In the Italian study, zinc was complemented with melatonin and magnesium; however, melatonin's effect on sleep patterns has been extensively studied and demonstrated [70] and it is unfortunate that the authors did not decipher how important each component was on sleep. Similarly, oysters contain a large amount of taurine, a γ-aminobutyric acid (GABA) receptor agonist known to promote sleep-like resting behavior in *D. melanogaster* [71]. Furthermore, environmental factors as well as emotional condition also have a major impact on how well and how long humans sleep, and can interfere with zinc's effects on sleep.

The most convincing study to date about the effect of zinc on sleep was obtained in our study published in 2015 [72]. In these experiments, we used a mouse model, which eliminates all the environmental and psychological factors that might have negatively influenced the previous human studies. Feeding the mice with zinc-containing yeast extract (equivalent to a dose of 10 to 160 mg/kg of elemental zinc) at the onset of the dark phase resulted in a drastic reduction of locomotor activity for a period of up to 6 h. Such an effect could not be observed if the mice were fed with a similar amount of yeast extract rich in the other divalent cations manganese, iron or copper, proving a specific effect for zinc. We could also precisely measure and characterize the zinc-induced sleep by recording the EEG and its power spectrum during the experiment. Yeast-zinc, orally administered at a dose of

40 or 80 mg/kg, dose-dependently and specifically increased the total amount of NREM sleep when administered at the onset of the dark phase, when the animal is most active (Figure 1B). Furthermore, the power spectrum of NREM sleep remained indistinguishable from that of physiological NREM sleep, demonstrating a good sleep quality.

However, the same doses of zinc-containing yeast extract had no significant effect on the amount of sleep when administered during daytime, when the animal is already mostly sleeping. Under basal conditions, a mouse sleeps an average of 20 min/h during nighttime and 40 min/h during daytime. When mice were fed with zinc at the onset of the dark phase, the total amount of sleep increased by up to 20–30 min/h, but, when they were fed during daytime, it remained at around the usual 40 min/h. In other words, zinc never induced sleep beyond the physiological level, contrary to more classical sleep-inducing molecules such as benzodiazepine, which also reduce the power density of NREM sleep and result in poor sleep quality. It is possible that zinc acts on circadian regulators and induces sleep when the animal is usually sleeping.

## 6. Solving the Mystery of Zinc-Induced Sleep

One may wonder how dietary zinc might act so quickly on the central nervous system (CNS) and regulate a function as essential as sleep. It is well accepted that the blood–brain barrier (BBB) has a very low permeability for zinc and the concentration of this ion remains extremely stable in the CSF regardless of serum zinc concentration [73]. However, a higher time-resolution measurement in rats revealed a rapid exchange of zinc between blood and brain during the first 30 min following intravenous administration, and zinc was not stored in the CSF but in an undetermined compartment of the brain [74,75]. A later study demonstrated the variable permeability of the BBB for zinc using an in vitro experimental model [76]. In this experiment, the BBB exhibited a constant zinc permeability when the plasma compartment concentration remained between 10 and 25 μM; however if this concentration increased further, BBB permeability for zinc increases suddenly and markedly. In our experiments, we found that zinc concentration increased in the serum up to 10-fold after the oral administration as compared to baseline. We therefore hypothesize that orally administered zinc reaches some specific compartment of the CNS after rapidly increasing in the blood, thus activating a signaling pathway that is responsible for the promotion of sleep. It would be unrealistic to think that, in physiological conditions, dietary zinc could be responsible for regulating sleep in animals and humans, especially since it seems unlikely that it would affect the concentration of zinc in the CNS. However, local zinc concentration may be less stable in the CNS than one would expect. Indeed, it was also reported that plasma zinc concentration, while being tightly controlled by diverse mechanisms, exhibits a circadian variation, with a minimum concentration measured in the evening while it reaches its peak in the morning [77]. Similarly, we hypothesize that the zinc concentration also varies in some specific regions of the CNS during the course of a day.

The idea that ions can regulate wake and sleep is actually not new. In 1927, Demole discovered that the injection of $CaCl_2$ into the pituitary of cats increased sleep for several hours. More recently, Nedergaard's lab thoroughly studied the interaction between sleep/wake status and the CSF concentration of three ions: $K^+$, $Ca^{2+}$ and $Mg^{2+}$ [78]. They reported that the concentration of extracellular $K^+$ increases in the CSF of mice during wakefulness, while those of $Ca^{2+}$ as well as $Mg^{2+}$ decreases. Moreover, the authors assessed whether the sleep/wake status was responsible for the change in ion concentrations or if the change in ion concentrations was physiologically responsible for the promotion of sleep and wakefulness. They demonstrated that infusion of artificial CSF mimicking the concentration of ions in wake or sleep reversed the neuronal activity and the behavioral state. Involvement of $Ca^{2+}$ in the control of sleep was also recently demonstrated in mammals [79]. These results shed new light on the importance of the non-neurotransmitter type of communication within the brain, involving ions such as $Ca^{2+}$, $Mg^{2+}$ and also $Zn^{2+}$, to control even the most critical physiological functions.

The specific mechanism of action of zinc in the CNS to promote sleep remains elusive. Most studies looking at the activity of $Zn^{2+}$ in the brain have focused on its interaction with the glutamatergic receptors, because $Zn^{2+}$ exists predominantly in the presynaptic vesicles of glutamatergic neurons to be co-released with glutamate. Some other receptors also interact with zinc. $Zn^{2+}$ is also found not only in glutamatergic axon terminals but also in some inhibitory axon terminals, potentially glycinergic, of the cerebellum and in the spinal cord [80]. Glycinergic neurons project to orexin neurons in the lateral hypothalamus (well characterized neurons that are involved in the maintenance of wakefulness), and can inhibit their activity [81]. The glycinergic receptor (GlyR) exhibits an atypical reaction in the presence of zinc. First, the synaptic activity of the $\alpha 1\beta$ GlyR isoform is increased in vitro even in the presence of a very low concentration of zinc (10 nM–1 $\mu$M). However, at a higher concentration (3–300 $\mu$M), zinc exhibits a dose-dependent inhibition of GlyR excitability. A recent in vivo study revealed that the free $Zn^{2+}$ concentration in the glycinergic synaptic cleft could rises to at least 1 $\mu$M following a single presynaptic stimulation, which is higher than the concentration in glutamatergic synapses [41]. Furthermore, selective mutation in the $\alpha 1$ subunit of the glycine receptor identified $Zn^{2+}$ as an essential endogenous modulator of glycinergic transmission, leading to the development of a hyperekplexia phenotype in mice, and demonstrated that zinc is essential for the proper functioning of the glycinergic system [82]. However, it remains unclear whether zinc is released directly from presynaptic glycinergic neurons or from neighboring zinc-containing glutamatergic neurons.

As well as acting as a cofactor for various receptors, zinc has its own receptor called G protein-coupled receptor 39 (GPR39) [83], which is a $G_{\alpha s}$ protein-coupled receptor to activate adenylate cyclase and cAMP-dependent signaling. In the brain, it is mainly expressed in the amygdala, the hippocampus and the auditory cortex, as well as in many other regions to a lesser extent [84]. The administration of zinc to hippocampal slices activated GPR39 in the CA3 region and regulated neuronal activity by inducing intracellular release of calcium, as well as phosphorylation of extracellular-regulated kinase and $Ca^{2+}$/calmodulin kinase II [85]. Deletion of GPR39 in mice led to the development of a depression-like behavior [86,87] as well as an increased risk of Alzheimer's disease [88], two well-characterized pathologies related to sleep disturbance [89]. To the best of our knowledge, nobody has yet checked the potential involvement of GPR39 in the regulation of sleep and wakefulness, let alone the potential role of zinc in this physiological process.

## 7. Conclusions

The role of zinc in the CNS has become increasingly important, since we first recognized its central role in the regulation of essential functions such as memory and, now, sleep. However, much work remains to comprehend properly the key functions of zinc in glutamatergic transmission and other types of neurotransmission. Although zinc ion was of little interest to the scientific community for a long time, accumulating evidence proves that endogenous zinc as well as available dietary zinc is of high importance, not only as an enzyme cofactor but also as a signaling molecule. One of the most unexpected functions of zinc to date may be in the regulation of sleep, an essential physiological function shared by the entire animal kingdom. While the mechanisms by which zinc regulates sleep remain unclear, rapid progress towards their elucidation is to be anticipated.

**Acknowledgments:** This work was supported by JSPS KAKENHI Grant Numbers JP16K18698 (Yoan Cherasse) and JP16H01881 (Yoshihiro Urade), and by the World Premier International Research Center Initiative (WPI) from MEXT.

**Conflicts of Interest:** The authors declare no conflict of interest.

## Abbreviations

| | |
|---|---|
| CNS | central nervous system |
| EEG | electroencephalogram |
| NREM | non-rapid eye movement |
| REM | rapid eye movement |
| LC | locus coeruleus |
| TMN | tuberomammillary nucleus |
| PB | parabrachial nucleus |
| VLPO | ventrolateral pre-optic area |
| DNA | deoxyribonucleic acid |
| GABA | γ-aminobutyric acid |
| NMDAR | N-methyl-D-aspartate receptor |
| AMPA | α-amino-3-hydroxy-5-methyl-4-isoxazolepropionic acid receptor |
| HC | high carbohydrate |
| LF | low fat |
| LC | low carbohydrate |
| HF | high fat |
| CPSQI | Chinese version of the Pittsburgh Sleep Quality Index |
| IDA | iron deficiency anaemia |
| BBB | blood–brain barrier |
| GlyR | glycinergic receptor |
| GPR39 | G protein-coupled receptor 39 |
| p.o. | per os |

## References

1. Cirelli, C.; Tononi, G. Is sleep essential? *PLoS Biol.* **2008**, *6*, e216. [CrossRef] [PubMed]
2. Bellesi, M.; de Vivo, L.; Chini, M.; Gilli, F.; Tononi, G.; Cirelli, C. Sleep loss promotes astrocytic phagocytosis and microglial activation in mouse cerebral cortex. *J. Neurosci.* **2017**, *37*, 5263–5273. [CrossRef] [PubMed]
3. Kripke, D.F.; Garfinkel, L.; Wingard, D.L.; Klauber, M.R.; Marler, M.R. Mortality associated with sleep duration and insomnia. *Arch. Gen. Psychiatry* **2002**, *59*, 131–136. [CrossRef] [PubMed]
4. Tamakoshi, A.; Ohno, Y. Self-reported sleep duration as a predictor of all-cause mortality: Results from the JACC study, Japan. *Sleep* **2004**, *27*, 51–54. [PubMed]
5. Patel, S.R.; Ayas, N.T.; Malhotra, M.R.; White, D.P.; Schernhammer, E.S.; Speizer, F.E.; Stampfer, M.J.; Hu, F.B. A prospective study of sleep duration and mortality risk in women. *Sleep* **2004**, *27*, 440–444. [CrossRef] [PubMed]
6. Diekelmann, S.; Born, J. The memory function of sleep. *Nat. Rev. Neurosci.* **2010**, *11*, 114–126. [CrossRef] [PubMed]
7. Inoue, S.; Honda, K.; Komoda, Y. Sleep as neuronal detoxification and restitution. *Behav. Brain Res.* **1995**, *69*, 91–96. [CrossRef]
8. Xie, L.; Kang, H.; Xu, Q.; Chen, M.J.; Liao, Y.; Thiyagarajan, M.; O'Donnell, J.; Christensen, D.J.; Nicholson, C.; Iliff, J.J.; et al. Sleep drives metabolite clearance from the adult brain. *Science* **2013**, *342*, 373–377. [CrossRef] [PubMed]
9. Sherin, J.E.; Shiromani, P.J.; McCarley, R.W.; Saper, C.B. Activation of ventrolateral preoptic neurons during sleep. *Science* **1996**, *271*, 216–219. [CrossRef] [PubMed]
10. Saper, C.B.; Fuller, P.M.; Pedersen, N.P.; Lu, J.; Scammell, T.E. Sleep state switching. *Neuron* **2010**, *68*, 1023–1042. [CrossRef] [PubMed]
11. Anaclet, C.; Lin, J.S.; Vetrivelan, R.; Krenzer, M.; Vong, L.; Fuller, P.M.; Lu, J. Identification and characterization of a sleep-active cell group in the rostral medullary brainstem. *J. Neurosci.* **2012**, *32*, 17970–17976. [CrossRef] [PubMed]
12. Anaclet, C.; Ferrari, L.; Arrigoni, E.; Bass, C.E.; Saper, C.B.; Lu, J.; Fuller, P.M. The GABAergic parafacial zone is a medullary slow wave sleep-promoting center. *Nat. Neurosci.* **2014**, *17*, 1217–1224. [CrossRef] [PubMed]

13. Lazarus, M.; Shen, H.Y.; Cherasse, Y.; Qu, W.M.; Huang, Z.L.; Bass, C.E.; Winsky-Sommerer, R.; Semba, K.; Fredholm, B.B.; Boison, D.; et al. Arousal effect of caffeine depends on adenosine A2A receptors in the shell of the nucleus accumbens. *J. Neurosci.* **2011**, *31*, 10067–10075. [CrossRef] [PubMed]

14. Oishi, Y.; Xu, Q.; Wang, L.; Zhang, B.J.; Takahashi, K.; Takata, Y.; Luo, Y.J.; Cherasse, Y.; Schiffmann, S.N.; de Kerchove d'Exaerde, A.; et al. Slow-wave sleep is controlled by a subset of nucleus accumbens core neurons in mice. *Nat. Commun.* **2017**, *8*, 734. [CrossRef] [PubMed]

15. Yuan, X.S.; Wang, L.; Dong, H.; Qu, W.M.; Yang, S.R.; Cherasse, Y.; Lazarus, M.; Schiffmann, S.N.; d'Exaerde, A.K.; Li, R.X.; et al. Striatal adenosine A2A receptor neurons control active-period sleep via parvalbumin neurons in external globus pallidus. *eLife* **2017**, *6*. [CrossRef] [PubMed]

16. Prasad, A.S. Discovery of human zinc deficiency: Its impact on human health and disease. *Adv. Nutr.* **2013**, *4*, 176–190. [CrossRef] [PubMed]

17. Klug, A. Zinc finger peptides for the regulation of gene expression. *J. Mol. Biol.* **1999**, *293*, 215–218. [CrossRef] [PubMed]

18. Pauzaite, T.; Thacker, U.; Tollitt, J.; Copeland, N.A. Emerging roles for Ciz1 in cell cycle regulation and as a driver of tumorigenesis. *Biomolecules* **2016**, *7*. [CrossRef] [PubMed]

19. Laity, J.H.; Lee, B.M.; Wright, P.E. Zinc finger proteins: New insights into structural and functional diversity. *Curr. Opin. Struct. Biol.* **2001**, *11*, 39–46. [CrossRef]

20. Kim, S.; Yu, N.K.; Kaang, B.K. CTCF as a multifunctional protein in genome regulation and gene expression. *Exp. Mol. Med.* **2015**, *47*, e166. [CrossRef] [PubMed]

21. Loef, M.; von Stillfried, N.; Walach, H. Zinc diet and Alzheimer's disease: A systematic review. *Nutr. Neurosci.* **2012**, *15*, 2–12. [CrossRef] [PubMed]

22. Hara, T.; Takeda, T.-A.; Takagishi, T.; Fukue, K.; Kambe, T.; Fukada, T. Physiological roles of zinc transporters: Molecular and genetic importance in zinc homeostasis. *J. Physiol. Sci.* **2017**, *67*, 283–301. [CrossRef] [PubMed]

23. Harbison, S.T.; Carbone, M.A.; Ayroles, J.F.; Stone, E.A.; Lyman, R.F.; Mackay, T.F.C. Co-regulated transcriptional networks contribute to natural genetic contribute variation in drosophila sleep. *Nat. Genet.* **2009**, *41*, 371–375. [CrossRef] [PubMed]

24. Groth, C.; Sasamura, T.; Khanna, M.R.; Whitley, M.; Fortini, M.E. Protein trafficking abnormalities in Drosophila tissues with impaired activity of the ZIP7 zinc transporter Catsup. *Development* **2013**, *140*, 3018–3027. [CrossRef] [PubMed]

25. Haug, F.M. Electron microscopical localization of the zinc in hippocampal mossy fibre synapses by a modified sulfide silver procedure. *Histochem. Histochem. Histochim.* **1967**, *8*, 355–368. [CrossRef]

26. Peters, S.; Koh, J.; Choi, D.W. Zinc selectively blocks the action of N-methyl-D-aspartate on cortical neurons. *Science* **1987**, *236*, 589–593. [CrossRef] [PubMed]

27. Westbrook, G.L.; Mayer, M.L. Micromolar concentrations of $Zn^{2+}$ antagonize NMDA and GABA responses of hippocampal neurons. *Nature* **1987**, *328*, 640–643. [CrossRef] [PubMed]

28. Vogt, K.; Mellor, J.; Tong, G.; Nicoll, R. The actions of synaptically released zinc at hippocampal mossy fiber synapses. *Neuron* **2000**, *26*, 187–196. [CrossRef]

29. Rassendren, F.A.; Lory, P.; Pin, J.P.; Nargeot, J. Zinc has opposite effects on NMDA and non-NMDA receptors expressed in xenopus oocytes. *Neuron* **1990**, *4*, 733–740. [CrossRef]

30. Xie, X.; Gerber, U.; Gahwiler, B.H.; Smart, T.G. Interaction of zinc with ionotropic and metabotropic glutamate receptors in rat hippocampal slices. *Neurosci. Lett.* **1993**, *159*, 46–50. [CrossRef]

31. Frederickson, C.J.; Koh, J.Y.; Bush, A.I. The neurobiology of zinc in health and disease. *Nat. Rev. Neurosci.* **2005**, *6*, 449–462. [CrossRef] [PubMed]

32. Rosati, A.M.; Traversa, U. Mechanisms of inhibitory effects of zinc and cadmium ions on agonist binding to adenosine A1 receptors in rat brain. *Biochem. Pharmacol.* **1999**, *58*, 623–632. [CrossRef]

33. Schetz, J.A.; Sibley, D.R. Zinc allosterically modulates antagonist binding to cloned D1 and D2 dopamine receptors. *J. Neurochem.* **1997**, *68*, 1990–1997. [CrossRef] [PubMed]

34. Gill, C.H.; Peters, J.A.; Lambert, J.J. An electrophysiological investigation of the properties of a murine recombinant 5-HT3 receptor stably expressed in HEK 293 cells. *Br. J. Pharmacol.* **1995**, *114*, 1211–1221. [CrossRef] [PubMed]

35. Vandenberg, R.J.; Mitrovic, A.D.; Johnston, G.A. Molecular basis for differential inhibition of glutamate transporter subtypes by zinc ions. *Mol. Pharmacol.* **1998**, *54*, 189–196. [PubMed]

36. Richfield, E.K. Zinc modulation of drug binding, cocaine affinity states, and dopamine uptake on the dopamine uptake complex. *Mol. Pharmacol.* **1993**, *43*, 100–108. [PubMed]
37. Busselberg, D.; Evans, M.L.; Rahmann, H.; Carpenter, D.O. $Zn^{2+}$ blocks the voltage activated calcium current of *Aplysia* neurons. *Neurosci. Lett.* **1990**, *117*, 117–122. [CrossRef]
38. Erdelyi, L. Zinc blocks the A-type potassium currents in Helix neurons. *Acta Physiol. Hung.* **1993**, *81*, 111–120. [PubMed]
39. Gilly, W.F.; Armstrong, C.M. Slowing of sodium channel opening kinetics in squid axon by extracellular zinc. *J. Gen. Physiol.* **1982**, *79*, 935–964. [CrossRef] [PubMed]
40. Kajita, H.; Whitwell, C.; Brown, P.D. Properties of the inward-rectifying Cl- channel in rat choroid plexus: Regulation by intracellular messengers and inhibition by divalent cations. *Pflugers Arch. Eur. J. Physiol.* **2000**, *440*, 933–940. [CrossRef]
41. Zhang, Y.; Keramidas, A.; Lynch, J.W. The free zinc concentration in the synaptic cleft of artificial glycinergic synapses rises to at least 1 muM. *Front. Mol. Neurosci.* **2016**, *9*, 88. [CrossRef] [PubMed]
42. Kodirov, S.A.; Takizawa, S.; Joseph, J.; Kandel, E.R.; Shumyatsky, G.P.; Bolshakov, V.Y. Synaptically released zinc gates long-term potentiation in fear conditioning pathways. *Proc. Natl. Acad. Sci. USA* **2006**, *103*, 15218–15223. [CrossRef] [PubMed]
43. Frederickson, C.J.; Giblin, L.J.; Krezel, A.; McAdoo, D.J.; Mueller, R.N.; Zeng, Y.; Balaji, R.V.; Masalha, R.; Thompson, R.B.; Fierke, C.A.; et al. Concentrations of extracellular free zinc (pZn)e in the central nervous system during simple anesthetization, ischemia and reperfusion. *Exp. Neurol.* **2006**, *198*, 285–293. [CrossRef] [PubMed]
44. Vergnano, A.M.; Rebola, N.; Savtchenko, L.P.; Pinheiro, P.S.; Casado, M.; Kieffer, B.L.; Rusakov, D.A.; Mulle, C.; Paoletti, P. Zinc dynamics and action at excitatory synapses. *Neuron* **2014**, *82*, 1101–1114. [CrossRef] [PubMed]
45. Li, Y.; Hough, C.J.; Suh, S.W.; Sarvey, J.M.; Frederickson, C.J. Rapid translocation of $Zn^{2+}$ from presynaptic terminals into postsynaptic hippocampal neurons after physiological stimulation. *J. Neurophysiol.* **2001**, *86*, 2597–2604. [PubMed]
46. Hancock, S.M.; Finkelstein, D.I.; Adlard, P.A. Glia and zinc in ageing and Alzheimer's disease: A mechanism for cognitive decline? *Front. Aging Neurosci.* **2014**, *6*, 137. [CrossRef] [PubMed]
47. Greer, S.M.; Goldstein, A.N.; Walker, M.P. The impact of sleep deprivation on food desire in the human brain. *Nat. Commun.* **2013**, *4*, 2259. [CrossRef] [PubMed]
48. Markwald, R.R.; Melanson, E.L.; Smith, M.R.; Higgins, J.; Perreault, L.; Eckel, R.H.; Wright, K.P., Jr. Impact of insufficient sleep on total daily energy expenditure, food intake, and weight gain. *Proc. Natl. Acad. Sci. USA* **2013**, *110*, 5695–5700. [CrossRef] [PubMed]
49. McEown, K.; Takata, Y.; Cherasse, Y.; Nagata, N.; Aritake, K.; Lazarus, M. Chemogenetic inhibition of the medial prefrontal cortex reverses the effects of REM sleep loss on sucrose consumption. *eLife* **2016**, *5*. [CrossRef] [PubMed]
50. Catterson, J.H.; Knowles-Barley, S.; James, K.; Heck, M.M.; Harmar, A.J.; Hartley, P.S. Dietary modulation of Drosophila sleep-wake behaviour. *PLoS ONE* **2010**, *5*, e12062. [CrossRef] [PubMed]
51. Hasegawa, T.; Tomita, J.; Hashimoto, R.; Ueno, T.; Kume, S.; Kume, K. Sweetness induces sleep through gustatory signalling independent of nutritional value in a starved fruit fly. *Sci. Rep.* **2017**, *7*, 14355. [CrossRef] [PubMed]
52. Frank, S.; Gonzalez, K.; Lee-Ang, L.; Young, M.C.; Tamez, M.; Mattei, J. Diet and sleep physiology: Public health and clinical implications. *Front. Neurol.* **2017**, *8*, 393. [CrossRef] [PubMed]
53. Phillips, F.; Chen, C.N.; Crisp, A.H.; Koval, J.; McGuinness, B.; Kalucy, R.S.; Kalucy, E.C.; Lacey, J.H. Isocaloric diet changes and electroencephalographic sleep. *Lancet* **1975**, *2*, 723–725. [CrossRef]
54. Lin, H.H.; Tsai, P.S.; Fang, S.C.; Liu, J.F. Effect of kiwifruit consumption on sleep quality in adults with sleep problems. *Asia Pac. J. Clin. Nutr.* **2011**, *20*, 169–174. [PubMed]
55. Masaki, M.; Aritake, K.; Tanaka, H.; Shoyama, Y.; Huang, Z.L.; Urade, Y. Crocin promotes non-rapid eye movement sleep in mice. *Mol. Nutr. Food Res.* **2012**, *56*, 304–308. [CrossRef] [PubMed]
56. Liu, Z.; Xu, X.H.; Liu, T.Y.; Hong, Z.Y.; Urade, Y.; Huang, Z.L.; Qu, W.M. Safranal enhances non-rapid eye movement sleep in pentobarbital-treated mice. *CNS Neurosci. Ther.* **2012**, *18*, 623–630. [CrossRef] [PubMed]

57. Qu, W.M.; Yue, X.F.; Sun, Y.; Fan, K.; Chen, C.R.; Hou, Y.P.; Urade, Y.; Huang, Z.L. Honokiol promotes non-rapid eye movement sleep via the benzodiazepine site of the GABA(A) receptor in mice. *Br. J. Pharmacol.* **2012**, *167*, 587–598. [CrossRef] [PubMed]

58. Chen, C.R.; Zhou, X.Z.; Luo, Y.J.; Huang, Z.L.; Urade, Y.; Qu, W.M. Magnolol, a major bioactive constituent of the bark of Magnolia officinalis, induces sleep via the benzodiazepine site of GABA(A) receptor in mice. *Neuropharmacology* **2012**, *63*, 1191–1199. [CrossRef] [PubMed]

59. Cho, S.; Yoon, M.; Pae, A.N.; Jin, Y.H.; Cho, N.C.; Takata, Y.; Urade, Y.; Kim, S.; Kim, J.S.; Yang, H.; et al. Marine polyphenol phlorotannins promote non-rapid eye movement sleep in mice via the benzodiazepine site of the GABAA receptor. *Psychopharmacology* **2014**, *231*, 2825–2837. [CrossRef] [PubMed]

60. Nakamura, Y.; Midorikawa, T.; Monoi, N.; Kimura, E.; Murata-Matsuno, A.; Sano, T.; Oka, K.; Sugafuji, T.; Uchiyama, A.; Murakoshi, M.; et al. Oral administration of Japanese sake yeast (Saccharomyces cerevisiae sake) promotes non-rapid eye movement sleep in mice via adenosine A2A receptors. *J. Sleep Res.* **2016**, *25*, 746–753. [CrossRef] [PubMed]

61. Monoi, N.; Matsuno, A.; Nagamori, Y.; Kimura, E.; Nakamura, Y.; Oka, K.; Sano, T.; Midorikawa, T.; Sugafuji, T.; Murakoshi, M.; et al. Japanese sake yeast supplementation improves the quality of sleep: A double-blind randomised controlled clinical trial. *J. Sleep Res.* **2016**, *25*, 116–123. [CrossRef] [PubMed]

62. Kaushik, M.K.; Kaul, S.C.; Wadhwa, R.; Yanagisawa, M.; Urade, Y. Triethylene glycol, an active component of Ashwagandha (Withania somnifera) leaves, is responsible for sleep induction. *PLoS ONE* **2017**, *12*, e0172508. [CrossRef] [PubMed]

63. Grandner, M.A.; Jackson, N.; Gerstner, J.R.; Knutson, K.L. Dietary nutrients associated with short and long sleep duration. Data from a nationally representative sample. *Appetite* **2013**, *64*, 71–80. [CrossRef] [PubMed]

64. Zhang, H.Q.; Li, N.; Zhang, Z.; Gao, S.; Yin, H.Y.; Guo, D.M.; Gao, X. Serum zinc, copper, and zinc/copper in healthy residents of Jinan. *Biol. Trace Elem. Res.* **2009**, *131*, 25–32. [CrossRef] [PubMed]

65. Song, C.H.; Kim, Y.H.; Jung, K.I. Associations of zinc and copper levels in serum and hair with sleep duration in adult women. *Biol. Trace Elem. Res.* **2012**, *149*, 16–21. [CrossRef] [PubMed]

66. Ji, X.; Liu, J. Associations between blood zinc concentrations and sleep quality in childhood: A cohort study. *Nutrients* **2015**, *7*, 5684–5696. [CrossRef] [PubMed]

67. Kordas, K.; Siegel, E.H.; Olney, D.K.; Katz, J.; Tielsch, J.M.; Kariger, P.K.; Khalfan, S.S.; LeClerq, S.C.; Khatry, S.K.; Stoltzfus, R.J. The effects of iron and/or zinc supplementation on maternal reports of sleep in infants from Nepal and Zanzibar. *J. Dev. Behav. Pediatr.* **2009**, *30*, 131–139. [CrossRef] [PubMed]

68. Rondanelli, M.; Opizzi, A.; Monteferrario, F.; Antoniello, N.; Manni, R.; Klersy, C. The effect of melatonin, magnesium, and zinc on primary insomnia in long-term care facility residents in Italy: A double-blind, placebo-controlled clinical trial. *J. Am. Geriat. Soc.* **2011**, *59*, 82–90. [CrossRef] [PubMed]

69. Saito, H.; Cherasse, Y.; Suzuki, R.; Mitarai, M.; Ueda, F.; Urade, Y. Zinc-rich oysters as well as zinc-yeast- and astaxanthin-enriched food improved sleep efficiency and sleep onset in a randomized controlled trial of healthy individuals. *Mol. Nutr. Food Res.* **2016**. [CrossRef] [PubMed]

70. Auld, F.; Maschauer, E.L.; Morrison, I.; Skene, D.J.; Riha, R.L. Evidence for the efficacy of melatonin in the treatment of primary adult sleep disorders. *Sleep Med. Rev.* **2016**. [CrossRef] [PubMed]

71. Lin, F.J.; Pierce, M.M.; Sehgal, A.; Wu, T.; Skipper, D.C.; Chabba, R. Effect of taurine and caffeine on sleep-wake activity in *Drosophila melanogaster*. *Nat. Sci. Sleep* **2010**, *2*, 221–231. [CrossRef] [PubMed]

72. Cherasse, Y.; Saito, H.; Nagata, N.; Aritake, K.; Lazarus, M.; Urade, Y. Zinc-containing yeast extract promotes nonrapid eye movement sleep in mice. *Mol. Nutr. Food Res.* **2015**, *59*, 2087–2093. [CrossRef] [PubMed]

73. Blair-West, J.R.; Denton, D.A.; Gibson, A.P.; McKinley, M.J. Opening the blood-brain barrier to zinc. *Brain Res.* **1990**, *507*, 6–10. [CrossRef]

74. Pullen, R.G.; Franklin, P.A.; Hall, G.H. 65zinc uptake from blood into brain and other tissues in the rat. *Neurochem. Res.* **1990**, *15*, 1003–1008. [CrossRef] [PubMed]

75. Kasarskis, E.J. Zinc metabolism in normal and zinc-deficient rat brain. *Exp. Neurol.* **1984**, *85*, 114–127. [CrossRef]

76. Bobilya, D.J.; Guerin, J.L.; Rowe, D.J. Zinc transport across an in vitro blood-brain barrier model. *J. Trace Elem. Exp. Med.* **1997**, *10*, 9–18. [CrossRef]

77. Markowitz, M.E.; Rosen, J.F.; Mizruchi, M. Circadian variations in serum zinc (Zn) concentrations: Correlation with blood ionized calcium, serum total calcium and phosphate in humans. *Am. J. Clin. Nutr.* **1985**, *41*, 689–696. [PubMed]

78. Ding, F.; O'Donnell, J.; Xu, Q.; Kang, N.; Goldman, N.; Nedergaard, M. Changes in the composition of brain interstitial ions control the sleep-wake cycle. *Science* **2016**, *352*, 550–555. [CrossRef] [PubMed]

79. Tatsuki, F.; Sunagawa, G.A.; Shi, S.; Susaki, E.A.; Yukinaga, H.; Perrin, D.; Sumiyama, K.; Ukai-Tadenuma, M.; Fujishima, H.; Ohno, R.; et al. Involvement of Ca$^{2+}$-dependent hyperpolarization in sleep duration in mammals. *Neuron* **2016**, *90*, 70–85. [CrossRef] [PubMed]

80. Danscher, G.; Stoltenberg, M. Zinc-specific autometallographic in vivo selenium methods: Tracing of zinc-enriched (ZEN) terminals, ZEN pathways, and pools of zinc ions in a multitude of other ZEN cells. *J. Histochem. Cytochem.* **2005**, *53*, 141–153. [CrossRef] [PubMed]

81. Hondo, M.; Furutani, N.; Yamasaki, M.; Watanabe, M.; Sakurai, T. Orexin neurons receive glycinergic innervations. *PLoS ONE* **2011**, *6*, e25076. [CrossRef] [PubMed]

82. Hirzel, K.; Muller, U.; Latal, A.T.; Hulsmann, S.; Grudzinska, J.; Seeliger, M.W.; Betz, H.; Laube, B. Hyperekplexia phenotype of glycine receptor alpha1 subunit mutant mice identifies Zn$^{2+}$ as an essential endogenous modulator of glycinergic neurotransmission. *Neuron* **2006**, *52*, 679–690. [CrossRef] [PubMed]

83. McKee, K.K.; Tan, C.P.; Palyha, O.C.; Liu, J.; Feighner, S.D.; Hreniuk, D.L.; Smith, R.G.; Howard, A.D.; van der Ploeg, L.H. Cloning and characterization of two human G protein-coupled receptor genes (GPR38 and GPR39) related to the growth hormone secretagogue and neurotensin receptors. *Genomics* **1997**, *46*, 426–434. [CrossRef] [PubMed]

84. Jackson, V.R.; Nothacker, H.P.; Civelli, O. GPR39 receptor expression in the mouse brain. *Neuroreport* **2006**, *17*, 813–816. [CrossRef] [PubMed]

85. Besser, L.; Chorin, E.; Sekler, I.; Silverman, W.F.; Atkin, S.; Russell, J.T.; Hershfinkel, M. Synaptically released zinc triggers metabotropic signaling via a zinc-sensing receptor in the hippocampus. *J. Neurosci.* **2009**, *29*, 2890–2901. [CrossRef] [PubMed]

86. Mlyniec, K.; Budziszewska, B.; Holst, B.; Ostachowicz, B.; Nowak, G. GPR39 (zinc receptor) knockout mice exhibit depression-like behavior and CREB/BDNF down-regulation in the hippocampus. *Int. J. Neuropsychopharmacol.* **2014**, *18*. [CrossRef] [PubMed]

87. Mlyniec, K.; Doboszewska, U.; Szewczyk, B.; Sowa-Kucma, M.; Misztak, P.; Piekoszewski, W.; Trela, F.; Ostachowicz, B.; Nowak, G. The involvement of the GPR39-Zn(2+)-sensing receptor in the pathophysiology of depression. Studies in rodent models and suicide victims. *Neuropharmacology* **2014**, *79*, 290–297. [CrossRef] [PubMed]

88. Khan, M.Z. A possible significant role of zinc and GPR39 zinc sensing receptor in Alzheimer disease and epilepsy. *Biomed. Pharmacother.* **2016**, *79*, 263–272. [CrossRef] [PubMed]

89. Burke, S.L.; Maramaldi, P.; Cadet, T.; Kukull, W. Associations between depression, sleep disturbance, and apolipoprotein E in the development of Alzheimer's disease: Dementia. *Int. Psychogeriatr.* **2016**, *28*, 1409–1424. [CrossRef] [PubMed]

International Journal of
*Molecular Sciences*

MDPI

*Article*

# ZnT3 Gene Deletion Reduces Colchicine-Induced Dentate Granule Cell Degeneration

Bo Young Choi, Dae Ki Hong and Sang Won Suh *

Department of Physiology, College of Medicine, Hallym University, Chuncheon 24252, Korea;
bychoi@hallym.ac.kr (B.Y.C.); zxnm01220@gmail.com (D.K.H.)
* Correspondence: swsuh@hallym.ac.kr; Tel.: +82-10-8573-6364

Received: 29 September 2017; Accepted: 17 October 2017; Published: 19 October 2017

**Abstract:** Our previous study demonstrated that colchicine-induced dentate granule cell death is caused by blocking axonal flow and the accumulation of intracellular zinc. Zinc is concentrated in the synaptic vesicles via zinc transporter 3 (ZnT3), which facilitates zinc transport from the cytosol into the synaptic vesicles. The aim of the present study was to identify the role of ZnT3 gene deletion on colchicine-induced dentate granule cell death. The present study used young (3–5 months) mice of the wild-type (WT) or the $ZnT3^{-/-}$ genotype. Colchicine (10 µg/kg) was injected into the hippocampus, and then brain sections were evaluated 12 or 24 h later. Cell death was evaluated by Fluoro-Jade B; oxidative stress was analyzed by 4-hydroxy-2-nonenal; and dendritic damage was detected by microtubule-associated protein 2. Zinc accumulation was detected by $N$-(6-methoxy-8-quinolyl)-para-toluenesulfonamide (TSQ) staining. Here, we found that $ZnT3^{-/-}$ reduced the number of degenerating cells after colchicine injection. The $ZnT3^{-/-}$-mediated inhibition of cell death was accompanied by suppression of oxidative injury, dendritic damage and zinc accumulation. In addition, $ZnT3^{-/-}$ mice showed more glutathione content than WT mice and inhibited neuronal glutathione depletion by colchicine. These findings suggest that increased neuronal glutathione by ZnT3 gene deletion prevents colchicine-induced dentate granule cell death.

**Keywords:** ZnT3; colchicine; axonal transport; zinc; neuron death; oxidative stress; glutathione

## 1. Introduction

Colchicine, a potent neurotoxin derived from plants of the genus *Colchicum autumnale*, is well known to cause selective loss of dentate granule cells in the hippocampus [1,2], and to cause cognitive dysfunction [3] resulting from cytoskeletal alterations and impaired axonal transport, followed by progressive neuronal loss. The neuronal cytoskeleton is a system of highly complex structures that consist of microtubules, neurofilaments and microfilaments. These components are responsible for the supportive shape of neurons, as well as for crucial processes such as transport of materials in the axon [4]. It has been found that colchicine injection decreases the soluble tubulin pool and inhibits microtubule polymerization by binding tightly to tubulin, the major structural protein of microtubules [5]. Kumar et al. demonstrated that intracerebral administration of colchicine induces excessive free radical generation and consequently oxidative damage [6,7]. The central nervous system (CNS) is highly vulnerable to oxidative stress because of its increased oxygen consumption for lipid peroxidation and is relatively deficient in antioxidant systems [8,9]. Therefore, generation of free radicals and oxidative stress can cause neuronal death.

Glutathione (GSH) is an intracellular non-protein thiol that plays a central role in antioxidant defense against free radical production, especially reactive oxygen species (ROS) production. Excess of ROS may result in GSH depletion [10–12]. The onset of cell death is associated with a reduction of intracellular GSH levels in various cellular systems [13,14]. Therefore, attenuation of oxidative stress by increasing antioxidant defense through modulation of GSH can be a useful tool in the management of neurodegeneration processes [15–17].

The zinc ($Zn^{2+}$) ion, one of the most abundant trace metals in the CNS, is enriched in the human body and has been known to be important in the control of physiological and pathological functions in the brain [18–21]. Anterograde and retrograde zinc transporters between the cell body and the axon terminal are important for maintaining neuronal function [22,23]. Previously, our laboratory demonstrated that blocking of axonal flow by colchicine administration and then accumulation of intracellular free zinc caused dentate granule cell death [24]. Recently, an interesting study demonstrated that genetic deletion of zinc transporter 3 (*ZnT3*), a putative transporter of zinc into synaptic vesicles, increases free zinc levels in the cytosol of neurons [25]. In addition, numerous studies have argued that $Zn^{2+}$ has antioxidant properties in most systems, indicating a positive correlation between $Zn^{2+}$ and GSH content [26–29]. Parat et al. reported that these antioxidant properties can be related to various actions, the most commonly described being $Zn^{2+}$ interference with the absorption of other metals, such as Cu or Fe, and metal-catalyzed oxidation reactions, $Zn^{2+}$ involvement in cooper/zinc superoxide dismutase (CuZnSOD) stability and the protection of thiol groups by $Zn^{2+}$ [28].

In concert with the above findings, we propose the hypothesis that increased intracellular free zinc levels by *ZnT3* gene deletion may increase neuronal GSH levels, thereby preventing colchicine-induced oxidative injury in the dentate granule cell.

## 2. Results

### 2.1. Colchicine-Induced Dentate Granule Cell Degeneration Is Reduced in ZnT3$^{-/-}$ Mice

It has previously been shown that the dentate granule cell is particularly vulnerable to colchicine and that this may contribute to subsequent cognitive impairment [30,31]. We first investigated whether genetic deletion of *ZnT3* influenced dentate granule cell death at 24 h after colchicine injection. Colchicine-induced dentate granule cell degeneration was analyzed by Fluoro-Jade B (FJB) staining to detect dying cells. Sham-operated groups didn't show any FJB (+) cells. Intrahippocampal colchicine injection induced several FJB (+) cells in the dentate gyrus (DG). However, we found that *ZnT3*$^{-/-}$ (KO) mice had a significantly reduced number of FJB (+) cells in the DG after colchicine injection, compared to the colchicine-injected WT mice (WT, $428.3 \pm 84.06$; KO, $55.5 \pm 19.93$; an 87% reduction). These results indicated that the lack of *ZnT3* suppressed the colchicine-induced dentate granule cell degeneration (Figure 1A,B).

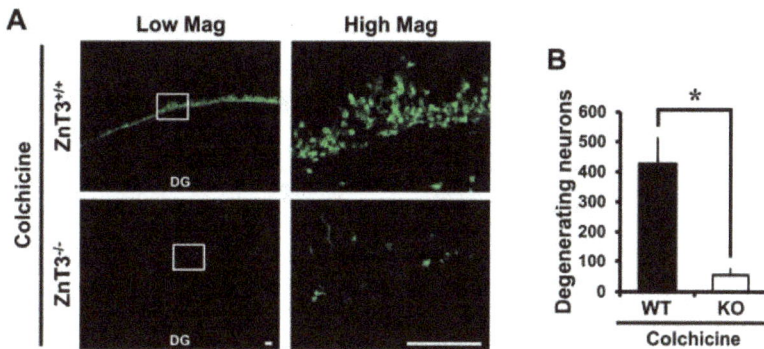

**Figure 1.** *Cont.*

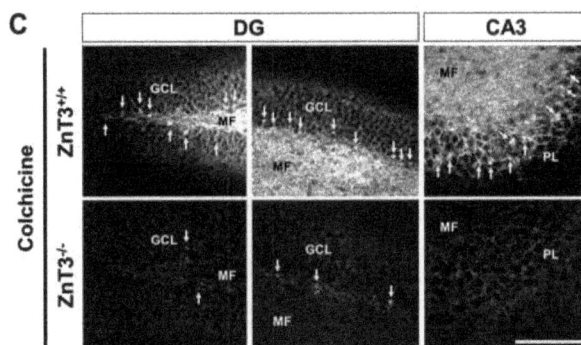

**Figure 1.** $ZnT3^{-/-}$ mice exhibit reduced dentate granule cell death after colchicine injection. Brain sections obtained from WT and $ZnT3^{-/-}$ (KO) mice at 24 h after colchicine injection (WT-colchicine, $n = 4$; KO-colchicine, $n = 5$) were analyzed by Fluoro-Jade B (FJB) staining to measure the degree of neurodegeneration. Representative images (**A**) and quantification (**B**) for the degree of neurodegeneration are shown as the number of degenerating cells of dentate gyrus (DG). Scale bar = 50 μm. Data are the mean ± SEM, * $p < 0.05$ versus WT mice; (**C**) Fluorescence photomicrographs show zinc-specific TSQ staining in the hippocampal DG and cornus ammonis 3 (CA3) at 24 h after colchicine injection. The dark holes represent the normal appearance of cell bodies, and the bright white fluorescence in the cell bodies (marked by a white arrow) indicates abnormal zinc accumulation. Scale bar = 100 μm. MF: mossy fiber. GCL: granular cell layer. PL: pyramidal cell layer.

## 2.2. ZnT3 Gene Deletion Prevents Intracellular Zinc Accumulation in the Dentate Granule Cells after Colchicine Injection

Next, we assessed if the $ZnT3$ gene deletion can prevent intracellular zinc accumulation in the granule cells of DG after colchicine injection. Colchicine-induced intracellular zinc accumulation was detected by the zinc-specific stain TSQ. Hippocampal sections harvested 24 h after colchicine injection showed an intense fluorescence signal in the cell bodies of dentate granule cells, indicative of labile zinc accumulation in these cells. However, the number of TSQ (+) neurons in the DG and cornus ammonis 3 (CA3) was significantly reduced in $ZnT3^{-/-}$ mice, compared to the colchicine-injected WT mice, indicating that deletion of the $ZnT3$ gene prevented colchicine-induced intracellular zinc accumulation (Figure 1C).

## 2.3. ZnT3 Gene Deletion Showed Less Oxidative Injury after Colchicine Injection

We next examined whether reduced neuronal death after colchicine injection in mice lacking the $ZnT3$ was related to less oxidative injury in the DG. To assess oxidative injury, brain sections were immunohistochemically stained with 4HNE at 12 h after colchicine injection. One of the major generators of oxidative stress, 4-hydroxynonenal (4HNE), has been widely considered as a bioactive marker of lipid peroxidation [32,33]. There were almost no 4HNE-stained granule cells in sham-operated mice of either WT or $ZnT3^{-/-}$ (WT, 32.5 ± 2.77; KO, 33.6 ± 1.99; average gray scale intensities). The intensity of 4HNE-immunoreactivity (IR) was remarkably increased in DG of WT mice after colchicine injection. However, colchicine-injected $ZnT3^{-/-}$ mice showed markedly less intensity of 4HNE-IR in the granule cell of DG, compared to the colchicine-injected WT mice (WT, 147.1 ± 7.49; KO, 91.4 ± 4.75; average gray scale intensities, a 38% reduction). These results suggested that the resistance of $ZnT3^{-/-}$ mice to colchicine-induced dentate granule cell degeneration might be related to the reduced oxidative injury in the DG of hippocampus (Figure 2).

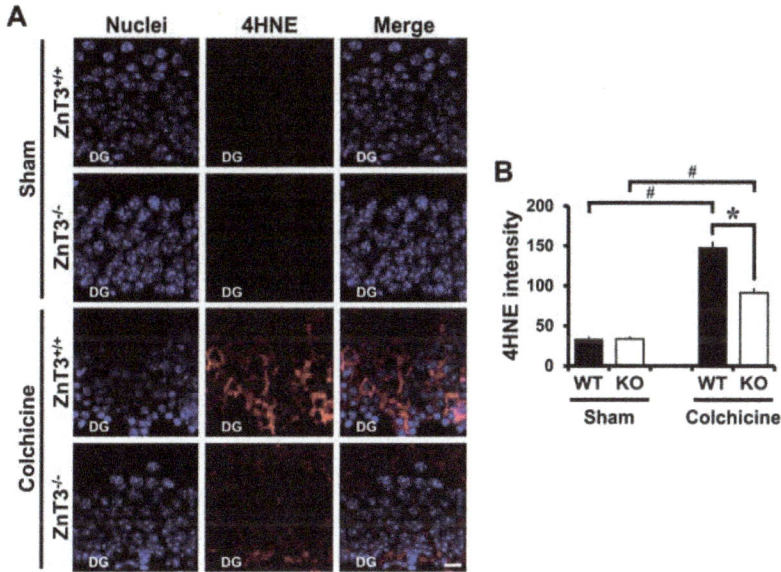

**Figure 2.** Colchicine-induced oxidative injury is reduced in $ZnT3^{-/-}$ mice. WT and $ZnT3^{-/-}$ mice were either sham-operated (WT-sham, $n$ = 3; KO-sham, $n$ = 3) or colchicine-injected (WT-colchicine, $n$ = 4; KO-colchicine, $n$ = 5). Brain sections were immunohistochemically stained with anti-4-hydroxynonenal (4HNE) to detect oxidative injury. (**A**) Representative images reveal 4HNE-labeled cells in the hippocampal DG from either WT or $ZnT3^{-/-}$ mice at 12 h after sham surgery or colchicine injection. Scale bar = 10 μm; (**B**) The bar graph shows the intensity of 4HNE-immunoreactivity (IR) in the granule cell of DG from sham-operated and colchicine-injected mice of either WT or $ZnT3^{-/-}$. Data are the mean ± SEM, * $p < 0.05$ versus WT mice, # $p < 0.05$ versus sham-operated mice.

### 2.4. ZnT3 Gene Deletion Reduced Dendritic Damage after Colchicine Injection

We also determined whether deletion of *ZnT3* reversed colchicine-induced dendritic damage in the hippocampus. MAP2 is exclusively expressed by dendrites of neurons where it binds to tubulin [34] and is thought to be involved in microtubule assembly, acting to stabilize microtubules [35]. It is considered that the loss of MAP2 protein is a characteristic of dendritic damage [36]. There were no significant differences in the distribution of MAP2-IR between the sham-operated WT and $ZnT3^{-/-}$ mice (cornus ammonis 1 (CA1): WT, 106.5 ± 5.10; KO, 108.9 ± 3.99; cornus ammonis 3 (CA3): WT, 137.3 ± 2.22; KO, 140.3 ± 5.07; DG: WT, 65.2 ± 4.54; KO, 68.7 ± 5.11; average gray scale intensities). In both cases, MAP2-IR was observed in the apical dendrites of the CA1 and CA3, as well as the dendrites of granule cells within the DG. Compared with the sham-operated group, expression levels of MAP2 in the hippocampal CA1, CA3 and DG were significantly reduced in WT mice 24 h after colchicine injection. In contrast, colchicine-injected $ZnT3^{-/-}$ mice revealed highly increased IR to MAP2, compared to the colchicine-injected WT mice (CA1: WT, 30.6 ± 4.03; KO, 58.6 ± 4.53, a 91% increase; CA3: WT, 35.2 ± 6.42; KO, 74.3 ± 6.25, a 111% increase; DG: WT, 35.9 ± 3.70; KO, 48.0 ± 4.02, a 34% increase; average gray scale intensities). These results suggested that the lack of *ZnT3* inhibited the colchicine-induced dendritic damage in the hippocampus (Figure 3).

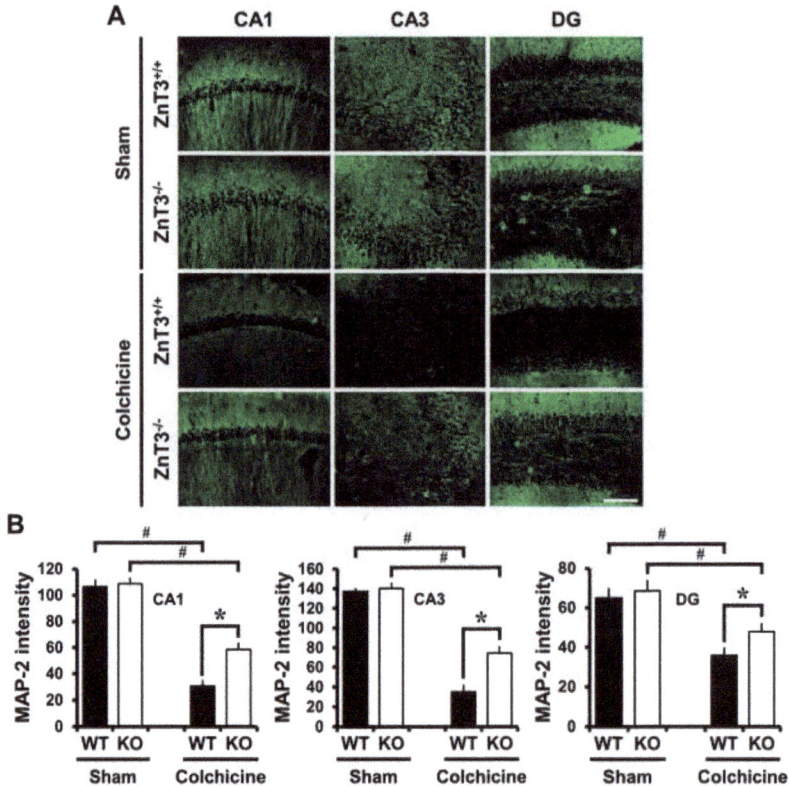

**Figure 3.** Genetic deletion of *ZnT3* reduces colchicine-induced dendritic damage. WT and *ZnT3*$^{-/-}$ mice were either sham-operated (WT-sham, *n* = 3; KO-sham, *n* = 3) or colchicine-injected (WT-colchicine, *n* = 4; KO-colchicine, *n* = 5). Brain sections were immunohistochemically stained with 4HNE to detect oxidative injury. (**A**) Representative images reveal MAP2-IR in the hippocampal CA1, CA3 and DG from either WT or *ZnT3*$^{-/-}$ mice at 24 h after sham surgery or colchicine injection. Scale bar = 100 μm; (**B**) The graph represents the intensity of MAP2-IR in the CA1, CA3 and DG from sham-operated and colchicine-injected mice of either WT or *ZnT3*$^{-/-}$. Data are the mean ± SEM, * *p* < 0.05 versus WT mice, $^{\#}$ *p* < 0.05 versus sham-operated mice.

## 2.5. Colchicine-Induced Neuronal GSH Depletion Is Prevented in ZnT3$^{-/-}$ Mice

To evaluate whether *ZnT3* gene deletion affected neuronal GSH, sections were histologically analyzed by probing for GSH-N-NEM adducts at 24 h after colchicine injection. Neurons in the hippocampal regions including subiculum, CA1, CA2, CA3 and DG were immunoreactive to the GS-NEM antibody, as can be seen in Figure 4A. GS-NEM IR in the subiculum, CA1 and CA2 of *ZnT3*$^{-/-}$ mice was similar to that in the WT mice, but was higher in CA3 and DG (subiculum: WT, 44.0 ± 3.17; KO, 46.0 ± 4.00; CA1: WT, 44.5 ± 3.43; KO, 48.8 ± 5.39; CA2: WT, 49.4 ± 3.57; KO, 52.6 ± 4.17; CA3: WT, 36.9 ± 9.02; KO, 58.9 ± 1.83, a 60% increase; DG: WT, 27.6 ± 1.89; KO, 49.8 ± 3.73, a 81% increase). Colchicine injection caused a decrease in the level of GS-NEM IR in all hippocampal regions. In addition to that, *ZnT3*$^{-/-}$ mice subjected to colchicine injection showed a significant increase of GS-NEM IR in all hippocampal regions, compared to colchicine-injected WT mice (subiculum: WT, 6.2 ± 0.35; KO, 17.9 ± 4.98, a 189% increase; CA1: WT, 7.5 ± 2.14; KO, 20.0 ± 5.66, a 167% increase; CA2: WT, 5.5 ± 1.81; KO, 20.4 ± 3.12, a 274% increase; CA3: WT, 6.3 ± 0.74; KO, 17.7 ± 4.31, a 181% increase; DG: WT, 7.4 ± 1.01; KO, 16.4 ± 2.68, a 123% increase). These results indicate that $ZnT3^{-/-}$ mice have increased neuronal glutathione and decreased vulnerability to oxidants induced by colchicine injection (Figure 4).

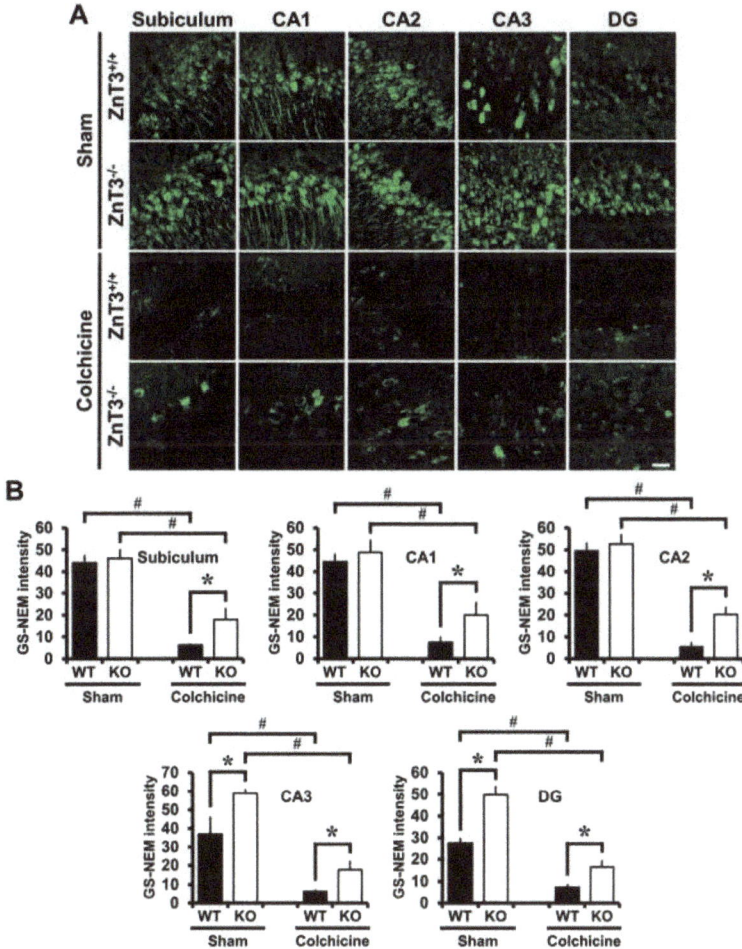

**Figure 4.** $ZnT3^{-/-}$ mice have increased neuronal glutathione and reversed a reduction of neuronal glutathione (GSH) content after colchicine injection. Brain sections obtained from WT and $ZnT3^{-/-}$ mice at 24 h after sham surgery or colchicine injection (WT-sham, $n = 3$; KO-sham, $n = 3$; WT-colchicine, $n = 4$; KO-colchicine, $n = 5$) were stained by GSH-N-ethylmaleimide (NEM) adduct (GS-NEM) antibody to detect the reduced form of GSH in the neurons. Representative immunofluorescence images (**A**) and quantification (**B**) for the degree of neuronal GSH levels are shown as the intensity of GS-NEM in the individual neurons of hippocampal regions including subiculum, CA1, CA2, CA3 and DG. Scale bar = 20 μm. Data are the mean ± SEM, * $p < 0.05$ versus WT mice, # $p < 0.05$ versus sham-operated mice.

## 3. Discussion

In the present study, we found that $ZnT3^{-/-}$ mice exhibited reduced dentate granule cell death in the hippocampus after colchicine injection compared to WT mice. The reduction of colchicine-induced dentate granule cell death by $ZnT3$ gene deletion is accompanied by suppression of oxidative injury, dendritic damage and zinc accumulation in the hippocampus. In addition, $ZnT3^{-/-}$ mice showed a higher GSH level than WT mice in the hippocampus and showed reduced neuronal GSH depletion after colchicine injection. These findings suggest that the increased intracellular free zinc level by $ZnT3$ gene deletion increased the neuronal GSH levels, thereby preventing the colchicine-induced dentate granule cell death.

Microtubules are one of the crucial cytoskeletal components in neurons, having an important role in stabilizing neuronal morphology because they provide platforms for intracellular transport that are involved in a variety of cellular cascades, including the movement of organelles, secretory vesicles and intracellular macromolecular assemblies, as well as controlling local signaling events [37]. Colchicine inhibits polymerization of microtubules by binding to β-tubulin [5], which is essential for cellular mitosis. It is also well-known to effectively act as a mitotic inhibitor used for cancer treatment. Since one of the characteristics of cancer cells is an increased rate of mitosis, cancer cells are more vulnerable to colchicine toxicity than normal cells. However, the therapeutic value of colchicine against cancer is limited by its toxicity against normal cells. Numerous studies indicated that colchicine causes the selective destruction of granule cells in the DG of the hippocampus [1,2]. In addition, our previous study has shown that cytoskeleton-disrupting agents such as colchicine or vincristine induce dentate granule cell death by blocking axonal zinc flow and intracellular zinc accumulation [24]. Therefore, these results suggest a new mechanism of neuronal death that arises by the blockade of the axonal zinc transport and subsequent intracellular zinc accumulation as intermediary steps in colchicine-induced dentate granule cell death.

It has been well established that the generation of free radicals and subsequent oxidative stress occur prior to neuronal death and play important roles in the pathogenesis of neurodegenerative diseases [15–17]. A previous study reported that central administration of colchicine is associated with an increase in free radical generation, and subsequent oxidative stress leads to cognitive dysfunction [3]. It has been reported that colchicine induces oxidative damage, possibly by the following mechanisms: (i) it causes an increase of the glutamate/γ-aminobutyric acid (GABA) ratio in the cortex of mice [38], and this relative increase in glutamate activity exhibits neurotoxic effects by generating hydroxyl free radicals [39]; (ii) colchicine also leads to increased production of nitric oxide (NO) and inducible nitric oxide synthase (iNOS) in the brain [40]; the generated NO can induce the peroxynitrite-mediated formation of free radicals by interacting with the superoxide anion; (iii) the generated NO also causes nitrosylation of diverse enzymes, thereby inhibiting glycolysis and inducing brain damage [41]. Our lab also demonstrated that NO increased vesicular zinc release and subsequent intracellular free zinc accumulation [42].

In the present study, intrahippocampal injection of colchicine caused significant injury of dentate granule cells as evidenced by increased oxidative stress, dendritic damage, zinc accumulation and neuronal death. Furthermore, the colchicine-induced dentate granule cell death was significantly reduced in $ZnT3^{-/-}$ mice. $ZnT3$ is mainly localized at $Zn^{2+}$-containing synaptic glutamatergic vesicles in the hippocampus, cortex and olfactory bulb. Yoo et al. have demonstrated that the free $Zn^{2+}$ level is higher in neuronal somata of $ZnT3^{-/-}$ mice than those of wild-type mice, and metallothioneins 1 and 2 ($Mt1/2$), which are known to be induced by increases in cytosolic free $Zn^{2+}$ levels, were also substantially increased in $ZnT3^{-/-}$ mice [25]. Glutathione is a tripeptide composed of glutamate, glycine and cysteine. It plays a major role as an endogenous antioxidant present in the reduced form within cells [12]. It has been shown to react with free radicals and prevents the generation of hydroxyl free radicals [14]. The decreased GSH level and glutathione S-transferase (GST) activity after colchicine injection suggest that there was an increase in free radical generation and a depletion of the GSH-dependent antioxidant system during oxidative stress. Several studies have argued that $Zn^{2+}$ status affects glutathione concentrations in tissues [26,27,29,43]. Parat et al. demonstrated a positive correlation between intracellular zinc deprivation and GSH depletion in cells treated with the zinc chelator $NNN'N'$-tetrakis(2-pyridylmethyl)ethylenediamine (TPEN) [28]. In the present study, results showed that $ZnT3^{-/-}$ mice had more reduced glutathione (GSH) content than WT mice in the hippocampal CA3 and DG. Furthermore, $ZnT3^{-/-}$ mice showed prevention of colchicine-induced neuronal GSH depletion in the hippocampal subiculum, CA1, CA2, CA3 and DG. Although the present study suggests that $ZnT3$ gene deletion causes an increased intracellular free zinc level and increased neuronal GSH levels, the exact mechanism of how $ZnT3$ gene deletion provides neuroprotective effects is not clear. The present study is seemingly paradoxical insofar that $ZnT3^{-/-}$ mice showed increased intracellular levels of zinc as presented in the previous study [25], but the proposed diagram in Figure 5 suggests that there is less accumulation of zinc. Thus, we hypothesize that this is in relation to the effect that the accumulation of zinc in response to colchicine is due to impaired axonal transport specifically at 24 h, whereas $ZnT3^{-/-}$ mice might have alternative mechanisms for dealing with a normally elevated level of zinc.

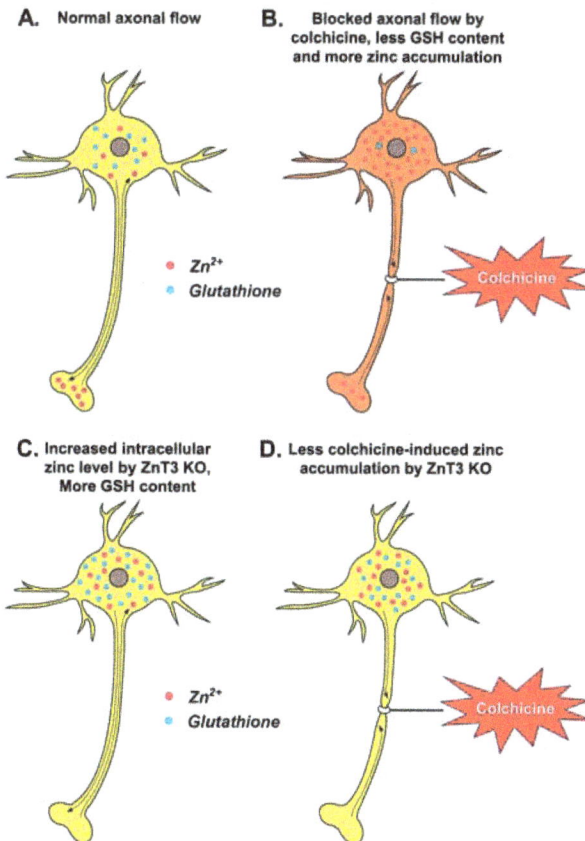

**Figure 5.** Proposed mechanism by which $ZnT3$ knockout reduces colchicine-induced dentate granule cell degeneration. This schematic drawing indicates several chain reactions that are thought to occur after colchicine injection in $ZnT3^{-/-}$ mice. (**A**) Cytoplasmic and vesicular zinc are normally moved by the axonal flow in WT mice. Zinc transport is essential for axonal flow in neurons. In addition, the concentration of GSH is much greater in the neuronal cytoplasm; (**B**) Blocked axonal flow by colchicine induces depletion of GSH content and an increase of intracellular zinc accumulation, thereby causing neuronal death; (**C**) $ZnT3$ gene deletion causes an increased intracellular free zinc level and increased neuronal GSH levels; (**D**) Genetic deletion of $ZnT3$ attenuates colchicine-induced zinc accumulation and dentate granule cell death.

Taken together, we suggest that the increased intracellular free zinc level by *ZnT3* gene deletion is correlated with increased neuronal GSH levels, which in part provide neuroprotective effects in the colchicine-induced oxidative injury in dentate granule cell.

## 4. Materials and Methods

### 4.1. Mouse Colonies

The animal care protocol and experimental procedures were approved by the committee on animal use for research and education at Hallym University (Protocol # Hallym 2014-28), in accordance with NIH guidelines. This manuscript was written in compliance with the guidelines of ARRIVE (animal research: reporting in vivo experiments) [44]. Wild-type (WT) and *ZnT3* KO male mice (background strains, C57BL/6 and Sv129 hybrid), aged 8 weeks, were propagated and maintained in the facility of Hallym University, College of Medicine. Mice were housed in a regulated environment ($22 \pm 2$ °C, $55 \pm 5\%$ humidity, 12:12 h light:dark cycle with lights on at 8:00 a.m.) and received a standard diet by Purina (Purina, Gyeonggi, Korea). Food and water were accessed ad libitum. As described previously [45], PCR genotyping was performed with a primer set to amplify WT (5′-GGT ATC CAT GCC CTT CCT CTA GAG-3′), or common (5′-ATA GTC ACT GGC ATC CTC CTG TAC C-3′), or the KO allele (5′-CCT GTG CTC TAG TAG CTT TAC GG-3′) prior to all experiments. The WT band is 650 bp in size, and the KO band is 400 bp in size.

### 4.2. Colchicine Intrahippocampal Injection and Experimental Design

To assess the role of *ZnT3* on colchicine-induced dentate granule cell degeneration, colchicine (10 µg/kg; Sigma, St. Louis, MO, USA) was intrahippocampally injected as previously described [24]. Briefly, mice were deeply anesthetized with isoflurane (1–2% for maintenance; 3% for induction) in a 70:30 mixture of nitrous oxide and oxygen using an isoflurane vaporizer (VetEquip Inc., Livermore, CA, USA) and positioned in a stereotaxic apparatus (David-Kopf Instruments, Tujunga, CA, USA). A burr hole was made in the skull, and a 29-gauge needle was inserted through the burr hole into the hippocampus (coordinates: 1.4 mm lateral from the midline, 1.5 mm posterior to the bregma and 1.8 mm below the cortical surface). We then slowly injected colchicine and removed it after placement for another 5 min. The burr hole was sealed with bone wax. Following the suture of the skin incision, anesthetics were discontinued. When mice showed spontaneous respiration, they were returned to a recovery room maintained at 37 °C. Core temperature was kept at 36.5–37.5 °C with a homoeothermic blanket control unit (Harvard apparatus, Holliston, MA, USA). Sham-operated mice received the same skin incision under isoflurane anesthesia, but they were administered with 1 µL sterile saline into the hippocampus. Mice were divided into four groups for histological evaluation: (1) sham-operated WT mice (WT-sham, $n = 6$), (2) sham-operated $ZnT3^{-/-}$ mice (KO-sham, $n = 6$), (3) colchicine-induced WT mice (WT-colchicine, $n = 8$) and (4) colchicine-induced $ZnT3^{-/-}$ mice (KO-colchicine, $n = 10$).

### 4.3. Tissue Preparation

Mice were perfused with 0.9% saline followed by 4% paraformaldehyde (PFA) under urethane (1.5 g/kg, i.p.) anesthesia. The brains were post-fixed with 4% PFA in phosphate-buffered saline (PBS) for 1 h and then immersed in 30% sucrose until subsided for cryo-protection. Thereafter, the entire brain was frozen and coronally sectioned at a 30 µm thickness using a cryostat (CM1850, Leica, Wetzlar, Germany).

### 4.4. Detection of Cell Death

Cell death after colchicine injection was evaluated after a 24 h survival period. To identify degenerating cells, Fluoro-Jade B (FJB, Histo-Chem, Jefferson, AR, USA) staining was used [46]. Five coronal sections (180 µm intervals, collected from 1.2–2.1 mm caudal to bregma) were analyzed from each mouse. A blinded experimenter counted the total number of FJB-positive cells in the hippocampal dentate gyrus (DG) from the ipsilateral hemisphere. Data were represented as the average number of degenerating cells per each region.

### 4.5. Detection of Neuronal Glutathione (GSH)

To detect the reduced form of GSH in the brain sections, we probed for GSH-$N$-ethylmaleimide (NEM) adducts on the free-floating coronal sections [47–49]. Brain sections were incubated with 10 mM NEM for 4 h at 4 °C, washed and incubated with mouse anti-GS-NEM (diluted 1:100, Millipore, Billerica, MA, USA). After washing, the sections were incubated with Alexa Fluor 488-conjugated goat anti-mouse IgG (diluted 1:250, Invitrogen, Carlsbad, CA, USA) for 2 h. The sections were mounted on gelatin-coated slides, and fluorescence signals were detected using a Zeiss LSM 710 confocal imaging system (Carl Zeiss, Oberkochen, Germany). Stacks of images (1024 × 1024 pixels) from consecutive slices of 0.9–1.2 μm in thickness were obtained by averaging eight scans per slice and were processed with ZEN 2010 (Carl Zeiss, Oberkochen, Germany). To quantify GSH intensity, individual neurons from the brain section images were selected as regions of interest (ROIs) and measured using ImageJ (NIH, Bethesda, MA, USA). Briefly, to quantify the GSH intensity, the image was loaded into ImageJ v. 1.50f and converted into an 8-bit image through the menu option Image → Type → 8-bit. Then, regions comprising individual neurons in the subiculum, CA1, CA2, CA3 and DG images were selected as ROIs. The resulting image was then binarized and restricted to the region of measurement for individual neurons. To measure this area, the menu option Analyze → Measure was selected, and then the signal from individual neurons was expressed as the mean gray value [49–51].

### 4.6. Immunofluorescence Staining

The immunolabeling procedures were performed as per routine immunostaining protocols. To block endogenous peroxidase activity, sections were immersed in 3% hydrogen peroxide for 15 min at room temperature. After washing in PBS, the sections were incubated with the following antibodies as primary antibodies: a rabbit polyclonal anti-4-hydroxynonenal (4HNE, recognizing oxidative stress, diluted 1:500, Alpha Diagnostic International, San Antonio, TX, USA) or a mouse monoclonal anti-microtubule-associated protein 2 (MAP2, recognizing a neuron-specific cytoskeletal protein, diluted 1:200, Millipore, Billerica, MA, USA). For visualization of antibody binding, Alexa Fluor 488- and 594-conjugated antibodies were applied at a dilution of 1:250 as secondary antibodies. Between incubations, the sections were washed with PBS 3 times for 10 min each and then mounted on gelatin-coated slides. According to the method modified from the above-described method, 4HNE or MAP2 intensity was quantified using ImageJ. The image was converted to 8 bit through the menu options (Image/Type/8-bit). Next, the image was thresholded as follows (Image/Adjust/Threshold): the type was set to black and white and the bottom slider moved to a value sufficient to show only the 4HNE or MAP2 immunoreactive area. The thresholded image was binary and only represented 4HNE or MAP2 immunoreactivity. The selected part in the whole image was sorted, and then the intensity of 4HNE or MAP2 was expressed as the mean gray value.

### 4.7. Free Zinc Staining

For analyzing the vesicular and intraneuronal free zinc, the fresh frozen brain sections were stained with $N$-(6-methoxy-8-quinolyl)-para-toluenesulfonamide (TSQ, Molecular Probes, Eugene, OR, USA) fluorescent probes as previously described [52]. TSQ is a membrane-permeant $Zn^{2+}$ indicator. Mice were anesthetized with 3–5% isoflurane in oxygen. Brain tissue was harvested without perfusion, quickly frozen on powdered dry ice and then stored at −80 °C. Unfixed fresh frozen brains were coronally sectioned with a 25 μm thickness. Sections were immersed for 1 min in a solution of 4.5 μM TSQ, 140 mM sodium barbital and 140 mM sodium acetate (pH 10.5–11), then rinsed for 1 min in 0.9% normal saline. TSQ–zinc binding was evaluated using an Olympus upright fluorescence microscope with an excitation/emission wavelength of 360/490 nm and photographed using a charge-coupled device (CCD) cooled digital color camera (Hamamatsu Co., Bridgewater, NJ, USA) with Infinity 3 (Lumenera Co., Ottawa, ON, Canada).

*4.8. Statistical Analysis*

Comparisons between experimental groups were conducted using repeated measures analysis of variance (ANOVA) followed by the Student–Newman–Keuls post hoc test. Data were presented as the mean $\pm$ SEM, and differences were considered significant at $p < 0.05$.

**Acknowledgments:** We would like to express our special thanks to Ms. Tae Yul Kim for the schematic illustrations. This study was supported by the National Research Foundation of Korea (NRF) (NRF-2017R1C1B1004226) to Bo Young Choi. This work was also supported by the Brain Research Program through the NRF funded by the Ministry of Science, ICT (Information & Communication Technology) & Future Planning (NRF-2017M3C7A1028937) to Sang Won Suh.

**Author Contributions:** Bo Young Choi researched the data, reviewed and edited the manuscript. Dae Ki Hong researched the data, and Sang Won Suh contributed to the discussion and wrote/reviewed and edited the manuscript. Sang Won Suh is the person who takes full responsibility for the manuscript and its originality. All authors read and approved the final manuscript.

**Conflicts of Interest:** The authors declare no competing financial interests.

## References

1. Goldschmidt, R.B.; Steward, O. Preferential neurotoxicity of colchicine for granule cells of the dentate gyrus of the adult rat. *Proc. Natl. Acad. Sci. USA* **1980**, *77*, 3047–3051. [CrossRef] [PubMed]
2. Muller, G.J.; Geist, M.A.; Veng, L.M.; Willesen, M.G.; Johansen, F.F.; Leist, M.; Vaudano, E. A role for mixed lineage kinases in granule cell apoptosis induced by cytoskeletal disruption. *J. Neurochem.* **2006**, *96*, 1242–1252. [CrossRef] [PubMed]
3. Veerendra Kumar, M.H.; Gupta, Y.K. Intracerebroventricular administration of colchicine produces cognitive impairment associated with oxidative stress in rats. *Pharmacol. Biochem. Behav.* **2002**, *73*, 565–571. [CrossRef]
4. Hirokawa, N. The neuronal cytoskeleton: Roles in neuronal morphogenesis and organelle transport. *Prog. Clin. Biol. Res.* **1994**, *390*, 117–143. [PubMed]
5. Uppuluri, S.; Knipling, L.; Sackett, D.L.; Wolff, J. Localization of the colchicine-binding site of tubulin. *Proc. Natl. Acad. Sci. USA* **1993**, *90*, 11598–11602. [CrossRef] [PubMed]
6. Kumar, A.; Seghal, N.; Naidu, P.S.; Padi, S.S.; Goyal, R. Colchicines-induced neurotoxicity as an animal model of sporadic dementia of Alzheimer's type. *Pharmacol. Rep.* **2007**, *59*, 274–283. [PubMed]
7. Kumar, A.; Dogra, S.; Prakash, A. Neuroprotective effects of centella asiatica against intracerebroventricular colchicine-induced cognitive impairment and oxidative stress. *Int. J. Alzheimers Dis.* **2009**. [CrossRef] [PubMed]
8. Muller, D.P. Neurological disease. *Adv. Pharmacol.* **1997**, *38*, 557–580. [PubMed]
9. Friedman, J. Why is the nervous system vulnerable to oxidative stress? In *Oxidative Stress and Free Radical Damage in Neurology. Oxidative Stress in Applied Basic Research and Clinical Practice*; Gadoth, N., Göbel, H., Eds.; Humana Press: Clifton, NJ, USA, 2011; pp. 19–27.
10. Buttke, T.M.; Sandstrom, P.A. Oxidative stress as a mediator of apoptosis. *Immunol. Today* **1994**, *15*, 7–10. [CrossRef]
11. Schulz, J.B.; Lindenau, J.; Seyfried, J.; Dichgans, J. Glutathione, oxidative stress and neurodegeneration. *Eur. J. Biochem.* **2000**, *267*, 4904–4911. [CrossRef] [PubMed]
12. Dringen, R. Metabolism and functions of glutathione in brain. *Prog. Neurobiol.* **2000**, *62*, 649–671. [CrossRef]
13. Beaver, J.P.; Waring, P. A decrease in intracellular glutathione concentration precedes the onset of apoptosis in murine thymocytes. *Eur. J. Cell Biol.* **1995**, *68*, 47–54. [PubMed]
14. Bains, J.S.; Shaw, C.A. Neurodegenerative disorders in humans: The role of glutathione in oxidative stress-mediated neuronal death. *Brain Res. Brain Res. Rev.* **1997**, *25*, 335–358. [CrossRef]
15. Markesbery, W.R. Oxidative stress hypothesis in Alzheimer's disease. *Free Radic. Biol. Med.* **1997**, *23*, 134–147. [CrossRef]

16. Uttara, B.; Singh, A.V.; Zamboni, P.; Mahajan, R.T. Oxidative stress and neurodegenerative diseases: A review of upstream and downstream antioxidant therapeutic options. *Curr. Neuropharmacol.* **2009**, *7*, 65–74. [CrossRef] [PubMed]

17. Li, J.; O, W.; Li, W.; Jiang, Z.G.; Ghanbari, H.A. Oxidative stress and neurodegenerative disorders. *Int. J. Mol. Sci.* **2013**, *14*, 24438–24475. [CrossRef] [PubMed]

18. Frederickson, C.J.; Suh, S.W.; Silva, D.; Thompson, R.B. Importance of zinc in the central nervous system: The zinc-containing neuron. *J. Nutr.* **2000**, *130*, 1471S–1483S. [PubMed]

19. Vogt, K.; Mellor, J.; Tong, G.; Nicoll, R. The actions of synaptically released zinc at hippocampal mossy fiber synapses. *Neuron* **2000**, *26*, 187–196. [CrossRef]

20. Maret, W.; Sandstead, H.H. Zinc requirements and the risks and benefits of zinc supplementation. *J. Trace Elem. Med. Biol.* **2006**, *20*, 3–18. [CrossRef] [PubMed]

21. Deshpande, A.; Kawai, H.; Metherate, R.; Glabe, C.G.; Busciglio, J. A role for synaptic zinc in activity-dependent Abeta oligomer formation and accumulation at excitatory synapses. *J. Neurosci.* **2009**, *29*, 4004–4015. [CrossRef] [PubMed]

22. Cull, R.E. Role of axonal transport in maintaining central synaptic connections. *Exp. Brain Res.* **1975**, *24*, 97–101. [CrossRef] [PubMed]

23. Takeda, A.; Kodama, Y.; Ohnuma, M.; Okada, S. Zinc transport from the striatum and substantia nigra. *Brain Res. Bull.* **1998**, *47*, 103–106. [CrossRef]

24. Choi, B.Y.; Lee, B.E.; Kim, J.H.; Kim, H.J.; Sohn, M.; Song, H.K.; Chung, T.N.; Suh, S.W. Colchicine induced intraneuronal free zinc accumulation and dentate granule cell degeneration. *Metallomics* **2014**, *6*, 1513–1520. [CrossRef] [PubMed]

25. Yoo, M.H.; Kim, T.Y.; Yoon, Y.H.; Koh, J.Y. Autism phenotypes in *ZnT3* null mice: Involvement of zinc dyshomeostasis, MMP-9 activation and BDNF upregulation. *Sci. Rep.* **2016**, *6*, 28548. [CrossRef] [PubMed]

26. Mills, B.J.; Lindeman, R.D.; Lang, C.A. Effect of zinc deficiency on blood glutathione levels. *J. Nutr.* **1981**, *111*, 1098–1102. [PubMed]

27. Fernandez, M.A.; O'Dell, B.L. Effect of zinc deficiency on plasma glutathione in the rat. *Proc. Soc. Exp. Biol. Med.* **1983**, *173*, 564–567. [CrossRef] [PubMed]

28. Parat, M.O.; Richard, M.J.; Beani, J.C.; Favier, A. Involvement of zinc in intracellular oxidant/antioxidant balance. *Biol. Trace Elem. Res.* **1997**, *60*, 187–204. [CrossRef] [PubMed]

29. Iszard, M.B.; Liu, J.; Klaassen, C.D. Effect of several metallothionein inducers on oxidative stress defense mechanisms in rats. *Toxicology* **1995**, *104*, 25–33. [CrossRef]

30. Emerich, D.F.; Walsh, T.J. Cholinergic cell loss and cognitive impairments following intraventricular or intradentate injection of colchicine. *Brain Res.* **1990**, *517*, 157–167. [CrossRef]

31. Nakagawa, Y.; Nakamura, S.; Kase, Y.; Noguchi, T.; Ishihara, T. Colchicine lesions in the rat hippocampus mimic the alterations of several markers in Alzheimer's disease. *Brain Res.* **1987**, *408*, 57–64. [CrossRef]

32. Esterbauer, H.; Schaur, R.J.; Zollner, H. Chemistry and biochemistry of 4-hydroxynonenal, malonaldehyde and related aldehydes. *Free Radic. Biol. Med.* **1991**, *11*, 81–128. [CrossRef]

33. Zarkovic, N. 4-hydroxynonenal as a bioactive marker of pathophysiological processes. *Mol. Aspects Med.* **2003**, *24*, 281–291. [CrossRef]

34. Izant, J.G.; McIntosh, J.R. Microtubule-associated proteins: A monoclonal antibody to MAP2 binds to differentiated neurons. *Proc. Natl. Acad. Sci. USA* **1980**, *77*, 4741–4745. [CrossRef] [PubMed]

35. Huber, G.; Matus, A. Differences in the cellular distributions of two microtubule-associated proteins, MAP1 and MAP2, in rat brain. *J. Neurosci.* **1984**, *4*, 151–160. [PubMed]

36. Park, J.S.; Bateman, M.C.; Goldberg, M.P. Rapid alterations in dendrite morphology during sublethal hypoxia or glutamate receptor activation. *Neurobiol. Dis.* **1996**, *3*, 215–227. [CrossRef] [PubMed]

37. Vale, R.D. The molecular motor toolbox for intracellular transport. *Cell* **2003**, *112*, 467–480. [CrossRef]

38. Yu, Z.; Cheng, G.; Hu, B. Mechanism of colchicine impairment on learning and memory, and protective effect of CGP36742 in mice. *Brain Res.* **1997**, *750*, 53–58. [CrossRef]

39. Hammer, B.; Parker, W.D., Jr.; Bennett, J.P., Jr. NMDA receptors increase OH radicals in vivo by using nitric oxide synthase and protein kinase C. *Neuroreport* **1993**, *5*, 72–74. [CrossRef] [PubMed]

40. Dufourny, L.; Leroy, D.; Warembourg, M. Differential effects of colchicine on the induction of nitric oxide synthase in neurons containing progesterone receptors of the guinea pig hypothalamus. *Brain Res. Bull.* **2000**, *52*, 435–443. [CrossRef]

41. Wallis, R.A.; Panizzon, K.L.; Henry, D.; Wasterlain, C.G. Neuroprotection against nitric oxide injury with inhibitors of ADP-ribosylation. *Neuroreport* **1993**, *5*, 245–248. [CrossRef] [PubMed]
42. Frederickson, C.J.; Cuajungco, M.P.; LaBuda, C.J.; Suh, S.W. Nitric oxide causes apparent release of zinc from presynaptic boutons. *Neuroscience* **2002**, *115*, 471–474. [CrossRef]
43. Song, Y.; Leonard, S.W.; Traber, M.G.; Ho, E. Zinc deficiency affects DNA damage, oxidative stress, antioxidant defenses, and DNA repair in rats. *J. Nutr.* **2009**, *139*, 1626–1631. [CrossRef] [PubMed]
44. Kilkenny, C.; Browne, W.J.; Cuthill, I.C.; Emerson, M.; Altman, D.G. Improving bioscience research reporting: The ARRIVE guidelines for reporting animal research. *PLoS Biol.* **2012**, *8*, e1000412. [CrossRef] [PubMed]
45. Cole, T.B.; Wenzel, H.J.; Kafer, K.E.; Schwartzkroin, P.A.; Palmiter, R.D. Elimination of zinc from synaptic vesicles in the intact mouse brain by disruption of the *ZnT3* gene. *Proc. Natl. Acad. Sci USA* **1999**, *96*, 1716–1721. [CrossRef] [PubMed]
46. Schmued, L.C.; Hopkins, K.J. Fluoro-Jade B: A high affinity fluorescent marker for the localization of neuronal degeneration. *Brain Res.* **2000**, *874*, 123–130. [CrossRef]
47. Liblau, R.S.; Tisch, R.; Shokat, K.; Yang, X.; Dumont, N.; Goodnow, C.C.; McDevitt, H.O. Intravenous injection of soluble antigen induces thymic and peripheral T-cells apoptosis. *Proc. Natl. Acad. Sci. USA* **1996**, *93*, 3031–3036. [CrossRef] [PubMed]
48. Polman, C.H.; O'Connor, P.W.; Havrdova, E.; Hutchinson, M.; Kappos, L.; Miller, D.H.; Phillips, J.T.; Lublin, F.D.; Giovannoni, G.; Wajgt, A.; et al. A randomized, placebo-controlled trial of natalizumab for relapsing multiple sclerosis. *N. Engl. J. Med.* **2006**, *354*, 899–910. [CrossRef] [PubMed]
49. Choi, B.Y.; Kim, J.H.; Kim, H.J.; Yoo, J.H.; Song, H.K.; Sohn, M.; Won, S.J.; Suh, S.W. Pyruvate administration reduces recurrent/moderate hypoglycemia-induced cortical neuron death in diabetic rats. *PLoS ONE* **2013**, *8*, e81523. [CrossRef] [PubMed]
50. Tang, X.N.; Berman, A.E.; Swanson, R.A.; Yenari, M.A. Digitally quantifying cerebral hemorrhage using Photoshop and Image J. *J. Neurosci. Methods* **2010**, *190*, 240–243. [CrossRef] [PubMed]
51. Choi, B.Y.; Kim, J.H.; Kim, H.J.; Lee, B.E.; Kim, I.Y.; Sohn, M.; Suh, S.W. EAAC1 gene deletion increases neuronal death and blood brain barrier disruption after transient cerebral ischemia in female mice. *Int. J. Mol. Sci.* **2014**, *15*, 19444–19457. [CrossRef] [PubMed]
52. Frederickson, C.J.; Kasarskis, E.J.; Ringo, D.; Frederickson, R.E. A quinoline fluorescence method for visualizing and assaying the histochemically reactive zinc (bouton zinc) in the brain. *J. Neurosci. Methods* **1987**, *20*, 91–103. [CrossRef]

International Journal of
*Molecular Sciences*

MDPI

*Article*

# Zinc Transporter 3 (ZnT3) in the Enteric Nervous System of the Porcine Ileum in Physiological Conditions and during Experimental Inflammation

Sławomir Gonkowski [1,*], Maciej Rowniak [2] and Joanna Wojtkiewicz [3,4,5]

[1] Department of Clinical Physiology, Faculty of Veterinary Medicine, Oczapowskiego 13,
University of Warmia and Mazury, 10-718 Olsztyn, Poland
[2] Department of Comparative Anatomy, Faculty of Biology, Plac Łódzki 3, University of Warmia and Mazury,
10-727 Olsztyn, Poland; maciek@matman.uwm.edu.pl
[3] Department of Pathophysiology, Faculty of Medical Sciences, Warszawska 30, University of Warmia and
Mazury, 10-082 Olsztyn, Poland; asiawoj@uwm.edu.pl
[4] Laboratory for Regenerative Medicine, Faculty of Medical Sciences, University of Warmia and Mazury,
Olsztyn, 10-082 Olsztyn, Poland
[5] Foundation for Nerve Cells Regeneration, Warszawska 30, 10-082 Olsztyn, Poland
[*] Correspondence: slawomir.gonkowski@uwm.edu.pl; Tel.: +48-89-523-4376

Academic Editor: Toshiyuki Fukada
Received: 25 November 2016; Accepted: 3 February 2017; Published: 7 February 2017

**Abstract:** Zinc transporter 3 (ZnT3) is a member of the solute-linked carrier 30 (SLC 30) zinc transporter family. It is closely linked to the nervous system, where it takes part in the transport of zinc ions from the cytoplasm to the synaptic vesicles. ZnT3 has also been observed in the enteric nervous system (ENS), but its reactions in response to pathological factors remain unknown. This study, based on the triple immunofluorescence technique, describes changes in ZnT3-like immunoreactive (ZnT3-LI) enteric neurons in the porcine ileum, caused by chemically-induced inflammation. The inflammatory process led to a clear increase in the percentage of neurons immunoreactive to ZnT3 in all "kinds" of intramural enteric plexuses, i.e., myenteric (MP), outer submucous (OSP) and inner submucous (ISP) plexuses. Moreover, a wide range of other active substances was noted in ZnT3-LI neurons under physiological and pathological conditions, and changes in neurochemical characterisation of ZnT3$^+$ cells in response to inflammation depended on the "kind" of enteric plexus. The obtained results show that ZnT3 is present in the ENS in a relatively numerous and diversified neuronal population, not only in physiological conditions, but also during inflammation. The reasons for the observed changes are not clear; they may be connected with the functions of zinc ions and their homeostasis disturbances in pathological processes. On the other hand, they may be due to adaptive and/or neuroprotective processes within the pathologically altered gastrointestinal tract.

**Keywords:** zinc transporter 3 (ZnT3); enteric nervous system (ENS); inflammation; pigs

## 1. Introduction

The enteric nervous system (ENS) is localised in the wall of the gastrointestinal (GI) tract and takes part in all regulatory processes connected with digestive actions, such as intestinal motility and excretion [1]. It is characterised by significant independence from the central nervous system, as well as complex conformation. The ENS is made up of millions of neurons grouped in intramural ganglionated plexuses interconnected with a very dense network of nerves. The number and form of these plexuses clearly depend both on the animal species and the fragment of the GI tract [1–3]. In the porcine intestine, the ENS is built of three intramural plexuses: the myenteric plexus (MP)—located between the longitudinal and circular muscle layers, the outer submucous plexus (OSP)—in the inner side of the circular muscle layer, and the inner submucous plexus (ISP)—between the muscularis mucosa and lamina propria [4]. Enteric neurons vary in terms of their conformation, functions, electrophysiological properties and neurochemical coding [2]. Besides acetylcholine, which is the main transmitter in enteric neurons [5,6], several dozen other neuronal active substances have been described within the ENS [2,3]. The most important of these include, among others, vasoactive intestinal peptide (VIP), galanin (GAL), neuronal isoform of nitric oxide synthase (nNOS) and substance P (SP) [7–9]. Most of the abovementioned factors act as neuromediators and/or neuromodulators, but the functions of some neuronal substances described in the ENS remain unknown. One of these is zinc transporter 3 (ZnT3) [10].

ZnT3 is one of the solute-linked carrier 30 (SLC 30) protein family of zinc transporters, which in mammals consists of 10 members (ZnT1 to ZnT10) [11]. These molecules allow zinc ions to permeate from the cytoplasm to the intercellular space, as well as intracellular organelles [12]. From among the ZnT family of zinc transporters, only ZnT3 is closely linked to neuronal cells, where it is responsible for transport of zinc ions from the cytoplasm to synaptic vesicles, and thereby may influence neuronal conduction [12]. ZnT3 has been described in the brain, spinal cord and autonomic peripheral nervous system [13–15]. Within the central nervous system, this protein may be responsible for both sensory conduction and secretory activity and is also considered to be a marker of zinc-enriched nerves (ZEN), which have an inhibitory function [16]. Furthermore, it is known that ZnT3 can take part in some pathological processes within the central nervous system, such as epilepsy, mechanical damage or ischemia [15,17,18].

In contrast to the brain and spinal cord, knowledge about the distribution and functions of ZnT3 in the peripheral autonomic nervous system is very sparse [14]. This is particularly visible with reference to the ENS. Admittedly, ZnT3 has been described in the human descending colon [10], as well as in the porcine duodenum and jejunum [19,20], but some aspects of the functions of this peptide within the GI tract, especially during pathological states, remain within the realm of conjecture.

The roles of ZnT3 are probably closely linked to the functions of zinc in a living organism. This metal is known to be a very important factor, as it is a component of various enzymatic systems and takes part in the stabilisation of cellular membranes, DNA synthesis, cell division and the correct functioning of the immune system [21]. Relatively high levels of zinc have been described in the central nervous system, where this metal is indispensable to normal brain development and neuronal functioning [22]. Moreover, "free pool" reactive zinc ions can act as neurotransmitters and/or neuromodulators [23]. Zinc also plays an important role within the GI tract, where first of all it stimulates the absorption of ions in enterocytes by modulating intracellular cAMP concentration, and therefore it is used as a drug to decrease the severity and duration of diarrhoea [24]. Moreover, zinc as a component of zinc finger E-box-binding homeobox (Zeb) 2 protein takes part in the formation, migration and specification of cells in the neural crest, from which the ENS arises, and is also necessary for the correct functioning of enteric neurons [25].

On the other hand, it is relatively well known that the ENS regulates the functions of the intestinal mucosa, which is the first barrier against various pathological factors [1]. Therefore, enteric neurons may undergo changes caused by adaptive and/or neuroprotective processes during many intestinal and extra-intestinal diseases, and these changes mainly manifest themselves in modifications of neuronal chemical coding [25–29].

The aim of the present study was to investigate the changes in ZnT3-like immunoreactive (ZnT3-LI) enteric neurons in the porcine ileum during experimental chemically induced inflammation. The choice of the pig as the experimental animal during the present study was determined by the fact that this species more and more often appears to be an optimal animal model for pathological processes in the human organism, due to similarities in anatomical, histological and physiological properties between humans and pigs [30–32].

## 2. Results

In the present study, neurons immunoreactive to ZnT3 were observed within the porcine ileal enteric nervous system, both under physiological conditions and during inflammation (Table 1). All ZnT3-positive cells noted in the ganglionated plexuses of the ENS were also immunoreactive to protein gene product 9.5 (PGP 9.5—used as a panneronal marker). Moreover, during the present investigation, ZnT3-LI cells were noted outside of the enteric plexuses. These cells were scattered in various parts of the intestinal wall and were not PGP 9.5-positive.

### 2.1. ZnT3 in the ENS under Physiological Conditions

Under physiological conditions, the percentage of these neurons was relatively high, levelised in all "kinds" of plexuses, and amounted to 42.3% ± 4.7% in the MP (Figure 1), 43.5% ± 6.8% in the OSP (Figure 2) and 48.6% in the ISP (Figure 3). In individual enteric ganglion, most often three or more ZnT3-LI cells were noted (Figure 2c), but ganglia with one or two such perikarya were also observed (Figure 1a). There are no differences between particular "kinds" of plexuses in the distribution of ZnT3$^+$ cells in individual enteric ganglia.

#### 2.1.1. Neurochemical Characterisation of ZnT3$^+$ Cholinergic Enteric Neurons

A wide range of neuronal active substances was investigated in ZnT3-positive enteric neurons under physiological conditions (Table 1). A significant percentage of cells were immunoreactive to ZnT3, regardless of whether the "kind" of enteric plexus was cholinergic (positive to vesicular acetylcholine transporter—VAChT) (Figure 1a,c,d, Figures 2a–c and 3b–d). These neurons accounted for 85.0% ± 5.8%, 82.0% ± 5.9% and 90.0% ± 7.2% of all ZnT3$^+$ perikarya in the MP, OSP and ISP, respectively. Neurochemical coding of cholinergic neurons immunoreactive to ZnT3 was variable in different "kinds" of enteric plexuses. Some ZnT3$^+$/VAChT$^+$ neuronal cells in the MP and OSP were also immunoreactive to somatostatin (SOM), VIP and SP, but the degree of co-localisation of ZnT3 with the abovementioned particular substances was slight and did not exceed 7.5% (Table 1).

A completely different situation was observed in the ISP. In this plexus, a much higher percentage of cholinergic ZnT3-LI neurons were immunoreactive to SOM, VIP and/or SP. These values amounted to 24.1% ± 1.2%, 33.0% ± 3.6% and 55.1% ± 1.3%, respectively (Table 1). Moreover, 65.2% ± 1.2% of cholinergic ZnT3$^+$ cells in the ISP were also immunoreactive to GAL, contrary to MP and OSP, where ZnT3$^+$/VAChT$^+$/GAL$^+$ neurons were not observed at all. During the present study, immunoreactivity to nNOS, leu-enkephalin (LENK), neuropeptide Y (NPY) and calcitonin gene-related peptide (CGRP) was not observed in cholinergic ZnT3$^+$ neurons in any enteric plexus.

#### 2.1.2. Neurochemical Characterisation of ZnT3$^+$ Non-Cholinergic Enteric Neurons

Non-cholinergic neurons in the MP amounted to 15.0% ± 5.6% of all ZnT3$^+$ cells (Figure 1a,c,d). In the OSP and ISP, these values stood at 18.0% ± 5.9% and 10.0% ± 7.2%, respectively (Figures 2c and 3b). The degree of co-localisation of ZnT3 with other substances studied in non-cholinergic cells clearly depended on the "kind" of plexus (Table 1). The most visible differences were observed in the event of nNOS and GAL. In the MP, 17.1% ± 1.1% of non-cholinergic ZnT3$^+$ cells were also nNOS-positive, whereas in the OSP, this value amounted to only 7.4% ± 1.3%, and in the ISP, such cells were not observed at all. Cells simultaneously immunoreactive to ZnT3 and GAL comprised 11.9% ± 1.5% of all ZnT3-LI non-cholinergic neurons in the MP, but such perikarya were not noted in the OSP, and in the ISP, they amounted to only 2.8% ± 0.3%. The degree of co-localisations of ZnT3 with VIP, SOM and SP was levelised in all "kinds" of plexuses and was rather slight, because they did not exceed 5% of all non-cholinergic ZnT3-LI cells regardless of the "kind" of plexus. An especially low level of co-localisation of ZnT3 with LENK was observed. Cells simultaneously immunopositive to ZnT3 and LENK were observed only in the MP, and the percentage of these amounted to barely 0.5% ± 0.2% of all non-cholinergic ZnT3$^+$ neurons. During the present study, the co-localisations of ZnT3 and NPY, as well as ZnT3 and CGRP, were not observed in non-cholinergic ZnT3-LI cells in any plexus of the ileal enteric nervous system under physiological conditions.

**Table 1.** Zinc transporter 3-like immunoreactive (ZnT3-LI) neurons in the enteric nervous system (ENS) of the porcine ileum under physiological conditions (C group) and during chemically-induced inflammation (I group).

| Enteric Plexus | MP | | OSP | | ISP | |
|---|---|---|---|---|---|---|
| | C Group | I Group | C Group | I Group | C Group | I Group |
| PGP⁺/ZnT3⁺ [1] | 42.3 ± 4.7 | 84 ± 3.9 | 43.5 ± 6.8 | 85.6 ± 2.0 | 48.6 ± 4.8 | 79.0 ± 3.2 |
| ZnT3⁺/VAChT⁺ [2] | 85.0 ± 5.8 | 41.0 ± 2.7 | 82.0 ± 5.9 | 31.3 ± 2.8 | 90.0 ± 7.2 | 34.0 ± 3.0 |
| ZnT3⁺/VAChT⁺/CGRP⁺ [3] | 0 | 0 | 0 | 0 | 0 | 0 |
| ZnT3⁺/VAChT⁺/GAL⁺ [3] | 0 | 5.4 ± 1.5 | 0 | 4.7 ± 0.9 | 65.2 ± 1.2 | 4.7 ± 1.3 |
| ZnT3⁺/VAChT⁺/LENK⁺ [3] | 0 | 0 | 0 | 0 | 0 | 0 |
| ZnT3⁺/VAChT⁺/NOS⁺ [3] | 0 | 0 | 0 | 0 | 0 | 0 |
| ZnT3⁺/VAChT⁺/NPY⁺ [3] | 0 | 0 | 0 | 0 | 0 | 0 |
| ZnT3⁺/VAChT⁺/SOM⁺ [3] | 2.9 ± 1.4 | 0 | 7.4 ± 1.4 | 1.1 ± 0.8 | 24.1 ± 1.2 | 4.1 ± 0.8 |
| ZnT3⁺/VAChT⁺/SP⁺ [3] | 1.2 ± 0.5 | 4.2 ± 1.2 | 4.6 ± 1.6 | 2.8 ± 1.4 | 55.1 ± 1.3 | 6.8 ± 1.5 |
| ZnT3⁺/VAChT⁺/VIP⁺ [3] | 2.9 ± 1.4 | 11.2 ± 1.9 | 1.4 ± 0.8 | 3.8 ± 1.4 | 33.0 ± 3.6 | 7.3 ± 1.6 |
| ZnT3⁺/VAChT⁻ [2] | 15.0 ± 5.6 | 59.0 ± 2.7 | 18.0 ± 5.9 | 68.7 ± 2.8 | 10.0 ± 7.2 | 66.0 ± 3.0 |
| ZnT3⁺/VAChT⁻/CGRP⁺ [3] | 0 | 0 | 0 | 0 | 0 | 0 |
| ZnT3⁺/VAChT⁻/GAL⁺ [3] | 11.9 ± 1.5 | 30.2 ± 3.4 | 0 | 36.3 ± 7.5 | 2.8 ± 0.1 | 54.7 ± 7.4 |
| ZnT3⁺/VAChT⁻/LENK⁺ [3] | 0.5 ± 0.2 | 41.2 ± 2.7 | 0 | 0 | 0 | 0 |
| ZnT3⁺/VAChT⁻/NOS⁺ [3] | 17.1 ± 1.1 | 57.0 ± 4.0 | 7.4 ± 1.3 | 10.3 ± 2.4 | 0 | 0 |
| ZnT3⁺/VAChT⁻/NPY⁺ [3] | 0 | 0 | 0 | 0 | 0 | 0 |
| ZnT3⁺/VAChT⁻/SOM⁺ [3] | 3.4 ± 1.3 | 0 | 1.4 ± 0.6 | 43.0 ± 3.6 | 3.7 ± 1.0 | 38.6 ± 1.5 |
| ZnT3⁺/VAChT⁻/SP⁺ [3] | 1.1 ± 0.6 | 32.2 ± 0.4 | 4.7 ± 1.8 | 9.5 ± 2.4 | 4.1 ± 1.2 | 36.9 ± 1.5 |
| ZnT3⁺/VAChT⁻/VIP⁺ [3] | 3.9 ± 0.3 | 52.1 ± 2.8 | 4.6 ± 1.6 | 20.7 ± 9.7 | 2.9 ± 1.3 | 51.4 ± 3.5 |

MP: myenteric plexus; OSP: outer submucous plexus; ISP: inner submucous plexus; [1] The relative frequency of ZnT3-LI neuronal cells is presented as % (mean ± SEM) in relation to all neurons counted within the enteric ganglionated plexuses stained for PGP 9.5 (PGP 9.5-LI cells were treated as 100%); [2] The percentage of VAChT⁺ or VAChT⁻ neurons is presented as % (mean ± SEM) of all neurons counted within the ganglionated plexuses stained for ZnT3; [3] The relative frequency of neurons immunoreactive to particular substances is presented as % (mean ± SEM) of all ZnT3⁺/VAChT⁺ or ZnT3⁺/VAChT⁻ neurons counted within the ganglionated plexuses. Statistically significant data ($p \leq 0.05$) are marked by a different colour (red—increase, green—decrease of the percentage of a particular neuronal population in I group compared to C group).The main groups of cells (general percentage of ZnT3⁺, cholinergic ZnT3⁺ and non-cholinergic ZnT3⁺ neurons) are indicated with bold font.

A completely different situation was observed in the ISP. In this plexus, a much higher percentage of cholinergic ZnT3-LI neurons were immunoreactive to SOM, VIP and/or SP. These values amounted to 24.1% ± 1.2%, 33.0% ± 3.6% and 55.1% ± 1.3%, respectively (Table 1). Moreover, 65.2% ± 1.2% of cholinergic ZnT3⁺ cells in the ISP were also immunoreactive to GAL, contrary to MP and OSP, where ZnT3⁺/VAChT⁺/GAL⁺ neurons were not observed at all. During the present study, immunoreactivity to nNOS, leu-enkephalin (LENK), neuropeptide Y (NPY) and calcitonin gene-related peptide (CGRP) was not observed in cholinergic ZnT3-positive neurons in any enteric plexus.

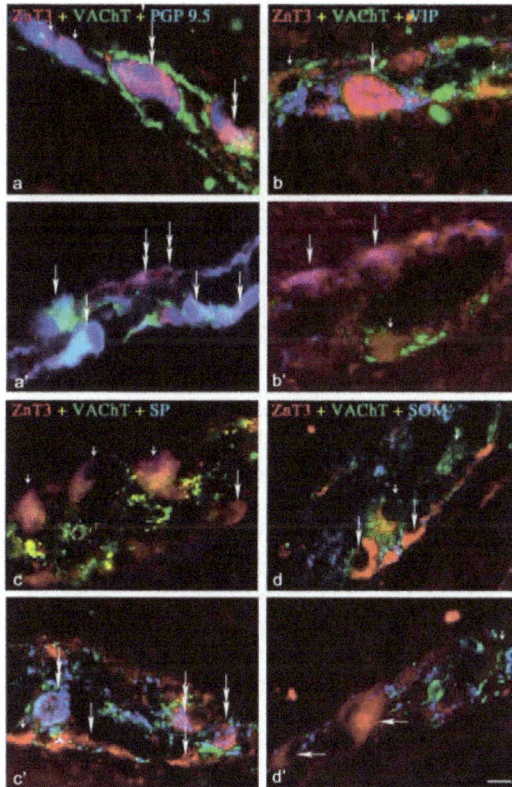

**Figure 1.** Representative images of ZnT3$^+$ neurons located in the myenteric plexus (MP) of the porcine ileum: myenteric ganglia. All images are composites of merged images taken separately from blue, red and green fluorescent channels. Control (C) group: (**a**) ZnT3$^+$/VAChT$^+$/PGP9.5$^+$ neurons are indicated with double-headed arrows, ZnT3$^-$/VAChT$^-$/PGP9.5$^+$ neurons are indicated with small arrows; (**b**) ZnT3$^+$/VAChT$^+$/VIP$^+$ neurons are indicated with an arrow, ZnT3$^+$/VAChT$^+$/VIP$^-$ neurons are indicated with small arrows; (**c**) ZnT3$^+$/VAChT$^+$/SP$^-$ neurons are indicated with small arrows; ZnT3$^+$/VAChT$^-$/SP$^-$ neurons are is indicated with arrows; (**d**) ZnT3$^+$/VAChT$^+$/SOM$^-$ neuron is indicated with a small arrow; ZnT3$^+$/VAChT$^-$/SOM$^-$ neurons are indicated with arrows. Inflammatory (I) group (**a′**) ZnT3$^+$/VAChT$^+$/PGP9.5$^+$ neurons are indicated with arrows, ZnT3$^+$/VAChT$^-$/PGP9.5$^+$ neurons are indicated with double-headed arrows; (**b′**) ZnT3$^+$/VAChT$^-$/VIP$^+$ neurons are indicated with arrows; ZnT3$^+$/VAChT$^-$/VIP$^-$ neurons are indicated with a small arrow; (**c′**) ZnT3$^+$/VAChT$^+$/SP$^-$ neuron is indicated with a small arrow; ZnT3$^+$/VAChT$^-$/SP$^+$ neurons are indicated with double-headed arrows; ZnT3$^+$/VAChT$^-$/SP$^-$ neurons are indicated with arrows; Control group (**d′**) ZnT3$^+$/VAChT$^+$/SOM$^-$ neuron is indicated with a small arrow; ZnT3$^+$/VAChT$^-$/SOM$^-$ neurons are indicated with arrows. Scale bar 25 μm.

## 2.2. ZnT3 in the ENS during the Inflammatory Process

Experimentally induced colitis changed both the percentage of ZnT3-positive cells and the degree of co-localisation of ZnT3 with the other active factors studied (Table 1). The observed modifications clearly depended on the "kind" of enteric plexus, as well as the type of neurochemical substance.

During the inflammatory process, an evident increase in the percentage of ZnT3$^+$ neurons (in relation to all protein gene product (PGP 9.5)-positive cells) was observed within all enteric plexuses. In the MP and OSP, these values were approx. two-fold higher than in the control group and amounted to 84.1% ± 3.9% and 85.6% ± 2.0%, respectively. Changes in the ISP were less pronounced, but also clearly visible (an increase from 48.6% ± 4.8% to 79.0% ± 3.2%).

Changes in Neurochemical Characterisation of ZnT3$^+$ Enteric Neurons during Inflammation

Myenteric Plexus

Inflammation influenced the neurochemical characterisation of both cholinergic and non-cholinergic enteric neurons immunoreactive to ZnT3. In the MP (Figure 1, Table 1), an increase of immunoreactivity in the majority of substances studied was observed in both classes of neurons, and the magnitude of these changes was most visible in non-cholinergic (VAChT$^-$) cells. The highest percentage of ZnT3$^+$/VAChT$^-$ neuronal cells during inflammation was also immunopositive to nNOS (57.0% ± 4.0%), VIP (52.1% ± 2.8%) and/or LENK (41.2% ± 2.7%). ZnT3-LI cholinergic cells (VAChT$^+$) contained only VIP (11.2% ± 1.9%), GAL (5.4% ± 1.5%) and SP (4.2% ± 1.1%). Contrary to the majority of substances studied, the number of ZnT3-positive cells (both cholinergic and non-cholinergic) immunoreactive to SOM within MP dropped to zero.

Outer Submucous Plexus

In the OSP (Figure 2), an increase in the percentage of cholinergic ZnT3-positive cells simultaneously immunoreactive to VIP (from 1.4% ± 0.8% to 3.8% ± 1.4%) and/or GAL (from 0% to 4.7% ± 0.9%) was noted, contrary to neurons ZnT3$^+$/VAChT$^+$/SOM$^+$ and ZnT3$^+$/VAChT$^+$/SP$^+$, where inflammation caused a decrease from 7.4% ± 1.4% to 1.1% ± 0.8% and 4.6% ± 1.6% to 2.8% ± 1.4%, respectively. Moreover, in the population of OSP non-cholinergic neurons (just as in the MP), the observed changes were most visible. The percentage of ZnT3$^+$/VaChT$^-$ neurons immunoreactive to the majority of substances studied was higher during inflammation. The highest number of ZnT3$^+$ non-cholinergic cells was also immunopositive to SOM (43.0% ± 3.6%) and/or GAL (36.3% ± 7.5%) (Table 1).

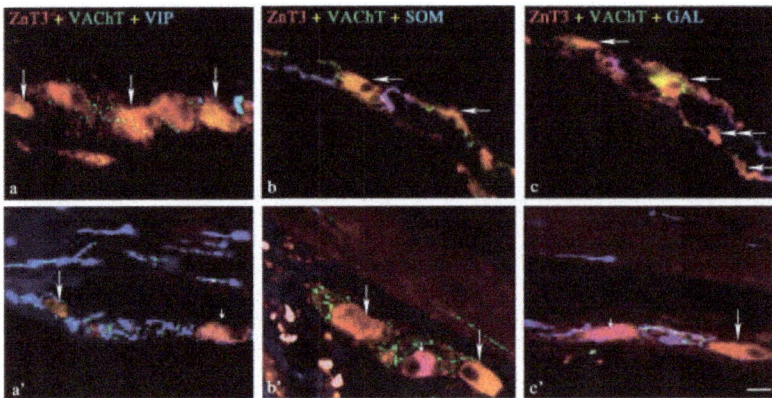

**Figure 2.** Representative images of ZnT3$^+$ neurons located in the outer submucous plexus (OSP) of the porcine ileum. All images are composites of merged images taken separately from blue, red and green fluorescent channels. Control (C) group: (a) ZnT3$^+$/VAChT$^+$/VIP$^-$ neurons are indicated with arrows; (b) ZnT3$^+$/VAChT$^+$/SOM$^-$ neurons are indicated with arrows; (c) ZnT3$^+$/VAChT$^+$/GAL$^-$ neurons are indicated with arrows; ZnT3$^+$/VAChT$^-$/GAL$^-$ neuron is indicated with a double-headed arrow. Inflammatory (I) group: (a′) ZnT3$^+$/VAChT$^+$/VIP$^-$ neuron is indicated with an arrow; ZnT3$^-$/VAChT$^-$/VIP$^+$ neuron is indicated with a small arrow; (b′) ZnT3$^+$/VAChT$^-$/SOM$^+$ neuron is indicated with a small arrow; ZnT3$^+$/VAChT$^+$/VIP$^-$ neurons are indicated with arrows; (c′) ZnT3$^+$/VAChT$^-$/GAL$^+$ neuron is indicated with a small arrow; ZnT3$^+$/VAChT$^+$/GAL$^-$ neuron is indicated with an arrow. Scale bar 25 μm.

Inner Submucous Plexus

During inflammation, the degree of co-localisation of ZnT3 with the majority of substances studied in cholinergic neurons within the ISP (Figure 3) was several times lower than in control animals. The inflammatory process caused a decrease in the percentage of neurons immunoreactive to ZnT3/VAChT/SOM (from 24.1% ± 1.2% to 4.1% ± 08%), ZnT3/VAChT/VIP (from 33.0% ± 3.6% to 7.3% ± 1.6%), ZnT3/VAChT/SP (from 55.1% ± 1.3% to 6.8% ± 1.5%) and ZnT3/VAChT/GAL (from 65.2% ± 1.2% to 4.7% ± 1.3%) (Table 1). Contrary to cholinergic neurons, an increase of the percentage of cells immunoreactive to the majority of all substances studied was observed in non-cholinergic ZnT3-positive perikarya. During inflammation, most of the ZnT3$^+$/VAChT$^-$ neurons were also immunopositive to GAL (54.7% ± 7.4%) and/or nNOS (51.4% ± 3.5%) (Table 1).

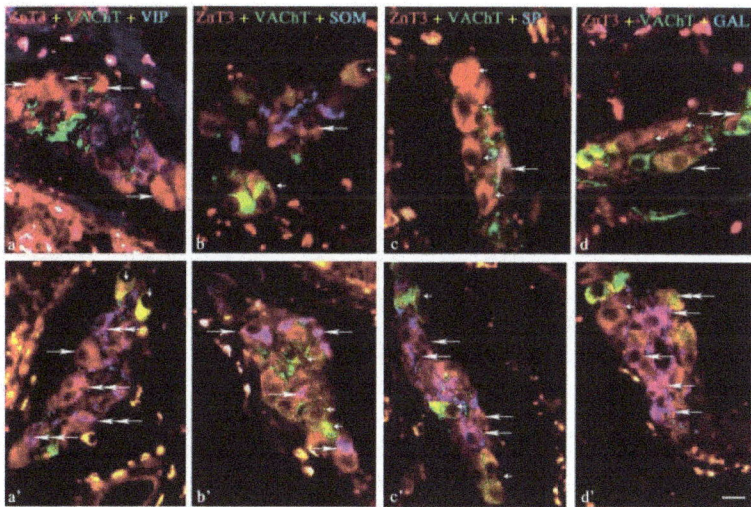

**Figure 3.** Representative images of ZnT3+ neurons located in the inner submucous plexus (ISP) of porcine ileum. All images are composites of merged images taken separately from blue, red and green fluorescent channels. Control (C) group: (**a**) ZnT3$^+$/VAChT$^-$/VIP$^-$ neurons are indicated with arrows; I group (**b**) ZnT3$^+$/VAChT$^+$/SOM$^-$ neurons are indicated with small arrows; ZnT3$^+$/VAChT$^-$/SOM$^-$ neuron is indicated with an arrow; (**c**) ZnT3$^+$/VAChT$^-$/SP$^-$ neurons are indicated with small arrows; ZnT3$^+$/VAChT$^+$/SP$^+$ neuron is indicated with an arrow; (**d**) ZnT3$^+$/VAChT$^-$/GAL$^-$ neurons are indicated with small arrows; ZnT3$^+$/VAChT$^+$/GAL$^-$ neuron is indicated with an arrow; ZnT3$^-$/VAChT$^+$/GAL$^-$ neuron is indicated with a double-headed arrow; Inflammatory (I) group: (**a′**) ZnT3$^+$/VAChT$^+$/VIP$^-$ neurons are indicated with small arrows; ZnT3$^+$/VAChT$^-$/VIP$^+$ neurons are indicated with double-headed arrows; ZnT3$^+$/VAChT$^-$/VIP$^-$ neuron is indicated with an arrow; (**b′**) ZnT3$^+$/VAChT$^-$/SOM$^+$ neurons are indicated with arrows; ZnT3$^+$/VAChT$^+$/SOM$^-$ neurons are indicated with small arrows; (**c′**) ZnT3$^+$/VAChT$^-$/SP$^+$ neurons are indicated with arrows; ZnT3$^+$/VAChT$^+$/SP$^-$ neurons are indicated with small arrows; (**d′**) ZnT3$^+$/VAChT$^+$/GAL$^-$ neuron is indicated with a double-headed arrow; ZnT3$^+$/VAChT$^-$/GAL$^-$ neuron is indicated with a small arrow; ZnT3$^+$/VAChT$^-$/GAL$^+$ neurons are indicated with arrows. Scale bar 25 μm.

## 3. Discussion

The results obtained during the present study show that ZnT3-positive neurons are relatively numerous in the ENS of the porcine ileum, which is in accordance with previous studies on humans [10] and other parts of the porcine digestive tract [19,20]. In terms of the numerical account of ZnT3-positive neurons, the ENS is very different from other previously studied parts of the peripheral autonomic nervous system, where such perikarya were rather sparse [14]. These observations strongly suggest the important role of ZnT3 within enteric neurons, but the factors that affect the expression of this peptide in enteric neurons under physiological conditions, as well as the exact functions of ZnT3 in the GI tract, are unknown. During the present study, all ZnT3-LI cells observed in the enteric plexuses were also immunoreactive to PGP 9.5. This fact shows that ZnT3 in the ENS is present solely in neuronal cells, which is in agreement with previous studies [10,19,20]. On the other hand, ZnT3-positive cells scattered in the intestinal wall outside of the enteric plexuses were not neurons (PGP 9.5-negative). This observation confirms previous studies describing the presence of ZnT3 in pancreatic β cells [33]. Presently, ZnT3-LI non-neuronal cells have not been observed in the intestine, and the character and functions of such cells are completely unknown.

One of the more interesting aspects of the occurrence of ZnT3 in the GI tract is the dependence of its level on the amount of zinc in food. Previous studies showed that a diet deficient in zinc caused changes in mRNA levels for various ZnT family peptides in the pancreas, digestive tract and central nervous system of rodents [34–36], but at present, the influence of food on ZnT3 in the ENS has not been studied. The role of this zinc transporter in the GI tract also remains unknown, and all assumptions concerning this problem are based on an analogy with the central nervous system, where the functions of ZnT3 are better known. Namely, this zinc transporter could take part in the conduction of sensory stimuli and regulate neuronal secretion [37,38]. Moreover, ZnT3 is regarded as a marker of zinc-enriched terminals, which are associated with inhibitory processes [16]. The present study, in which many other neuroactive substances were observed in ZnT3+ neurons, suggests that this zinc transporter is present in varied classes of enteric neurons and, consequently, could play various roles not only in the central nervous system, but also in the ENS. Proof of this seems to be the presence of ZnT3 in both cholinergic and non-cholinergic nNOS+ enteric neurons, observed during the present study. Acetylcholine, the main neuromediator of the ENS, present in different functional classes of enteric neurons [2], is responsible for the activation of intestinal motility [5,6] and the secretory functions of the GI tract [39]. In contrast, nitric oxide (NO), which is characteristic mainly for non-cholinergic neuronal cells, may act as an inhibitory factor. It is known that NO has a relaxatory effect on intestinal muscles, suppressing the exudation of electrolytes and hormones by the GI tract and also regulating blood flow in the gut by blood vessel relaxation [40–42]. Moreover, differences in neurochemical characterisation between ZnT3-positive enteric neurons in the ileum (this study) and the jejunum [19] or duodenum [20] may suggest that the exact functions of ZnT3 in the ENS depend on the fragment of the digestive tract.

The influence of experimental colitis on ZnT3-positive enteric neurons has been observed during the present study. These observations confirm the relatively well-known capabilities of the ENS to changes under various pathological factors [26] and strongly suggest that ZnT3 can take part in adaptive, neuroprotective and/or regenerative reactions within the GI tract. Previous studies on the central nervous system described the increase in mRNA levels for various ZnT family peptides under cerebral ischemia [43], which is probably connected with the augmentation of the amount of zinc in neuronal tissue during this type of pathological process [44,45]. It should be pointed out that zinc in non-physiological conditions can act on neurons in two ways. On the one hand, zinc exhibits neuroprotective activity and reduces neuronal damage [46], but on the other hand, it is known that this metal can be deposited in the cytoplasm and present neurotoxic effects [44]. In this case, the increase in the expression of ZnT transporters is intended to protect neuronal cells. Similar mechanisms probably occur in the GI tract and are the basis of the changes observed during the present study. This is all the more likely as the neuroprotective role of ZnT3 has been described in the central nervous system during Alzheimer's disease [47], and interactions between ZnT transporters and NO in neuroprotective activity have also been observed. The high degree of co-localisation of ZnT3 with nNOS, noted in the present study especially during inflammation, seems to confirm these interactions. Moreover, other substances (VIP, GAL) that co-localised with ZnT3+ enteric neurons observed during this experiment are known as factors that have neuroprotective properties [48–50].

Another argument for the neuroprotective role of ZnT3 in the ENS is the fact that pathological processes cause an increase in the expression of active factors that exhibit neuroprotective effects with a simultaneous decline in the amount of other substances [49]. This situation has been observed in the present study. Namely, a decrease in the percentage of ZnT3-positive cholinergic neurons has been noted, and acetylcholine is the main neuromediator in the ENS under physiological conditions [5,6]. At the same time, an increase in the degree of co-localisation of ZnT3 with other substances, such as nNOS, VIP, SP, SOM, LENK and/or GAL, has generally been observed.

The co-localisation of ZnT3 with the abovementioned neurochemical factors suggests that this zinc transporter can play similar roles (besides neuroprotective) for these neurochemical factors. ZnT3 are possibly inhibitory factors in the ENS, like VIP or nNOS [40,41,51], and can also take part in the conduction of sensitive stimuli, like SP or LENK [9,52,53]. These functions of ZnT3 have also been described in the central nervous system [16].

On the other hand, it should be pointed out that the functions of ZnT3 in the central nervous system, although important (by analogy) to explain its role in the ENS, do not clarify all doubts. For example, ZnT3-positive nerve fibres, known as zinc-enriched nerves, which can play an inhibitory role, have been studied in the central nervous system [16], where both the present experiment and previous studies [10,19,20] did not reveal nerve fibres to be immunoreactive to this zinc transporter in the intestinal wall. Moreover, the exact reasons for changes in immunoreactivity to ZnT3 observed during the present study are difficult to explain, as they may arise from modifications in various stages of peptide synthesis, such as transcription or translation, as well as during post-translational modifications and even shifts in the transport of ZnT3 within neuronal cells.

The changes observed during the present investigations are most likely related to disturbances in zinc homeostasis. It is relatively well established that zinc, as one of the essential micronutrients, plays important roles during diseases localised in various organs and systems. Previous studies showed that the deregulation of zinc homeostasis can be involved in processes connected with Alzheimer's disease, asthma and diabetes [47,54,55], as well as pathological processes within the GI tract, including inflammatory bowel disease, ulcerative colitis and Crohn's disease [56,57].

On the other hand, zinc transporters are key factors implicated in the regulation of zinc distribution in a living organism. There are two main families of zinc transporters: the ZnT proteins (SLC30) and the Zip (Zrt- and Irt-like proteins) family (solute-linked carrier 39—SLC39) [58]. The first of these transport zinc ions from the cytoplasm outside of cells or into the intracellular vesicle, thereby reducing intracellular zinc concentration. The second group has the opposite action, as they promote zinc transport from extracellular space into the cytoplasm.

ZnT3, which is mainly localised in neuronal cells, transports zinc ions from the cytoplasm into synaptic vesicles [12], and the correlation between the expression of this zinc transporter and synaptic $Zn^{2+}$ levels has been described [12,59].

The exact roles in the maintenance of synaptic zinc ions homeostasis in the ENS remain unknown, but they are probably similar to those observed within the CNS. It is known that synaptic $Zn^{2+}$ in the brain and spinal cord plays a modulatory role in synaptic transmission, and its homeostasis determines the correct functioning of the nervous system [59]. Both synaptic $Zn^{2+}$ deficiency and excess causes disturbances in postsynaptic neurons. In the first case, dysfunctions of postsynaptic neurons have been observed, and in the second neurodegeneration has been observed [59,60]. Given this context, $Zn^{2+}$ is regarded as a potentially neurotoxic factor that is involved in neuronal loss and plays some role in neurodegenerative diseases, including Huntington's disease, Parkinson's disease and amyotrophic lateral sclerosis [60]. On the other hand, it is known that both ZnT3 and synaptic $Zn^{2+}$ are inherent to correct brain development [61,62].

The increase of ZnT3-like immunoreactivity observed in the present study was probably connected with synaptic zinc ions excess. This effect could be due to the direct neurodegenerative influence of the inflammatory process. On the other hand, it has been connected with neuroprotective and/or adaptive processes in enteric neurons. Namely, inflammation may have been responsible for the dysfunction of postsynaptic neurons, which is often connected with a deficiency of synaptic $Zn^{2+}$ [59]. For this reason, presynaptic neurons enhanced the expression of ZnT3 in order to maintain zinc homeostasis. The increase in ZnT3-like immunoreactivity could also be the result of an excessive concentration of intracellular cytoplasmic zinc. This is more likely as previous studies described the activation of ZIP transporters (responsible for the transport of zinc ions from extracellular fluid into the cytoplasm) during inflammatory processes [63].

The observed changes could also be connected with other functions of zinc in the intestine, including participation in the maintenance of intestinal mucosal layer integrity, controlling leukocyte infiltration, as well as excretion of pro-inflammatory factors [64–66]. Moreover, it is known that ZnT3 probably takes part in sensory stimuli conduction [67], and fluctuations in this transporter's levels observed during the present study were caused by pain processes during inflammation.

## 4. Materials and Methods

The present study was performed on 10 immature female pigs of the Large White Polish breed at the age of approximately eight weeks. The animals were kept under standard laboratory conditions. All actions connected with the experiment were carried out in compliance with the regulations of the Local Ethical Committee in Olsztyn (Poland)—(decision numbers: 90/2007 from 20 November 2007—submission 90/2007/N and 27/2009 from 18 March 2009—submission 26/2009/DTN).

Sows were randomly divided into two experimental groups: control (C group; $n = 5$) without any surgical operations, and animals in which experimental acute colitis and visceral pain were induced according to the method described previously by Miampamba et al. [68] and modified by Gonkowski et al. [69] (Inflammatory—I group; $n = 5$).

The pigs of I group were subjected to the following experimental procedures: (a) premedication with Stressnil (Janssen, Beerse, Belgium, 75 µL/kg of body weight given intravenously) 15 min before the application of the main anaesthetic; (b) general anaesthesia using sodium thiopental (Thiopental, Sandoz, Kundl-Rakúsko, Austria; 20 mg/kg of body weight given intravenously); and (c) median laparotomy. During laparotomy, the sows were injected with 80 µL of 10% formalin solution (microinjections of 5–8 µL) into the wall of the ileum five centimetres before the ileocecal valve. The group of "sham" operated animals (injections of saline solution instead of formalin) was deliberately abandoned during the experiment due to the fact that previous studies clearly showed that this type of manipulation of the intestine does not influence the neurochemical coding of enteric neurons [9,27,70]. Reducing the number of animals was in accordance with the regulations of the Local Ethical Committee.

After five days, all animals (C and I groups) were euthanized by an overdose of sodium thiopental and then perfused transcardially with 4% buffered paraformaldehyde (pH 7.4) prepared ex tempore.

The same parts (about 2 cm long) of the ileum (from the area where injections of the formalin solution were made and inflammatory changes were observed in animals of I group) were collected and post-fixed by immersion in the same solution used during transcardial perfusion for 20 min. Then tissues were rinsed in a phosphate buffer (0.1 M; pH 7.4; 4 °C) for three days with the buffer changing every day and, finally, stored in 18% sucrose (at least 10 days). After this period, the fragments of ileum were frozen at −25 °C and cut into 10 µm-thick cryostat sections, which were subjected to standard triple-labelling immunofluorescence according to the method described previously by Wojtkiewicz et al. [19]. In short, the immunofluorescence technique was performed as follows: slices with tissue sections were air-dried at room temperature (rt) for 45 min and incubated with a blocking solution (10% normal goat serum, 0.1% bovine serum albumin, 0.01% NaN₃, Triton x-100 and thimerozal in PBS) for 1 h (rt). After this period, tissues were incubated overnight in a humid chamber (rt) with a mixture of three primary antibodies: (1) anti-protein gene product 9.5 (PGP 9.5—used here as a pan-neuronal marker); (2) anti-vesicular acetylcholine transporter (VAChT—used here as a marker of cholinergic neurons); and (3) directed against one of the following substances: calcitonin gene-related peptide (CGRP), galanin (GAL), leu-enkephalin (LENK), neuronal isoform of nitric oxide synthase

(Nnos—used here as a marker of nitrergic neurons), neuropeptide Y (NPY), somatostatin (SOM), substance P (SP) and vasoactive intestinal polypeptide (VIP). The following day, slices were incubated (1 h, rt) with species-specific secondary antisera conjugated to 7-amino-4-methylcoumarin-3-acetic acid (AMCA), fluorescein isothiocyanate (FITC) or biotin, which was then visualised by a streptavidin-CY3 complex (1 h; rt). The specification of primary and secondary antibodies used in the present study is shown in Table 2. Each stage of immunofluorescence labelling was followed by rinsing the slices with PBS (3 × 10 min, pH 7.4). The specificity of antisera was tested by standard controls, including pre-absorption of antibodies with the appropriate antigen, omission test and replacement of antisera by non-immune sera. The abovementioned controls completely eliminated labelling in the tissue.

The overall percentage of ZnT3-LI neurons was defined by the examination of at least 700 PGP-9.5-labelled cell bodies for ZnT3-like immunoreactivity in each "kind" of enteric plexus (i.e., muscular, outer submucous and inner submucous plexuses), and the number of neurons immunoreactive to PGP 9.5 was treated as 100%. In the case of the study on the co-localisation of ZnT3 with acetylcholine, at least 300 ZnT3-positive cell bodies in particular types of enteric ganglia were examined for immunoreactivity to VAChT. In these studies, ZnT3-positive neurons were considered as representing 100%. To determine the chemical coding of cholinergic and non-cholinergic neurons immunoreactive to ZnT3, at least 300 $ZnT3^+/VAChT^+$ and $ZnT3^+/VAChT^-$ were studied for immunoreactivity against particular substances. Evaluation of immunopositive cells and the counting of neurons were performed by two independent investigators. During the present study, only double- or triple-labelled neuronal cell bodies with a clearly visible nucleus were determined under an Olympus BX51 microscope equipped with epi-fluorescence and appropriate filter sets, pooled and presented as mean ± SEM. The section of ileum that was the subject of the study was located at least 100 μm apart to prevent double counting of neurons. All pictures were captured with a digital camera connected to a PC. Statistical analysis was performed with the Anova test using Statistica 10 software (StatSoft Inc., Tulsa, OK, USA). The differences were considered statistically significant at $p \leq 0.05$.

**Table 2.** Specification of immune reagents used in the study: PGP9.5—protein gene product 9.5, ZnT3—zinc transporter 3, NOS—nitric oxide synthase, VIP—vasoactive intestinal peptide, SOM—somatostatin, VAChT—Vesicular acetylcholine transporter, NPY—neuropeptide Y, GAL—galanin, CGRP—calcitonin-gene related peptide, FITC—fluorescein isothiocyanate, AMCA—7-amino-4-methylcoumarin-3-acetic acid, H heavy chain, L light chain.

| Primary Antibody | | | | |
|---|---|---|---|---|
| Antisera | Code | Host Species | Dilution | Supplier |
| PGP9.5 | 7863-2004 | Mouse | 1:2000 | Biogenesis Inc., Poole, UK; www.biogenesis.co.uk |
| ZnT3 | - | Rabbit | 1:600 | Gift from prof. Palmiter, University of Washington Seattle, WA, USA |
| NOS | N2280 | Mouse | 1: 2000 | Sigma-Aldrich, Saint Louis, MS, USA; www.sigma-aldrich.com |
| VIP | 9535-0504 | Mouse | 1: 2000 | Biogenesis Inc. |
| SP | 8450-0505 | Rat | 1:300 | Biogenesis Inc. |
| SOM | 8330-0009 | Rat | 1: 100 | Biogenesis Inc. |
| LENK | 4140-0355 | Mouse | 1: 1000 | Biogenesis Inc. |
| VAChT | H-V007 | Goat | 1: 2000 | Phoenix, Pharmaceuticals, INC., Belmont, CA, USA; www.phoenixpeptide.com |
| NPY | NZ1115 | Rat | 1:300 | Biomol Research Laboratories Inc., Plymouth, PA, USA |
| GAL | T-5036 | Guinea pig | 1:1000 | Peninsula Labs., San Carlos, CA, USA; see Bachem AG; www.bachem.com |
| CGRP | T-5027 | Guinea pig | 1:1000 | Peninsula Labs. |

Table 2. *Cont.*

| Secondary Antibodies | | |
|---|---|---|
| Reagent | Dilution | Supplier |
| FITC-conjugated donkey-anti-mouse IgG (H+L) | 1:800 | Jackson, 715-095-151, West Grove, PA, USA |
| FITC-conjugated donkey-anti-rat IgG (H+L) | 1:800 | Jackson, 712-095-153 |
| FITC-conjugated donkey-anti-guinea pig IgG (H+L) | 1:1000 | Jackson, 706-095-148 |
| FITC-conjugated donkey-anti-goat IgG (H+L) | 1:1000 | Jackson, 705-096-147 |
| Biotinylated goat anti-rabbit immunoglobulins | 1:1000 | DAKO, E 0432, Carpinteria, CA, USA |
| Biotin conjugated F(ab)′ fragment of affinity Purified anti-rabbit IgG (H+L) | 1:1000 | BioTrend, 711-1622, Cologne, Germany |
| AMCA-conjugated donkey-anti-mouse IgG (H+L) | 1:50 | Jackson, 715-155-151 |
| AMCA-conjugated donkey-anti-rat IgG (H+L) | 1:50 | Jackson, 715-155-153 |
| AMCA-conjugated donkey-anti-goat IgG (H+L) | 1:50 | Jackson, 705-156-147 |
| CY3-conjugated Streptavidin | 1:9000 | Jackson, 016-160-084 |

## 5. Conclusions

In summary, the present investigation shows that ZnT3 is present in relatively numerous populations of enteric neurons in the porcine ileum, both under physiological conditions and during experimental inflammation in cholinergic and non-cholinergic cells. The wide range of active substances (including SP, VIP, SOM, nNOS and GAL) that co-localise with ZnT3 suggests that this zinc transporter occurs in various classes of enteric neurons and may take part in different regulatory processes within the intestinal wall. Meanwhile, changes in ZnT3-like immunoreactivity during experimental inflammation could denote the functions of ZnT3 in adaptive and/or regenerative reactions in the ENS. ZnT3 is probably present in enteric neurons, which use zinc ions as a neuromodulator, and fluctuations in ZnT3 expression are connected with disturbances in zinc ions homeostasis. Moreover, it is known that, on the one hand, zinc is essential for correct development and functioning of the ENS [25], and during inflammation or nerve damage, ZnT3 can play a neuroprotective role. On the other hand, zinc ions can exhibit neurodegenerative activity [59]. Therefore, a lot of aspects connected with the role of ZnT3 within the ENS remain unknown and require further investigations.

**Acknowledgments:** Publication supported by KNOW (Leading National Research Centre) Scientific Consortium "Healthy Animal—Safe Food", decision of Ministry of Science and Higher Education No. 05-1/KNOW2/2015 and grant No. NN401178639 from the State Committee for Science Research of Poland.

**Author Contributions:** Sławomir Gonkowski and Joanna Wojtkiewicz conceived and designed this experiment; Sławomir Gonkowski and Joanna Wojtkiewicz performed the experimental procedures; Maciej Rowniak analysed data in the statistical program and took part in formulating the figures; Sławomir Gonkowski wrote the manuscript, with grammatical editing from Joanna Wojtkiewicz.

**Conflicts of Interest:** The authors declare no conflict of interest.

## References

1. Furness, J.B.; Callaghan, B.P.; Rivera, L.R.; Cho, H.J. The enteric nervous system and gastrointestinal innervation: Integrated local and central control. *Adv. Exp. Med. Biol.* **2014**, *817*, 39–71. [PubMed]
2. Furness, J.B. The enteric nervous system and neurogastroenterology. *Nat. Rev. Gastroenterol. Hepatol.* **2012**, *9*, 286–294. [CrossRef] [PubMed]
3. Wojtkiewicz, J.; Gonkowski, S.; Bladowski, M.; Majewski, M. Characterization of cocaine- and amphetamine-regulated transcript-like immunoreactive (CART-LI) enteric neurons in the porcine small intestine. *Acta Vet. Hung.* **2012**, *60*, 371–381. [CrossRef] [PubMed]

4.    Gonkowski, S.; Burlinski, P.; Calka, J. Proliferative enteropathy (PE)-induced changes in galanin—Like immunoreactivity in the enteric nervous system of the porcine distal colon. *Acta Vet. Beograd.* **2009**, *59*, 321–330.

5.    Porter, A.J.; Wattchow, D.A.; Brookes, S.J.; Schemann, M.; Costa, M. Choline acetyltransferase immunoreactivity in the human small and large intestine. *Gastroenterology* **1996**, *111*, 401–408. [CrossRef] [PubMed]

6.    Porter, A.J.; Wattchow, D.A.; Brookes, S.J.; Costa, M. Cholinergic and nitrergic interneurones in the myenteric plexus of the human colon. *Gut* **2002**, *51*, 70–75. [CrossRef] [PubMed]

7.    Kaleczyc, J.; Klimczuk, M.; Franke-Radowiecka, A.; Sienkiewicz, W.; Majewski, M.; Łakomy, M. The distribution and chemical coding of intramural neurons supplying the porcine stomach—The study on normal pigs and on animals suffering from swine dysentery. *Anat. Histol. Embryol.* **2007**, *36*, 186–193. [CrossRef] [PubMed]

8.    Pidsudko, Z.; Kaleczyc, J.; Wasowicz, K.; Sienkiewicz, W.; Majewski, M.; Zajac, W.; Lakomy, M. Distribution and chemical coding of intramural neurons in the porcine ileum during proliferative enteropathy. *J. Comp. Pathol.* **2008**, *138*, 23–31. [CrossRef] [PubMed]

9.    Gonkowski, S. Substance P as a neuronal factor in the enteric nervous system of the porcine descending colon in physiological conditions and during selected pathogenic processes. *Biofactors* **2013**, *39*, 542–551. [CrossRef] [PubMed]

10.   Gonkowski, S.; Kaminska, B.; Landowski, P.; Skobowiat, C.; Burlinski, P.; Majewski, M.; Calka, J. A population of zinc transporter 3-like immunoreactive neurons is present in the ganglia of human descending colon. *Adv. Clin. Exp. Med.* **2009**, *18*, 243–248.

11.   Palmiter, R.D.; Huang, L. Efflux and compartmentalization of zinc by members of the SLC30 family of solute carriers. *Pflug. Arch.* **2004**, *447*, 744–751. [CrossRef] [PubMed]

12.   Palmiter, R.D.; Cole, T.B.; Quaife, C.J.; Findley, S.D. ZnT-3, a putative transporter of zinc into synaptic vesicles. *Proc. Natl. Acad. Sci. USA* **1996**, *93*, 14934–14939. [CrossRef] [PubMed]

13.   Wenzel, H.J.; Cole, T.B.; Born, D.E.; Schwartzkroin, P.A.; Palmiter, R.D. Ultrastructural localization of zinc transporter-3 (ZnT3) to synaptic vesicle membranes within mossy fiber boutons in the hippocampus of mouse and monkey. *Proc. Natl. Acad. Sci. USA* **1997**, *94*, 12676–12681. [CrossRef] [PubMed]

14.   Wang, Z.Y.; Danscher, G.; Dahlstrom, A.; Li, J.Y. Zinc transporter 3 and zinc ions in the rodent superior cervical ganglion neurons. *Neuroscience* **2003**, *120*, 605–616. [CrossRef]

15.   Kaneko, M.; Noguchi, T.; Ikegami, S.; Sakurai, T.; Kakita, A.; Toyoshima, Y.; Kambe, T.; Yamada, M.; Inden, M.; Hara, H.; et al. Zinc transporters ZnT3 and ZnT6 are downregulated in the spinal cords of patients with sporadic amyotrophic lateral sclerosis. *J. Neurosci. Res.* **2015**, *93*, 370–379. [CrossRef] [PubMed]

16.   Jo, S.M.; Won, M.H.; Cole, T.B.; Jansen, M.S.; Palmiter, R.D.; Danscher, G. Zinc-enriched (ZEN) terminals in mouse olfactory bulb. *Brain Res.* **2000**, *865*, 227–236. [CrossRef]

17.   Takeda, A. Movement of zinc and its functional significance in the brain. *Brain Res. Rev.* **2000**, *34*, 137–148. [CrossRef]

18.   Molnar, P.; Nadler, J.V. Lack of effects of mossy fiber—Released zinc on granule cell (GABA$_A$ receptors in the pilocarine model of epilepsy. *J. Neurophysiol.* **2001**, *85*, 1932–1940. [PubMed]

19.   Wojtkiewicz, J.; Równiak, M.; Crayton, R.; Majewski, M.; Gonkowski, S. Chemical coding of zinc-enriched neurons in the intramural ganglia of the porcine jejunum. *Cell Tissue Res.* **2012**, *350*, 215–223. [CrossRef] [PubMed]

20.   Wojtkiewicz, J.; Gonkowski, S.; Równiak, M.; Crayton, R.; Majewski, M.; Jałyński, M. Neurochemical characterization of zinc transporter 3-like immunoreactive (ZnT3+) neurons in the intramural ganglia of the porcine duodenum. *J. Mol. Neurosci.* **2012**, *48*, 766–776. [CrossRef] [PubMed]

21.   Vallee, B.L.; Falchuk, K.H. The biochemical basis of zinc physiology. *Physiol. Rev.* **1993**, *73*, 79–118. [PubMed]

22.   Dineley, K.E.; Votyakova, T.V.; Reynolds, I.J. Zinc inhibition of cellular energy production: Implications for mitochondria and neurodegenaration. *J. Neurochem.* **2003**, *85*, 563–570. [CrossRef] [PubMed]

23.   Barañano, D.E.; Ferris, C.D.; Snyder, S.H. Atypical neural messengers. *Trends. Neurosci.* **2001**, *24*, 99–106. [CrossRef]

24. Bhutta, Z.A.; Bird, S.M.; Black, R.E.; Brown, K.H.; Gardner, J.M.; Hidayat, A.; Khatun, F.; Martorell, R.; Ninh, N.X.; Penny, M.E.; et al. Therapeutic effects of oral zinc in acute and persistent diarrhea in children in developing countries: Pooled analysis of randomized controlled trials. *Am. J. Clin. Nutr.* **2000**, *72*, 1516–1522. [PubMed]

25. Hegarty, S.; Sullivan, A.M.; O'Keeffe, G.W. Zeb2: A multifunctional regulator of nervous system development. *Prog. Neurobiol.* **2015**, *132*, 81–95. [CrossRef] [PubMed]

26. Vasina, V.; Barbara, G.; Talamonti, L.; Stanghellini, V.; Corinaldesi, R.; Tonini, M.; de Ponti, F.; de Giorgio, R. Enteric neuroplasticity evoked by inflammation. *Auton. Neurosci.* **2006**, *126–127*, 264–272. [CrossRef] [PubMed]

27. Gonkowski, S.; Calka, J. Changes in the somatostatin (SOM)-like immunoreactivity within nervous structures of the porcine descending colon under various pathological factors. *Exp. Mol. Pathol.* **2010**, *88*, 416–423. [CrossRef] [PubMed]

28. Gonkowski, S.; Obremski, K.; Calka, J. The influence of low doses of zearalenone on distribution of selected active substances in nerve fibers within the circular muscle layer of porcine ileum. *J. Mol. Neurosci.* **2015**, *56*, 878–886. [CrossRef] [PubMed]

29. Makowska, K.; Gonkowski, S.; Zielonka, L.; Dabrowski, M.; Calka, J. T2 toxin-induced changes in cocaine- and amphetamine-regulated transcript (CART)—Like immunoreactivity in the enteric nervous system within selected fragments of the porcine digestive tract. *Neurotox. Res.* **2016**. [CrossRef] [PubMed]

30. Brown, D.R.; Timmermans, J.P. Lessons from the porcine enteric nervous system. *Neurogastroenterol. Motil.* **2004**, *16*, 50–54. [CrossRef] [PubMed]

31. Litten-Brown, J.C.; Corson, A.M.; Clarke, L. Porcine models for the metabolic syndrome, digestive and bone disorders: A general overview. *Animal* **2010**, *4*, 899–920. [CrossRef] [PubMed]

32. Verma, N.; Rettenmeier, A.W.; Schmitz-Spanke, S. Recent advances in the use of *Sus scrofa* (pig) as a model system for proteomic studies. *Proteomics* **2011**, *11*, 776–793. [CrossRef] [PubMed]

33. Smidt, K.; Larsen, A.; Brønden, A.; Sørensen, K.S.; Nielsen, J.V.; Praetorius, J.; Martensen, P.M.; Rungby, J. The zinc transporter ZNT3 co-localizes with insulin in INS-1E pancreatic β cells and influences cell survival, insulin secretion capacity, and ZNT8 expression. *Biometals* **2016**, *29*, 287–298. [CrossRef] [PubMed]

34. Liuzzi, J.P.; Bobo, J.A.; Lichten, L.A.; Samuelson, D.A.; Cousins, R.J. Responsive transporter genes within the murine intestinal—Pancreatic axis form a basis of zinc homeostasis. *Proc. Natl. Acad. Sci. USA* **2004**, *101*, 14355–14360. [CrossRef] [PubMed]

35. Chowanadisai, W.; Kelleher, S.L.; Lonnerdal, B. Zinc defficiency is associated with increased brain zinc import and LIV-1 expression and decreased ZnT-1 expression in neonatal rats. *J. Nutr.* **2005**, *135*, 1002–1007. [PubMed]

36. Pfaffl, M.W.; Windisch, W. Influence of zinc deficiency on the mRNA expression of zinc transporters in adult rats. *J. Trace Elem. Med. Biol.* **2003**, *17*, 97–106. [CrossRef]

37. Danscher, G.; Jo, S.M.; Varea, E.; Wang, Z.; Cole, T.B.; Schrøder, H.D. Inhibitory zinc-enriched terminals in mouse spinal cord. *Neuroscience* **2001**, *105*, 941–947. [CrossRef]

38. Danscher, G.; Wang, Z.; Kim, Y.K.; Kim, S.J.; Sun, Y.; Jo, S.M. Immunocytochemical localization of zinc transporter 3 in the ependyma of the mouse spinal cord. *Neurosci. Lett.* **2003**, *342*, 81–84. [CrossRef]

39. Bader, S.; Diener, M. Novel aspects of cholinergic regulation of colonic ion transport. *Pharmacol. Res. Perspect.* **2015**, *3*, e00139. [CrossRef] [PubMed]

40. Kuwahara, A.; Kuramoto, H.; Kadowaki, M. 5-HT activates nitric oxide-generating neurons to stimulate chloride secretion in guinea pig distal colon. *Am. J. Physiol.* **1998**, *275*, G829–G834. [PubMed]

41. Schleiffer, R.; Raul, F. Nitric oxide and the digestive system in mammals and non-mammalian vertebrates. *Comp. Biochem. Physiol.* **1997**, *118A*, 965–974. [CrossRef]

42. Page, A.J.; O'Donnell, T.A.; Cooper, N.J.; Young, R.L.; Blackshaw, L.A. Nitric oxide as an endogenous peripheral modulator of visceral sensory neuronal function. *J. Neurosci.* **2009**, *29*, 7246–7255. [CrossRef] [PubMed]

43. Aguilar-Alonso, P.; Martinez-Fong, D.; Pazos-Salazar, N.G.; Brambila, E.; Gonzalez-Barrios, J.A.; Mejorada, A.; Flores, G.; Millan-Perez Peña, L.; Rubio, H.; Leon-Chavez, B.A. The increase in zinc levels and upregulation of zinc transporters are mediated by nitric oxide in the cerebral cortex after transient ischemia in the rat. *Brain Res.* **2008**, *1200*, 89–98. [CrossRef] [PubMed]

44. Frederickson, C.J.; Maret, W.; Cuajungco, M.P. Zinc and excitotoxic brain injury: A new model. *Neuroscientist* **2004**, *10*, 18–25. [CrossRef] [PubMed]

45. Frederickson, C.J.; Giblin, L.J.; Krezel, A.; McAdoo, D.J.; Muelle, R.N.; Zeng, Y.; Balajim, R.V.; Masalha, R.; Thompson, R.B.; Fierke, C.A.; et al. Concentrations of extracellular free zinc (pZn)$_e$ in the central nervous system during simple anesthetization, ischemia and reperfusion. *Exp. Neurol.* **2006**, *198*, 285–293. [CrossRef] [PubMed]

46. Yeiser, E.C.; Vanlandingham, J.W.; Levenson, C.W. Moderate zinc deficiency increases cell death after brain injury in the rat. *Nutr. Neurosci.* **2002**, *51*, 345–352. [CrossRef] [PubMed]

47. Devirgiliis, C.; Zalewski, P.D.; Perozzi, G.; Murgia, C. Zinc fluxes and zinc transporters genes in chronic diseases. *Mutation Res.* **2007**, *622*, 84–93. [CrossRef] [PubMed]

48. Lin, Z.; Sandgren, K.; Ekblad, E. Increased expression of nitric oxide synthase in cultured neurons from adult rat colonic submucous ganglia. *Auton. Neurosci.* **2004**, *114*, 29–38. [CrossRef] [PubMed]

49. Arciszewski, M.B.; Ekblad, E. Effects of vasoactive intestinal petide and galanin on survival of cultured porcine myenteric neurons. *Regul. Pept.* **2005**, *125*, 185–192. [CrossRef] [PubMed]

50. Holmes, F.E.; Mahoney, S.A.; Wynick, D. Use of genetically engineered transgenic mice to investigate the role of galanin in the peripheral nervous system after injury. *Neuropeptides* **2005**, *39*, 191–199. [CrossRef] [PubMed]

51. Kasparek, M.S.; Fatima, J.; Iqbal, C.W.; Duenes, J.A.; Sarr, M.G. Role of VIP and Substance P in NANC innervation in the longitudinal smooth muscle of the rat jejunuminfluence of extrinsic denervation. *J. Surg. Res.* **2007**, *141*, 22–30. [CrossRef] [PubMed]

52. Brehmer, A.; Lindig, T.M.; Schrödl, F.; Neuhuber, W.; Ditterich, D.; Rexer, M.; Rupprecht, H. Morphology of enkephalin-immunoreactive myenteric neurons in the human gut. *Histochem. Cell Biol.* **2005**, *123*, 131–138. [CrossRef] [PubMed]

53. Shimizu, Y.; Matsuyama, H.; Shiina, T.; Takewaki, T.; Furness, J.B. Tachykinins and their functions in the gastrointestinal tract. *Cell. Mol. Life Sci.* **2008**, *65*, 295–311. [CrossRef] [PubMed]

54. Chimienti, F. Zinc, pancreatic islet cell function and diabetes: New insights into an old story. *Nutr. Res. Rev.* **2013**, *26*, 1–11. [CrossRef] [PubMed]

55. Zalewski, P.D.; Truong-Tran, A.Q.; Grosser, D.; Jayaram, L.; Murgia, C.; Ruffin, R.E. Zinc metabolism in airway epithelium and airway inflammation: Basic mechanisms and clinical targets: A review. *Pharmacol. Ther.* **2005**, *105*, 127–149. [CrossRef] [PubMed]

56. Kruis, W.; Phuong Nguyen, G. Iron deficiency, zinc, magnesium, vitamin deficiencies in Crohn's disease: Substitute or not? *Dig. Dis.* **2016**, *34*, 105–111. [CrossRef] [PubMed]

57. Ananthakrishnan, A.N.; Khalili, H.; Song, M.; Higuchi, L.M.; Richter, J.M.; Chan, A.T. Zinc intake and risk of Crohn's disease and ulcerative colitis: A prospective cohort study. *Int. J. Epidemiol.* **2015**, *44*, 1995–2005. [CrossRef] [PubMed]

58. Kambe, T.; Tsuji, T.; Hashimoto, A.; Itsumura, N. The physiological, biochemical, and molecular roles of zinc transporters in zinc homeostasis and metabolism. *Physiol. Rev.* **2015**, *95*, 749–784. [CrossRef] [PubMed]

59. Takeda, A.; Nakamura, M.; Fujii, H.; Tamano, H. Synaptic Zn$^{2+}$ homeostasis and its significance. *Metallomics* **2013**, *5*, 417–423. [CrossRef] [PubMed]

60. Weiss, J.H.; Sensi, S.L.; Koh, J.Y. Zn$^{2+}$: A novel ionic mediator of neural injury in brain disease. *Trends Pharmacol. Sci.* **2000**, *21*, 395–401. [CrossRef]

61. Marger, L.; Schubert, C.R.; Bertrand, D. Zinc: An underappreciated modulatory factor of brain function. *Biochem. Pharmacol.* **2014**, *91*, 426–435. [CrossRef] [PubMed]

62. Takeda, A. Significance of Zn$^{2+}$ signaling in cognition: Insight from synaptic Zn$^{2+}$ dyshomeostasis. *J. Trace Elem. Med. Biol.* **2014**, *28*, 393–396. [CrossRef] [PubMed]

63. Troche, C.; Aydemir, T.B.; Cousins, R.J. Zinc transporter Slc39a14 regulates inflammatory signaling associated with hypertrophic adiposity. *Am. J. Physiol. Endocrinol. Metab.* **2016**, *310*, E258–E268. [CrossRef] [PubMed]

64. Sturniolo, G.C.; Fries, W.; Mazzon, E.; di Leo, V.; Barollo, M.; D'inca, R. Effect of zinc supplementation on intestinal permeability in experimental colitis. *J. Lab. Clin. Med.* **2002**, *139*, 311–315. [CrossRef] [PubMed]

65. Finamore, A.; Massimi, M.; Conti Devirgiliis, L.; Mengheri, E. Zinc deficiency induces membrane barrier damage and increases neutrophil transmigration in Caco-2 cells. *J. Nutr.* **2008**, *138*, 1664–1670. [PubMed]

66. Ranaldi, G.; Ferruzza, S.; Canali, R.; Leoni, G.; Zalewski, P.D.; Sambuy, Y.; Perozzi, G.; Murgia, C. Intracellular zinc is required for intestinal cell survival signals triggered by the inflammatory cytokine TNFα. *J. Nutr. Biochem.* **2013**, *24*, 967–976. [CrossRef] [PubMed]

67. Jo, S.M.; Danscher, G.; Schrøder, H.D.; Suh, S.W. Depletion of vesicular zinc in dorsal horn of spinal cord causes increased neuropathic pain in mice. *Biometals* **2008**, *21*, 151–158. [CrossRef] [PubMed]

68. Miampamba, M.; Chery-Croze, S.; Gorry, F.; Berger, F.; Chayvialle, J.A. Inflammation of the colonic wall induced by formalin as a model of acute visceral pain. *Pain* **1994**, *57*, 327–334. [CrossRef]

69. Gonkowski, S.; Burliński, P.; Skobowiat, C.; Majewski, M.; Całka, J. Inflammation- and axotomy-induced changes in galanin-like immunoreactive (GAL-LI) nerve structures in the porcine descending colon. *Acta Vet. Hung.* **2010**, *58*, 91–103. [CrossRef] [PubMed]

70. Burliński, P.J. Inflammation- and axotomy-induced changes in cocaine- and amphetamine-regulated transcript peptide-like immunoreactive (CART-LI) nervous structures in the porcine descending colon. *Pol. J. Vet. Sci.* **2012**, *15*, 517–524. [PubMed]

International Journal of
*Molecular Sciences*

MDPI

*Article*

# Zinc as a Signal to Stimulate Red Blood Cell Formation in Fish

**Yen-Hua Chen [1,2], Jhe-Ruei Shiu [2], Chia-Ling Ho [2] and Sen-Shyong Jeng [2,\***

[1]   Center of Translational Medicine, Department of Basic Medicine, Xiamen Medical College, Xiamen 361023, China; yanhua09123@hotmail.com
[2]   Department of Food Science, College of Life Sciences, National Taiwan Ocean University, Keelung 20224, Taiwan; syujhe.7@gmail.com (J.-R.S.); tina41048@gmail.com (C.-L.H.)
\*   Correspondence: jengss@mail.ntou.edu.tw; Tel.: +886-2-2463-2781

Academic Editors: Toshiyuki Fukada and Taiho Kambe
Received: 20 October 2016; Accepted: 6 January 2017; Published: 11 January 2017

**Abstract:** The common carp can tolerate extremely low oxygen levels. These fish store zinc in a specific zinc-binding protein presented in digestive tract tissues, and under low oxygen, the stored zinc is released and used as a signal to stimulate erythropoiesis (red blood cell formation). To determine whether the environmental supply of zinc to other fish species can serve as a signal to induce erythropoiesis as in the common carp, head kidney cells of four different fish species were cultured with supplemental $ZnCl_2$. Zinc stimulated approximately a three-fold increase in immature red blood cells (RBCs) in one day. The stimulation of erythropoiesis by zinc was dose-dependent. $ZnSO_4$ solution was injected into an experimental blood loss tilapia model. Blood analysis and microscopic observation of the blood cells indicated that, in vivo, the presence of additional zinc induced erythropoiesis in the bled tilapia. In the fish species studied, zinc could be used as a signal to stimulate erythropoiesis both in vitro and in vivo. The present report suggests a possible approach for the induction of red blood cell formation in animals through the supply of a certain level of zinc through either diet or injection.

**Keywords:** zinc; erythropoiesis; red blood cells; fish; erythropoietin

## 1. Introduction

In humans and other mammals, erythropoiesis, the process of red blood cell production, occurs in the bone marrow; in fish, the head kidney is the main erythropoietic organ [1]. The cellular composition of the teleost head kidney resembles that of mammalian bone marrow [2]. Erythropoiesis in fish is similar to that in other vertebrates [3,4]. Erythropoiesis in mammals is regulated by the hormone erythropoietin (EPO) [5,6]. Fish and mammalian erythropoietic systems have similar responses to hypoxia because erythropoiesis is influenced by EPO in fish [7]. The human hepatoma cell lines HepG2 and Hep3B have been utilized to study the regulation of EPO. Studies of Hep3B cells revealed that $Co^{2+}$, $Ni^{2+}$, and $Mn^{2+}$ can stimulate EPO synthesis, whereas $Zn^{2+}$ is ineffective [8]. However, Chen et al. [1] observed zinc stimulation of erythropoiesis in vitro when common carp head kidney cells were cultured with $ZnCl_2$ supplementation. In vivo, an increase in the rate of erythropoiesis in the common carp was triggered and facilitated by zinc in the head kidney when the common carp were exposed to air [9]. In nature, stimulation of erythropoiesis in common carp by zinc occurs via the high concentration of zinc stored in a specific 43 kDa zinc-binding protein present in the digestive tract tissue of the common carp [10]. When needed, such as under conditions of anoxia, the zinc in the 43-kDa zinc-binding protein is released [11] and used as a signal to stimulate the formation of new red blood cells (RBCs) in the head kidney of the common carp [1,9]. It is of great interest to determine whether the supply of additional zinc to other fish species from the environment can serve as a signal to induce erythropoiesis as in the common carp. In this report, the effects of supplementation of $ZnCl_2$ to cultures of the head kidney cells of four fish species were determined: (1) crucian carp, *Carassius carassius*; (2) grass carp, *Ctenopharyngodon idella*; (3) silver carp, *Aristichthys nobilis*; and (4) tilapia, *Oreochromis aureus*. In addition to this in vitro study, $ZnSO_4$ solution was injected into an experimental blood loss tilapia model to determine whether zinc serves as a signal to stimulate erythropoiesis in fish in vivo. Our results indicated that zinc could be used as a signal to induce RBC formation in fish both in vitro and in vivo.

## 2. Results

### 2.1. Suspension Culture of the Head Kidney Cells from Four Fish Species with or without ZnCl₂

In teleost fish, the maturation of erythrocytes involves a progressive increase in cell size [12]. For common carp, immature RBCs can be separated from mature RBCs through two-fraction separation with discontinuous Percoll density centrifugation. The immature RBCs are separated into fraction 1 ($p = 1.020$ g/mL) due to their smaller cell size, whereas the mature RBCs and other larger cells are separated into fraction 2 ($p = 1.070$ g/mL) [1]. The morphology of the head kidney cells of crucian carp, grass carp, silver carp and tilapia were found to be similar as in other species of fish, the major cells were lymphocytes, granulocytes, and erythrocytes [1–4]. After Percoll density centrifugation, it was found that the immature RBCs were separated into fraction 1 the same as that reported before [1]. As shown in Figure 1, after suspension culture of head kidney cells with or without $ZnCl_2$, the total cell density changed only slightly during the four-day period for all four fish species (Figure 1A-a,B-a,C-a,D-a). However, an increase in the number of cells in fraction 1 and a decrease in the number of cells in fraction 2 were observed for all four fish species (Figure 1A-b,B-b,C-b,D-b). In fraction 1, for all four fish species, the increased cell density was significantly higher in the $ZnCl_2$ groups than in the groups without $ZnCl_2$ after day one and entered the plateau phase from day two to four (Figure 1A-b-1,B-b-1,C-b-1,D-b-1). At day one, the ratio of cell numbers in fraction 1 were approximately three-fold higher in the $ZnCl_2$ groups than in groups without $ZnCl_2$ for all four fish species. These results demonstrate that zinc stimulated the growth of the fraction 1 cells in all four fish species in one day, with an approximately three-fold increase in the proliferation of new immature RBCs.

### 2.2. Effects of ZnCl₂ Levels on the Growth of Fraction 1 Cells

For crucian carp, grass carp, silver carp, and tilapia, significant activation of fraction 1 cell growth was observed at 0.6, 0.2, 0.3, and 1.2 mM $ZnCl_2$, and maximal cell growth was approximately 500%, 520%, 660%, and 290%, respectively (Figure 2).

### 2.3. Characteristics of the Cultured Head Kidney Cells of the Crucian Carp

As shown in Figure 3A-b, when crucian carp head kidney cells were cultured with supplemental $ZnCl_2$ for one day, many new cells proliferated that were identified as erythrocytes at different stages of development. Immunofluorescence staining of the transferrin receptor was only observed in RBCs at stages 1, 2, and 3, and not in RBCs at stages 4 and 5 (Figure 3B-b). This result indicates that immunofluorescence staining of the transferrin receptor can be used to identify immature RBCs (RBC stages 1, 2, and 3).

### 2.4. Characteristics of the Cultured Head Kidney Cells of Grass Carp, Silver Carp and Tilapia

Immunofluorescent staining of the transferrin receptor revealed the proliferation of new immature RBCs when the three fish species (grass carp, silver carp, and tilapia) were supplemented in culture with $ZnCl_2$ for one day (Figure 4).

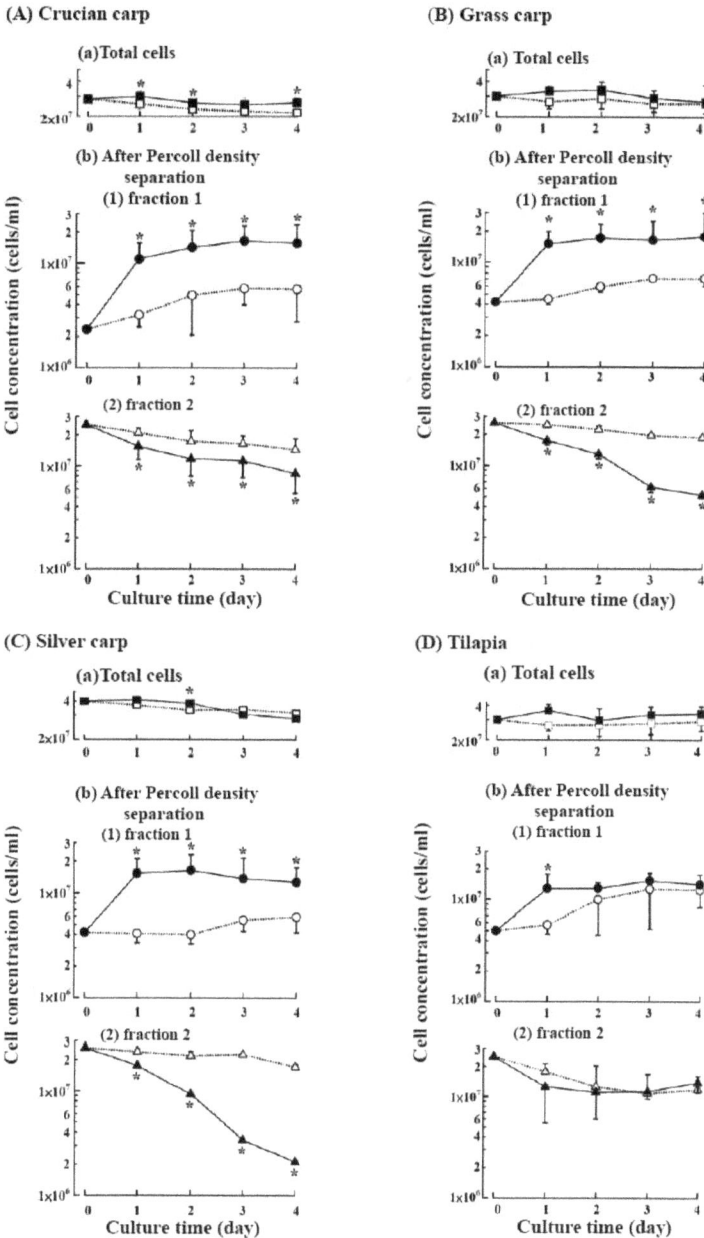

**Figure 1.** Suspension cultures of head kidney cells from the four fish species with or without $ZnCl_2$ supplementation. (**A**) Crucian carp; (**B**) grass carp; (**C**) silver carp; and (**D**) tilapia. All cells remained in suspension for four days. The growth curve of the total cells is shown in (**a**). The total cells were further separated by Percoll density centrifugation into fractions 1 and 2 ($p = 1.020$ and $1.070$ g/mL), and the growth curves are shown in **b(1)** and **b(2)**, respectively. Filled symbols and continuous lines represent cultures supplemented with $ZnCl_2$; open symbols and broken lines represent cultures without $ZnCl_2$. The data are expressed as the means ± SDs of four independent experiments. * Statistically significant differences with $p < 0.05$ between the groups supplemented with or without $ZnCl_2$.

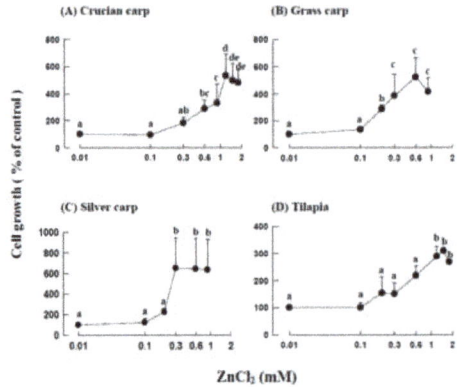

**Figure 2.** Effects of ZnCl$_2$ concentrations on erythropoiesis in fish head kidney cells: (**A**) crucian carp, (**B**) grass carp, (**C**) silver carp, and (**D**) tilapia. The fish head kidney cells were suspension cultured with 0.01 (control), 1.0, or 1.8 mM ZnCl$_2$ in the presence of 10% fish serum. Measurement of the cell growth is described in Materials and Methods. The results are the means ± SDs of six independent experiments. Values with different letter superscripts are significantly different at $p < 0.05$.

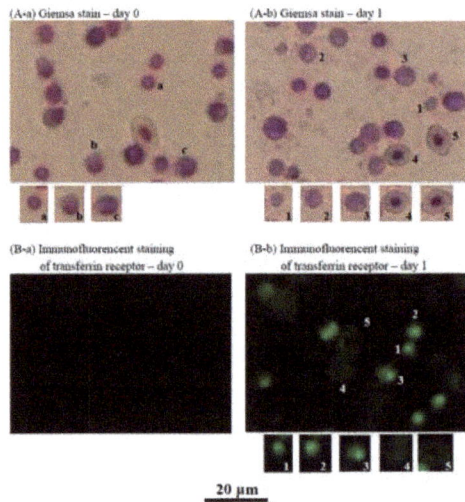

**Figure 3.** Cells collected from the suspension cultures of crucian carp head kidney cells grown in medium supplemented with ZnCl$_2$. (**A-a**) Giemsa-stained cells collected at day zero. The major cells were lymphocytes (**a**) (approximately 70%); other cells, including neutrophilic progranulocytes (**b**) and basophilic granulocytes (**c**) were also observed; (**A-b**) Giemsa-stained cells collected at day one. Various newly proliferated cells (approximately 42%) were identified as erythrocytes of different development stages: 1. lymphoid hemoblasts, 2. early erythroblasts, 3. polychromatophilic erythroblasts, 4. orthochromatic erythroblasts, and 5. erythrocytes; (**B-a**) Immunofluorescent staining of the transferrin receptor of the cells at day zero; no cells were stained; and (**B-b**) Immunofluorescent staining of the transferrin receptor of the cells at day one. Comparing the cell size and cell morphology in (**A-b**) and (**B-b**), the transferrin receptor immunofluorescent-stained cells were RBC stages 1, 2, and 3. RBC stages 4 and 5 and cells other than RBCs were not immunofluorescently stained.

(A-a) Grass carp – day 0

(A-b) Grass carp – day 1

(B-a) Silver carp – day 0

(B-b) Silver carp – day 1

(C-a) Tilapia – day 0

(C-b) Tilapia – day 1

**20 μm**

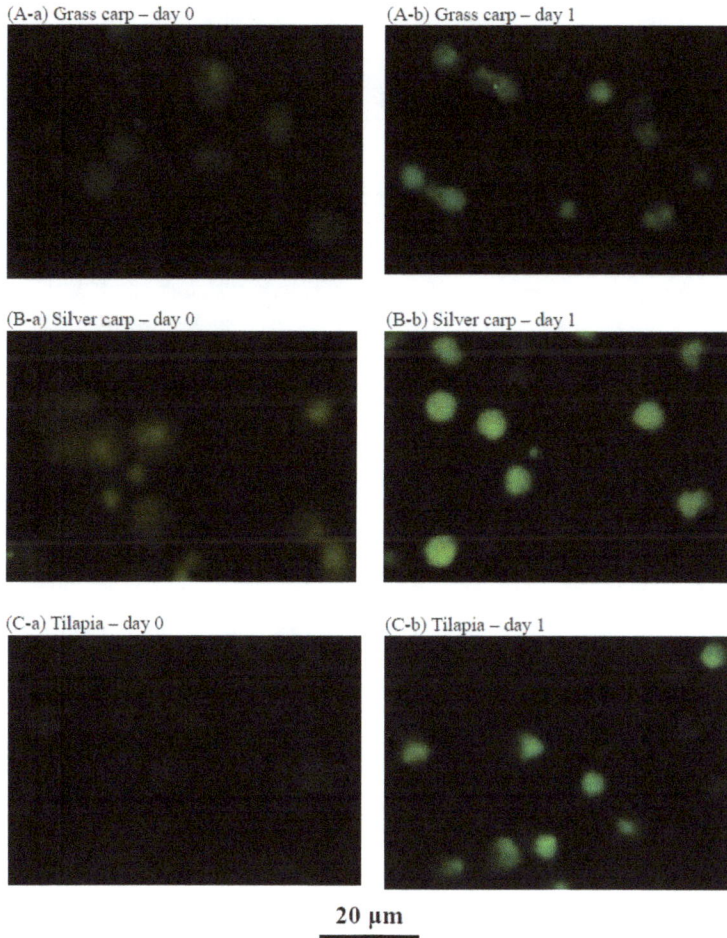

**Figure 4.** Immunofluorescent staining of the transferrin receptor of cells collected from suspension cultures of head kidney cells from (**A**) grass carp, (**B**) silver carp, and (**C**) tilapia at day zero and day one. At day 0, virtually no cells were immunofluorescently stained in the three fish species, as shown in (**A-a**), (**B-a**), and (**C-a**). At day one, various cells were immunofluorescently stained, representing newly proliferated RBC cells in stages 1, 2, and 3, as shown in (**A-b**), (**B-b**), and (**C-b**).

*2.5. Isolation of the Active Substance in Fish Serum That Stimulates the Proliferation of Fraction 1 Cells in the Head Kidney Cells of the Four Fish Species*

When common carp head kidney cells were cultured with supplemental $ZnCl_2$ in the presence of carp serum, transferrin was identified as the active substance in fish serum that stimulates the proliferation of immature RBCs [1]. As shown in Figure 5, transferrin was confirmed as the active substance for all four fish species in the present study.

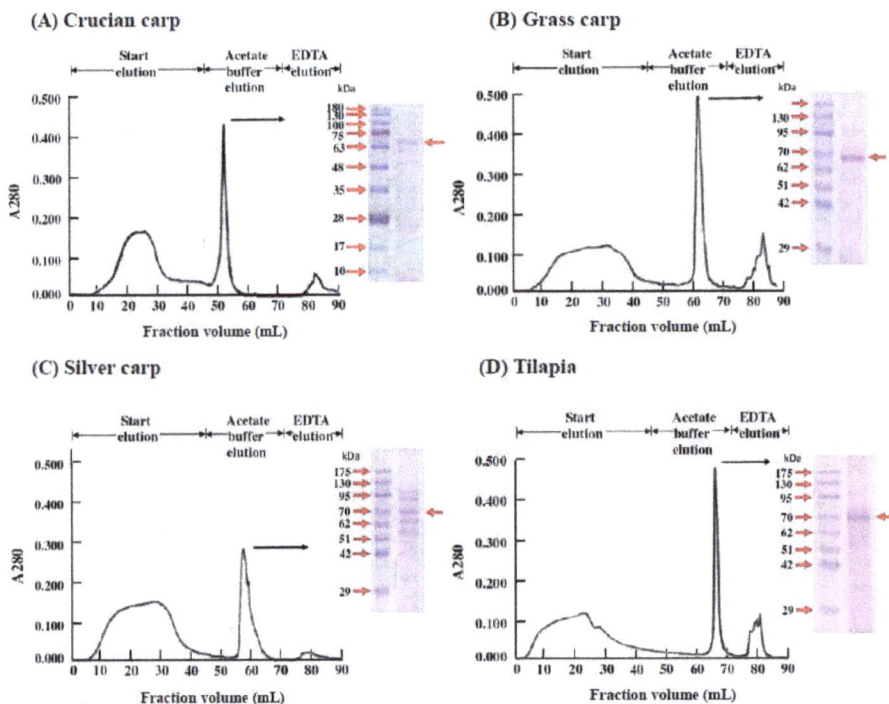

**Figure 5.** Isolation of the fish serum component stimulating the proliferation of new RBCs in head kidney cells of (**A**) crucian carp, (**B**) grass carp, (**C**) silver carp, and (**D**) tilapia. The head kidney cell suspension was supplemented with 10% serum from each fish species and ZnCl$_2$ and then cultured for three days before being harvested. The harvested cells were separated into fractions 1 and 2 using a Percoll density gradient. The cells collected from fraction 1 were extracted using a detergent buffer, and the detergent extract was purified by Zn$^{2+}$-IMAC. The acetate buffer eluates were pooled and analyzed by SDS-PAGE under reducing conditions, followed by staining with Coomassie blue to visualize the protein bands. The arrowhead denotes the position of the protein identified by nano-LC-MS/MS analysis as transferrin in crucian carp, grass carp, silver carp, and tilapia, respectively.

*2.6. Effects of the Injection of ZnSO$_4$ on Erythropoiesis in a Blood Loss Tilapia Model*

Two hours after the loss of approximately 20% of the blood in the tilapia, the RBC counts, hematocrit, and hemoglobin levels of the fish decreased to approximately 55%–65% of the baseline values (Figure 6). However, the group of tilapia injected with ZnSO$_4$ exhibited significantly higher RBC counts, hematocrit values, and hemoglobin levels than the control group (injection of saline) after three days and until 12 days. Microscopic observation of the blood cells revealed few immature RBCs in the group injected with saline throughout the experimental period (Figure 7A-a,A-b,A-c). In contrast, immature RBCs of different stages were observed three and six days after injection of ZnSO$_4$ (Figure 7B-b,B-c). These results indicate that excess in vivo zinc induced erythropoiesis in fish in three days. At day six, the blood cells from the fish injected with ZnSO$_4$ included more stage 4 RBCs (orthochromatic erythroblasts) (Figure 7B-c).

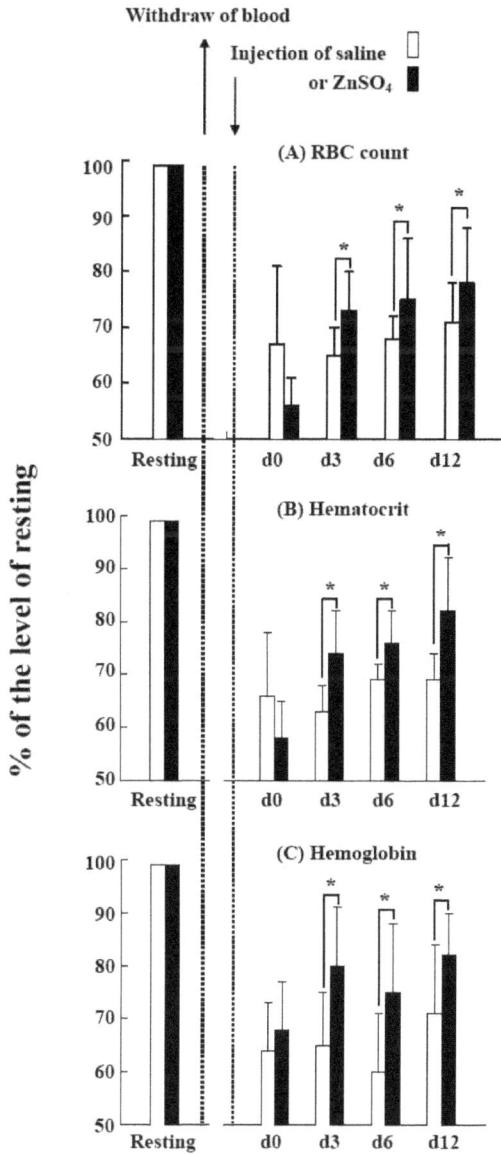

**Figure 6.** Effects of the injection of ZnSO$_4$ on erythropoiesis in the blood loss tilapia model. At day zero, blood was withdrawn from the fish, and saline or ZnSO$_4$ was injected. At day three, day six, and day 12, blood was sampled from the treated fish and analyzed. The data are expressed as the ratios of the RBC count (**A**), hematocrit (**B**) and hemoglobin (**C**) level at different days to the values at baseline. The results are the means ± SDs of three independent experiments. * Significant difference ($p < 0.05$) between the saline and ZnSO$_4$ treatments.

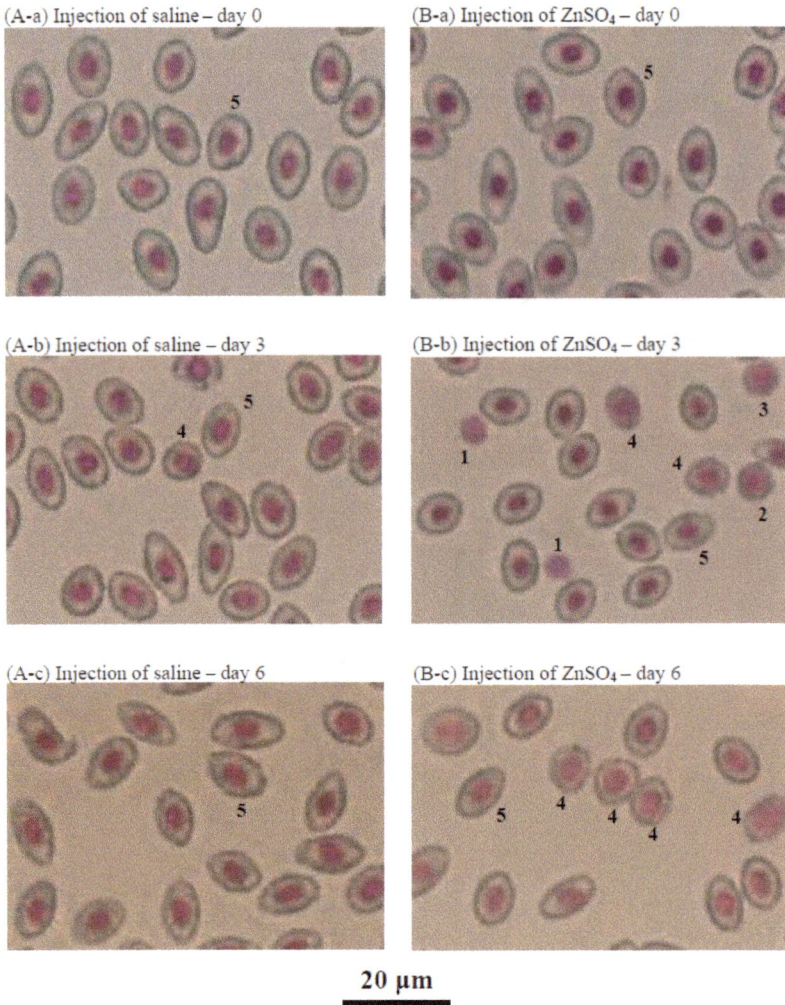

(A-a) Injection of saline – day 0          (B-a) Injection of ZnSO₄ – day 0

(A-b) Injection of saline – day 3          (B-b) Injection of ZnSO₄ – day 3

(A-c) Injection of saline – day 6          (B-c) Injection of ZnSO₄ – day 6

**20 μm**

**Figure 7.** Giemsa staining of the blood cells sampled from tilapia subjected to blood loss and injected with saline (**A**) or ZnSO₄ (**B**). At day zero, approximately 95% of the blood cells were stage 5 RBCs, erythrocytes (**A-a,B-a**). At day three and day six, the blood cell composition of the fish injected with saline (**A-b,A-c**) was similar to that on day zero; that is, approximately 95% stage 5 RBCs and 3% stage 4 RBCs (orthochromatic erythroblasts). However, for fish injected with ZnSO₄, newly proliferated cells were observed on day three (**B-b**), such as stage 1 RBCs (lymphoid hemoblasts), stage 2 RBCs (early erythroblasts), and stage 3 RBCs (polychromatophilic erythroblasts). At day six, the blood cells from the fish injected with ZnSO₄ included more stage 4 RBCs (orthochromatic erythroblasts) (**B-c**).

## 3. Discussion

The common carp has extraordinarily high zinc levels in digestive tract tissues (~300 μg/g tissue) compared to other fish (~20 μg/g tissue), as first reported more than 40 years ago [13]. The physiological relevance of this high level of zinc was not clear until recently [1,9,11,14]. At low oxygen concentrations, zinc is released from the digestive tract tissue of the common carp and used as a signal to stimulate erythropoiesis in the head kidney of the fish [1]. Whether the use of zinc as a signal to induce erythropoiesis was unique to this fish species was unclear, but the present report indicates that this type of "zinc signaling" pathway also exists in other fish if a sufficient amount of zinc is sent to the head kidney cells.

Despite the variety of the four fish studied (crucian carp, grass carp, and silver carp belong to the *Cyprinidae* family, but the tilapia are in a completely different family, *Cichlidae*) and the different zinc levels (crucian carp have zinc levels as high as ~500 µg/g fresh tissue, but grass carp, silver carp, and tilapia all have zinc levels of ~20 µg/g fresh tissue), the stimulation of erythropoiesis by zinc was dose-dependent, with a significant activation concentration of 0.2–1.2 mM and maximum cell growth of 300%–600% (Figure 2). In vitro, zinc could be used as a signal to induce blood cell formation in fish. In the blood, the zinc is carried by transferrin (Figure 5).

An experimental blood loss model has been established in several fish species [15–17]. The response to blood loss appears to depend on the fish species: some species recovered in several days [15], whereas others require weeks [16,17]. In the present study, the bled tilapia exhibited no significant stimulation of erythropoiesis in 12 days (Figure 6, control group). However, the ZnSO$_4$ injection group exhibited increased RBC, hematocrit and hemoglobin levels. These values decreased to 55%–65% of baseline in the control group but only 70%–80% of baseline in the ZnSO$_4$ group at three days post-bleeding (Figure 6, ZnSO$_4$ group). Microscopic observation of the blood of the ZnSO$_4$-injected fish confirmed that many new immature RBCs proliferated (Figure 7). In vivo, the presence of additional zinc induced erythropoiesis in the bled tilapia.

In recent years, "zinc signaling" was reported to play a role in brain function, immunology, inflammation, growth control and bone diseases, and cancer [18]. To date, the use of zinc as a signal to stimulate erythropoiesis appears to only have been observed in the common carp [1,9,11]. Although it is not clear whether this mechanism also exists in other animals, the present report suggests a possible approach for the induction of erythropoiesis in animals through the supply of a certain additional level of zinc through either diet or injection. If zinc can act as a signal to trigger erythropoiesis in many fish species, this mechanism might also exist in a broader range of species. Mammalian red cells lack a nucleus and differ fundamentally from the red blood cells of fish, birds, and reptiles. Erythropoiesis occurs in the head kidney in teleost fish but in the bone marrow in mammals. We recently studied whether zinc could be used as a signal to induce red blood formation in rat bone marrow cells, and our results suggest that at certain levels, zinc could trigger erythropoiesis in mammals (unpublished paper). Anemia is a global human health concern, and the use of zinc as a signal to induce red blood cell formation in patients through zinc supplementation might be a possible treatment. Thus, the special feature of using zinc as a signal to stimulate red blood formation in the common carp might have implications for human health.

## 4. Materials and Methods

### 4.1. Fish

Cultured live fish (crucian carp, grass carp. silver carp, and tilapia) were obtained from a local fish farm and housed in a polyethylene tank in the laboratory for 2 months before the experiment [11]. The total length and total mass of the fish were: crucian carp, 22 ± 3 cm, 207 ± 69 g, (*n* = 44); grass carp, 41 ± 2 cm, 680 ± 101 g, (*n* = 6); silver carp, 39 ± 3 cm, 556 ± 73 g, (*n* = 6); and tilapia, 27 ± 1 cm, 362 ± 43 g, (*n* = 15). Animal use was reviewed and approved by the institutional animal care and use committee at National Taiwan Ocean University (105-J29803; License No. 102072).

### 4.2. Preparation of Head Kidney Cell Suspensions and Fish Serum

Head kidney cell suspensions and fish serum were prepared from each different fish species as described previously [1].

*4.3. Suspension Culture of Head Kidney Cells with or without ZnCl$_2$ Supplementation*

The head kidney cell suspension (approximately $60 \times 10^6$ cells/mL) from each different fish species was cultured with or without ZnCl$_2$ as described previously [1]. The cultures were incubated at 27 °C in an atmosphere of 5% CO$_2$. After 0, 1, 2, 3, and 4 days, cells were harvested to measure the total cell number and subjected to separation by Percoll density centrifugation. Four independent experiments were performed.

*4.4. Separation of Harvested Head Kidney Cells into Two Fractions by Discontinuous Percoll Density Centrifugation*

The harvested head kidney suspension was separated by discontinuous Percoll centrifugation (2.0 mL of 1.020 g/mL Percoll and 6.0 mL of 1.070 g/mL Percoll, both in 0.15 M NaCl solution) into 2 fractions as reported previously [9]. The cell numbers of fractions 1 and 2 were measured with an electronic cell counter.

*4.5. Effects of ZnCl$_2$ on the Growth of Fraction 1 Cells from the Head Kidneys of Different Fish Species*

Aliquots of 0.1 mL of head kidney cell suspension ($60 \times 10^6$ cells/mL) from different fish species were added to equal volumes of DMEM/F12 medium supplemented with 20% fish serum from the respective fish species and different concentrations of ZnCl$_2$. Zinc concentrations in the control groups were determined, and found to be about 0.01 mM. The cells were cultured as described above. At day 0 and 1, the cells were harvested and separated by Percoll density centrifugation. The cell numbers in fraction 1 were counted. Cell growth was expressed as the ratio of the fraction 1 cell number to that at the start of culture relative to the control. Six independent experiments were performed.

*4.6. Microscopic Observation of Head Kidney Cells or Tilapia Blood by Giemsa Staining*

The harvested head kidney cells or tilapia blood were directly fixed on slides, stained with the Giemsa method [19], and subjected to microscopic observation. RBCs were identified according to previous reports [3,12,20].

*4.7. Immunofluorescence Staining of the Transferrin Receptor in Fish Head Kidney Cells*

The fish head kidney cells were subjected to immunofluorescence staining of the transferrin receptor as previously reported [1] and were observed through fluorescence microscopy.

*4.8. Isolation of the Active Substance in Fish Serum That Stimulates the Proliferation of Fraction 1 Cells in Fish Head Kidneys*

Preparation of proliferated cells, detergent extraction of zinc-binding proteins from fraction 1 cells, immobilized metal affinity chromatography (IMAC) purification of zinc-binding proteins, electrophoresis, and nano-LC-MS/MS were performed as reported previously [1].

*4.9. Experimental Blood Loss in Tilapia and Injection of ZnSO$_4$ Solution*

In each experiment, 54 acclimated tilapias were used. Before the experiment, 6 fish were removed, and 0.2 mL of blood was sampled from resting fish as the baseline. The remaining 48 fish were subjected to experimental blood loss by withdrawal (caudal venipuncture) of 3.0 mL blood from each fish, which represented approximately 20% blood loss because preliminary experiments indicated that the extractable blood of the tilapia was approximately 15 mL. After bleeding, the fish were allocated to two groups of 24 fish. Group A (control group) was injected with saline solution (0.25 mL), and Group B (experimental group) was injected with (1.47 mg/mL, or 5.1 mM) ZnSO$_4$ solution (0.25 mL). The fish were returned to the culture tank, and no feed was given during the experimental period. Two hours after the injection of saline or ZnSO$_4$ solution, 0.2 mL of blood was sampled from each fish in each group (6 fish) as blood from day 0. At days 3, 6, and 12, 0.2 mL of blood was sampled from

6 fish in each group. The collected blood was used for blood analysis and microscopic observation. Three different experiments were performed.

### 4.10. Blood Analysis

At each sampling time, the hematocrit, hemoglobin, and total RBC counts were determined using an auto hematology analyzer (Excell 500; Danam Electronics, Dallas, TX, USA) as reported previously [9]. Blood smears were microscopically observed.

### 4.11. Data Analysis

The data are expressed as the means ± standard deviations (SDs). The statistical significance of experimental results was calculated by one-way analysis of variance, followed by the least significant difference post hoc test, using SPSS 10.0 (SPSS Inc., Chicago, IL, USA).

## 5. Conclusions

Zinc is an essential element in organism, it has many function in life. In nature, the common carp has a unique way to use it as a signal to stimulate red blood cell formation by having a specific zinc-binding protein. Many fish do have no this specific zinc-binding protein, but the present report indicated that supplying a specific additional level of zinc exogenously could also induce erythropoiesis. It is very possible that this mechanism also exist in other animals including human beings. The potential of using this mechanism to treat anemia in human beings is high. However, how zinc stimulates erythropoiesis in fish (and may be in other animals) must be further investigated.

**Acknowledgments:** The authors wish to express their gratitude to Pang-Hung Hsu for providing the nano-LC-MS/MS sample analysis and to Cai-Yi Wu and Hui-Lin Feng for their technical assistance. This research was supported by the Center of Excellence for the Oceans, Project no. 105-J29803, National Taiwan Ocean University.

**Author Contributions:** Sen-Shyong Jeng contributed to the conception and instruction of the study; Yen-Hua Chen designed the experiments, performed the experiments, and analyzed the data. Jhe-Ruei Shiu and Chia-Ling Ho performed the experiments; Sen-Shyong Jeng wrote the paper.

**Conflicts of Interest:** The authors declare no conflict of interest.

## Abbreviations

EPO       Erythropoietin
IMAC      Immobilized metal affinity chromatography

## References

1. Chen, Y.H.; Fang, S.W.; Jeng, S.S. Zinc transferrin stimulates red blood cell formation in the head kidney of common carp (*Cyprinus carpio*). *Comp. Biochem. Physiol. A Mol. Integr. Physiol.* **2013**, *166*, 1–7. [CrossRef] [PubMed]
2. Fänge, R. Blood cells, haemopoiesis and lymphomyeloid tissues in fish. *Fish Shellfish Immunol.* **1994**, *4*, 405–411. [CrossRef]
3. Fijan, N. Composition of main haematopoietic compartments in normal and bled channel catfish. *J. Fish Biol.* **2002**, *60*, 1142–1154. [CrossRef]
4. Fijan, N. Morphogenesis of blood cell lineages in channel catfish. *J. Fish. Biol.* **2002**, *60*, 999–1014. [CrossRef]
5. Bunn, H.F. Erythropoietin. *Cold Spring Harb. Perspect. Med.* **2013**, *3*, a011619. [CrossRef] [PubMed]
6. Jelkmann, W. Physiology and pharmacology of erythropoietin. *Transfus. Med. Hemother.* **2013**, *40*, 302–309. [CrossRef] [PubMed]
7. Lai, J.C.; Kakuta, I.; Mok, H.O.; Rummer, J.L.; Randall, D. Effects of moderate and substantial hypoxia on erythropoietin levels in rainbow trout kidney and spleen. *J. Exp. Biol.* **2006**, *209*, 2734–2738. [CrossRef] [PubMed]

8.  Goldberg, M.A.; Dunning, S.P.; Bunn, H.F. Regulation of the erythropoietin gene: Evidence that the oxygen sensor is a heme protein. *Science* **1988**, *242*, 1412–1415. [CrossRef] [PubMed]
9.  Chen, Y.H.; Chen, H.H.; Jeng, S.S. Rapid renewal of red blood cells in the common carp following prolonged exposure to air. *Fish. Sci.* **2015**, *81*, 255–265. [CrossRef]
10. Jeng, S.S.; Wang, M.S. Isolation of a Zn-binding protein mediating cell adhesion from common carp. *Biochem. Biophys. Res. Commun.* **2003**, *309*, 733–742. [CrossRef] [PubMed]
11. Lin, T.Y.; Chen, Y.H.; Liu, C.L.; Jeng, S.S. Role of high zinc levels in the stress defense of common carp. *Fish. Sci.* **2011**, *77*, 557–574. [CrossRef]
12. Catton, W.T. Blood cell formation in certain teleost fishes. *Blood* **1951**, *6*, 39–60. [PubMed]
13. Jeng, S.S.; Lo, H.W. High zinc concentration in common carp viscera. *Bull. Jpn. Soc. Sci. Fish.* **1974**, *40*, 509. [CrossRef]
14. Jeng, S.S.; Lin, T.Y.; Wang, M.S.; Chang, Y.Y.; Chen, C.Y.; Chang, C.C. Anoxia survival in common carp and crucian carp is related to high zinc concentration in tissues. *Fish. Sci.* **2008**, *74*, 627–634. [CrossRef]
15. Montero, D.; Tort, L.L.; Izquierdo, M.S.; Socorro, J.; Vergara, J.M.; Robaina, L.; Fernández-Palacios, H. Hematological recovery in *Sparus aurata* after bleeding. A time course study. *Rev. Esp. Fisiol.* **1995**, *51*, 219–226. [PubMed]
16. Kondera, E.; Dmowska, A.; Rosa, M.; Witeska, M. The effect of bleeding on peripheral blood and head kidney hematopoietic tissue in common carp (*Cyprinus carpio*). *Turk. J. Vet. Anim. Sci.* **2012**, *36*, 169–175.
17. Fazio, F.; Piccione, G.; Arfuso, F.; Faggio, C. Peripheral blood and head kidney haematopoietic tissue response to experimental blood loss in mullet (*Mugil cephalus*). *Mar. Biol. Res.* **2015**, *11*, 197–202. [CrossRef]
18. Fukada, T.; Kambe, T. (Eds.) *Zinc Signals in Cellular Functions and Disorders*; Springer: Tokyo, Japan, 2014.
19. Freshney, R.I. *Culture of Animal Cells*, 4th ed.; Wiley-Liss: New York, NY, USA, 2000.
20. Kondera, E. Haematopoiesis in the head kidney of common carp (*Cyprinus carpio* L.): A morphological study. *Fish Physiol. Biochem.* **2011**, *37*, 355–362. [CrossRef] [PubMed]

International Journal of
*Molecular Sciences*

MDPI

*Article*

# Zinc Up-Regulates Insulin Secretion from β Cell-Like Cells Derived from Stem Cells from Human Exfoliated Deciduous Tooth (SHED)

**Gyuyoup Kim [1], Ki-Hyuk Shin [2] and Eung-Kwon Pae [3,\*]**

[1] School of Medicine Department of Obstetrics, Gynecology and Reproductive Sciences, University of Maryland, Baltimore, MD 21201, USA; gykim@fpi.umaryland.edu
[2] UCLA School of Dentistry, Los Angeles, CA 90095, USA; kshin@dentistry.ucla.edu
[3] Department of Orthodontics and Pediatric Dentistry, School of Dentistry, University of Maryland, Baltimore, MD 21201, USA
\* Correspondence: epae@umaryland.edu; Tel.: +1-443-478-6025

Academic Editor: Toshiyuki Fukada
Received: 29 September 2016; Accepted: 6 December 2016; Published: 13 December 2016

**Abstract:** Stem cells from human exfoliated deciduous tooth (SHED) offer several advantages over other stem cell sources. Using SHED, we examined the roles of zinc and the zinc uptake transporter ZIP8 (Zrt- and irt-like protein 8) while inducing SHED into insulin secreting β cell-like stem cells (i.e., SHED-β cells). We observed that ZIP8 expression increased as SHED differentiated into SHED-β cells, and that zinc supplementation at day 10 increased the levels of most pancreatic β cell markers—particularly Insulin and glucose transporter 2 (GLUT2). We confirmed that SHED-β cells produce insulin successfully. In addition, we note that zinc supplementation significantly increases insulin secretion with a significant elevation of ZIP8 transporters in SHED-β cells. We conclude that SHED can be converted into insulin-secreting β cell-like cells as zinc concentration in the cytosol is elevated. Insulin production by SHED-β cells can be regulated via modulation of zinc concentration in the media as ZIP8 expression in the SHED-β cells increases.

**Keywords:** dental pulp; insulin; stem cells; zinc; Zrt- and irt-like protein (ZIP)

## 1. Introduction

Type-1 diabetes (T1D), a major and difficult-to-manage health problem [1], results from dysfunctional and/or insufficient pancreatic β cells, leading to a loss of insulin as well as glucose homeostasis as the disease advances [2]. Interventions include insulin injection, which isoften accompanied by significant side-effects (e.g.,hypoglycemia), particularly in children. A more amenable approach provides an alternative, supplemental source for the generation of functional insulin-producing β cells using mesenchymal stem cells which can overcome the shortage of donated islets and limitations in the ability to expand β cells in vitro. Stem cells from human exfoliated deciduous tooth (SHED) offer advantages over other stem cell sources for conversion into insulin secreting cells [3,4], including circumvention of ethical issues, noninvasiveness, long-term banking, abundance of tissue source, availability for autologous cells to avoid immunosuppressive regimen for graft rejection responses, and no chance of potential transfection from donors.

Several studies were performed to evaluate "multipotency" and "stemness" of SHED-derived insulin-secreting cell aggregates [5–7]; however, few studies focused on what factors control insulin secretion. We previously reported that β cells start losing the capability to secrete insulin before signs of inflammatory cell infiltration or conspicuous cytokine expression emerged in [8], and this notion was supported by such observations in a recent clinical report [9]. We also observed that a lack of zinc results in a significant decline of insulin secretion in conjunction with decreased ZIP8 (Zrt- and irt-like protein 8, a zinc uptake transporter) expression in Sprague-Dawley rats [10]. In accordance with these previous results and other studies [8,10], we presumed that zinc supplementation would up-regulate insulin secretion from insulin-secreting β cell-like cells (or SHED-β cells) via up-regulation of ZIP8. Based on our previous observation and published data, we hypothesized that without zinc accumulation in the cytosol, SHED cannot fully differentiate into insulin-producing SHED-β cells, and therefore, the existance of zinc uptake transporters and maintaining an optimum level of intracellular zinc are critical for maximizing insulin production.

For this preliminary study, we successfully generated SHED-β cells from SHED, based on a published protocol [5]. Using these SHED-β cells, we tested a set of hypotheses: first, if zinc is an essential factor in the regulationprocess of insulin production and secretion. Second, if SHED-β cells contain ZIP8, a zinc uptake transporter. Third, if ZIP8 is involved in the insulin secretion mechanism in SHED-β cells. The purposes of this study were: (1) to demonstrate β cell markers which show SHED differentiating to β cell lineage to become SHED-β cells; (2) to show that ZIP8 expression increasesas SHED-β cell differentiates; (3) to demonstrate the effect of zinc supplementation on insulin secretion.

## 2. Results

### 2.1. SHED Express a Strong "Stemness" Compared to Dental Pulp and Periodontal Ligament Originated Stem Cells (DPSC and PDLSC)

Results of real-time quantitative PCR (qPCR) assays showed significantly strong expressions of pluripotent stem cell markers in SHED compared to dental pulp stem cells (DPSC) and periodontal ligament originated stem cells (PDLSC) (see Figure 1). This result prompted us to focus on SHED rather than using other stem cell origins. Error bars and P values were calculated from triplicates of each measurements.

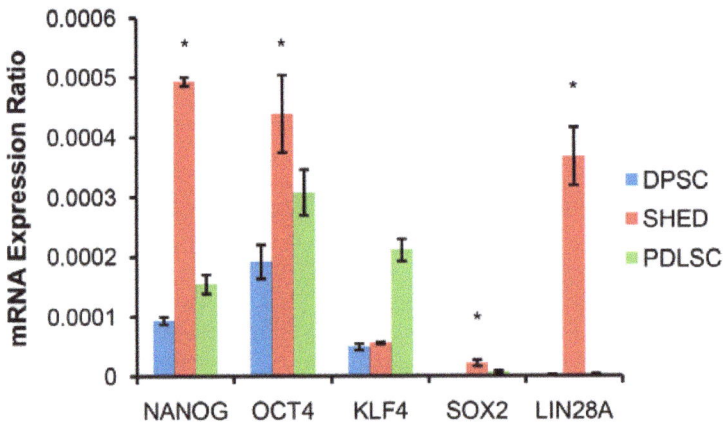

**Figure 1.** Expression of pluripotent transcription factors determined by quantitative PCR (qPCR) normalized with the level of GAPDH. DPSC: dental pulp stem cells; PDLSC: periodontal ligament stem cells; SHED: stem cells from human exfoliated deciduous tooth. SHED show a significantly stronger expression than other (asterisk * $p < 0.05$) markers, except KLF4. Gene Descriptions: NANOG, Nanog homeobox; OCT4, Octamer-binding protein 4; KLF4, Kruppel like factor 4; SOX2, SRY (sex determining region Y)-box 2; LIN28A, lin-28 homolog A; GAPDH, glyceraldehydes-3-phosphate dehydrogenase.

## 2.2. Genetic Markers in Pancreatic β Cell Lineage Expressed in SHED Indicate a Successful Conversion of SHED to SHED-β Cells

SHED-β cells differentiated through the three-step media treatment as previously offered by Govindasamy et al. [5] expressed biomarkers in the pancreatic lineage, such as pancreatic and duodenal homeobox 1 (PDX1), neurogenin 3 (NEUROG3), NK6 homeobox 1 (NKX6.1), paired box 4 (Pax4), and aristaless related homeobox (ARX), which were markedly increased between d5 and d10 (see Figure 2).

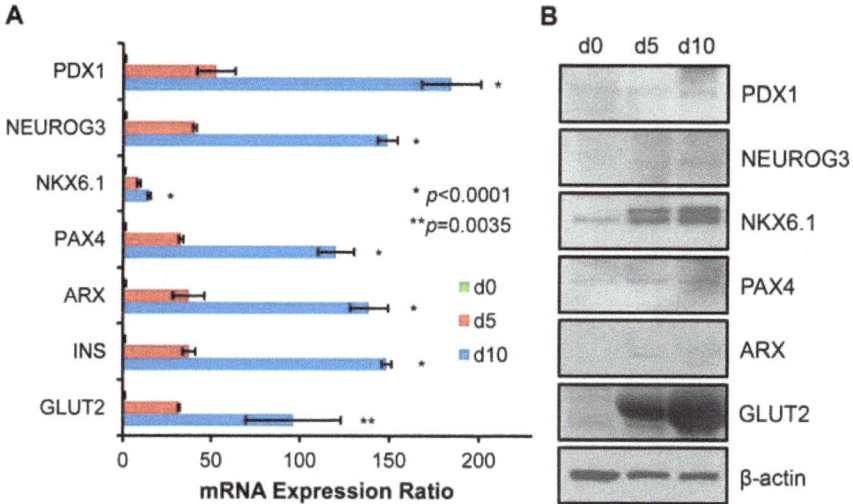

**Figure 2.** Differentiation of biomarkers in pancreatic lineage shown in the levels of (**A**) mRNA; and (**B**) protein. Every day 5 (or d5) and day 10 (d10) marker expression differed from one another and from d0 in mRNA. $n = 3$. Gene Descriptions: PDX1, pancreatic and duodenal homeobox 1; NEUROG3, neurogenin 3; NKX6.1, NK6 homeobox 1; PAX4, paired box 4; ARX, aristaless related homeobox; INS, insulin; GLUT2, glucose transporter 2.

Particularly, mRNA expressions for insulin (INS) and glucose transporter2 (GLUT2, a main transporter importing glucose molecules into the β-cell cytosol) increased significantly at both mRNA and protein levels. Compared to Day-5 (d5), d10 cells show significantly stronger expression of every gene (see Figure 2A).

## 2.3. ZIP8 Expression and Zinc Supplementation Effects on the Markers in β Cell Lineage

To examine the hypothesized one-to-one relationship between zinc and the markers expressed in β cell lineage, we measured ZIP8 expression changes because we assumed that the ZIP8 transporter mediates the insulin production and secretion in SHED-β cells (Figure 3). The transcriptional activity of the ZIP8 gene increased significantly as differentiation proceeded (See Figure 3A).

Zinc (50 μM) was added to the media on day 10. This supplementation of zinc increased not only the genes in the β cell lineage, but also INS and GLUT2. This may indicate that zinc supplementation could augment insulin secretion.

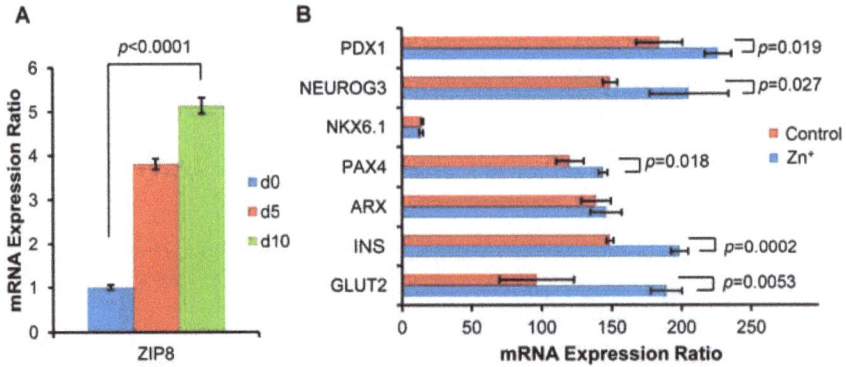

**Figure 3.** mRNA level changes in Zrt- and irt-like protein 8 (ZIP8) expression and β cell differentiation markers. (**A**) ZIP8 expression increased between d0 and d10; (**B**) Zinc supplementation at d10 increased the levels of most markers, particularly Insulin and GLUT2. $n = 3$.

### 2.4. Levels of Zinc in Conjunction with ZIP8 in the Cytosol of SHED-β Cells Modulate Insulin Secretion

To investigate the roles of zinc in the cytosol of SHED-β cells in association with ZIP8 expression, we augmented the zinc concentration of the media by 50 µM, and then we compared before and after supplementation. Figure 4 contrasts the changes in SHED-β cells in the images obtained by immunofluorescence staining. Figure 4A demonstrates a marked increase in the number of insulin-containing β cell-like stem cells after zinc supplementation. Insulin molecules in the cytosolare shown in brighter red color after the augmentation (Figure 4A). Figure 4B demonstrates a significant increase of ZIP8 protein after zinc supplementation. Lastly, supplementation of 50 µM of zinc significantly increased insulin secretion by approximately 25% ($9.8 \pm 0.025$ vs. $12.39 \pm 0.035$ ng/mL) compared to control media, as shown in Figure 4C.

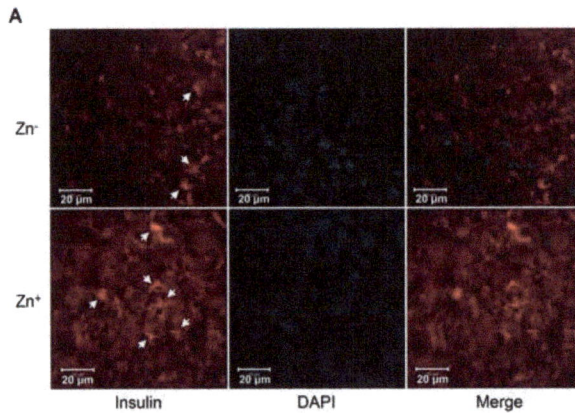

**Figure 4.** *Cont.*

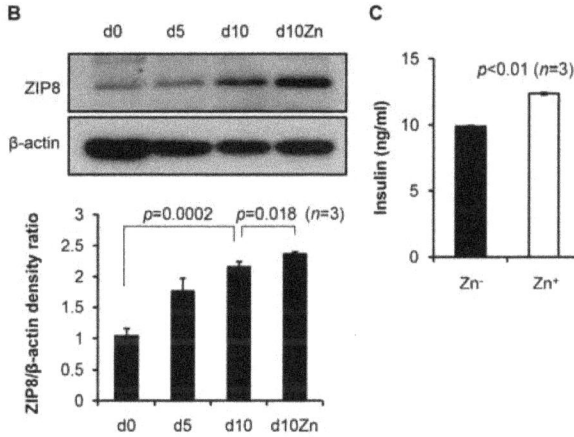

**Figure 4.** Zn supplementation affects the production and secretion of insulin and the expression of ZIP8 in SHED-β cells. (**A**) Immunofluorescence staining of insulin (indicated by red color and white arrow heads) in SHED-β cells with or without zinc supplementation. Note the more prominent expression of insulin after zinc supplementation in the merged panel; (**B**) Western blot analysis of ZIP8 expression in different time points with or without zinc supplementation; (**C**) Constitutive secretion of insulin from SHED-β cells evaluated by ELISA on $n = 3$. In these assays, 50 μM of zinc was supplemented.

## 3. Discussion

Several previous studies thoroughly demonstrated the process by which mesenchymal stem cells from various sources could convert to insulin-secreting cells [3,4,6]. However, most studies focused on proving the pluripotency of their cells or showing β-cell-like phenotypes of their cells. A scarce number of studies have reported what factors play a key role in influencing quantity and quality of insulin secretion. Based on previous experience and understanding, our study focused on what factors play a key role in the augmentation of insulin secretion. The results of this study support a new concept that zinc and zinc uptake transporters (particularly ZIP8) may affect the proliferation of SHED-β cells. Zinc is an essential element involved in many basic cellular biochemical processes for human physiology. Zinc plays its roles as a part of protein structure, signaling process, and enzymatic regulation [11,12]. For instance, when zinc and osteocalcin are lacking during perinatal growth, brittle bones in rats [13] and a short height [14] may result. Conversely, Zinc supplementation dampens osteoclastic activity [15] and upregulates osteoprotegerin [16] in diabetic rodents and cell lines. Zinc in calcium phosphate is known to modulate bone induction [17] and mineralization [18]; therefore, zinc plays a significant role in bone formation. Bones in patients with type 1 diabetes often suffer from a lack of bone (i.e., osteopenia and osteoporosis) [19–21]. Indeed, a lack of bone in diabetes is associated with zinc deficiency [22,23].

Zinc homeostasis in the cytosol is tightly regulated via ubiquitous zinc-bound intracellular proteins such as metallothionein as well as zinc pool in intracellular organelles such as Golgi apparatus. Due to these reservoirs inside the cells, pathologic symptoms of zinc depletion cannot be easily recognized [24]. Particularly, insulin-producing pancreatic β cells are specified to contain a high level of zinc because insulin hexamer molecules are structured around zinc ions [25]. In addition, β cells are exposed to oxidative stress due to reactive oxygen species (ROS) produced during the cleaving process from proinsulin to insulin and during glucose metabolism, which consume anti-oxidants such as zinc containing metallothionein [26,27]. Thus, zinc is an essential metal for synthesizing insulin.

Previous studies have shown that lack of zinc in pancreatic islets is associated with reduced insulin secretion [8,10]. Our current results indicate that zinc supplementation assists the process of converting SHED into SHED-β cells. In addition, this augmentation is strongly associated with the increased expressions of ZIP8 in SHED-β cells. A recent publication revealed that their culture media contained 10 μM $Zn^{2+}$ [4]. Then, since we used 50 μM, a question would merge if 10 μM of zinc is optimal.

Zinc transporters are cell structures involved in regulating zinc homeostasis. Therefore, questions must start from which zinc transporter(s) play a role in zinc homeostasis. We reported that one of the zinc uptake transporters—Zip8—plays a major role in insulin secretion [10] using primary β cells harvested from rats. More recently, a research group investigated a role of Zip4 in insulin secretion; however, Zip4 zinc uptake transporters were not directly associated with insulin secretion in mice [28]. Another independent study concluded that ZIP6 and ZIP7 are functionally important [29] for maintaining zinc homeostasis in human islets and β cell lines; however, they did not measure changes in insulin secretion in pancreatic cells in response to ZIP expression changes. Therefore, ZIP8 may be an unknown zinc uptake transporter involved in the regulation of insulin production and secretion. We confirmed that ZIP8 may be an important functional transporter in SHED-β cells essential for optimal production of insulin via the regulation of intracellular zinc levels. We found that ZIP8 up-regulation for increased accumulation of zinc is an important factor in the differentiation process of SHED into SHED-β cells.

Finally, our findings could develop an ex vivo study model determining what important factors play a major role as zinc and ZIP8 augmented SHED-β cells are transplanted to human β cells aggregates. Glucose-stimulated insulin secretion studies on SHED-β cells with and without zinc supplementation would confirm our conclusion. Future experiments to determine the optimized conditions for the mechanism involved in zinc regulation for the optimal insulin production in β cells is a critical point for understanding clinical implications in not only Type-1 but Type-2 diabetes mellitus [30].

## 4. Methods

### 4.1. Cell Cultures

Three types of human dental mesenchymal stem cells were cultured and used for the study. Primary SHED cells were isolated from normal exfoliated deciduous incisors. Dental pulp stem cells (DPSCs) were isolated from extracted teeth. Periodontal ligament stem cells (PDLSCs) were scraped from harvested third molars. All three cell types were cultured in α-minimum essential media (or α-MEM)medium (Invitrogen, Carlsbad, CA, USA) supplemented with 10% fetal bovine serum (Invitrogen), 5 μg/mL gentamicin sulfate (Gemini Bio-Products, West Sacramento, CA, USA), and 20 mmol/L L-glutamine (Invitrogen).

### 4.2. In Vitro SHED-β Cell Differentiation

Collected SHED cells were amplified to passage 5 in the media, as described above. SHED-β cells were differentiated using the previously established protocol with slight modifications [5]. After trypsin-treated and centrifuged, the cells were resuspended in Medium-A (containing Dulbecco's modified Eagle's medium/F-12 Knock-out (DMEM-KO), 17.5 mM Glucose, 1% bovine serum albumin Cohn Fraction V (BSA-CF; fatty acid free), 1× insulin-transferrin selenium (ITS), 4 nM Activin A, 1 mM Sodium Butyrate, 50 μM 2-Mercaptoethanol, and 5 μg/mL gentamicin sulfate) and plated in sterilized borosilicate glass plates (60 mm, Corning, Keene, NH, USA) without serum. After the cultures were incubated for 48 h, the medium was changed to Medium-B (containing DMEM-KO, 17.5 mM Glucose, 1% BSA-CF, 1× ITS, 0.3 mM taurine, and 5 μg/mL gentamicin sulfate), and was substituted with Medium-C (DMEM-KO, 17.5 mM Glucose, 1.5% BSA-CF, 1× ITS, 3 mM Taurine, 100 nM glucagon-like peptide-1, 1× non-essential amino acids, and 5 μg/mL gentamicin sulfate) on the fifth day. Gravity-downed aggregated cell pellets were gently aspirated, and the medium was replaced with fresh Medium-C every 2 days for the next 5 days. To make the zinc-rich environment for the differentiation step, 50 μM zinc chloride was added in every differentiation medium. All reagents were purchased from Sigma-Aldrich, St. Louis, MO, USA; Invitrogen, Carlsbad, CA, USA; R&D Systems, Minneapolis, MN, USA.

### 4.3. Real-Time qPCR

Stem cells and differentiated SHED-β cells were harvested, and total RNAs were purified using RNeasy Mini Kit (QIAGEN Sciences, Germantown, MD, USA) in accordance with the manufacturer's protocol. First-strand cDNA was synthesized from 1 µg of RNAs using the High Capacity cDNA Reverse Transcription Kits (Applied Biosystems, Foster City, CA, USA) primed with a mixture of random primers. A 2 µL volume of cDNA template was used on the mixture of 25 µL volume of SYBR Green master mix (Applied Biosystems, Carlsbad, CA, USA) with 5 pmol of primers. The primer sequences (5 to 3 primes) for each gene are the following:

NANOG, F: AGATGCCTCACACGGAGACT; R: TCTCTGCAGAAGTGGGTTGTT.
OCT4, F: GAAAACCCACACTGCAGATCA; R: CGGTTACAGAACCACACTCG.
KLF4, F: GGGAGAAGACACTGCGTCA; R: GGAAGCACTGGGGGAAGT.
SOX2, F: TCTCATGATGTTCAACCATTCAC; R: CACATTTACATTCAAAGCACCAG.
LIN28A, F: GAAGCGCAGATCAAAAGGAG; R: GCTGATGCTCTGGCAGAAGT.
GAPDH, F: GGTGTGAACCATGAGAAGTATGA; R: GAGTCCTTCCACGATACCAAAG.
PDX1, F: GGGTGACCACTAAACCAAAGA; R: GGTCATACTGGCTCGTGAATAG.
NEUROG3, F: GCTGCTCATCGCTCTCTATTC; R: GGCAGGTCACTTCGTCTTC.
NKX6.1, F: GAAGAGGACGACGACTACAATAAG; R: CTGCTGGACTTGTGCTTCT.
PAX4, F: TGGGAAGGAGATGGCATAGA; R: ATCACAGGAAGGAGGAAGGA.
ARX, F: GGCAAGGAGGTGTGCTAAA; R: GCTGGTCCTCTGTTTCCATT.
INS, F: CTGGAGAACTACTGCAACTAGAC; R: TGCTGGTTCAAGGGCTTTAT.
GLUT2, F: CCGCTGAGAAGATTAGACTTGG; R: GACTAGCTCCTGCCTGTTTATT.
SLC39A8, F: GCTGGCTATTGGGACTCTTT; R: GCAACTGCCTTCTCAACATAAC.
ACTB, F: GGATCAGCAAGCAGGAGTATG; R: AGAAAGGGTGTAACGCAACTAA.

Quantitative PCR reactions were triplicated for each sample with the Eppendorf Realplex System (Eppendorf, Hamburg, Germany), and the threshold cycle ($C_t$) for each reaction was normalized ($\Delta C_t$) by the value of the β-ACTIN (ACTB) housekeeping gene. The value of $\Delta C_T$ was further normalized to exhibit the comparative expression levels with respect to the mean value.

### 4.4. Western-Blots

Collected SHED and SHED-β cells were lysed in ice-cold RIPA buffer containing protease inhibitor cocktails (Roche Applied Science, Indianapolis, IN, USA). The same amount of proteins was resolved on the SDS-PAGE and transferred onto a polyvinylidene fluoride (PVDF) membrane using an electroblot. After blocking with 5% milk TBS-T, the membrane was incubated with anti-PDX1, anti-β-actin (Cell Signaling Technology Inc., Beverly, MA, USA), anti-NEUROG3, anti-ARX, anti-GLUT2 (Sigma-Aldrich, St. Louis, MO, USA), anti-NKX6.1 (R&D Systems, Minneapolis, MN, USA), anti-PAX4 (GeneTex Inc., Irvine, CA, USA), and anti-ZIP8 (Thermo Scientific Inc., Rockford, IL, USA) antibodies. A horseradish peroxidase-conjugated secondary antibody was added, and chemiluminescent reagents were used to detect immunoreactive proteins on X-ray films. A density measurement was performed on Multi Gauge V3.0 (Fujifilm, Tokyo, Japan), and the quantities were calculated by subtraction between ZIP8 and β-actin bands.

### 4.5. Immunofluorescence Assay

On the ninth day of differentiation, aggregated SHED-β cells were transferred on the Lab-Tek II CC2 Chamber Slide (Thermo Scientific Inc., New York, NY, USA) and incubated for 24 h to allow proper attachment on the surface. The cells were fixed with CytoCell Fixative solution

(Biocare Medical, Concord, CA, USA) for 20 min, and incubated in blocking solution (CAS-BLOCK; Invitrogen, Halethorpe, MD, USA) for 15 minutes at room temperature. SHED-β cells were stained with anti-Insulin (GenScript, Piscataway, NJ, USA) antibody at room temperature for 2 hours, washed three times with PBS, and then incubated with Alexa Fluor 594-conjugated secondary antibody (Invitrogen, Eugene, OR, USA) for 1 h. After being washed with PBS, the slide was mounted in Vectashield mounting medium containing DAPI (4′,6-diamidino-2-phenylindole) (Vector Laboratories, Burlingame, CA, USA). Fluorescent images of the samples were obtained using a Zeiss LSM530 META Confocal Microscope (Carl Zeiss, Thornwood, NY, USA).

*4.6. ELISA Assay for Insulin*

On day 10 of the differentiation of SHED-β cells, incubation medium was refreshed and maintained for 2 h to allow insulin secretion. The media were collected to quantify insulin levels using Insulin ELISA Kit (EMD Millipore Corp., Billerica, MA, USA). An equal amount of sample was incubated on the anti-insulin antibody-coated plate with biotinylated capture antibody for 90 min. After the plate was washed, a horseradish peroxidase-conjugated streptavidin was added, and TMB (3,3′,5,5′-tetramethylbenzidine) substrate and stop solution were mixed for color developing reaction. Absorbance was measured at 450 nm using a spectrophotometer (BioTek Instrument Inc., Winooski, VT, USA).

*4.7. Statistical Analysis*

For group comparison, Student's $t$-tests were performed with SPSS v. 21 (IBM Co., Armonk, NY, USA). Significance level of $p < 0.05$ was adopted for inference tests on mRNA and protein measurements. Samples collected from multiple teeth were used for triplicated measurements.

## 5. Conclusions

(1) β cell-differentiation markers emerge when SHED convert into SHED-β cells; (2) ZIP8 expression increases as SHED differentiate into SHED-β cells; (3) zinc supplementation enhances insulin secretion by SHED-β cells; (4) insulin secretion is augmented by ZIP8 elevation.

**Author Contributions:** Gyuyoup Kim—Performed assays/Analyzed and interpreted the data/Prepared figures; Ki-Hyuk Shin—Performed assay/Collated and interpreted the data; Eung-Kwon Pae—Concenptualized the project/Designed the study/Analyzed and interpreted the data/Wrote and revised the manuscript. This work was supported by a seed grant to EP.

**Conflicts of Interest:** The authors declare no conflict of interest.

## References

1. Vogel, G. Biomedicine. Stem cell recipe offers diabetes hope. *Science* **2014**. [CrossRef] [PubMed]
2. Atkinson, M.A. The pathogenesis and natural history of type 1 diabetes. *Cold Spring Harb. Perspect. Med.* **2012**, *2*, a007641. [CrossRef] [PubMed]
3. Kanafi, M.M.; Mamidi, M.K.; Sureshbabu, S.K.; Shahani, P.; Bhawna, C.; Warrier, S.R.; Bhonde, R. Generation of islet-like cell aggregates from human non-pancreatic cancer cell lines. *Biotechnol. Lett.* **2015**, *37*, 227–233. [CrossRef] [PubMed]
4. Carnevale, G.; Riccio, M.; Pisciotta, A.; Beretti, F.; Maraldi, T.; Zavatti, M.; Cavallini, G.M.; la Sala, G.B.; Ferrari, A.; de Pol, A. In vitro differentiation into insulin-producing β-cells of stem cells isolated from human amniotic fluid and dental pulp. *Dig. Liver Dis.* **2013**, *45*, 669–676. [CrossRef] [PubMed]
5. Govindasamy, V.; Ronald, V.S.; Abdullah, A.N.; Nathan, K.R.; Ab Aziz, Z.A.; Abdullah, M.; Musa, S.; Kasim, N.H.; Bhonde, R.R. Differentiation of dental pulp stem cells into islet-like aggregates. *J. Dent. Res.* **2011**, *90*, 646–652. [CrossRef] [PubMed]
6. Wang, X.; Sha, X.J.; Li, G.H.; Yang, F.S.; Ji, K.; Wen, L.Y.; Liu, S.Y.; Chen, L.; Ding, Y.; Xuan, K. Comparative characterization of stem cells from human exfoliated deciduous teeth and dental pulp stem cells. *Arch. Oral Biol.* **2012**, *57*, 1231–1240. [CrossRef] [PubMed]
7. Bhonde, R.R.; Sheshadri, P.; Sharma, S.; Kumar, A. Making surrogate β-cells from mesenchymal stromal cells: Perspectives and future endeavors. *Int. J. Biochem. Cell Biol.* **2014**, *46*, 90–102. [CrossRef] [PubMed]

8.  Pae, E.K.; Ahuja, B.; Kim, M.; Kim, G. Impaired glucose homeostasis after a transient intermittent hypoxic exposure in neonatal rats. *Biochem. Biophys. Res. Commun.* **2013**, *441*, 637–642. [CrossRef] [PubMed]
9.  Koskinen, M.K.; Helminen, O.; Matomäki, J.; Aspholm, S.; Mykkänen, J.; Mäkinen, M.; Simell, V.; Vähä-Mäkilä, M.; Simell, T.; Ilonen, J.; et al. Reduced β-cell function in early preclinical type 1 diabetes. *Eur. J. Endocrinol.* **2016**, *174*, 251–259. [CrossRef] [PubMed]
10. Pae, E.K.; Kim, G. Insulin production hampered by intermittent hypoxia via impaired zinc homeostasis. *PLoS ONE* **2014**, *9*, e90192. [CrossRef] [PubMed]
11. King, J.C. Zinc: An essential but elusive nutrient. *Am. J. Clin. Nutr.* **2011**, *94*, 679S–684S. [CrossRef] [PubMed]
12. Vallee, B.L.; Auld, D.S. Zinc coordination, function, and structure of zinc enzymes and other proteins. *Biochemistry* **1990**, *29*, 5647–5659. [CrossRef] [PubMed]
13. Kim, G.; Elnabawi, O.; Shin, D.; Pae, E.K. Transient Intermittent Hypoxia Exposure Disrupts Neonatal Bone Strength. *Front. Pediatr.* **2016**, *4*, 15. [CrossRef] [PubMed]
14. Doménech, E.; Díaz-Gómez, N.M.; Barroso, F.; Cortabarria, C. Zinc and perinatal growth. *Early Hum. Dev.* **2001**, *65*, S111–S117. [CrossRef]
15. Iitsuka, N.; Hie, M.; Tsukamoto, I. Zinc supplementation inhibits the increase in osteoclastogenesis and decrease in osteoblastogenesis in streptozotocin-induced diabetic rats. *Eur. J. Pharmacol.* **2013**, *714*, 41–47. [CrossRef] [PubMed]
16. Liang, D.; Yang, M.; Guo, B.; Cao, J.; Yang, L.; Guo, X. Zinc upregulates the expression of osteoprotegerin in mouse osteoblasts MC3T3-E1 through PKC/MAPK pathways. *Biol. Trace Elem. Res.* **2012**, *146*, 340–348. [CrossRef] [PubMed]
17. Luo, X.; Barbieri, D.; Davison, N.; Yan, Y.; de Bruijn, J.D.; Yuan, H. Zinc in calcium phosphate mediates bone induction: In vitro and in vivo model. *Acta Biomater.* **2014**, *10*, 477–485. [CrossRef] [PubMed]
18. Nagata, M.; Lönnerdal, B. Role of zinc in cellular zinc trafficking and mineralization in a murine osteoblast-like cell line. *J. Nutr. Biochem.* **2011**, *22*, 172–178. [CrossRef] [PubMed]
19. Saito, M.; Marumo, K. Bone quality in diabetes. *Front. Endocrinol.* **2013**, *14*, 72. [CrossRef] [PubMed]
20. Hamilton, E.J.; Rakic, V.; Davis, W.A.; Chubb, S.A.; Kamber, N.; Prince, R.L.; Davis, T.M. Prevalence and predictors of osteopenia and osteoporosis in adults with Type 1 diabetes. *Diabet. Med.* **2009**, *26*, 45–52. [CrossRef] [PubMed]
21. Liu, T.M.; Lee, E.H. Transcriptional regulatory cascades in Runx2-dependent bone development. *Tissue Eng. B Rev.* **2013**, *19*, 254–263. [CrossRef] [PubMed]
22. Fukada, T.; Hojyo, S.; Furuichi, T. Zinc signal: A new player in osteobiology. *J. Bone Miner. Metab.* **2013**, *31*, 129–135. [CrossRef] [PubMed]
23. Alcantara, E.H.; Lomeda, R.A.; Feldmann, J.; Nixon, G.F.; Beattie, J.H.; Kwun, I.S. Zinc deprivation inhibits extracellular matrix calcification through decreased synthesis of matrix proteins in osteoblasts. *Mol. Nutr. Food Res.* **2011**, *55*, 1552–1560. [CrossRef] [PubMed]
24. Lowe, N.M.; Fekete, K.; Decsi, T. Methods of assessment of zinc status in humans: A systematic review. *Am. J. Clin. Nutr.* **2009**, *89*, 2040S–2051S. [CrossRef] [PubMed]
25. Dodson, G.; Steiner, D. The role of assembly in insulin's biosynthesis. *Curr. Opin. Struct. Biol.* **1998**, *8*, 189–194. [CrossRef]
26. Chausmer, A.B. Zinc, insulin and diabetes. *J. Am. Coll. Nutr.* **1998**, *17*, 109–115. [CrossRef] [PubMed]
27. Fujimoto, S.; Mukai, E.; Inagaki, N. Role of endogenous ROS production in impaired metabolism-secretion coupling of diabetic pancreatic β cells. *Prog. Biophys. Mol. Biol.* **2011**, *107*, 304–310. [CrossRef] [PubMed]
28. Hardy, A.B.; Prentice, K.J.; Froese, S.; Liu, Y.; Andrews, G.K.; Wheeler, M.B. Zip4 mediated zinc influx stimulates insulin secretion in pancreatic beta cells. *PLoS ONE* **2015**, *10*, e0119136. [CrossRef] [PubMed]
29. Liu, Y.; Batchuluun, B.; Ho, L.; Zhu, D.; Prentice, K.J.; Bhattacharjee, A.; Zhang, M.; Pourasgari, F.; Hardy, A.B.; Taylor, K.M.; et al. Characterization of Zinc Influx Transporters (ZIPs) in Pancreatic β Cells: Roles in regulating cytosolic zinc homeostasis and insulin secretion. *J. Biol. Chem.* **2015**, *290*, 18757–18769. [CrossRef] [PubMed]
30. Potter, K.J.; Westwell-Roper, C.Y.; Klimek-Abercrombie, A.M.; Warnock, G.L.; Verchere, C.B. Death and dysfunction of transplanted β-cells: Lessons learned from type 2 diabetes? *Diabetes* **2014**, *63*, 12–19. [CrossRef] [PubMed]

International Journal of
*Molecular Sciences*

MDPI

*Discussion*

# Pan-Domain Analysis of ZIP Zinc Transporters

**Laura E. Lehtovirta-Morley [1,2], Mohammad Alsarraf [1] and Duncan Wilson [1,\*]**

[1]  Aberdeen Fungal Group, Medical Research Council Centre for Medical Mycology, Institute of Medical Sciences, University of Aberdeen, Aberdeen AB25 2ZD, UK; L.Lehtovirta-Morley@uea.ac.uk (L.E.L.-M.); r01mjaa@abdn.ac.uk (M.A.)

[2]  School of Biological Sciences, University of East Anglia, Norwich NR4 7TJ, UK

\*  Correspondence: Duncan.Wilson@abdn.ac.uk

Received: 9 November 2017; Accepted: 1 December 2017; Published: 6 December 2017

**Abstract:** The ZIP (Zrt/Irt-like protein) family of zinc transporters is found in all three domains of life. However, little is known about the phylogenetic relationship amongst ZIP transporters, their distribution, or their origin. Here we employed phylogenetic analysis to explore the evolution of ZIP transporters, with a focus on the major human fungal pathogen, *Candida albicans*. Pan-domain analysis of bacterial, archaeal, fungal, and human proteins revealed a complex relationship amongst the ZIP family members. Here we report (i) a eukaryote-wide group of cellular zinc importers, (ii) a fungal-specific group of zinc importers having genetic association with the fungal zincophore, and, (iii) a pan-kingdom supercluster made up of two distinct subgroups with orthologues in bacterial, archaeal, and eukaryotic phyla.

**Keywords:** ZIP; zinc; transport; transporter; evolution; pathogenic fungi; *Candida albicans*

## 1. Introduction

Zinc is an essential micronutrient for all living organisms. This is because many proteins (particularly enzymes and transcription factors) require zinc to function. In fact, 9% of eukaryotic proteins are predicted to interact with this metal. As well as acting as an essential cofactor for proteins involved in a large number of cellular processes, zinc can also be toxic if present in excess. Therefore, zinc acquisition, homeostasis, and detoxification is crucial for cell survival and proliferation. However, pathogenic microbes have to deal with extremes in zinc bioavailability due to the action of nutritional immunity. This term describes a variety of host processes which manipulate microbial exposure to trace metals, particularly zinc, iron, manganese, and copper. For example, following phagocytosis by macrophages, bacterial cells can face potential zinc and copper toxicity [1]. However, most examples of nutritional immunity involve host-driven metal sequestration together with microbial starvation [2].

Many bacterial pathogens utilise an ABC (ATP-binding cassette) transporter (ZnuABC) for high-affinity zinc uptake during infection. These systems consist of a substrate-binding protein ZnuA, permease ZnuB, and ATPase (ZnuC). An increasing body of literature is demonstrating an important role for ZnuABC-mediated zinc assimilation in bacterial pathogenicity. For a recent review on bacterial zinc assimilation, readers are directed to Capdevila et al. [3].

In contrast, with the exception of some recent studies in fungi, far less is known about how zinc homeostasis in eukaryotic pathogens influences their virulence. Zinc import via a ZnuABC-like system has not been reported in eukaryotes. Rather, they appear to predominantly employ ZIP-type transporters for cellular zinc import [4]. The name derives from fungal Zrt (zinc regulated transporter) [5] and plant Irt (iron-regulated transporter) [6] proteins.

However, the understanding of this ZIP-mediated zinc transport is complicated by the architecture of the eukaryotic cell. Unlike most bacteria and archaea, where cellular import only occurs at the plasma membrane, eukaryotic ZIPs can also deliver zinc from various intracellular organelles into the cytoplasm. Metal promiscuity may also confound phylogenetic interpretation: certain bacterial ZIPs have been shown to transport multiple metals. For example, *Salmonella enterica* ZupT transports both zinc and manganese [7]. In eukaryotes, different ZIP transporters can transport different metals. For example, *Saccharomyces cerevisiae* Atx2 is implicated in Golgi manganese homeostasis [8]. However, most eukaryotic ZIPs are implicated in zinc transport [4].

*S. cerevisiae* is one of the best understood models of eukaryotic ZIP-mediated zinc transport. This yeast encodes five ZIP transporters: two plasma membrane importers and three intracellular organellar transporters. There is an emerging and important role for ZIP transporters in the pathogenicity of human fungal pathogens [9–12], but little is known about the phylogeny, distribution, or origin of these transporters either within fungal pathogens or throughout different domains of life. Here we take advantage of OrthoMCL, to investigate the evolution of ZIP-type zinc transporters, with a focus on the medically important fungus *Candida albicans*.

## 2. Phylogenetic Analysis of ZIP Transporter

In order to investigate their phylogenetic relationships, ZIP orthologue groups were constructed using OrthoMCL (Available online: http://orthomcl.org/orthomcl/) [13]. OrthoMCL generates orthologue superclusters from 36 Bacteria (including six Firmicutes and 19 Proteobacteria), 16 Archaea, 9 Euglenozoa (Trypanosomes, *Leishmania*), 4 Amoebae, 11 Viridiplantae (plants, algae), 15 Alveolates (e.g., Apicomplexa such as *Plasmodium*), 24 Fungi (including 4 Basidiomycetes, 3 Microsporidia, and 17 Ascomycetes), 29 Metazoa, and 6 miscellaneous Eukaryotes (Oomycete, *Giardia*) genomes.

This approach resulted in 38 orthologue groups with ZIP zinc transporter Pfam annotations. The majority of these contained only few poorly connected ZIP proteins, and several represented likely expansions in metazoans or Viridiplantae; that is, likely associated with the development of multicellularity. However, several superclusters of interest were identified, which we discuss below.

## 3. Conserved Plasma Membrane Importer Cluster (OG5_126707)

OG5_126707 member ZIP transporters were exclusively eukaryotic and contained transporters from all studied eukaryotic groups. Figure 1 shows the cluster graph of OG5_126707. With the exception of the three Microsporidian species, fungal orthologues clustered together, as did Viridiplantae, Euglenozoa parasites (e.g., Trypanosomes), and to a lesser extent, Alveolates. Amoebic orthologues (blue circles) distributed between parasite clusters. The large cluster to the central right of Figure 1 includes Metazoan ZIP transporters. Not only were all eukaryotic groups represented in this supercluster, but most analysed species were also present. For example, 23 of the 24 analysed fungal species were represented.

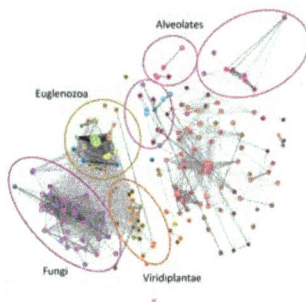

**Figure 1.** Conserved plasma membrane importer Cluster (OG5_126707). Clustering performed using OrthoMCL. Orthologues of *C. albicans* Zrt2 are conserved within eukaryotes. All characterised members of the clusters are implicated in plasma membrane zinc import.

Several members of the OG5_126707 supercluster have already been characterised. These include human ZIP1 and ZIP3, *S. cerevisiae* Zrt1 and Zrt2, *S. pombe* Zrt1, *A. fumigatus* ZrfA and ZrfB, *C. neoformans* Zip1 and Zip2, *C. albicans* Zrt2, and *Leishmania infantum*. Notably, all 11 of these transporters are implicated in cellular zinc import at the plasma membrane [5,11,14–19].

## 4. The Fungal Zincophore Locus Cluster (OG5_141027)

This cluster was unique to fungi (with the exception of two very loosely connected Trypanosomal sequences). However, unlike OG5_126707 which included 23 of the 24 fungal species, this cluster only contained 10 species. These included *C. albicans* (Zrt1) and *A. fumigatus* (ZrfC). Interestingly, both of these transporters have been reported to be up-regulated specifically at neutral/alkaline pH, and in the case of *C. albicans* Zrt1, to act as a cell surface docking protein for the secreted zincophore, Pra1 [20,21]. We have previously reported that the Zrt1 and Pra1 encoding genes are syntenic, not only in *C. albicans*, but in multiple fungal species [21,22]. Moreover, for *C. albicans* and *A. fumigatus*, the gene pairs are known to be co-expressed in response to zinc limitation [20,21]. Based on these observations, we have proposed that the *ZRT1/PRA1* locus may function as a conserved zincophore/receptor in multiple fungal species [22]. We therefore interrogated the genetic loci of those species identified in cluster OG5_141027. Gene order analysis revealed that seven of the 10 species here have maintained a syntenic relationship between orthologues of *C. albicans PRA1* and *ZRT1*. Two species—*Yarrowia lipolytica* and *Neurospora crassa*—have lost *PRA1*, and in one (*Gibberella zeae*, also called *Fusarium graminearum*), *PRA1-ZRT1* synteny has broken. One of the *PRA1*-negative species, *Yarrowia lipolytica*, has undergone duplication and divergence of the Zrt1 orthologue (Figure 2).

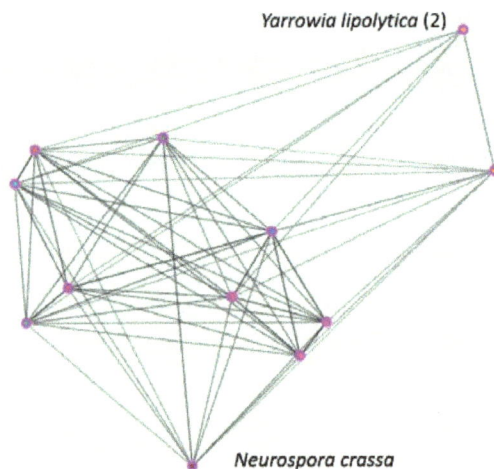

**Figure 2.** The fungal zincophore locus cluster (OG5_141027). Clustering performed using OrthoMCL. Orthologues of the zincophore-associated ZIP, Zrt1 (zinc regulated transporter 1) in *C. albicans* are specific to the fungal kingdom.

These observations are similar to our previous study—of 16 selected species analysed in Citiulo et al., 10 encoded *PRA1* orthologues and, of these 10 species, six maintained synteny with a *ZRT1* orthologue. To examine how widespread the syntenic relationship is, we interrogated the NCBI database. Of 102 species analysed, we identified Pra1 orthologues in 87 (85.3%) species and, of the Pra1[+] species, 61 (70.1%) have maintained a syntenic relationship between *PRA1* and *ZRT1* (Table S1).

The fact that only ascomycete ZIPs were identified within this OrthoMCL cluster is probably due to the low number of basidiomycete species present in this database. In fact, BLASTp analysis of

*C. albicans* Zrt1 against non-ascomycetes identified numerous ZIPs which reciprocally hit *C. albicans* Zrt1. Moreover, both ascomycete and basidiomycete species exhibit synteny of zincophore and ZIP orthologues (see [21] and Table S1).

While it should be pointed out that both *PRA1* [21] and *ZRT1* [22] orthologues have been lost multiple times throughout the fungal kingdom, this indicates that, when present, the genes tend to share a syntenic relationship. This most likely serves to simplify modular co-expression. The observations reported here support our earlier conclusion that *PRA1-ZRT1* synteny represents an ancient and highly successful adaption within the fungal kingdom [21].

## 5. The ZupT/ZIP11/Zrt3 Pan-Domain Supercluster (OG5_127397)

The OG5_127397 supercluster (Figure 3) was the only cluster to contain ZIP proteins from all three domains of life. In fact, all phyla, with the exception of Alveolate and Euglenozoa parasites were represented.

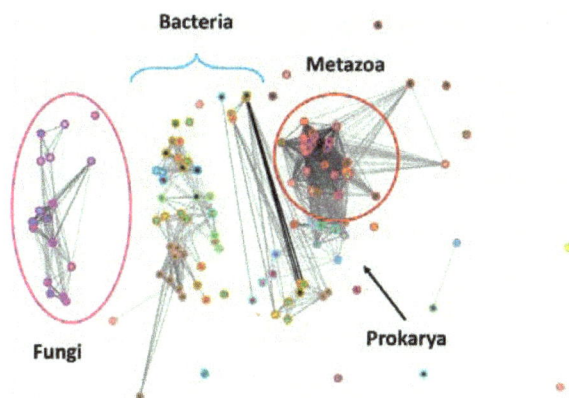

**Figure 3.** Fungal Zrt3, Prokaryote ZupT, Metazoan Zip11 pan-domain supercluster (OG5_127397). Clustering performed using OrthoMCL. Note the separation of eukaryotic (Fungi and Metazoan) subclusters by prokaryotic proteins.

A number of bacterial (16) and archaeal (5 or 6) members were present in the OG5_127397 supercluster. No archaeal ZIP transporters have been studied to-date. In bacteria, the Zip transporter, ZupT, has been characterised in *Escherichia coli*, *Cupriavidus metallidurans*, and *Salmonella enterica*. In all three species, a role in zinc import has been described [7,23,24]. *E. coli* ZupT appears to transport several other cations in addition to zinc [25], and *S. enterica* ZupT imports both zinc and manganese [7].

Orthologues were present throughout the fungal kingdom, but are absent from the Microsporidia and Basidiomycota. The *S. cerevisiae* member, Zrt3, has been shown to transport zinc out of the fungal vacuole [26], and our own work indicates that the *C. albicans* orthologue plays a similar role [27]. The human member, ZIP11, has been implicated in Golgi zinc transport [28].

Based on similarity between human ZIP11 and bacterial ZIP (ZupT) proteins, Yu et al. [27] have proposed that this family represents the most ancient ZIP [27], present in the last universal common ancestor.

Similarly, the identification of *S. cerevisiae* Zrt3 led to the recognition of prokaryotic ZIP-type transporters in the first place, as Zrt3 (but not the previously characterised yeast Zrt1 and Zrt2) shared sequence similarity with bacterial and archaeal proteins [26].

In this context, the position of metazoan ZIP11, fungal Zrt3, and prokaryotic ZupT within the same supercluster is in line with an ancient origin [26,29].

Surprisingly, however, Fungal (Zrt3) and Metazoan (ZIP11) clusters were very distinct, and both had higher similarity to prokaryotic ZIPs than to each other (Figure 3). Furthermore, direct alignments showed that fungal Zrt3 and human ZIP11 shared only 28% sequence identity (e-value $3.1 \times 10^{-2}$). This was unexpected, as within the Eukarya, fungi and metazoans are very closely related [30]. This suggests that fungal Zrt3 and metazoan ZIP11 may not be closely related.

We therefore compared fungal (Zrt3-type) and metazoan (ZIP11-type) with more bacterial and archaeal sequences.

In order to capture bacterial and archaeal diversity as broadly as possible, fungal (*C. albicans* Zrt3) and metazoan (human ZIP11) sequences were subjected to individual BLASTp searches against Firmicutes, Proteobacteria, Actinobacteria, Spirochetes, Euryarchaeota, and Crenarchaeota species available at NCBI. These analyses identified predicted ZIP transporters in all six prokaryotic phyla.

Sequence similarities between fungal Zrt3 and prokaryotic best hits were 30–40% (e-value $10^{16}$–$10^{15}$) for bacteria, and even lower, $\leq 30\%$ identity (e-value $\sim 10^{-12}$), for archaea. Sequence similarity between fungal Zrt3 and bacterial proteins was limited to the C-terminal 200 amino acids. Indeed, when we repeated BLASTp analysis with the C-terminal 200 amino acids alone, we identified greater similarity (e-value $6 \times 10^{-19}$). No sequence identity was observed for the N-terminal 391 amino acids out with the fungal kingdom.

We observed a higher degree of sequence similarity between Hs ZIP11 and bacterial and archaeal species ($\sim 45\%$ identity, e-value $10^{-81}$–$10^{-60}$).

Next, we aligned fungal, mammalian, bacterial, and archaeal species' ZIP sequences using Phylo.fr [31,32]. The resulting tree formed two distinct branches: one containing the fungal Zrt3 and the other, metazoan ZIP11 (Figure 4). Remarkably, both branches contained ZIP transporters from Firmicutes, Proteobacteria, Actinobacteria, Spirochete, and Euryarchaeota species, whilst the Human Zip11 branch rooted against the two identified Crenarchaeota. This suggests that prokaryotes have two different ZIP transporters: one related to fungal Zrt3 and the other to metazoan ZIP11.

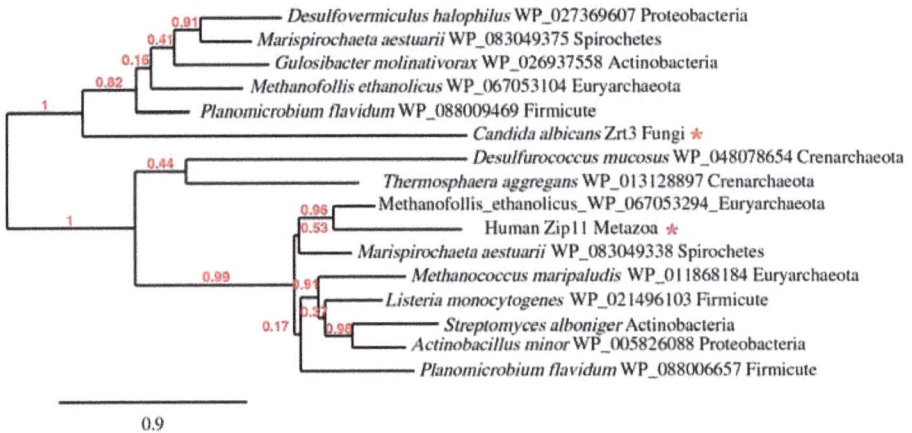

**Figure 4.** Pan-domain supercluster phylogeny. Phylogenetic tree generated using Phylogeny.fr (Available online: http://www.phylogeny.fr/). Note the presence of two branches with prokaryotic ZIPs related to both Fungal Zrt3 and Metazoan Zip11.

If this is the case, we may anticipate the existence of extant prokaryotic species with both types. Indeed, the respective best hits of fungal Zrt3 and metazoan ZIP11 against Spirochetes were two independent ZIP transporters in the same species: *Marispirochaeta aestuarii* (Figure 4).

We therefore subjected human ZIP11 to BLASTp analysis against prokaryotic species which were identified in the fungal Zrt3 search, and vice versa. In most cases, we identified the same ZIP as in the previous search round, or the sequence similarity was too low to return a subject. However, when we queried fungal Zrt3 against the Firmicute *Planomicrobium flavidum* and the Euryarchaeota *Methanofollis ethanolicus* (two species which had metazoan ZIP11 orthologues), we identified independent ZIP transporters. The ZIP pairs from these three species clustered on the two distinct branches of the tree (Figure 4). Therefore, it would appear that these three prokaryotic species encode two independent ZIP transporters.

This is interesting because it demonstrates the existence of two distinct classes of ZIP transporter in multiple prokaryotic phyla.

Although fungal Zrt3 and metazoan ZIP11 were identified as belonging to the same orthologue supercluster, their similarity was very low (identity 28%, e-value $3.1 \times 10^{-2}$). Moreover, their relationship to distinct prokaryotic proteins (Figures 3 and 4) is not suggestive of a close phylogenetic relationship.

We therefore performed BLASTp analysis of fungal Zrt3 excluding the fungal kingdom (NCBI). Intriguingly, outside of the fungal kingdom, Zrt3 shares highest similarity with bacterial sequences and not with other Eukaryotes, as would be expected.

We therefore systematically analysed ZupT from *Desulfovermiculus halophilus* (which was one of the bacterial ZIPs with highest similarity to fungal Zrt3) against the major eukaryotic phyla.

*D. halophilus* ZupT did not share sequence identity with any proteins within the Parabasalia, Diplomonadida, Ciliophora, or Euglenozoa. Only one species within the Heterolobosea (the Apicomplexa), and a handful of *Dictyostelium* and *Acytostelium* species within the Mycetozoa had proteins with similarity to *D. halophilus* ZupT (not shown).

We retrieved a large number of hits from within the Heterokonta (e-value $7 \times 10^{-60}$ [47% identity] to e-value $4 \times 10^{-26}$ [28% identity]) and Viridiplantae ($10^{-50}$, 43% identity) and, of those top hits analysed, they aligned to the fungal Zrt3 branch of Figure 4. Within the Metazoa, we did identify ZIP transporters with sequence similarity to *D. halophilus* ZupT, but (with the exception of *Oikopleura dioica*) these aligned to the human ZIP11 branch of the tree (not shown).

Therefore, it appears that the origin of fungal Zrt3 is complex. It is possible that the gene was inherited vertically into the fungi, and that it has been lost multiple times within extant eukaryotic lineages. However, given the absence of Zrt3 orthologues from basal eukaryotes, its acquisition via horizontal gene transfer may represent an alternative explanation.

We note that the observed similarities of ZIP11 and Zrt3 with prokaryotic proteins are in agreement with the conclusions of both MacDiarmid (2000) and Yu (2013) [26,29], that these proteins may represent ancient ZIP transporters in metazoans and in fungi, respectively. However, the diversity of bacterial and archaeal protein sequences within this orthologue supercluster (Figure 4) suggests that they arose from distinct genes.

In summary, our analysis of fungal ZIP transporters indicates that there are three major orthologue groups with different degrees of conservation within and outside of the eukaryotes.

(i) A conserved group of eukaryotic proteins (OG5_126707) encompassing fungal, metazoan, and parasite plasma membrane importers; (ii) A fungal-specific group of zinc importers (OG5_141027), genetically associated with the fungal zincophore; (iii) A pan-domain supercluster (OG5_127397), formed of two distinct groups with orthologues in all three domains of life.

At this stage, it is unclear whether eukaryotic members of this supercluster were inherited vertically or horizontally. However, our analyses indicate the presence of two relatively distinct groups of ZIP transporters in extant bacterial and archaeal species. Interestingly, the fungal members of this group appear to be involved in organellar (vacuolar) zinc export, rather than plasma membrane import.

Since the emergence of the Eukarya, ZIP transporter genes have clearly undergone multiple rounds of expansion. This is presumably to meet the requirements of an organellar lifestyle and, in the case of metazoans (humans for example have 14 ZIP family members), multicellularity.

Because zinc can be highly limited during infection due to the action of nutritional immunity, understanding the nature of pathogen (and host) zinc transporters may help inform future therapeutic or diagnostic strategies.

**Supplementary Materials:** Supplementary materials can be found at www.mdpi.com/1422-0067/18/12/2631/s1. Table S1. ZIP and zincophore gene synteny. Table lists the presence (denoted by respective accession numbers) or absence (N) of *PRA1* and *ZRT1* orthologues in 102 fungal species (NCBI). In species where both genes are present, column 4 indicates syntenic (Y), or non-syntenic (N) relationship.

**Author Contributions:** Duncan Wilson and Laura E. Lehtovirta-Morley conceived the study. Duncan Wilson and Mohammad Alsarraf acquired and analysed the data. Duncan Wilson and Laura E. Lehtovirta-Morley drafted and critically revised the manuscript. All authors approved the final submitted version and agree to accountability for all aspects of the work.

**Conflicts of Interest:** The authors declare no conflict of interest.

**Funding:** Duncan Wilson is supported by a Sir Henry Dale Fellowship jointly funded by the Wellcome Trust and the Royal Society (102549/Z/13/Z), a Wellcome Trust ISSF seed corn grant (RG12723 14), the Medical Research Council and University of Aberdeen (MR/N006364/1) and a Wellcome Trust Strategic Award for Medical Mycology and Fungal Immunology (097377/Z/11/Z). Laura E. Lehtovirta-Morley is supported by a Royal Society Dorothy Hodgkin Fellowship (DH150187). Mohammad Alsarraf is funded by the Public Authority for Applied Education and Training, Kuwait.

## References

1.  Botella, H.; Peyron, P.; Levillain, F.; Poincloux, R.; Poquet, Y.; Brandli, I.; Wang, C.; Tailleux, L.; Tilleul, S.; Charriere, G.M.; et al. Mycobacterial P$_1$-type atpases mediate resistance to zinc poisoning in human macrophages. *Cell Host Microbe* **2011**, *10*, 248–259. [CrossRef] [PubMed]

2.  Hood, M.I.; Skaar, E.P. Nutritional immunity: Transition metals at the pathogen-host interface. *Nat. Rev. Microbiol.* **2012**, *10*, 525–537. [CrossRef] [PubMed]

3.  Capdevila, D.A.; Wang, J.; Giedroc, D.P. Bacterial strategies to maintain zinc metallostasis at the host-pathogen interface. *J. Biol. Chem.* **2016**, *291*, 20858–20868. [CrossRef] [PubMed]

4.  Eide, D.J. Zinc transporters and the cellular trafficking of zinc. *Biochim. Biophys. Acta* **2006**, *1763*, 711–722. [CrossRef] [PubMed]

5.  Zhao, H.; Eide, D. The yeast *zrt1* gene encodes the zinc transporter protein of a high-affinity uptake system induced by zinc limitation. *Proc. Natl. Acad. Sci. USA* **1996**, *93*, 2454–2458. [CrossRef] [PubMed]

6.  Eide, D.; Broderius, M.; Fett, J.; Guerinot, M.L. A novel iron-regulated metal transporter from plants identified by functional expression in yeast. *Proc. Natl. Acad. Sci. USA* **1996**, *93*, 5624–5628. [CrossRef] [PubMed]

7.  Karlinsey, J.E.; Maguire, M.E.; Becker, L.A.; Crouch, M.L.; Fang, F.C. The phage shock protein pspa facilitates divalent metal transport and is required for virulence of *Salmonella enterica* sv. Typhimurium. *Mol. Microbiol.* **2010**, *78*, 669–685. [CrossRef] [PubMed]

8.  Lin, S.J.; Culotta, V.C. Suppression of oxidative damage by *Saccharomyces cerevisiae atx2*, which encodes a manganese-trafficking protein that localizes to golgi-like vesicles. *Mol. Cell. Biol.* **1996**, *16*, 6303–6312. [CrossRef] [PubMed]

9.  Amich, J.; Calera, J.A. Zinc acquisition: A key aspect in *Aspergillus fumigatus* virulence. *Mycopathologia* **2014**, *178*, 379–385. [CrossRef] [PubMed]

10. Schneider Rde, O.; Diehl, C.; Dos Santos, F.M.; Piffer, A.C.; Garcia, A.W.; Kulmann, M.I.; Schrank, A.; Kmetzsch, L.; Vainstein, M.H.; Staats, C.C. Effects of zinc transporters on *Cryptococcus gattii* virulence. *Sci. Rep.* **2015**, *5*, 10104. [CrossRef] [PubMed]

11. Do, E.; Hu, G.; Caza, M.; Kronstad, J.W.; Jung, W.H. The zip family zinc transporters support the virulence of *Cryptococcus neoformans*. *Med. Mycol.* **2016**, *54*, 605–615. [CrossRef] [PubMed]

12. Dade, J.; DuBois, J.C.; Pasula, R.; Donnell, A.M.; Caruso, J.A.; Smulian, A.G.; Deepe, G.S., Jr. Hczrt2, a zinc responsive gene, is indispensable for the survival of *Histoplasma capsulatum* in vivo. *Med. Mycol.* **2016**, *54*, 865–875. [CrossRef] [PubMed]

13. Fischer, S.; Brunk, B.P.; Chen, F.; Gao, X.; Harb, O.S.; Iodice, J.B.; Shanmugam, D.; Roos, D.S.; Stoeckert, C.J., Jr. Using orthomcl to assign proteins to orthomcl-db groups or to cluster proteomes into new ortholog groups. *Curr. Protoc. Bioinform.* **2011**. [CrossRef]

14. Gaither, L.A.; Eide, D.J. The human zip1 transporter mediates zinc uptake in human k562 erythroleukemia cells. *J. Biol. Chem.* **2001**, *276*, 22258–22264. [CrossRef] [PubMed]

15. Wang, F.; Dufner-Beattie, J.; Kim, B.E.; Petris, M.J.; Andrews, G.; Eide, D.J. Zinc-stimulated endocytosis controls activity of the mouse zip1 and zip3 zinc uptake transporters. *J. Biol. Chem.* **2004**, *279*, 24631–24639. [CrossRef] [PubMed]

16. Zhao, H.; Eide, D. The zrt2 gene encodes the low affinity zinc transporter in *Saccharomyces cerevisiae*. *J. Biol. Chem.* **1996**, *271*, 23203–23210. [CrossRef] [PubMed]

17. Vicentefranqueira, R.; Moreno, M.A.; Leal, F.; Calera, J.A. The *zrfa* and *zrfb* genes of *Aspergillus fumigatus* encode the zinc transporter proteins of a zinc uptake system induced in an acid, zinc-depleted environment. *Eukaryot. Cell* **2005**, *4*, 837–848. [CrossRef] [PubMed]

18. Carvalho, S.; Barreira da Silva, R.; Shawki, A.; Castro, H.; Lamy, M.; Eide, D.; Costa, V.; Mackenzie, B.; Tomas, A.M. Lizip3 is a cellular zinc transporter that mediates the tightly regulated import of zinc in *Leishmania infantum* parasites. *Mol. Microbiol.* **2015**, *96*, 581–595. [CrossRef] [PubMed]

19. Crawford, A.; Lehtovirta-Morley, L.E.; Alamir, O.; Niemiec, M.J.; Alawfi, B.; Alsarraf, M.; Skrahina, V.; Costa, A.C.B.P.; Anderson, A.; Yellagunda, S.; et al. Biphasic zinc compartmentalisation in a human fungal pathogen. Unpublished work. 2017.

20. Amich, J.; Vicentefranqueira, R.; Leal, F.; Calera, J.A. *Aspergillus fumigatus* survival in alkaline and extreme zinc-limiting environments relies on the induction of a zinc homeostasis system encoded by the *zrfc* and *aspf2* genes. *Eukaryot. Cell* **2010**, *9*, 424–437. [CrossRef] [PubMed]

21. Citiulo, F.; Jacobsen, I.D.; Miramon, P.; Schild, L.; Brunke, S.; Zipfel, P.; Brock, M.; Hube, B.; Wilson, D. *Candida albicans* scavenges host zinc via pra1 during endothelial invasion. *PLoS Pathog.* **2012**, *8*, e1002777. [CrossRef] [PubMed]

22. Wilson, D. An evolutionary perspective on zinc uptake by human fungal pathogens. *Metallomics* **2015**, *7*, 979–985. [CrossRef] [PubMed]

23. Grass, G.; Wong, M.D.; Rosen, B.P.; Smith, R.L.; Rensing, C. Zupt is a zn(II) uptake system in *Escherichia coli*. *J. Bacteriol.* **2002**, *184*, 864–866. [CrossRef] [PubMed]

24. Herzberg, M.; Bauer, L.; Nies, D.H. Deletion of the zupt gene for a zinc importer influences zinc pools in *Cupriavidus metallidurans* ch34. *Metallomics* **2014**, *6*, 421–436. [CrossRef] [PubMed]

25. Grass, G.; Franke, S.; Taudte, N.; Nies, D.H.; Kucharski, L.M.; Maguire, M.E.; Rensing, C. The metal permease zupt from *Escherichia coli* is a transporter with a broad substrate spectrum. *J. Bacteriol.* **2005**, *187*, 1604–1611. [CrossRef] [PubMed]

26. MacDiarmid, C.W.; Gaither, L.A.; Eide, D. Zinc transporters that regulate vacuolar zinc storage in *Saccharomyces cerevisiae*. *EMBO J.* **2000**, *19*, 2845–2855. [CrossRef] [PubMed]

27. Alamir, O.; Wilson, D. Zrt3 mediates vacuolar zinc efflux in *Candida albicans*. Unpublished work, 2017.

28. Kelleher, S.L.; Velasquez, V.; Croxford, T.P.; McCormick, N.H.; Lopez, V.; MacDavid, J. Mapping the zinc-transporting system in mammary cells: Molecular analysis reveals a phenotype-dependent zinc-transporting network during lactation. *J. Cell. Physiol.* **2012**, *227*, 1761–1770. [CrossRef] [PubMed]

29. Yu, Y.; Wu, A.; Zhang, Z.; Yan, G.; Zhang, F.; Zhang, L.; Shen, X.; Hu, R.; Zhang, Y.; Zhang, K.; et al. Characterization of the gufa subfamily member SLC39A11/Zip11 as a zinc transporter. *J. Nutr. Biochem.* **2013**, *24*, 1697–1708. [CrossRef] [PubMed]

30. Baldauf, S.L.; Roger, A.J.; Wenk-Siefert, I.; Doolittle, W.F. A kingdom-level phylogeny of eukaryotes based on combined protein data. *Science* **2000**, *290*, 972–977. [CrossRef] [PubMed]

31. Dereeper, A.; Guignon, V.; Blanc, G.; Audic, S.; Buffet, S.; Chevenet, F.; Dufayard, J.F.; Guindon, S.; Lefort, V.; Lescot, M.; et al. Phylogeny.Fr: Robust phylogenetic analysis for the non-specialist. *Nucleic Acids Res.* **2008**, *36*, W465–W469. [CrossRef] [PubMed]

32. Dereeper, A.; Audic, S.; Claverie, J.M.; Blanc, G. Blast-explorer helps you building datasets for phylogenetic analysis. *BMC Evol. Biol.* **2010**, *10*, 8. [CrossRef] [PubMed]

MDPI

St. Alban-Anlage 66

4052 Basel, Switzerland

Tel. +41 61 683 77 34

Fax +41 61 302 89 18

http://www.mdpi.com

*IJMS* Editorial Office

E-mail: ijms@mdpi.com

http://www.mdpi.com/journal/ijms

www.ingramcontent.com/pod-product-compliance
Lightning Source LLC
Chambersburg PA
CBHW051719210326
41597CB00032B/5536